MATHEMATICAL ANALYSIS OF PHYSICAL PROBLEMS

Philip R. Wallace
McGill University

DOVER PUBLICATIONS, INC.
New York

Published in Canada by General Publishing Company, Ltd., 30 Lesmill
Road, Don Mills, Toronto, Ontario.
Published in the United Kingdom by Constable and Company, Ltd., 10
Orange Street, London WC2H 7EG.

This Dover edition, first published in 1984, is an unabridged and corrected
republication of the work first published by Holt, Rinehart and Winston,
Inc., N.Y., in 1972.

Manufactured in the United States of America
Dover Publications, Inc., 31 East 2nd Street, Mineola, N.Y. 11501

Library of Congress Cataloging in Publication Data

Wallace, Philip R. (Philip Russell) 1915-
Mathematical analysis of physical problems.

Reprint. Originally published: New York : Holt, Rinehart and Winston,
1972.
Bibliography: p.
Includes index.
1. Mathematical physics. I. Title.
QC20.W27   1984   530.1'55   83-20640
ISBN 0-486-64676-9

To my wife Jean, for her patience and understanding.

And to the memory of two men who brought both dignity and excitement to the study of physics: Leopold Infeld and Georges Placzek.

PREFACE

"After having spent years trying to be accurate, we must spend as many more in discovering when and how to be inaccurate"

Ambrose Bierce

In writing a book on mathematical physics, the author is from the beginning faced with the fact that he has not chosen a uniquely defined subject. The task would have been easier in the nineteenth century, when physics was somewhat more stable and mathematics and physics lived in closer association with each other. Today, the words "mathematical physics" have, by fairly general consent, in North America at least, taken on a special and rather conventional meaning, though there is still some fuzziness at the boundaries with other areas. As exemplified by the American Physical Society's journal of that name, the phrase signifies the investigation of mathematical techniques current in physics. The emphasis is on methodology rather than on physical content. In other places and other times, however, the separation between form and content is, and has been, less marked.

It is important, first, to understand the difference in purpose of a book such as this and a book on mathematics *per se*. Modern mathematics is primarily concerned with axiomatic systems and formal deductions made within these systems. In this sense, it consists of games: The axiomatic system defines the rules of the game, and one then plays according to these rules. The rules need not have any obvious relation to the world in which we live.

The physicist, on the other hand, is concerned with understanding and describing the physical aspects of the world in which he lives. It is this which determines and governs his "mathematics." For him mathematics is a language, a shorthand, for *describing* and coordinating his comprehension of that world. The mathematician is concerned with rigor and with the completeness and logical consistency of his systems. The "rules of the game" must be fully defined and adhered to. The criterion which the physicist applies to his "mathematics" is its conformity to nature. Mathematical symbols represent the magnitudes of physical quantities, and "mathematical" formulas express the relations between these quantities. His systems are, and can be, only as complete as his comprehension. As for consistency, it is assured by the correspondence between his mathematical models and physical reality; he is prepared to assume the internal consistency of nature! Thus, as has been pointed out by Landau (and many others), mathematical rigor has no relevance to physics.

This must be the major feature distinguishing a book on mathematics from one on mathematical physics. The one proceeds axiomatically, governed by the requirements of rigor. The other tries to construct workable (and necessarily approximate and incomplete) models of aspects of physical reality. To attempt to judge either by the standards of the other is therefore inappropriate and mistaken.

When, in the present book, mathematical structures are described (as in chapter 2 on Linear Vector Spaces), they should be considered as the description of a common framework within which theories can be constructed of a variety of different sorts of physical phenomenon. The important questions are, whether the mathematical framework serves as an effective description of the physics, and whether it provides us with answers to physical problems which can be tested by physical observation. No greater mathematical generality is relevant than is required to describe the phenomena with which we are concerned.

Given this general definition of our aims, a few words should be said about the particular material which we have chosen to include in this book. Clearly, any book of this sort could not be "complete" without being encyclopaedic; it must represent an arbitrary selection of topics for discussion. Our selection has been chosen to link classical and modern physics through common techniques and concepts. The first chapter is on vibrating strings, in other words, on problems of one-dimensional wave propagation. Aside from recognizing the wide importance of wave phenomena in physics, this subject provides us with a testing ground for a wide range of concepts and methods with wider relevance, and does so in a simple and familiar physical context. Thus, when similar problems are met in newer and less familiar physical contexts, it will not be necessary to cope with technical difficulties as well as conceptual ones.

Chapter 2, on linear vector spaces, provides the thread which ties together most of the rest of the book. This is, of course, the basic conceptual framework of quantum mechanics; it also unifies our treatment of the problems of classical physics and provides us with some essential mathematical tools.

In chapter 3 we introduce the problems of potentials and the Laplace and Poisson differential equations. Since these are three-dimensional problems, we are led to introduce the method of separation of variables. An important feature is the introduction of spherical harmonics, which reappear later in other problems, including that of angular momentum in quantum mechanics. We have also tried, in this chapter, to illustrate how physical considerations can be used to provide the motivation for the development of mathematical techniques.

The fourth chapter is again primarily a "mathematical" one. It is concerned with the methods of Laplace and Fourier transforms and the relation between them. The methods are illustrated with a selection of useful examples.

Chapters 5 and 6 deal with important classes of problems in classical physics. Chapter 5 is concerned with problems of wave propagation in three dimensions, and chapter 6 with problems primarily of a diffusive character. Problems of the propagation of electromagnetic waves are considered at some length. Our intention is not to replace a more extensive and detailed treatment such as given in Jackson's book; chapter 5 *does* (along with chapter 3) provide a somewhat more streamlined account of some of the basic problems of the field.

Chapter 7 is concerned with probabilistic methods in physics. Some basic methods and concepts are introduced, and the basis is provided of a more fundamental description of some of the diffusive processes dealt with in chapter 6.

The final three chapters are devoted to a discussion of the fundamentals of quantum mechanics. Any treatment of methods of mathematical physics which was arbitrarily confined to "classical" physics would be artificial, as well as foregoing the pedagogical advantage of exploiting the technical similarities of many classical and quantum problems. Our emphasis is on useful techniques in quantum mechanics. In chapter 8 we deal with problems of a quite general nature, including a fairly extensive treatment of the time evolution of quantum systems, and of perturbation theory. In chapter 9 we deal with the mathematical theory of some standard problems: the hydrogen atom, the harmonic oscillator in one, two, and three dimensionals, and angular momentum theory. Extensive use is made of ladder (or creation and annihilation) operators, which provide the framework for the introduction of elementary field theoretical methods in the last chapter. Chapter 10 provides an introduction to many-body problems using the occupation number representation, and introduces a number of fundamental problems: the Hartree–Fock method, the electron gas, and the theory of density matrices and linear response.

As for omissions, perhaps the most striking is that of group theory. This is partly arbitrary, since any such book must be selective in its coverage; an additional justification is, however, that it is a difficult subject in which to give an adequate and self-contained treatment of modest length, while for longer and more extensive treatments, there have appeared in recent years a very large number of excellent books on group-theoretical methods in physics.

It is evident that, given the rapid evolution of physics curricula at the present time, and the great variety of ways in which the sort of mathematical physics incorporated in this book is distributed among course units in different universities, it is almost impossible to write a "text book" which will correspond simultaneously to the patterns of a very large number of universities. It seems to the author, however, that the very notion of a course textbook at this level is unrealistic, and that the advanced undergraduate or beginning graduate student should use a number of books in any course. My purpose is, therefore, to produce a generally useful book; one which will be a valuable addition to the personal library of students of physics. If it fits the needs of particular courses, so much the better.

A final word about problems. We have departed from the usual practice of providing collections of problems at the ends of chapters, but have instead interspersed them throughout the text. Some of these, in fact, form an integral part of the text, and will serve to test the student's grasp of what he has read, and his ability to extend it. Others are exercises in the techniques dealt with in the text. An effort has been made to avoid "problems for problems' sake." On the other hand, it is impossible to overemphasize the importance of the student "doing" for himself, and not merely learning theory by memory. I have tried, in many of the problems, to raise the sort of question which a good student might ask himself. If he *is* encouraged to ask further questions, and to seek their answers, their purpose will have been served.

I should like to express my thanks to Dr. Robert Heck for carefully checking the manuscript.

P. R. Wallace

CONTENTS

xvi   Contents

REFERENCES

We list first a number of general references, largely though not entirely of a mathematical character.

Courant, R. and D. Hilbert, *Methods of Mathematical Physics.* New York: Wiley–Interscience, 1961.

Goertzel, G. H. and N. Tralli, *Some Mathematical Methods of Physics.* New York: McGraw-Hill, 1960.

Irving, J. and N. Mullineux, *Mathematics in Physics and Engineering.* New York: Academic, 1959.

Jeffreys, H. and B. S. Jeffreys, *Methods of Mathematical Physics.* London: Cambridge Press, 1956.

Margenau, H. and G. M. Murphy, *The Mathematics of Physics and Chemistry.* Princeton, New Jersey, Van Nostrand, 1961. (This book is more oriented toward physics than most, and, although old, is somewhat more in the spirit of the present book.)

Morse, P. M. and H. Feshbach, *Methods of Theoretical Physics.* New York: McGraw-Hill, 1953, 2 vols.

Sommerfeld, A., *Partial Differential Equations in Physics.* New York: Academic, 1949.

Webster, A. G., *Partial Differential Equations of Mathematical Physics.* New York: Dover, 1955.

Whittaker, E. T. and G. N. Watson, *A Course of Modern Analysis*. London: Cambridge Press, 1958.

There are also several useful references on topics some knowledge of which is assumed in the present book, for example, ordinary differential equations and complex variable theory, (the knowledge assumed is, however, far less than that covered by these books).

Byerly, W. E., *Introduction to the Calculus of Variations*. Cambridge, Mass.: Harvard University Press, 1933.
Heins, M., *Selected Topics in the Classical Theory of Functions of a Complex Variable*. New York: Holt, Rinehart and Winston, 1962.
Ince, E. L., *Integration of Ordinary Differential Equations*. New York: Wiley–Interscience, 1956.
MacRobert, T. M., *Functions of a Complex Variable*. London: Macmillan, 1950, 3rd ed.
Pennisi, L. L., *Elements of Complex Variables*. New York: Holt, Rinehart and Winston, 1963.
Smith, L. P., *Mathematical Methods for Engineers and Scientists*. Englewood Cliffs, N.J.: Prentice-Hall, 1953.

For the student wishing problems supplementary to those given in this book, there exist collections of problems, mostly with solutions:

Cronin, J. A., D. F. Greenberg, and V. Telegdi, *University of Chicago Graduate Problems in Physics*. Reading, Mass.: Addison-Wesley, 1967.
Goldman, J. J. and V. D. Krivchenkov, *Problems in Quantum Mechanics*. Reading, Mass.: Addison-Wesley, 1961.
Lebedev, N. N., I. P. Stalskaya and Y. S. Uflyand, *Worked Problems in Applied Mathematics*. New York: Dover, 1979.
Misyurkeyev, I. V., *Problems in Mathematical Physics*, New York: McGraw-Hill, 1966.
Smirnov, M. M., *Problems on the Equations of Mathematical Physics*. Groningen, Netherlands: P. Noordhoff, 1966.
ter Haar, D., *Selected Problems in Quantum Mechanics*. Infosearch, 1964.

Finally, there are various extremely useful books of numerical tables and tables of formulas:

Abramowitz, M. and I. A. Stegun, *Handbook of Mathematical Functions*. New York: Dover, 1965.
Dwight, H. B., *Tables of Integrals and Other Mathematical Data*. London: Macmillan, 1957.
Erdelyi, A. W., W. Magnus, F. Oberhettinger, and F. G. Tricomi, *Higher Transcendental Functions*. New York: McGraw-Hill, 1953, 3 vols.
Gradshteyn, I. S. and I. N. Ryazhik, *Tables of Integrals, Series, and Products*. New York and London: Academic, 1965.
Gröbner W., and N. Hofreiter, *Integraltafeln*. Berlin: Springer, 1957, 2 vols.

Jahnke, E., F. Emde, and F. Losch, *Tables of Higher Functions*. New York: McGraw-Hill, 1960.

Jolley, L. B. W., *Summation of Series*. New York: Dover, 1961.

Madelung, E., *Mathematical Tools for the Physicist*. New York: Dover, 1943.

Magnus, W. and F. Oberhettinger, *Formulae and Theorems for the Special Functions of Mathematical Physics*. New York: Chelsea, 1949.

Sneddon, I. N., *Special Functions of Mathematical Physics and Chemistry*. New York: Wiley–Interscience, 1956.

MATHEMATICAL
ANALYSIS OF
PHYSICAL PROBLEMS

PRELUDE TO CHAPTER 1

In this chapter as in subsequent ones, we shall provide
an outline of the topics with which the chapter deals.
In addition to providing a guide to the organization of
its contents, we indicate the mathematical background
assumed, and call attention to those methods and concepts which have an application in different physical
problems.

The main theme of Chapter 1 is *waves*. A useful
reference is the small book of Coulson (*Waves*[1]). We
deal here only with one-dimensional problems; the
theme is picked up again, in the context of *three* dimensions, in Chapter 5. Many of the concepts and methods
developed here are capable of generalization to the
three-dimensional case.

Because the physical context of our discussion is
that of the vibrating string, we first provide a derivation
of the equation of motion of a stretched elastic string.
We then discuss very general solutions of this equation
for an "infinitely long" string. We first obtain solutions
in terms of arbitrary functions, which are determined
by initial conditions. This leads to rather simple physical pictures of the motion.

It is then shown that these solutions may be
adapted to describe the "reflection" of a wave at an
end which is fixed. If there are *two* such fixed ends, it
is then shown that the solution is periodic, and is thus
expansible in a Fourier series (*an elementary knowledge of Fourier series is assumed*).

The string problem is next discussed from the
viewpoint of energy. Energy density (energy per unit
length) and energy flux are defined. Two important
results are obtained:

(1) The energies of waves traveling in opposite
directions are distinct in an infinite string; two such
waves can pass through each other and emerge un-

---

[1]Publishing information may be found in the list of references concluding the prelude to
each chapter for books not specifically footnoted throughout.

altered. That is, *the waves do not scatter each other.*
This is a consequence of the linear approximation. If
nonlinear terms were kept, they would in fact give rise
to mutual scattering of waves.

(2) The velocity of energy transport (group velocity) is the same as the phase velocity.

Turning to *harmonics* (Fourier components), it is
verified that each harmonic also has its own energy
content, determined by initial conditions; there is no
transfer of energy between harmonics (again, this is a
consequence of linearity).

The next problem treated is that of *scattering*,
which, in one dimension, is that of determining reflection and transmission at some point of irregularity. An
example is that of reflection at a point mass fixed to the
string. We have introduced into the problem the concept of the scattering matrix (*S* matrix), which plays an
essential role in modern quantum scattering theory,
but which here takes a particularly simple form and
interpretation.

We then consider the problem of regularly spaced
scatterers, a prototype of the "periodic lattice" type
of problem which arises in other areas, notably electrical networks and lattice vibrations ("phonons") in
solids [see Brillouin, *Wave Propagation in Periodic
Structures*[2]]. The existence of "forbidden bands" of
frequency is demonstrated.

The problem of scattering by a continuous scatterer (e.g., a portion of string of different density) is
also dealt with. This problem is analogous to that of
"potential barriers" in quantum mechanics, and is
mathematically almost identical for the one-dimensional problem.

The treatment, up to this point, has been confined
to the *uniform* string. Turning now to the inhomogeneous one, a different approach is adopted, – that of
separation of variables. This is the almost universal
method of dealing with partial differential equations
in mathematical physics. It leads to the so-called
Sturm–Liouville eigenvalue problem. In the case of the
homogeneous string, it leads again to the Fourier (har-

[2]New York: Dover, 1946.

monic) expansion. Solutions are sought such that every point of the string vibrates with the same frequency.

It is found that the boundary conditions can in general only be satisfied for specific discrete frequencies, which determine the *spectrum* of excitations. Each such solution has its characteristic pattern (normal mode). The *orthogonality* of the normal modes is proven; from this it follows (again, as in the homogeneous case) that each normal mode has its characteristic energy, and there is no mutual scattering.

Eigenvalue problems of the type dealt with here are the commonest form of problems met in quantum mechanics, as is seen in Chapters 8 and 9.

The next topic introduced is that of the variational principle of Rayleigh and Ritz. It is shown that the normal modes are those solutions which minimize a certain quantity, which is related to the frequency of the mode. This principle, which also has wide application in quantum mechanics (especially where exact solutions cannot conveniently be found), permits us to determine the normal frequencies (particularly the lowest ones) approximately. Examples are given.

It is shown that general solutions may be expressed as linear combinations of "normal modes." The variational principle is used in this demonstration.

The problem of a string acted upon by external forces is used to introduce the "Green's function" of the problem. The Green's function is the response of the string to a "unit impulse" or fundamental disturbance, spatially localized and instantaneous in time. The response to a general disturbance is then obtained by adding those of the constituent "unit impulses." This approach is used in *all* areas of mathematical physics, and is a very fruitful one. The particular case of a periodic disturbance, leading to the phenomenon of "resonance," is fully discussed.

Two useful approximate methods are then introduced, each of which is best known for its use in quantum mechanics. These methods are used when exact analytical solutions are unattainable. The first is the method of "perturbations," applicable when the problem differs only by a small amount from a soluble one. The other is the JWKB (Jeffreys–Wentzel–Kramers–

Brillouin) method, applicable when the properties of the string are slowly varying. Examples are given of both methods.

Finally, the string problem is treated by the Lagrangian and Hamiltonian methods of classical mechanics. This is a simple prototype of the problem of the dynamics of continuous fields. Since Hamiltonian mechanics can be used as the basis of the quantum mechanics of a problem, methods and concepts similar to those developed here (though of course more complicated in detail) are fundamental to such problems as quantum electrodynamics and quantum field theory.

REFERENCES

Coulson, C. A., *Waves*. Edinburgh: Oliver and Boyd, (5th ed.) 1948.

Goldstein, H., *Classical Mechanics*. Reading, Mass.: Addison-Wesley, 1957.

Lord Rayleigh, *Theory of Sound*. New York: Dover, 1945, 2 Vols.

Morse, P. M., *Vibration and Sound*. New York: McGraw-Hill, 1948, Chapter 3.

Schwartz, L., *Mathematics for the Physical Sciences*. Reading, Mass.: Addison-Wesley, 1966.

1
THE VIBRATING STRING

"Harp not on that string"

Shakespeare, Richard III

1. Introduction

We begin with a discussion of the stretched vibrating string for a number of reasons. In the first place, it is a simple one-dimensional system, and so does not lead to too many mathematical complications. Secondly, it gives us an opportunity, in this fairly simple context, to discuss the main features of wave propagation. Furthermore, it is a problem which may be linked rather directly with the methods of classical mechanics, which are usually first met for discrete systems but may here be extended to a continuous one. Finally, it provides a testing ground for many of the concepts and methods which are met later in more complicated systems.

2. Derivation of the Equation of Motion

We concentrate our attention on the purely transverse small vibrations of an elastic string, such as a violin string, under constant tension.

Transverse vibrations are defined as those in which each point of the string is displaced perpendicular to its length. We deal, later, with a purely *longitudinal* motion, in which each point is displaced *along* the string.

5

Let x be a coordinate measured along the length of the string, and $y(x, t)$, ($t =$ time), be the transverse displacement at point x and time t. Thinking in terms of conventional dynamics, the coordinate x enumerates the degrees of freedom of the system, which are here continuous. The dynamical variables are the y's for each value of x and they are, of course, functions of the time. The problem is idealized by taking the cross section of the string to be negligibly small.

Because the dynamical variables form a continuous set, a little care must be taken to set up the equations of motion. The variables are, of course, coupled to each other. The most convenient device to deal with this problem is to consider the motion of the center of mass of a very small portion of the string. The length of this portion may then be allowed ultimately to approach zero.

Figure 1.1

Let us imagine that we look at a very small portion of the string, between points x and $x + \Delta x$, under a magnifying glass (Figs. 1.1 and 1.2).

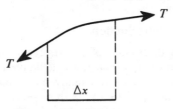

Figure 1.2

The only significant forces acting on this portion of string are assumed to be the forces of tension on the two ends. Since the slope of the string at any point is $\partial y/\partial x$, the component of tension in the y direction at the right-hand end is

$$ T \left[\frac{\partial y}{\partial x} \bigg/ \sqrt{1 + \left(\frac{\partial y}{\partial x}\right)^2} \right]_{x + \Delta x} $$

where the subscript $x + \Delta x$ indicates that the quantity in the bracket is evaluated at this point, and T is the force of tension in the string. The corresponding component at the left-hand end is

$$ - T \left[\frac{\partial y}{\partial x} \bigg/ \sqrt{1 + \left(\frac{\partial y}{\partial x}\right)^2} \right]_{x} $$

The net force in this direction is the difference of the above quantities. By the mean value theorem this difference is

$$T\frac{\partial}{\partial x}\left[\frac{\partial y}{\partial x}\left/\sqrt{1+\left(\frac{\partial y}{\partial x}\right)^2}\right.\right]\Delta x \tag{1.1}$$

the whole expression being evaluated at some point between x and $x+\Delta x$.

The mass times y component of acceleration for the segment of string is

$$\Delta x\left[\rho\frac{\partial^2 y}{\partial t^2}\right] \tag{1.2}$$

ρ being the density (mass per unit length) of the string. The quantity in the square bracket is a mean value on $(x, x+\Delta x)$; this, once again, is the value at some intermediate point.

If the quantities (1.1) and (1.2) are now equated to express the Newtonian law of motion for the element of string, and Δx is allowed to approach zero, the equation of motion is

$$\rho\frac{\partial^2 y}{\partial t^2}=T\frac{\partial}{\partial x}\left[\frac{\partial y}{\partial x}\left/\sqrt{1+\left(\frac{\partial y}{\partial x}\right)^2}\right.\right] \tag{1.3}$$

If, for some reason, the tension were dependent on position, represented by a function $T(x)$, it would appear under the derivative.

Equation (1.3) is nonlinear and is difficult to solve. The situation is simplified, however, if $\partial y/\partial x \ll 1$, in which case the square root may be replaced by unity. The equation then becomes

$$\rho\frac{\partial^2 y}{\partial t^2}=\frac{\partial}{\partial x}\left(T\frac{\partial y}{\partial x}\right) \tag{1.4}$$

This is the equation for small vibrations, by which we mean vibrations for which the slope $\partial y/\partial x$ is very small. It is this equation which we study in some detail.

Assuming T and ρ constant, and putting

$$T/\rho = c^2 \tag{1.5}$$

we obtain the one-dimensional wave equation

$$\frac{\partial^2 y}{\partial x^2}=\frac{1}{c^2}\frac{\partial^2 y}{\partial t^2} \tag{1.6}$$

A similar equation holds for purely longitudinal oscillations. We let y represent the displacement of a point x on the string from its equilibrium position. We define the *strain* in the string as the extension per unit length; this quantity varies from point to point. The portion of the string initially between x and $x+\Delta x$ is stretched by an amount

$$y(x+\Delta x)-y(x)$$

so that the strain is $\partial y/\partial x$. y is now the *longitudinal* displacement.

Let the "stress" be defined as the excess over its equilibrium value of the tension of the string at a point. It is then assumed that the stress is proportional to the strain, being equal to.

$$k\frac{\partial y}{\partial x}$$

where k is the Young's Modulus.

As in the case of transverse displacement, the net force acting on the element Δx is the excess of the stress at $x + \Delta x$ over that at x; thus, the force per unit length is

$$\frac{\partial}{\partial x}\left(k\frac{\partial y}{\partial x}\right)$$

k may, of course, be variable if the wire is not homogeneous. Since the mass per unit length times the acceleration is equal to the force per unit length, the equation of motion is

$$\rho\frac{\partial^2 y}{\partial t^2} = \frac{\partial}{\partial x}\left(k\frac{\partial y}{\partial x}\right) \tag{1.7}$$

which is formally the same as (1.4).

3. Solution of the Equation

It may be immediately verified that

$$y = f_1(x - ct) \tag{1.8}$$

and

$$y = f_2(x + ct) \tag{1.9}$$

are solutions of the equation, where f_1 and f_2 are arbitrary functions. Each of these solutions has a simple physical interpretation. In the case of (1.8), the displacement y at point x and time t is the same as that at point $x + c\Delta t$ at time $t + \Delta t$. Thus, the whole pattern of disturbance of the string moves in the direction of increasing x with a velocity c. c is therefore the "velocity of propagation" of the disturbance.

In the case of (1.9), on the other hand, the displacement at point x at time t is found, at time $t + \Delta t$, at the point with coordinate $x - c\Delta t$. This disturbance, therefore, travels in the direction of decreasing x with velocity c.

The general solution of (1.6) is the sum of solutions such as (1.8) and (1.9):

$$y = f_1(x - ct) + f_2(x + ct) \tag{1.10}$$

The functions f_1 and f_2 are determined by the initial conditions of the motion. We know from general dynamical principles that if the initial values of the coordinates and their time rates of change (velocities) are known, the motion is completely determined. Let these initial conditions be that

$$y = y_0(x) \quad \text{at} \quad t = 0 \tag{1.11}$$

and

$$\frac{\partial y}{\partial t} = v_0(x) \quad \text{at} \quad t = 0 \tag{1.12}$$

If they are imposed on the solution (1.10), f_1 and f_2 are expressed in terms of $y_0(x)$ and $v_0(x)$ by the equations

$$y_0 = f_1 + f_2 \tag{1.13}$$

and

$$v_0 = -cf_1' + cf_2' \tag{1.14}$$

The primes represent, in the first instance, derivatives of the functions f_1 and f_2 with respect to their arguments; however, at $t = 0$ the argument is simply x.

Equations (1.13) and (1.14) may be solved for f_1 and f_2:

$$f_1(x) = \tfrac{1}{2}\left[y_0(x) - \frac{1}{c}\int_{x_1}^{x} v_0(x')dx' \right] \tag{1.15}$$

and

$$f_2(x) = \tfrac{1}{2}\left[y_0(x) + \frac{1}{c}\int_{x_1}^{x} v_0(x')dx' \right] \tag{1.16}$$

The lower limit x_1 on the integrals is arbitrary; in any case, it does not appear in the complete solution, which is

$$y(x, t) = \tfrac{1}{2}\left[y_0(x - ct) + y_0(x + ct) + \frac{1}{c}\int_{x-ct}^{x+ct} v_0(x')\, dx' \right] \tag{1.17}$$

This solution has a simple physical interpretation, which is most easily understood if we consider first two particular cases, in which $v_0(x) = 0$ and $y_0(x) = 0$, respectively. The two particular solutions may subsequently be combined to give (1.17).

When $v_0 = 0$ the motion is given by

$$y = \tfrac{1}{2}[y_0(x - ct) + y_0(x + ct)] \tag{1.18}$$

This says that half of the initial disturbance travels in the positive direction and half in the negative direction. This situation is illustrated in Fig. 1.3.

Figure 1.3(a) shows the initial state. Figure 1.3(b) shows the displacement a little later, while the two displacements are still partially overlapping. Figure 1.3(c) shows the situation at a later time, when they are completely separated and traveling away from each other.

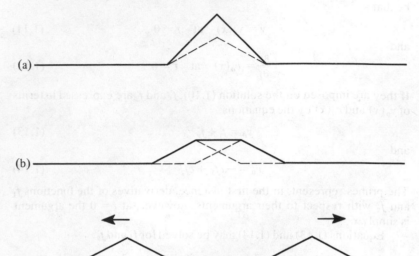

(a)

(b)

(c)

Figure 1.3

The case just discussed is that in which the string is given an initial displacement, and released from rest. Consider now the case in which there is no initial displacement, but an impulse represented by a velocity distribution $v_0(x)$. The solution in this case, which in general must be added to the previous one, is

$$y = \tfrac{1}{2}[y_1(x+ct) - y_1(x-ct)] \tag{1.19}$$

where

$$y_1(x) = \frac{1}{c}\int_{x_0}^{x} v_0(x')\,dx' \tag{1.20}$$

This solution again consists of two parts, representing disturbances which travel in opposite directions, but also in this instance have opposite signs. If $v_0 > 0$, the disturbance traveling to the left is positive, while that traveling to the right is negative. At $t = 0$, of course, they completely cancel. The situation is portrayed in Fig. 1.4. Figure 1.4(a) shows an initial *velocity* distribution of a simple sort. Figure 1.4(b) displays the functions y_1, $-y_1$. In Fig. 1.4(c), the disturbances have moved a small distance relative to each other. In Fig. 1.4(d), they have moved much further.

The preceding discussion deals with the propagation of disturbances on a string of indefinite length, that is, one for which the effect of end conditions may be neglected. Suppose, however, the string is finite, the ends being at $x = 0$ and $x = L$. Then, the functions $y_0(x)$ and $v_0(x)$ are defined only on this range. Clearly, however, (1.17) requires them to be defined for *all* values. The problem is, then, how to extend the definition of these functions to arbitrary values of the argument.

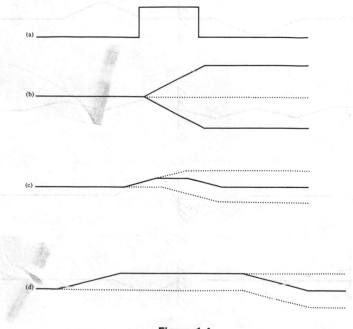

(a)

(b)

(c)

(d)

Figure 1.4

Again, for simplicity, consider first the case in which $v_0 = 0$, and in which there is a single fixed end at $x = 0$. If the solution (1.18) is to satisfy this condition it is necessary that, for *all* values of t,

$$y_0(-ct) = -y_0(ct) \tag{1.21}$$

The boundary condition then tells us how to extend the function into the nonphysical region; it is simply to be defined as an odd function with respect to the end point. Once this condition has been applied, the equation (1.18) may be used, and can be used to describe the reflection of a wave from the end. This is illustrated in Fig. 1.5. In (a) there is shown an initial disturbance, along with its "reflection" in the fixed end. Half of it travels in each direction, and in (b) the half which is traveling to the left is shown at the moment at which it first reaches the end. Thereafter, the part of the "reflection" which is traveling to the right begins to overlap it, and must be added to it. Successive stages of this interference are shown in (c) and (d). Finally, the original disturbance has passed completely into the unphysical region to the left, while that which was originally in that region travels on the string toward the right, as shown in (e).

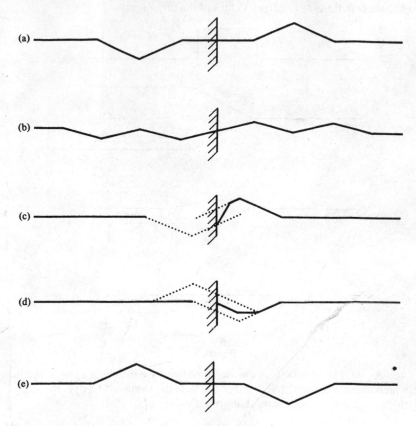

For clarity, (c), (d) and (e) are drawn on a larger scale than (a) and (b).

Figure 1.5

Physically, the process of reflection at the fixed end of the string is then seen as a reversal in direction of propagation of the disturbance which impinges on the end, as well as an inversion (reversal in the direction of displacement).

In the case of a motion resulting from an initial *velocity* distribution v_0 it is evident from Eq. (1.19) that the function $y_1(x)$ must be an *even* function; that is to say, that

$$\int_{x_0}^{x} v_0(x')dx' = \int_{x_0}^{-x} v_0(x')dx'$$

where we have again taken the fixed end at $x = 0$. Differentiation of both sides of the equation with respect to x gives

$$v_0(x) = -v_0(-x) \tag{1.22}$$

so that the definition of v_0, too, must be extended to make it an odd function. The lower limit in the integral defining y_1 can conveniently be taken to be zero. Although y_1 in the nonphysical region is of the same sign as in the physical region, it appears with a *minus* sign in the solution. Thus, again, one adds to the "real" disturbance its inverted reflection in the end of the string.

Suppose now that there are *two* fixed ends, at $x = 0$ and at $x = L$. In physical terms, we can envisage the motion as a propagation of the initial disturbance followed by successive reflections of its parts at the two ends; the reflected disturbances will then interfere with, that is to say, be added to, each other. Although the resulting disturbance may at first sight appear to be quite complicated, it may in fact be visualized in a rather simple way. Again, the most convenient device is to extend the definition of the displacement function $y(x)$ outside its original range of definition ($0 \leq x \leq L$). As already seen, it must be an *odd* function with respect to each end. In mathematical terms, this is expressed by the equations

$$y(-x) = -y(x) \tag{1.23}$$

and

$$y(-x+L) = -y(x+L) \tag{1.24}$$

Combining these two equations yields the interesting result that

$$y(x+L) = y(x-L) \tag{1.25}$$

so that the function must be extended in such a way as to be periodic with a period equal to *twice* the length of the string. The complete function can then be simply constructed as follows: The physical displacement on the range $(0, L)$ is inverted and reflected in the region $(0, -L)$, and then extended over successive regions of length $2L$ by periodicity. To state it differently, the function may be constructed from its value in $(0, L)$ by inverted reflections in its two ends, followed by *their* inverted reflections in the two ends, and so forth, thus producing an infinite sequence of "images."

To describe the motion, the resulting function must be considered as consisting of two equal parts, which move with velocity c in opposite directions, and interfere with each other.

Problem 1-1: Describe in detail the reflection at an end at which the boundary condition is $\partial y/\partial x = 0$. (This is an end which is free to move transversely, since the tension can do no work on the support.)

The fact of periodicity may be used to express the solution in a rather different form. For, if y_0 and v_0 are odd functions with period $2L$, they may be expressed in Fourier sine series on that interval:

$$y_0(x) = \sum_{n=1}^{\infty} a_n \sin \frac{n\pi x}{L} \tag{1.26}$$

$$v_0(x) = \sum_{n=1}^{\infty} b_n \sin \frac{n\pi x}{L} \tag{1.27}$$

where the coefficients a_n and b_n may be obtained from the standard formulas in terms of the given initial conditions:

$$a_n = \frac{2}{L} \int_0^L y_0(x') \sin \frac{n\pi x'}{L} \, dx' \tag{1.28}$$

and

$$b_n = \frac{2}{L} \int_0^L v_0(x') \sin \frac{n\pi x'}{L} \, dx' \tag{1.29}$$

Substituting the series (1.26) and (1.27) into the general solution (1.17) yields the solution in the form

$$
\begin{aligned}
y &= \frac{1}{2} \sum a_n \left[\sin \frac{n\pi(x-ct)}{L} + \sin \frac{n\pi(x+ct)}{L} \right] \\
&\quad + \frac{1}{2c} \sum b_n \frac{L}{n\pi} \left[\cos \frac{n\pi(x-ct)}{L} - \cos \frac{n\pi(x+ct)}{L} \right] \\
&= \sum \sin \frac{n\pi x}{L} \left(a_n \cos \frac{n\pi ct}{L} + \frac{L}{n\pi c} b_n \sin \frac{n\pi ct}{L} \right)
\end{aligned}
\tag{1.30}
$$

We see later that this solution is the one obtained by a quite different approach — that of separation of variables. This latter method is, however, more flexible, since it can be extended to the case of *variable* ρ and T. The present approach runs into difficulties in that case, in that it could only be pursued if one could write a general solution of the differential equation analogous to (1.10).

4. Energy of the String

Consider now the energy associated with the motion of the string. This is most naturally expressed in terms of an *energy density*, that is, energy per unit length. The calculation of kinetic and potential energies is essential to the formulation of the problem in terms of classical Lagrangian or Hamiltonian mechanics. (The problem is complicated, of course, by the fact that the degrees of freedom form a continuum. This, the usual *sums* over discrete coordinates will be replaced by *integrals* over the variable x which enumerates the present continuum of coordinates.)

The kinetic energy per unit length of string is clearly seen to be

$$\mathcal{K} = \frac{1}{2}\rho\left(\frac{\partial y}{\partial t}\right)^2$$

and the total kinetic energy

$$K = \frac{1}{2}\int \rho\left(\frac{\partial y}{\partial t}\right)^2 dx \tag{1.32}$$

the integral being taken over the whole length of the string, which we take to be $(0 \leqslant x \leqslant L)$.

The determination of the potential energy is only slightly more difficult. If a portion of the string of initial length dx is, when displaced, stretched to a length ds, the increase in length is

$$ds - dx = \left[\sqrt{1 + \left(\frac{\partial y}{\partial x}\right)^2} - 1\right]dx$$

In the approximation of "small vibrations" ($\partial y/\partial x \ll 1$) this becomes

$$\frac{1}{2}\left(\frac{\partial y}{\partial x}\right)^2 dx$$

Since this stretching takes place against a force of tension T, the potential energy gain, which is the work done against tension, is

$$\frac{1}{2}T\left(\frac{\partial y}{\partial x}\right)^2 dx$$

so that the potential energy density is

$$\mathcal{V} = \frac{1}{2}T\left(\frac{\partial y}{\partial x}\right)^2 \tag{1.33}$$

and the total potential energy is

$$V = \frac{1}{2}\int_0^L T\left(\frac{\partial y}{\partial x}\right)^2 dx \tag{1.34}$$

Problem 1-2: Derive the potential and kinetic energies for the small longitudinal vibrations of a wire.

The total energy is therefore

$$H = \int_0^L \mathcal{H}\, dx \qquad (1.35)$$

where the energy density $\mathcal{H}$ is

$$\mathcal{H} = \frac{1}{2}\rho\left(\frac{\partial y}{\partial t}\right)^2 + \frac{1}{2}T\left(\frac{\partial y}{\partial x}\right)^2 \qquad (1.36)$$

It is also possible to calculate the flux of energy past a given point x. This is the negative of the rate at which the y component of tension is doing work on the portion of the string to the left of x (thus, the flux is taken to be positive if energy is being transmitted from left to right, and vice versa). Since, as shown above, the y component of tension is $T(\partial y/\partial x)$, its rate of working is $(T(\partial y/\partial x))\partial y/\partial t$. Designating the rate of energy flux as J, it follows that

$$J = -T\frac{\partial y}{\partial x}\frac{\partial y}{\partial t} \qquad (1.37)$$

From (1.36) and (1.37), along with the equation of motion, it is possible to derive an equation of continuity for the energy. This equation expresses the fact that the rate of change of energy on a portion of the string

$$\frac{\partial}{\partial t}\int_{x_1}^{x_2} \mathcal{H}\, dx \qquad (1.38)$$

is equal to the rate at which energy flows *into* the section over the ends:

$$-T\frac{\partial y}{\partial x}\frac{\partial y}{\partial t}\bigg|_{x_1} + T\frac{\partial y}{\partial x}\frac{\partial y}{\partial t}\bigg|_{x_2} = -J(x_2) + J(x_1) \qquad (1.39)$$

For, calculating (1.38),

$$\frac{\partial}{\partial t}\int_{x_1}^{x_2}\left[\frac{1}{2}\rho\left(\frac{\partial y}{\partial t}\right)^2 + \frac{1}{2}T\left(\frac{\partial y}{\partial x}\right)^2\right]dx = \int_{x_1}^{x_2}\left[\rho\frac{\partial y}{\partial t}\frac{\partial^2 y}{\partial t^2} + T\frac{\partial y}{\partial x}\frac{\partial^2 y}{\partial x\partial t}\right]dx$$

Substituting for $(\rho(\partial^2 y/\partial t^2))$ from (1.4), this becomes

$$\frac{\partial}{\partial t}\int_{x_1}^{x_2}\mathcal{H}\, dx = \int_{x_1}^{x_2}\left[\frac{\partial y}{\partial t}\frac{\partial}{\partial x}\left(T\frac{\partial y}{\partial x}\right) + T\frac{\partial y}{\partial x}\frac{\partial}{\partial x}\left(\frac{\partial y}{\partial t}\right)\right]dx$$

$$= \frac{\partial y}{\partial t}T\frac{\partial y}{\partial x}\bigg|_{x_1}^{x_2}$$

$$= -J(x_2) + J(x_1)$$

as stated.

Alternatively, given the expression for the energy density, the physically obvious condition of continuity could be used to *define* the energy flux J.

It is instructive to calculate the energy density and energy flux for the general solution (1.10)

$$y = f_1(x-ct) + f_2(x+ct)$$

The kinetic energy density is

$$\mathscr{K} = \tfrac{1}{2}\rho c^2 (f_1' - f_2')^2 \tag{1.40}$$

where the primes indicate derivatives of the functions with respect to their arguments. The potential energy density is

$$\mathscr{V} = \tfrac{1}{2}T(f_1' + f_2')^2$$
$$= \tfrac{1}{2}\rho c^2 (f_1' + f_2')^2 \tag{1.41}$$

Thus the total energy density is

$$\mathscr{H} = \rho c^2 (f_1'^2 + f_2'^2) \tag{1.42}$$

The important thing to note is that the waves moving in opposite directions each have their own energies. While there are interference terms in the kinetic and potential energy densities individually, there are no such terms in the total energy. It is of course true also that the energy in each of the waves is constant, since each is an integral over all values of x, and therefore of the argument. This is true for a finite string because of the periodicity condition and the condition of oddness in a fixed end, which ensures that the integral of the square of the function is the same over every interval of length L. Thus, constancy of the integral over all x implies constancy of that on $(0, L)$.

Consider next the energy flux

$$J = -T(f_1' + f_2')(-cf_1' + cf_2')$$
$$= Tc(f_1'^2 - f_2'^2) \tag{1.43}$$

Thus, the two waves flow independently. In the case of the wave moving in the direction of positive x, the flux J_1 bears to the energy density $\mathscr{H}_1$ the ratio

$$\frac{J_1}{\mathscr{H}_1} = \frac{T}{\rho c} = c \tag{1.44}$$

so that c is the rate of propagation of energy as well as the phase velocity of the wave. Designating the energy flux and density of the wave traveling in the direction of negative x as J_2 and $\mathscr{H}_2$, respectively,

$$\frac{J_2}{\mathscr{H}_2} = -c \tag{1.45}$$

which says that energy in this component is transferred in the *negative* direction with a velocity c.

5. Energy in the Harmonics

In the preceding section we have an example of waves which interfere with each other so far as point by point displacement is concerned, but do not exchange energy with each other. This result is striking when one considers that two disturbances of finite extent and of the same shape but opposite in sign, and traveling in opposite directions, can momentarily extinguish each other in passing, and yet will, when they cease to interfere, each regain its original shape and propagate its characteristic energy.

An even more striking result, which we demonstrate in this section, is that the individual terms in the Fourier series solution (1.30) are similarly independent, in that each carries a fixed amount of energy, and this energy cannot be exchanged with the energies associated with the other terms. The terms in the Fourier series are designated as "normal modes" or harmonics, each of which represents a possible mode of motion of the string with a definite frequency. We subsequently discuss the question of normal modes in a more fundamental way and in a more general context, but it is interesting to demonstrate at this point the constancy of the energy of each normal mode.

The Fourier series solution is, as shown in (1.30),

$$y = \sum_{n=0}^{\infty} c_n \sin \frac{n\pi x}{L} \sin \left(\frac{n\pi ct}{L} - \alpha_n \right) \qquad (1.46)$$

where

$$c_n \cos \alpha_n = a_n \qquad (1.47)$$

and

$$c_n \sin \alpha_n = \frac{L}{n\pi c} b_n$$

In calculating the energy we make use of the "orthogonality" of the expansion functions, that is to say, of the fact that

$$\int_0^L \sin \frac{n\pi x}{L} \sin \frac{n'\pi x}{L} dx = \frac{L}{2} \delta_{nn'} \qquad (1.48)$$

where

$$\delta_{nn'} = 0, \quad n \neq n'$$
$$= 1, \quad n = n' \qquad (1.49)$$

and of a similar equation for the integral of two cosines of the same arguments. The kinetic energy in this case is

$$K = \frac{1}{2}\rho \int_0^L \left[\sum_n \frac{n\pi c}{L} c_n \cos\left(\frac{n\pi ct}{L} - \alpha_n\right) \sin\frac{n\pi x}{L} \right]^2 dx$$

$$= \frac{1}{4}\rho L \sum_n \left(\frac{n\pi c}{L}\right)^2 c_n{}^2 \cos^2\left(\frac{n\pi ct}{L} - \alpha_n\right) \tag{1.50}$$

the cross terms dropping out because of (1.48). Similarly the potential energy is

$$V = \frac{1}{2}T \int \left[\sum_n \frac{n\pi}{L} c_n \cos\frac{n\pi x}{L} \sin\left(\frac{n\pi ct}{L} - \alpha_n\right) \right]^2 dx$$

$$= \frac{1}{4}\rho L \sum_n \left(\frac{n\pi c}{L}\right)^2 c_n{}^2 \sin^2\left(\frac{n\pi ct}{L} - \alpha_n\right) \tag{1.51}$$

where we have used the fact that $T = \rho c^2$.

It is evident that there are no terms representing interactions of the normal modes in either the kinetic or potential energy. The potential and kinetic energies of a normal mode taken together give a constant; the total energy is

$$H = \frac{1}{4}\rho L \sum_n \left(\frac{n\pi c}{L}\right)^2 c_n{}^2 \tag{1.52}$$

In each mode, the energy oscillates between kinetic and potential forms, as the string itself oscillates. The periods of the oscillations of the string are

$$\tau_n = \frac{2L}{nc} \tag{1.53}$$

while those of the energy are half that great, corresponding to successive realizations of a given phase of the motion.

The decomposition of the energy of the string into contributions from distinct normal modes appears in this problem as a mathematical *tour de force* which is not at all physically evident. Only when we come to the development of quantum mechanics will this sort of phenomenon be given a specifically *physical* connotation, and thus take on a very fundamental significance.

6. The "Loaded" String

In the following sections we consider three special problems. First, we treat the problem of the reflection and transmission of a wave on the string by a mass M attached at a fixed point, which for convenience we take to be $x = 0$. The results of this calculation are then applied to propagation of waves on a string to which point masses are attached at regular intervals. Finally, we deal with the problem in which a mass is distributed uniformly over a finite length a, that is, in which such a portion of the string has a density different from that of the rest of the string. In each instance we consider a periodic wave of circular frequency ω.

7. Reflection and Transmission at a Fixed Mass

The effect of the mass may be formulated as a boundary condition to be applied at its position. The motion of the mass is determined by the condition that $M(\partial^2 y/\partial t^2)_{x=0}$ = net y component of tensional forces on the two sides of the mass. In the small-vibration approximation this is $T[(\partial y/\partial x)_{0+} - (\partial y/\partial x)_{0-}]$, where by 0+ we mean a point just to the right of $x = 0$ and by 0− a point just to the left of it. In other words, the effect of the mass is to produce a discontinuity in the slope of the string of amount

$$\left(\frac{\partial y}{\partial x}\right)_{0+} - \left(\frac{\partial y}{\partial x}\right)_{0-} = \frac{M}{T}\left(\frac{\partial^2 y}{\partial t^2}\right)_0 \tag{1.54}$$

Problem 1-3: A disturbance of arbitrary form $y = f(ct - x), (x < 0)$, travels along an infinite string in the direction of increasing x and impinges on a discrete mass M attached to the string at $x = 0$.

Show that the wave transmitted to the region of positive x is

$$y = \frac{2\rho}{M} \int_{-\infty}^{ct-x} f(x') e^{2(\rho/M)(x+x')} \, dx' \, e^{-2(\rho/M)ct}$$

$$= f_1(ct - x)$$

and that the reflected wave is

$$y = \frac{2\rho}{M} \int_{-\infty}^{ct+x} f(x') e^{-2(\rho/M)(x-x')} \, dx' \, e^{-2(\rho/M)ct} - f(ct + x)$$

Show that the fractions of the energy of the incident wave which are transmitted and reflected are

$$\tau = \frac{\int_0^\infty [f_1'(ct - x)]^2 \, dx}{\int_0^\infty [f'(ct - x)]^2 \, dx} \quad \text{and} \quad (1 - \tau)$$

respectively, $f(x)$ and $f_1(x) = 0$ for $x < 0$.

Problem 1-4:   Do the calculations explicitly for the case of an incident wave:

$$f(x) = e^{ikx}$$

What special problems arise from taking a *complex* function $f(x)$?

If the disturbance being propagated is periodic, its time behavior may be described by $e^{-i\omega t}$; it is, therefore,

$$y = A e^{ik(x-ct)} \tag{1.55}$$

where

$$k = \omega/c \tag{1.56}$$

The use of the complex exponential must of course be properly understood, since the displacement y is obviously a real quantity. An equation like (1.55) is to be understood to mean that the physical quantity is to be taken to be either the real or imaginary part of the expression on the right-hand side. If A is taken as complex

$$A = A_0 e^{i\alpha} \tag{1.57}$$

it does not matter whether the real or imaginary part is chosen; the imaginary part for a given α is the same as the real part for an α which is greater by $\pi/2$. Therefore, in general, we adopt the convention that the physical quantity is the *real* part of the complex function.

Consider a wave on the left-hand side of the string ($x < 0$), traveling toward the right, and therefore represented by (1.55). The reflected wave will be a wave of the same frequency on the range $x < 0$ and traveling to the left. Therefore the displacement on $x < 0$ is given by

$$y = A e^{ik(x-ct)} + B e^{-ik(x+ct)} \tag{1.58}$$

On the right there will be only a *transmitted* wave traveling to the right:

$$y = C e^{ik(x-ct)} \tag{1.59}$$

An additional term in $e^{-ik(x+ct)}$ on the right would give rise to a valid solution on that side, but would not correspond to the given conditions of the problem, since it would represent a wave *incident* from the right.

There are two boundary conditions, which serve to determine B and C in terms of the given incident amplitude A. Aside from (1.54), there is the obvious condition that y must have the same value whether calculated from (1.58) or (1.59). The two boundary equations are, then

$$A + B = C \tag{1.60}$$

and

$$ikC - ik(A - B) = -\frac{M}{T} k^2 c^2 C \tag{1.61}$$

These may be solved to give

$$B = A \frac{iMk/2}{\rho - iMk/2} \tag{1.62}$$

$$C = A \frac{\rho}{\rho - iMk/2} \tag{1.63}$$

If we introduce

$$\tan \beta = Mk/2\rho \tag{1.64}$$

the equations for B and C may be written as

$$B = A \sin \beta e^{i(\beta + \pi/2)} \tag{1.65}$$

and

$$C = A \cos \beta e^{i\beta} \tag{1.66}$$

As $M \to 0$, $\beta \to 0$ and $B \to 0$, so that the reflected wave ceases to exist; since $C = 1$ there is complete transmission. In general the transmitted wave suffers a phase change of β, while the reflected wave is further changed in phase by $\pi/2$.

The physical displacements are, on the left

$$y = A_0 \cos [k(x - ct) + \alpha] + A_0 \sin \beta \sin [k(x + ct) + \alpha + \beta] \tag{1.67}$$

and on the right

$$y = A_0 \cos \beta \cos [k(x - ct) + \alpha + \beta] \tag{1.68}$$

Since, as we have seen, the energies of waves traveling in opposite directions are distinct, we may use (1.42) to write the energy densities in each of the waves:

incident: $\rho \omega^2 A_0^2 \cos^2 [k(x - ct) + \alpha]$ $\tag{1.69}$

reflected: $\rho \omega^2 A_0^2 \sin^2 \beta \sin^2 [k(x + ct) + \alpha + \beta]$ $\tag{1.70}$

transmitted: $\rho \omega^2 A_0^2 \cos^2 \beta \cos^2 [k(x - ct) + \alpha + \beta]$ $\tag{1.71}$

The situation is somewhat simpler if we consider time averages, in which case each trigonometric function of time is replaced by $\frac{1}{2}$. It is evident then that the fraction of energy transmitted is

$$f_T = \cos^2 \beta \tag{1.72}$$

and the fraction reflected is

$$f_R = \sin^2 \beta \tag{1.73}$$

the sum being, as it should be, equal to unity.

If the mass were attached, not at $x = 0$ but at $x = x_0$, the above calculation would be altered only in that B would be multiplied by an extra factor e^{2ikx_0}.

It may be verified in a similar manner that if a wave $D e^{-ik(x+ct)}$ impinges from the right on a mass M at x_0, we have a transmitted wave

$$B = D \cos \beta e^{i\beta} \tag{1.74}$$

and a reflected wave

$$C = D \sin \beta e^{i(\beta + \pi/2)} e^{-2ikx_0} \tag{1.75}$$

In (1.65), (1.66), (1.74), and (1.75) we have four equations relating transmitted and reflected amplitudes to incident ones. These may be given a matrix formulation, which relates the matrix of the transmitted amplitudes, $\binom{C}{B}$, to that for the incident amplitudes on the two sides, $\binom{A}{D}$. This matrix is a simple example of a scattering matrix, and in this instance takes the form

$$S = e^{i\beta} \begin{pmatrix} \cos \beta & i \sin \beta\, e^{-2ikx_0} \\ i \sin \beta\, e^{2ikx_0} & \cos \beta \end{pmatrix} \tag{1.76}$$

If we write the transpose conjugate of this matrix

$$S^+ = e^{-i\beta} \begin{pmatrix} \cos \beta & -i \sin \beta\, e^{-2ikx_0} \\ -i \sin \beta\, e^{2ikx_0} & \cos \beta \end{pmatrix} \tag{1.77}$$

it is easily verified that

$$SS^+ = S^+S = 1 \tag{1.78}$$

Such a matrix is known as a *unitary* matrix, and is discussed in the chapter on linear vector spaces.

The utility of the scattering matrix lies in the fact that it gives, for any given wave incident on the scatterer, the amplitudes of the scattered waves, that is, of the waves emerging from the scatterer.

The unitary property of the matrix expresses conservation of energy —the fact that the total flux of energy in the scattered waves is equal to that in the incident wave. For the incident wave is completely specified by a matrix M, and the scattered one by a matrix M' related to the incident by

$$M' = SM \tag{1.79}$$

But by virtue of (1.37), the time average of energy flux in a wave is proportional to the magnitude squared of its amplitude (the proportionality factor being the same for all partial waves). Thus the incident energy is $M^+M = (A^*D^*)\binom{A}{D} = A^2 + D^2$, while the scattered energy is

$$M'^+M' = (M^+S^+)(SM) = M^+M \qquad (1.80)$$

by virtue of the unitarity of S.

It may be asked why the time average of the energy flux is used. The reason lies in the fact that in each wave the energy flux oscillates with time; what flows in flows out, but with a time or phase delay. Averaging over time eliminates this complication.

In the corresponding quantum problem, which is treated in a later chapter, this complication is not present, since the flux in that case is constant in time even though the "wave function" is oscillatory.

8.   Propagation on a String with Regularly Spaced Masses Attached

Let the attached masses be located at points $x = na$. In the section to the left of na, let the amplitudes of the waves moving to the right and to the left be designated r_n and l_n, respectively, while those to the right of na are r_{n+1} and l_{n+1}. These correspond to the quantities A, D, C, and B, respectively, of the previous problem. Thus they are related by the matrix equation

$$\binom{r_{n+1}}{l_n} = S_n\binom{r_n}{l_{n+1}} \qquad (1.81)$$

the subscript on the scattering matrix S indicating that the scattering takes place at $x_0 = na$. Relations like (1.81) connect the amplitudes in successive sections. A formal solution of these equations may be obtained by putting

$$r_n = e^{in(\theta - ka)}R \qquad (1.82)$$

and

$$l_n = e^{in(\theta + ka)}L \qquad (1.83)$$

in (1.81), using S from (1.76). Two simultaneous equations are obtained for R and L:

$$(e^{i(\theta - ka)} - \cos\beta\, e^{i\beta})R = ie^{i\beta}e^{i(\theta + ka)}\sin\beta \cdot L \qquad (1.84)$$

and

$$(1 - e^{i\beta}e^{i(\theta + ka)}\cos\beta)L = ie^{i\beta}\sin\beta \cdot R \qquad (1.85)$$

The condition for a nonzero solution is obtained by eliminating R and L. The resulting equation determines θ in terms of $k = \omega/c$ (or conversely).

After some simplification, it is found that

$$\cos \theta = \frac{\cos (\beta + ka)}{\cos \beta} \qquad (1.86)$$

It is evident from this equation that the disturbance is propagated without attenuation along the string only if the right side is of magnitude less than unity. If it is greater than 1, θ must be an imaginary quantity. If $\theta = i\theta'$, where $\theta' > 0$, r_n and l_n will be

$$r_n = e^{-n\theta'} e^{-inka} R \qquad (1.87)$$

$$l_n = e^{-n\theta'} e^{inka} L \qquad (1.88)$$

Thus the disturbance will be attenuated as it propagates to the right. If, on the other hand, $\theta' < 0$, these equations will contain positive rather than negative exponential, and it will be attenuated toward the left.

Consider, then, the conditions under which the magnitude of the right-hand side of (1.86) will be less than unity. Using (1.64) and writing

$$ka = x = \frac{\omega}{c} a \qquad (1.89)$$

and

$$\frac{M}{2\rho a} = \lambda \qquad (1.90)$$

it may be written as

$$f(x) = \cos x - \lambda x \sin x \qquad (1.91)$$

The general features of this function may be simply described. For small frequency ($\omega \ll c/a$) it will be less than 1. The second term in (1.91) will, however, as x becomes progressively larger, grow in amplitude, so that $f(x)$ will be less than 1 for intervals about the points $x = n\pi$, where n is an integer, which decrease in size as n becomes larger. Calling the intervals of ω for which $|f(x)| < 1$ "allowed regions" (representing the ranges of frequencies which can be propagated without attenuation), it is evident that sufficiently low frequencies are allowed, that "allowed" and "forbidden" ranges alternate, and that the allowed regions, which in any case become successively narrower as the frequency increases, do so more rapidly the larger the value of λ; that is, the larger the attached masses in relation to the mass of a section of the string. In Figs. 1.6(a), 1.6(b), and 1.6(c) are graphs of the function $f(x)$ for $\lambda = 0.2$, $\lambda = 1$, and $\lambda = 5$, respectively, the "allowed" regions being indicated by the sections of the x axis.

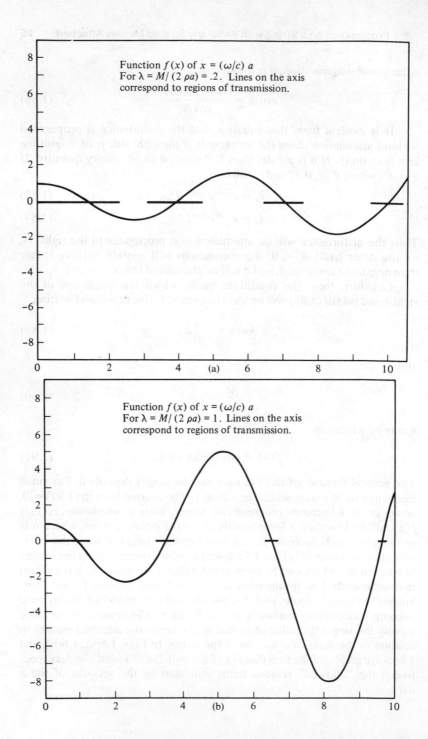

Function $f(x)$ of $x = (\omega/c)\,a$
For $\lambda = M/(2\,\rho a) = .2$. Lines on the axis
correspond to regions of transmission.

(a)

Function $f(x)$ of $x = (\omega/c)\,a$
For $\lambda = M/(2\,\rho a) = 1$. Lines on the axis
correspond to regions of transmission.

(b)

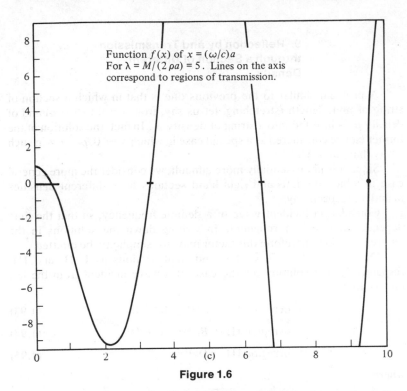

Function $f(x)$ of $x = (\omega/c)a$
For $\lambda = M/(2\rho a) = 5$. Lines on the axis
correspond to regions of transmission.

Figure 1.6

For very large λ, the allowed frequencies all lie close to the values

$$\omega = \frac{n\pi c}{a} \qquad (1.92)$$

In this case the portions of string between the masses vibrate effectively independently.

The greater the excess of $|f(x)|$ over 1, the larger will have to be the imaginary value of θ to satisfy (1.86), and the greater will be the rate of attenuation of the disturbances corresponding to the forbidden frequencies. Thus, the attenuation will be greater in the middle of the forbidden region than near its boundaries, and will in general increase with frequency.

The tendency of periodic systems to exhibit alternately allowed and forbidden frequency ranges is well known in various branches of physics, for example, in filter networks [Brillouin, *Wave Propagation in Periodic Structures*[1]], in the lattice vibrations, and in electronic spectra of solids [Ziman, *Theory of Solids*[2]].

[1]New York: Dover, 1946.
[2]London: Cambridge Press, 1964, Chapters 2 and 3.

9. Reflection by and Transmission through a Section of Different Density

A problem similar to the previous one is that in which a section of string of finite length (stretching, let us say, from $x = 0$ to $x = a$) and of density ρ_2 is inserted into a string of density ρ_1. In fact, the solution of the former can be considered as a special case in which $a \to 0$, $\rho_2 \to \infty$ in such a way that $\rho_2 a = M$.

Since it is not essentially more difficult, we consider the more general case in which the left- and right-hand sections have different densities ρ_1 and ρ_3, respectively.

Consider an incident wave of a definite frequency, so that the time dependence is $e^{-i\omega t}$ throughout. In writing down the solutions in the different section, therefore, this factor may, for simplicity, be omitted.

Designating the left, center, and right sections as I, II, and III, respectively, the solution for the case of a wave incident from the left only is, then

$$\text{in region I:} \quad e^{ik_1 x} + A e^{-ik_1 x} \tag{1.93}$$

$$\text{in region II:} \quad B e^{ik_2 x} + C e^{-ik_2 x} \tag{1.94}$$

$$\text{in region III:} \quad D e^{ik_3 x} \tag{1.95}$$

where

$$k_i = \frac{\omega}{c_i} = \omega \sqrt{\frac{\rho_i}{T}}, \quad i = 1, 2, 3 \tag{1.96}$$

The amplitude of the incident wave may be taken to be unity since all other amplitudes are proportional to it. One boundary condition is that the displacement y should be continuous at the boundaries of the regions. The other is that the slope $\partial y / \partial x$ should be continuous; this follows from (1.54) since there is no discrete mass at the boundaries.

The application of these boundary conditions at $x = 0$ and $x = a$ to the solutions (1.93)–(1.95) yields the equations

$$1 + A = B + C \tag{1.97}$$

$$ik_1(1 - A) = ik_2(B - C) \tag{1.98}$$

$$B e^{ik_2 a} + C e^{-ik_2 a} = D e^{ik_3 a} \tag{1.99}$$

and

$$ik_2(B e^{ik_2 a} - C e^{-ik_2 a}) = ik_3 D e^{ik_3 a} \tag{1.100}$$

The simplest method of solution is to solve (1.97) and (1.98) in terms of A, and (1.99) and (1.100) in terms of D. From the resulting equations we can solve for A (reflected amplitude) and D (transmitted amplitude). The results are

$$D = \frac{2k_1 k_2 e^{-ik_3 a}}{(k_1 + k_3)k_2 \cos k_2 a - i(k_2^2 + k_1 k_3) \sin k_2 a} \qquad (1.101)$$

and

$$A = \frac{k_2(k_1 - k_3) \cos k_2 a + i(k_2^2 - k_1 k_3) \sin k_2 a}{(k_1 + k_3)k_2 \cos k_2 a - i(k_2^2 + k_1 k_3) \sin k_2 a} \qquad (1.102)$$

Let these complex quantities be represented in terms of amplitude and phase:

$$A = A_0 e^{i\alpha} \qquad (1.103)$$

and

$$D = D_0 e^{i\delta} \qquad (1.104)$$

Then the real y's associated with the incident, reflected, and transmitted waves are

$$\text{incident:} \quad y = \cos (k_1 x - \omega t) \qquad (1.105)$$

$$\text{reflected:} \quad y = A_0 \cos (k_1 x + \omega t - \alpha) \qquad (1.106)$$

$$\text{transmitted:} \quad y = D_0 \cos (k_3 x - \omega t + \delta) \qquad (1.107)$$

Calculating the time averages of the energy flux we get $\frac{1}{2}T\omega k_1$, $\frac{1}{2}T\omega k_1 A_0^2$, and $\frac{1}{2}T\omega k_3 D_0^2$, respectively. Therefore the fraction of the energy which is reflected (reflection coefficient) is

$$f_R = A_0^2 = \frac{k_2^2(k_1 - k_3)^2 \cos^2 k_2 a + (k_2^2 - k_1 k_3)^2 \sin^2 k_2 a}{k_2^2(k_1 + k_3)^2 \cos^2 k_2 a + (k_2^2 + k_1 k_3)^2 \sin^2 k_2 a} \qquad (1.108)$$

and that which is transmitted (transmission coefficient) is

$$f_T = \frac{k_3}{k_1} D_0^2 = \frac{4k_1 k_3 k_2^2}{k_2^2(k_1 + k_3)^2 \cos^2 k_2 a + (k_2^2 + k_1 k_3)^2 \sin^2 k_2 a} \qquad (1.109)$$

It is easily verified that $f_R + f_T = 1$.

Problem 1-5: When $\rho_1 = \rho_3$, calculate the S matrix for scattering by the central section and verify that it is unitary.

Problem 1-6: Put $\rho_3 = \rho_1$ and verify that as $\rho_2 \to \infty$ and $a \to 0$, such that $\rho_2 a = M$, the results reduce to those of section 7.

Problem 1-7: An infinitely long string consists of two sections which have densities ρ_1 and ρ_2, respectively. A wave of unit amplitude incident on the junction of the sections from one side has reflected and transmitted amplitudes r and t, respectively. When the wave is incident from the other side the amplitudes are r' and t', respectively. Derive the relations

$$r^2 + tt' = 1, \qquad r' = -r$$

An interesting special result is the following: If the density in the central portion is intermediate between the densities of the end portions, the existence of that portion *increases* the transmission coefficient. On the other hand, if its density is either greater than or less than that of the rest of the string, it *decreases* the transmission coefficient. For the transmission coefficient when $a = 0$ is

$$f_T(0) = \frac{4k_1 k_3 k_2^2}{k_2^2 (k_1 + k_3)^2} \tag{1.110}$$

If this is subtracted from (1.109) the excess transmission due to the central portion is found to be

$$f_T - f_T(0) = \frac{4k_1 k_3 k_2^2}{[\ \ ][\ \ ]} \{k_2^2 (k_1 + k_3)^2 - (k_2^2 + k_1 k_3)^2\} \sin^2 k_2 a \tag{1.111}$$

where the empty brackets in the denominator represent the denominators of (1.109) and (1.110), respectively, both of which are positive. But the quantity in the curly brackets may be written as

$$-(k_2^2 - k_1^2)(k_2^2 - k_3^2) = -\frac{\omega^2}{T}(\rho_2 - \rho_1)(\rho_2 - \rho_3)$$

which is positive if ρ_2 is intermediate between ρ_1 and ρ_3 and negative otherwise.

10. Inhomogeneous String and the Method of Separation of Variables

The methods used up to this point do not lend themselves to obvious extension if more general or complicated problems are considered; if, for instance, the density or tension of the string varies with position x. The problem of variable density is a natural one to consider. The tension also may be variable, however, if, for instance, the string has weight which is not negligible, and vibrates about a *vertical* position (in this case, it is perhaps best to think of a flexible *chain* rather than a string), or is whirled with very high angular velocity about a vertical axis to which it is fixed. These problems are admittedly rather esoteric, but are of interest in that they lead to a type of equation which will appear repeatedly in other physical problems, and which is generally known as a Sturm–Liouville equation.

Consider the problem of solving the general equation (1.4)

$$\rho(x) \frac{\partial^2 y}{\partial t^2} = \frac{\partial}{\partial x}\left[T(x) \frac{\partial y}{\partial x}\right]$$

We approach this problem, as we do most other partial differential equations, by the "method of separation of variables." This method involves seeking special solutions which are products of functions of the different variables of the problem; in this case, solutions of the form

$$y = f(x)g(t) \tag{1.112}$$

If this is substituted into the equation, which is then divided through by $y = fg$ and by $\rho(x)$, the problem becomes that of solving

$$\frac{1}{g}\frac{d^2 g}{dt^2} = \frac{1}{\rho f}\frac{d}{dx}\left[T(x) \frac{df}{dx}\right] \tag{1.113}$$

Now it is observed that the left-hand side of this equation is a function of the variable t only, while the right-hand side is a function of x only. Yet the equation must be valid for *all* values of x and t. This is clearly only possible if the functions appearing on both sides are a constant. Thus we are led to the "separated" equations

$$\frac{1}{g}\frac{d^2 g}{dt^2} = \text{a constant } (-\lambda) \tag{1.114}$$

and

$$\frac{1}{\rho f}\frac{d}{dx}\left[T(x) \frac{df}{dx}\right] = -\lambda \tag{1.115}$$

both of which are *ordinary* differential equations.

Up to this point, there is nothing to prevent λ from having an *arbitrary* value. This situation is, however, changed by the application of *boundary* conditions. In the case of constant ρ and T, (1.115) becomes

$$c^2 \frac{d^2 f}{dx^2} = -\lambda f \tag{1.116}$$

If, as an example, the ends of the string at $x = 0$ and $x = L$ are fixed, y and therefore f must take the value zero at each of these points. If y were a positive constant, the solutions of (1.116) would be positive and negative exponentials (or alternatively hyperbolic sines and cosines), no combination of which can be made equal to zero at two points. Thus no negative value of λ is permissible if the boundary conditions are to be satisfied. [This is fortunate, of course, since the solutions of the time equation (1.114) would then also be exponentials, and we would obviously not be describing a vibration!]

If, on the other hand, λ is positive, a general solution of (1.116) is

$$f = A \sin \left(\frac{1}{c} \sqrt{\lambda} x + \phi \right) \tag{1.117}$$

where A and ϕ are constants. But this is zero at $x = 0$ only if $\phi = 0$, and at $x = L$ if

$$\frac{1}{c} \sqrt{\lambda} L = n\pi \tag{1.118}$$

where n is an integer. Thus

$$\lambda = \left(\frac{n\pi c}{L} \right)^2 \tag{1.119}$$

$$f = A \sin \frac{n\pi x}{L} \tag{1.120}$$

and

$$g = B \sin \left(\frac{n\pi ct}{L} - \alpha \right) \tag{1.121}$$

giving a solution which is one of the terms of (1.46).

Problem 1-8: A string is stretched between points $x = 0$ and $x = L$. At time $t = 0$ it is pulled sideways a distance y_0 at $x = a$ and released from rest. Show that the displacement at arbitrary time is

$$y = \frac{2y_0}{\pi^2} \frac{L^2}{a(L-a)} \sum_{n=1}^{\infty} \frac{\sin n\pi a/L}{n^2} \sin \frac{n\pi x}{L} \cos \frac{n\pi ct}{L}$$

c being the velocity of propagation of a wave on the string. Calculate the initial energy and verify directly conservation of energy.

The boundary conditions have, therefore, in this case, limited the permissible values of λ to a discrete set of values. These values are a particular case of what are in general designated eigenvalues; the corresponding solutions are known as "eigenfunctions."

Equation (1.115) is a slightly special case of an equation known as the Sturm–Liouville equation. The Sturm–Liouville equation is

$$\frac{d}{dx}\left[p(x) \frac{dy}{dx} \right] - q(x)y + \lambda r(x)y = 0 \tag{1.122}$$

In (1.115) the function corresponding to q is equal to zero. The function $r(x), \rho(x)$ in the case of the string, is known as the "weighting function," and is taken throughout to be a function that is positive for the range of x on which the problem is defined.

It is not difficult to show that any second-order linear differential equation containing a y term with an eigenvalue multiplier can be transformed into the Sturm–Liouville form. For consider

$$f_0(x)\frac{d^2y}{dx^2}+f_1(x)\frac{dy}{dx}+f_2(x)y = 0 \qquad (1.123)$$

Let us multiply this equation by $g(y)$, and choose g so that the coefficient of dy/dx is the derivative of that of d^2y/dx^2:

$$\frac{d}{dx}(gf_0) = gf_1 \qquad (1.124)$$

Putting $gf_0 = F$,

$$\frac{dF}{dx} = F\frac{f_1}{f_0} \qquad (1.125)$$

from which it follows that

$$F = A \exp \int_{x_0}^{x} \frac{f_1(x')}{f_0(x')}\,dx' \qquad (1.126)$$

A being an arbitrary constant. Therefore if we choose the "integrating factor"

$$g = \frac{A}{f_0(x)} \exp \int_{x_0}^{x} \frac{f_1(x')}{f_0(x')}\,dx' \qquad (1.127)$$

the differential equation may be written

$$Fy'' + F'y' + f_2gy = 0$$

(where $y' = dy/dx$, $y'' = d^2y/dx^2$, etc.), or

$$\frac{d}{dx}\left[F(x)\frac{dy}{dx}\right]+f_2gy = 0 \qquad (1.128)$$

Of course, it need not always be true that $f_2g > 0$; we, however, confine our attention to this case, the significance of which becomes evident when we discuss the question of orthogonality of eigenfunctions.

Problem 1-9: Write in Sturm–Liouville form the Bessel equation

$$\frac{d^2y}{dx^2}+\frac{1}{x}\frac{dy}{dx}+\left(\lambda-\frac{n^2}{x^2}\right)y = 0$$

---

11. Boundary Conditions and the Eigenvalue Problem

For each value of the eigenvalue quantity λ the Sturm–Liouville equation (1.122) has two independent solutions. One formal way to see that this is the case is to think of expanding the solution in a Taylor's series about any point x_0 at which the coefficients in the equation are all finite:

$$y = \sum_{n=0} \frac{(x-x_0)^n}{n!} y^{(n)}(x_0) \qquad (1.129)$$

where by $y^{(n)}(x_0)$ we mean the nth derivative of $y(x)$ calculated at $x = x_0$. If we choose arbitrary values for $y(x_0)$ and $y'(x_0)$ (the first derivative), the differential equation makes it possible to calculate the second derivative $y''(x_0)$. Then, successive differentiations of the equation will determine the higher derivatives. In this way, a formal solution is obtained in Taylor's series form. Furthermore, because of the linearity of the equation, the series has the form

$$y = y(x_0)y_1(x) + y'(x_0)y_2(x) \qquad (1.130)$$

where y_1 and y_2 are separately solutions.

In a physical problem in which the Sturm–Liouville equation arises (such as the problem of the vibrating string) the solution is normally subjected to boundary conditions. Consider as an example the case in which y must be zero at $x = 0$ and $x = L$, and suppose that we try to obtain the solution as a Taylor's series about $x = 0$. Then the first solution in (1.130) will not appear. The condition that $y_2(x)$ be zero at $x = L$ will then yield an equation for λ. Clearly, the two boundary conditions will only be satisfied for particular values of λ. In short, *the boundary conditions determine the eigenvalues*.

We shall see, however, that the boundary conditions are not always so clearly specified; one or more may simply require that the solution remain *finite* for a physically significant value of the variable. Again, however, this will only be the case for suitable values of λ.

If the boundary conditions are applied at finite points, $x = a$ and $x = b$, and the solutions are both finite at these points, boundary conditions

$$\alpha y' + \beta y = 0 \qquad (1.131)$$

which prescribe a linear relationship between the function and its derivative at each boundary lead in a similar manner to the determination of the eigenvalue. For if such a relation exists at $x = a$, and we expand the solution about this point, one of the constants in (1.130) may either be eliminated or expressed in terms of the other. In this case, again, only a suitable choice of λ will permit the satisfaction of the condition at the other boundary.

12. Orthogonality of Eigenfunctions

This section is concerned with a very important theorem, which says that, under conditions to be specified, the eigenfunctions of the Sturm–Liouville equation are orthogonal. This *orthogonality* is defined by the equation

$$\int_a^b y_m(x)y_n(x)r(x)\,dx = 0, \qquad m \neq n \tag{1.132}$$

where y_m and y_n are the eigenfunctions belonging to different eigenvalues λ_m and λ_n, and $r(x)$ is the weighting function.

The most general statement of the condition for the validity of the orthogonality equation is that

$$p(y_m y_n' - y_n y_m')\big|_a^b = 0 \tag{1.133}$$

the left-hand side being the difference of $p(y_m y_n' - y_n y_m')$ at the points b and a. This condition may be satisfied in various ways, the most important of which are

(a) that boundary conditions (1.131) hold at each end, or

(b) that y and y' remain finite while $p = 0$ at one or both of the ends, in place of the linear boundary condition.

A further possible condition would of course be that p, y_m, and y_n be periodic on the interval (a, b).

The orthogonality theorem is proven as follows: y_m and y_n satisfy

$$(py_m')' - qy_m + \lambda_m ry_m = 0 \tag{1.134}$$

and

$$(py_n')' - qy_n + \lambda_n ry_n = 0 \tag{1.135}$$

where the primes denote differentiation. Let us multiply (1.134) by y_n, (1.135) by y_m, subtract, and integrate from a to b. Now

$$\int_a^b [y_n(py_m')' - y_m(py_n')']\,dx = p(y_n y_m' - y_m y_n')\big|_a^b$$

on integration. Thus, by (1.133), it is zero. The q terms vanish, and one has left only

$$(\lambda_m - \lambda_n)\int_a^b y_m(x)y_n(x)r(x)\,dx = 0 \tag{1.136}$$

The orthogonality theorem follows, since $\lambda_m \neq \lambda_n$.

Since the y's are determined only to within a multiplicative constant, we may choose them to obey the "normalization condition."

$$\int_a^b r y_n^2(x)\,dx = 1 \tag{1.137}$$

This relation, along with (1.132), may be written

$$\int_a^b r y_n y_m dx = \delta_{nm} = 1, \qquad n = m$$
$$= 0, \qquad n \neq m \tag{1.138}$$

The functions y_n are known as an "orthonormal" (mutually orthogonal and normalized) set.

13. Rayleigh–Ritz Variational Principle

This section is concerned with a variational principle for the eigenfunctions of the Sturm–Liouville equation. Consider the quantity

$$I(y) = \int_a^b (p y'^2 + q y^2)\,dx \Big/ \int_a^b r y^2 dx \tag{1.139}$$

We can show that the functions y for which I has a stationary value are the eigenfunctions of the Sturm–Liouville equation. This statement has the following meaning. Suppose I is calculated for y equal to an eigenfunction y_n. Let us then calculate it for the function

$$y = y_n + \epsilon \phi \tag{1.140}$$

where ϵ is a parameter and ϕ is an arbitrary bounded function which is zero at both ends of the interval of integration, $x = a$ and $x = b$. Then I is stationary for $y = y_n$ if

$$\left(\frac{\partial I}{\partial \epsilon}\right)_{\epsilon=0} = 0 \tag{1.141}$$

Suppose (as will generally be the case in the problems we shall consider) that I is a local *minimum* in ϵ at $\epsilon = 0$. Then, our principle states that I will be *lower* when calculated for an eigenfunction than for any function slightly displaced from it.

It is convenient to introduce the following notation: Let us expand I as a series in ϵ

$$I(y) = I(y_n) + \epsilon \delta I + O(\epsilon^2) \tag{1.142}$$

δI then stands for the coefficient in the term in the expansion which is linear in the parameter ϵ. Equation (1.141) then may be written

$$\delta I = 0 \tag{1.143}$$

Before proving the variational principle, let us express it in another way. If we write

$$I(y) = N/D \tag{1.144}$$

N being the numerator and D the denominator of (1.139), the principle may equally well be written: For an eigenfunction

$$\delta N = 0 \tag{1.145}$$

subject to the auxiliary condition

$$D = \int_a^b ry^2 dx = 1 \tag{1.146}$$

Thus, of all *normalized* functions, the ones for which the numerator N is stationary are the eigenfunctions.

In either case, of course, the principle is stated in such a way as to ensure that the quantity being varied cannot be changed in value simply by multiplying the function by an arbitrary constant.

That (1.145) and (1.146) are equivalent to (1.139) can be shown as follows:

$$\delta I = \delta(N/D)$$

$$= \frac{D\delta N - N\delta D}{D^2} \tag{1.147}$$

If y is normalized, $D = 1$, $\delta D = 0$, and $\delta I = 0$ implies that $\delta N = 0$ and vice versa.

If y is *not* necessarily normalized, it may be seen from (1.147) that the variational principle may be expressed in the form

$$\delta N - I\delta D = 0 \tag{1.148}$$

Under suitable conditions, the value of I for $y = y_n$ may be shown to be equal to the eigenvalue λ_n. For if we multiply the eigenvalue condition (1.135) by y_n and integrate from $x = a$ to $x = b$ and then integrate the first term by parts, we get the equation

$$py_ny_n'|_a^b - \int_a^b (py_n'^2 + qy_n^2)\,dx + \lambda_n \int_a^b ry_n^2 dx = 0$$

Therefore, $\lambda_n = I$ provided

$$py_ny_n'|_a^b = 0 \tag{1.149}$$

This condition, it should be noted, is different from the boundary condition (1.133) which defines the Sturm–Liouville problem. Nevertheless, there are circumstances which will ensure the satisfaction of both, for example,

(a) one of p, y, or $y' = 0$ at each end, or

(b) periodicity of p, y, and y' on the interval (a, b).

In either of these circumstances,

$$I(y_n) = \lambda_n \qquad (1.150)$$

and the variational principle satisfied by the eigenvalues may be written as

$$\delta N - \lambda_n \delta D = 0 \qquad (1.151)$$

14. Approximate Calculation of Eigenvalues from the Variational Principle

The Rayleigh–Ritz principle may be used to estimate eigenvalues when exact calculation proves difficult. The essential fact is, that if we calculate

$$I = \frac{N}{D}$$

for a function which is a reasonably good approximation to an eigenfunction, the error being of order ϵ, the value of I will differ from the minimum only to order ϵ^2. If, as under the conditions specified in the previous section, I is the eigenvalue, we can thus make a quite accurate estimate of it. This is particularly useful if there is a lowest eigenvalue, in that we can in this way put an upper bound on that eigenvalue.

It is, of course, important that we find a "trial function" which can approximate reasonably well the true eigenfunction. The choice of good trial functions requires skill. Usually, one's physical intuition suggests its qualitative behavior. If one then chooses a function which can represent such a behavior, and containing one or more disposable parameters, one can minimize I as a function of those parameters. The minimizing values will then define the best possible approximation to the true eigenfunction of the form chosen.

It must, of course, be emphasized that the function obtained in this way need not be a particularly good approximation to the true eigenfunction, even if it gives an excellent approximation to the eigenvalue. This is, after all, the whole point of the variational method – that the discrepancy in the eigenvalue is only of second order relative to that in the eigenfunction.

It should be noted, finally, that one requirement of the trial function is that it should satisfy the boundary conditions of the problem.

To demonstrate the accuracy of the method, consider a simple problem for which the answer is known exactly, viz. the solution of the equation

$$y'' + \lambda y = 0$$

subject to $y = 0$ at $x = 0$ and $x = 1$. The eigenfunctions are of course $\sin n\pi x$, where n is an integer, and the eigenvalues are

$$\lambda = n^2 \pi^2$$

In this case

$$I = \frac{\int_0^1 y'^2 dx}{\int_0^1 y^2 dx} \tag{1.152}$$

A simple trial function, containing only one parameter, is

$$y = x(1-x)(1+ax) \tag{1.153}$$

The corresponding value of λ is

$$\lambda = 10 \frac{2/5a^2 + a + 1}{2/7a^2 + a + 1} \tag{1.154}$$

This is found to have zero derivative for $a = 0$ and $a = -2$, the corresponding values of λ being

$$\lambda_1 = 10 \quad \text{and} \quad \lambda_2 = 42$$

These values lie quite close to the lowest two eigenvalues,

$$\lambda_1 = \pi^2 = 9.870 \quad \text{and} \quad \lambda_2 = 4\pi^2 = 39.479$$

It may then be observed that with $a = 0$, (1.153) is an even function about the central point $x = \frac{1}{2}$ of the interval, while for $a = -2$ it is *odd*. In fact, if we put

$$\bar{x} = x - \tfrac{1}{2} \tag{1.155}$$

we obtain

$$y = (\bar{x}^2 - \tfrac{1}{4})[a\bar{x} + (\tfrac{1}{2}a + 1)] \tag{1.156}$$

Is this symmetry property of the eigenfunctions fortuitous, or is it of fundamental significance? It is not an accident at all. To see that this is so, let us define a symmetry (parity) operator P such that

$$Pf(\bar{x}) = f(-\bar{x}) \tag{1.157}$$

Now under the operation P both our differential equation and the boundary conditions are unaltered. Therefore, if $y_n(\bar{x})$ is an eigenfunction, $Py_n(\bar{x}) = y_n(-\bar{x})$ is an eigenfunction with the same eigenvalue. The same is then true of

$$y_n^{(e)}(\bar{x}) = \tfrac{1}{2}[y_n(\bar{x}) + y_n(-\bar{x})] \tag{1.158}$$

and

$$y_n^{(o)}(\bar{x}) = \tfrac{1}{2}[y_n(\bar{x}) - y_n(-\bar{x})] \tag{1.159}$$

which are, respectively, even and odd functions, and of any linear combination

$$k_1 y_n^{(e)}(\bar{x}) + k_2 y_n^{(o)}(\bar{x})$$

We may substitute this in (1.150); if we make use of the fact that the integral of the product of an even and odd function is zero, we find that

$$I = \frac{\displaystyle\int_{-1/2}^{1/2} y'^2(\bar{x})\,d\bar{x}}{\displaystyle\int_{-1/2}^{1/2} y^2(\bar{x})\,d\bar{x}}$$

$$= \frac{k_1^2 \displaystyle\int_{-1/2}^{1/2} [y_n^{(e)'}(\bar{x})]^2 d\bar{x} + k_2^2 \displaystyle\int_{-1/2}^{1/2} [y_n^{(o)'}(\bar{x})]^2 d\bar{x}}{k_1^2 \displaystyle\int_{-1/2}^{1/2} [y_n^{(e)}(\bar{x})]^2 d\bar{x} + k_2^2 \displaystyle\int_{-1/2}^{1/2} [y_n^{(o)}(\bar{x})]^2 d\bar{x}}$$

$$= \frac{k_1^2 A + k_2^2 B}{k_1^2 C + k_2^2 D}$$

Minimizing with respect to k_1 and k_2, it is seen that the solutions correspond either to $k_1 = 0$ or $k_2 = 0$. Thus, the eigenfunctions *must* be either even or odd.

This example illustrates the value of symmetry principles in the treatment of eigenvalue problems for linear operator equations. This point is dealt with more fully in Chapter 2 on the general theory of linear operators.

In the context of the present problem, symmetry considerations suggest how to get better approximations to the lower eigenfunctions. To improve the calculation of the lowest eigenvalue, we can choose an even function containing one arbitrary parameter,

$$y = (\tfrac{1}{4} - \bar{x}^2)(1 + A\bar{x}^2) \tag{1.160}$$

On calculation it is found that

$$I = 6 \cdot \frac{44A^2 + 28A + 35}{4A^2 + 12A + 21}$$

Minimization leads to the quadratic equation

$$52A^2 + 196A + 21 = 0$$

To approximate the lowest eigenvalue we must choose that root which leads to the smaller value of λ; this turns out to be

$$A = -0.1103749$$

The corresponding value of λ is

$$\lambda = 9.86975$$

compared with the correct value (to the same number of significant figures)

$$\lambda = 9.86959$$

Thus, with only one arbitrary parameter, if it is well chosen, it is possible to obtain the eigenvalue with very high accuracy.

What of the calculation of the higher eigenvalues of each symmetry? Of all functions orthogonal to the lowest eigenfunction,[3] that which yields the minimum value for the integral I is the second lowest. This remark, however, has limited practical value, since, unless one knows the lowest eigenfunction exactly, one cannot apply to trial functions the necessary orthogonality condition. It is, of course, possible to choose only functions orthogonal to the *approximate* form of the lowest eigenfunction. It may be possible, by choosing a trial function judiciously, to get a quite close estimate of the next eigenvalue. However, one cannot be sure, in this case, that the estimate will be *above* the true eigenvalue; it might, in fact, be either above or below it.

Problem 1-10: Using the trial function

$$y = \bar{x}(\tfrac{1}{4} - \bar{x}^2)(1 + B\bar{x}^2)$$

estimate the second lowest eigenvalue.

It should be emphasized in any case, that whether or not the variational principle yields an approximation to the lowest eigenvalue depends entirely on whether the trial function can, for suitable value of its parameters, approximate the lowest eigenfunction more closely than any other. This point may be emphasized by pointing out that if, by some accident, we were to choose a trial function which was orthogonal to the five lowest eigenfunctions, it could at best give us an approximation to the sixth eigenvalue.

Problem 1-11: Show that the trial functions given by the two values of a in (1.153) are mutually orthogonal.

Problem 1-12: Set up the Rayleigh-Ritz variational principle for the zeros of the Bessel function of order n, $J_n(kx)$, which

---

[3]We use the term "lowest eigenfunction" as an abbreviation for "eigenfunction corresponding to the lowest eigenvalue."

satisfies the differential equation

$$y'' + \frac{1}{x}y' + \left(k^2 - \frac{n^2}{x^2}\right)y = 0$$

and the boundary conditions: (i) at $x = 0$, J_n is an even or odd function according as n is even or odd, and (ii) $J_n(kx) = 0$ at $x = 1$.

For the zero-order Bessel function, use the trial wave function

$$y = (1-x^2)(1+ax^2)$$

to get approximate values for the two lowest zeros.

Show that the trial functions for the two values of a are mutually orthogonal.

Problem 1-13: A similar result was obtained for the oscillator equation. Can you find a general theorem of which these are particular examples?

Problem 1-14: Using $(1-x^2)$ times a quartic orthogonal to the approximate lowest eigenvalue, estimate the next eigenvalue of the same symmetry.

Problem 1-15: Solve the same problem for the Bessel function of order 1, using the trial function $y = x(1-x^2)(1+ax^2)$.

Answers: For J_0, the approximate values of the two lowest zeros, using the one-parameter trial function, are found to be 2.4050 and 6.0731, compared with correct values 2.4049 and 5.5201. With $y = (1-x^2)$ times a quartic, we get 5.5418. For J_1, the approximate values are 3.8343 and 8.0807, and the correct ones 3.8318 and 7.0155. Note that the higher zero is, in all these examples, given less accurately than the lower. Can you give a plausible explanation why this is so?

---

15. Expansion in Eigenfunctions

In the vibrating string problem, the method of separation of variables can provide solutions which are products of functions of the two variables, and which satisfy the boundary conditions. The spatial functions are the eigenfunctions of the Sturm–Liouville problem.

Because of the linearity of the original differential equation, any linear combination of these separated (product) solutions also satisfies both the equation and the boundary conditions. The most general such solution is

$$y = F(x, t) = \Sigma\, y_n(x) g_n(t) \qquad (1.161)$$

where the arbitrary constants are absorbed into g (we assume the y_n's to be normalized).

Now the *initial* conditions of the problem are in the form

$$y = f_0(x) \text{ at } t = 0 \qquad (1.162)$$

and

$$\frac{\partial y}{\partial t} = v_0(x) \text{ at } t = 0 \qquad (1.163)$$

We will be able to satisfy these conditions with the solution (1.161) if we can choose the two arbitrary constants in the g_n's to satisfy the equations

$$f_0(x) = \Sigma'\, g_n(0) y_n(x) \qquad (1.164)$$

and

$$v_0(x) = \Sigma\, g_n(0) y_n(x) \qquad (1.165)$$

where $\dot{g}_n = dg_n/dt$.

The question is, then, whether the somewhat arbitrary functions f_0 and v_0 can be represented as series in the eigenfunctions $y_m(x)$ of the Sturm–Liouville problem. We say "somewhat" arbitrary because physics certainly does not permit the functions which are to be expanded to be anything but a very well-behaved special set out of the totality of "mathematician's functions." They must, for instance, be continuous and bounded.

We assume "without proof" that the functions $y_m(x)$ form an infinite set, and also that the eigenvalues form a set tending to infinity. To talk of "proving" these statements is in fact not to put the problem correctly. In all of the physical problems with which we deal, it will be found, by actual calculation, that these conditions hold. It is therefore clearly possible to "prove" them under *some* conditions broad enough to encompass all physicists' problems. The mathematical question might then be inverted to ask, what are the most general conditions under which the statements *can* be proven? Put this way, however, the question is really not of much interest to a physicist. If all of the problems he turns up are of this type, he will be interested in whether the expansion theorem can be proven under these conditions, and this we show. If, perchance, he should one day meet a problem in which these conditions do *not* hold, he will, either using his own intelligence, or by consulting a mathematician, have to find other methods of proceeding.

One further remark should be made before proceeding further: We have made the assumption that the eigenfunctions and eigenvalues form a discrete set, which may then be enumerated by integers n. There *are*, however, physical problems in which the "spectrum" of eigenvalues is, at least in part, continuous. In this case the eigenvalues will be enumerated by a continuous variable α. At least part of the "sum" in the expansion theorem will then have to be replaced by an integral, and we have

$$y(x) = \sum_n c_n y_n(x) + \int_{\alpha_1}^{\alpha_2} c(\alpha) y(x;\alpha)\, d\alpha \qquad (1.166)$$

The sum may then be either a finite or an infinite one. The orthogonality theorem still holds, but we can see that it is necessary to modify somewhat the normalization condition. For suppose we multiply both sides of (1.166) by $r(x)y(x, \alpha')$ and integrate over x. Assume that we can interchange the order of the integrations. The integral

$$\int y(x;\alpha) y(x;\alpha') r(x)\, dx$$

is different from zero only for one value of α. Now the left-hand side of the equation will certainly not be zero, in general. The contribution from the *sum* on the right, on the other hand, is zero by orthogonality. Thus, the integral of $c(\alpha)$ with a function which is zero at all points except $\alpha = \alpha'$ must be finite. Such a "function" is known as a "Dirac δ function" and is written $\delta(\alpha - \alpha')$. Then $\delta(\alpha - \alpha') = 0$ except when $\alpha = \alpha'$, yet $\int c(\alpha)\delta(\alpha - \alpha')d\alpha$ is finite. We specifically define the "δ-function" so that

$$\int c(\alpha)\delta(\alpha - \alpha')d\alpha = c(\alpha') \qquad (1.167)$$

whenever the range of integration includes the value $\alpha = \alpha'$. Thus, the integral of the "δ function" itself is unity. The "equation"

$$\int y(x;\alpha) y(x;\alpha') r(x)\, dx = \delta(\alpha - \alpha') \qquad (1.168)$$

then represents a way of "normalizing" the functions y. Multiplying y by a constant A simply produces a multiple A^2 of the "δ function" on the right.

The terms "δ function," "equation," and "normalizing" in the preceding paragraphs have been put between quotation marks because, in the strict mathematical sense, the "δ function" is not a function at all, so that both the "equation" and the "normalization" have not quite the expected meanings. The "δ function" (which we refer to henceforth without its qualifying quotation marks) represents a generalization of the concept of function known to mathematicians as a "distribution".[4] Within the theory

---

[4]Lighthill, M. J., *Introduction to Fourier Analysis and Generalized Functions*. London: Cambridge University Press, 1958.

of distributions, the operations which we apply to δ functions in this book can be made mathematically valid and meaningful. In any case, δ functions were first used by physicists for the description of *physical* reality, and the validity of the operations carried out on them can be justified on physical grounds.

Continuous eigenvalue spectra most often arise in cases in which the dependent variable has an infinite range. But infinite ranges of variables are, from the point of view of physics, only convenient fictions. Normally, one can approximate actual physical conditions just as well by the application of boundary conditions at finite, if large, values. Consequently, it is usually possible to avoid continuous-spectrum problems. It is, however, not always technically convenient to do so. However, for the subsequent development, we confine our attention only to discrete spectra. When, in Chapter 3, we explore the transition from Fourier series to Fourier integrals, it will be seen what sort of considerations are involved in the transition from a finite to an infinite range.

Returning now to the question of expansion in eigenfunctions, suppose it is *assumed* that a function $f(x)$ can be so expanded;

$$f(x) = \sum_{n=1} a_n y_n(x) \tag{1.169}$$

where the y_n's are normalized. The a_n's may be found simply by multiplying the equation by any particular eigenfunction $y_m(x)$ and by the weighting function $r(x)$, and integrating between the values of x for which the boundary conditions are specified. Because of the orthogonality of the eigenfunctions, only one term survives on the right-hand side, and the coefficient is determined:

$$a_m = \int_{x_1}^{x_2} f(x) y_m(x) r(x) dx \tag{1.170}$$

Thus, if the expansion is valid, the coefficient is determined.

We may understand (1.170) better if we proceed in a different way. Suppose one tries to make a best weighted least-squares fit to $f(x)$ by a *finite* sum of the first N terms of the series. To do so we want to minimize

$$\Delta_N = \int \left[f(x) - \sum_{n=1}^{N} c_n y_n(x) \right]^2 r(x) dx \tag{1.171}$$

Using the orthonormality of the y_n's, this may be written

$$\Delta_N = \int [f(x)]^2 r(x) dx - 2 \sum_{n=1}^{N} c_n a_n + \sum_{n=1}^{N} c_n^2$$

where a_n is given by (1.170) and the integrals are over the range (x_1, x_2).

Δ_N is minimized by the condition

$$\frac{\partial \Delta_N}{\partial c_n} = 0$$

from which it is found that

$$c_n = a_n$$

Thus, the coefficients (1.170) give the best approximation to $f(x)$ by a *finite* series of eigenfunctions.

An obvious way of posing the question of the validity of the expansion theorem is, now, to ask whether $\Delta_N \rightarrow 0$ as $N \rightarrow \infty$. For if it does, since the integrand of Δ_N is never negative, it must be everywhere zero. Again, this need not be so for "mathematicians' functions"; $f(x)$ and the series could differ on a set of points of measure zero. This, however, does not represent a physically realizable situation, due to the well-behaved nature of the physicists' functions.

Consider the function

$$\phi_N(x) = f(x) - \sum_{n=1}^{N} a_n y_n(x) \tag{1.172}$$

This may immediately be shown to be orthogonal to all of the y_n's for $n \leq N$. This function may be normalized by dividing it by

$$\Delta_N = \int [\phi_N(x)]^2 r(x)\,dx \tag{1.173}$$

Let us now recall the variational theorem as expressed in (1.139) and (1.141). This theorem stated that the quantity $I(y)$ defined there was a minimum when y was the lowest eigenfunction of the Sturm–Liouville problem. It also followed that, of all functions orthogonal to the lowest eigenfunction, that which made y a minimum was the second lowest eigenfunction, and so on. Thus, of all functions orthogonal to the first N eigenfunctions, that which makes $I(y)$ a minimum is the $(N+1)$st eigenfunction, which gives it a value equal to the $(N+1)$st eigenvalue λ_{N+1}. Therefore,

$$I(\phi_N) \geq \lambda_{N+1}$$

or

$$\frac{1}{\Delta_N} \int_{x_1}^{x_2} (p\phi_N'^2 + q\phi_N^2)\,dx \geq \lambda_{N+1} \tag{1.174}$$

Therefore, when, as $N \rightarrow \infty$, $\lambda_{N+1} \rightarrow \infty$, subject only to the condition that the integral exists, it follows that $\Delta_N \rightarrow 0$, and the expansion theorem follows.

16. The Inhomogeneous Problem
for the Vibrating String

First, let us recapitulate the solution of the homogeneous equation. The results of separation of variables were shown in (1.114) and (1.115). The latter determine the eigenvalues $\lambda = \lambda_n$. Solution of (1.114) shows that $\sqrt{\lambda_n}$ is the circular frequency of the periodic time-dependent function g. We therefore write

$$\lambda_n = \omega_n{}^2 \tag{1.175}$$

Let us now consider the problem of the vibrating string (which may be inhomogeneous) under the influence of an external applied force. The latter will be described by a "force density" or force per unit length $F(x, t)$ which will of course also in general vary with time. The equation of motion is then

$$\rho \frac{\partial^2 y}{\partial t^2} = \frac{\partial}{\partial x}\left(T \frac{\partial y}{\partial x}\right) + F(x, t) \tag{1.176}$$

It is actually convenient to introduce, in place of the force per unit length F, the force per unit mass

$$E(x, t) = F(x, t)/\rho \tag{1.177}$$

A solution of (1.176) may be found by expanding both y and E in eigenfunctions of the associated Sturm–Liouville problem, the coefficients being functions of t;

$$y = \sum c_n(t) y_n(x) \tag{1.178}$$

$$E(x, t) = \sum e_n(t) y_n(x) \tag{1.179}$$

Substituting in (1.176), making use of the fact that y_n satisfies the eigenvalue equation

$$\frac{d}{dx}\left(T \frac{dy_n}{dx}\right) = -\rho \omega_n{}^2 y_n \tag{1.180}$$

and equating coefficients of each y_n, we obtain the equation

$$\frac{d^2 c_n}{dt^2} = -\omega_n{}^2 c_n + e_n \tag{1.181}$$

It is easily verified that the solution of this equation satisfying the initial conditions that $c_n(0) = \dot{c}_n(0) = 0$ is

$$c_n(t) = \frac{1}{\omega_n} \int_0^t e_n(t') \sin \omega_n(t - t')\,dt' \tag{1.182}$$

Substitution in (1.178) then gives the solution of the problem when the string is initially at rest and the force is applied starting at $t = 0$.

An interesting particular case is that in which the applied force has a definite frequency ω_0. This may be solved from (1.182) with

$$e_n(t) = e_n(0) \sin \omega_0 t \qquad (1.183)$$

Problem 1-16: Solve the problem of the uniform string for which the end at $x = 0$ is fixed and that at $x = L$ is jiggled according to the law

$$y(L) = a \cos \omega_0 t$$

and was at rest at time $t = 0$. [Hint: Introduce as dependent variable $z = y - a(x/L) \cos \omega_0 t$.]

Problem 1-17: Solve the problem of the motion of a string fixed at $x = 0$ and $x = L$ for the external force per unit mass

$$E(x, t) = \delta(x - x') \delta(t - t')$$

[The resulting solution $y = G(x, x'; t - t')$ is the "Green's function" of the problem.]

On substitution and evaluation of the integrals it is found that

$$c_n(t) = \frac{e_n(0)}{\omega_n^2 - \omega_0^2} \left(\frac{\omega_n \sin \omega_0 t - \omega_0 \sin \omega_n t}{\omega_n} \right) \qquad (1.184)$$

The energy absorbed by the string per unit time is

$$\frac{\partial W}{\partial t} = \int_0^L \rho(x') E(x', t) \frac{\partial y}{\partial t}(x', t) dx' \qquad (1.185)$$

Using the expansions (1.178) and (1.179) and the orthonormality of the y_n's, this becomes

$$\frac{\partial W}{\partial t} = \sum_n e_n(t) \frac{\partial c_n(t)}{\partial t} \qquad (1.186)$$

Substituting from (1.183) and (1.184) we find that

$$\frac{\partial W}{\partial t} = \sum_n \tfrac{1}{2} [e_n(0)]^2 \frac{\omega_0}{\omega_n^2 - \omega_0^2} [\sin 2\omega_0 t - \sin (\omega_n + \omega_0)t - \sin (\omega_0 - \omega_n)t] \qquad (1.187)$$

The average increase of energy of the string over a long period of time is zero unless $\omega_0 = \omega_n$, in which case all terms average to zero except the nth; from it, however,

$$\frac{\partial W}{\partial t} = \tfrac{1}{4} [e_n(0)]^2 t(1 - \cos 2\omega_n t) \qquad (1.188)$$

In this case the energy of the string increases with time. Mathematically, it becomes infinitely great; it must be remembered, however, that as soon as the oscillations cease to be small, the approximations used in setting up the problem become invalid.

What we have shown, of course, is that the string "resonates" to applied frequencies equal to any of its natural frequencies. This resonance occurs, however, only when $\omega_0 = \omega_n$ exactly; that is, the resonance has no *width*.

If there is a frictional or resistive force proportional to the velocity, such as would be provided in practice by air resistance, the equation of motion takes the form

$$\rho \frac{\partial^2 y}{\partial t^2} = \frac{\partial}{\partial x}\left(T \frac{\partial y}{\partial x}\right) - k\rho \frac{\partial y}{\partial t} + \rho E \tag{1.189}$$

where k is the resistive force constant per unit mass. Expanding both sides as series in the eigenfunctions $y_n(x)$ and equating coefficients of the two sides gives equations for the coefficients c_n:

$$\ddot{c}_n + \omega_n^2 c_n + k\dot{c}_n = e_n \tag{1.190}$$

If $e_n = e_n(0) \sin \omega_0 t = e_n(0) \operatorname{Im} e^{i\omega_0 t}$, where Im means "imaginary part of," putting $c_n = \operatorname{Im} \gamma_n e^{i\omega_0 t}$ yields a particular integral of (1.190):

$$c_n = \operatorname{Im} \frac{e_n(0)}{\omega_n^2 - \omega_0^2 + ik\omega_n} e^{i\omega_0 t}$$

$$= \frac{e_n(0)}{\sqrt{(\omega_n^2 - \omega_0^2)^2 + k^2 \omega_n^2}} \sin (\omega_0 t - \beta_n) \tag{1.191}$$

where

$$\tan \beta_n = \frac{k\omega_n}{\omega_n^2 - \omega_0^2} \tag{1.192}$$

The rate of energy absorption formula (1.186) now becomes

$$\frac{\partial W}{\partial t} = \sum_n [e_n(0)]^2 \frac{\omega_0}{\sqrt{(\omega_n^2 - \omega_0^2)^2 + k^2 \omega_n^2}} \sin \omega_0 t \cos (\omega_0 t - \beta_n) \tag{1.193}$$

The average rate is zero except for those terms in which the force and velocity are in phase. In that case we can average over a number of cycles.

$$\langle \sin \omega_0 t \cos (\omega_0 t - \beta_n) \rangle = \frac{1}{2} \sin \beta_n = \frac{1}{2} \frac{k\omega_n}{\sqrt{(\omega_n^2 - \omega_0^2)^2 + k^2 \omega_n^2}} \tag{1.194}$$

where the brackets $\langle \quad \rangle$ indicate a time average. Therefore

$$\left\langle \frac{\partial W}{\partial t} \right\rangle = \frac{1}{2} \sum_n \frac{k\omega_n \omega_0}{(\omega_n^2 - \omega_0^2)^2 + k^2 \omega_n^2} [e_n(0)]^2 \tag{1.195}$$

This shows resonant peaks for ω_0 in the neighborhood of the natural frequencies ω_n, but these peaks now have a "width." One frequently specifies the half-width at half-maximum, which is

$$|\omega_0 - \omega_n| = k\omega_n \qquad (1.196)$$

The "strength" of the resonance (value of $\partial W/\partial t$ at resonance) is

$$\omega_0/k\omega_n$$

and so decreases as the resonant frequency increases. As $k \rightarrow 0$, the strength of the resonance approaches infinity and its width approaches zero, as seen above.

17. Green's Function

The previous section has been concerned with the solution of the inhomogeneous problem; the determination of the response of a system to an external stimulus. The concept of a "Green's function" is a very useful one for this problem. In the present section we introduce a one-dimensional Green's function associated with the Sturm–Liouville problem, and show how it may be used to solve the problem of the vibrating string under the influence of an external force which is periodic in time but has an arbitrary spatial variation. In Chapter 5, we generalize the concept of Green's functions to two or more dimensions, one corresponding to each independent variable of the problem. We will then be able to discuss the problem of arbitrary time as well as space variation. However, this is conveniently done only after the development of the theory of Fourier transforms. For the present, the more limited problem will serve to introduce the concept and to illustrate the use of Green's functions.

If, in the string problem, we assume an external force with a periodic time variation $e^{i\omega t}$,

$$F(x, t) = \rho(x)w(x)e^{i\omega t} \qquad (1.197)$$

it is clearly possible to find a particular solution of the differential equation in which y has the same time variation

$$y = u(x)e^{i\omega t} \qquad (1.198)$$

The inhomogeneous differential equation for u is then

$$\frac{d}{dx}\left(T\frac{du}{dx}\right) + \omega^2\rho u = -\rho w u \qquad (1.199)$$

A solution may be found by expanding both sides of the equation as a series in the normalized eigenfunctions $y_n(x)$ of the homogeneous problem:

$$u = \sum c_n y_n(x) \tag{1.200}$$

$$w(x) = \sum a_n y_n(x) \tag{1.201}$$

The coefficients in (1.201) are

$$a_n = \int_{x_1}^{x_2} y_n w \rho \, dx \tag{1.202}$$

Substituting the expansions in (1.199), equating coefficients, and using the eigenvalue equation

$$\frac{d}{dx}\left(T \frac{dy_n}{dx}\right) + \omega_n^2 \rho y_n = 0 \tag{1.203}$$

we find that

$$c_n(\omega^2 - \omega_n^2) = -a_n$$

or

$$c_n = \frac{a_n}{\omega_n^2 - \omega^2} \tag{1.204}$$

Using (1.202) for a_n and substituting for c_n in (1.200) we obtain the solution

$$u(x) = \sum_n \frac{y_n(x)}{\omega_n^2 - \omega^2} \int_{x_1}^{x_2} \rho(x') y_n(x') w(x') \, dx' \tag{1.205}$$

This may be written in the form

$$u(x) = \int_{x_1}^{x_2} G(x, x') \rho(x') w(x') \, dx' \tag{1.206}$$

where

$$G(x, x') = \sum_n \frac{y_n(x') y_n(x)}{\omega_n^2 - \omega^2} \tag{1.207}$$

$G(x, x')$ is the "Green's function" of the problem.

Problem 1-18:   Prove that

$$y_n(x) = (\omega_n^2 - \omega^2) \int_{x_1}^{x_2} \rho(x') y_n(x') G(x, x') \, dx'$$

Let us now enumerate some properties of the Green's function.

(a) Since $\rho(x')w(x') \, dx'$ is the force exerted on the element of string $(x', x' + dx')$, $G(x, x')$ is the quantity by which one must multiply that force to get the resulting displacement at x. The total displacement is then the sum (integral) of the contributions from the different elements of force (a result which follows in any case from the linearity of the equation).

(b) $G(x, x_0)$ is the displacement at x due to a δ-function force density (force per unit length) at x_0. This follows on putting

$$\rho(x')w(x') = \delta(x' - x_0) \tag{1.208}$$

in (1.206).

(c) The eigenvalues of the problem are the values at which G, considered as a function of the eigenvalue variable, has singularities. This again reflects the resonance phenomenon mentioned earlier.

This last observation may in some instances provide a means of obtaining the eigenvalues. For it is sometimes possible to solve directly the Green's function equation

$$\frac{d}{dx}\left(T\frac{dG}{dx}\right) + \omega^2 G = -\delta(x - x') \tag{1.209}$$

Problem 1-19:   By calculating directly the Green's function for the case in which ρ and T are constant, determine the eigenvalues of the problem.

At all points except $x = x'$, G satisfies the homogeneous equation. The δ-function on the right simply provides a boundary condition at $x = x'$. For suppose we integrate the equation from $x = x' - \epsilon$ to $x = x' + \epsilon$ and then let $\epsilon \to 0$. The contribution from the second term approaches 0, and we get

$$T\frac{\partial G}{\partial x}\bigg|_{x'+} - T\frac{\partial G}{\partial x}\bigg|_{x'-} = -1 \tag{1.210}$$

If we were to integrate (1.209) from a fixed point, say x_1, to an arbitrary point x, we would get

$$T\frac{\partial G}{\partial x} - \left[T\frac{\partial G}{\partial x}\right]_{x_1} + \int_{x_1}^{x} w^2 \rho G \, dx' = 0 \qquad x < x'$$

$$= -1 \qquad x > x' \tag{1.211}$$

If this is divided by $T(x)$, and the resulting equation is integrated from $x' - \epsilon$ to $x' + \epsilon$, it follows that as $\epsilon \to 0$, $G(x, x')$ is continuous across the point $x = x'$. This condition, along with (1.210), then provides boundary conditions subject to which the homogeneous equation must be solved to give G.

18. Effect of a Perturbation of Density

Let us suppose that we are able to solve the string problem for a density distribution $\rho_0(x)$; in this section we show how we may solve approximately in the case

$$\rho(x) = \rho_0(x) + \lambda\rho_1(x) \qquad (1.212)$$

where $\lambda\rho_1 \ll \rho_0$ everywhere. It is useful to introduce the parameter as a small constant, so that ρ_1 may be comparable with ρ_0; however, aside from that, the choice of λ is not important, since only $\lambda\rho_1$ is physically significant. The string equation now takes the form

$$\frac{\partial}{\partial x}\left(T\frac{\partial y}{\partial x}\right) = (\rho_0 + \lambda\rho_1)\frac{\partial^2 y}{\partial t^2} \qquad (1.213)$$

The time variation may again be taken as periodic:

$$y = u(x)e^{i\omega t} \qquad (1.214)$$

Thus (1.213) becomes

$$\frac{d}{dx}\left(T\frac{du}{dx}\right) + (\rho_0 + \lambda\rho_1)\omega^2 u = 0 \qquad (1.215)$$

The method of solution is now the following; we expand u and ω^2 as power series in λ:

$$u = \sum \lambda^n u^{(n)} \qquad (1.216)$$

$$\omega^2 = \sum \lambda^n \omega^{(n)2} \qquad (1.217)$$

We then substitute in (1.215) and equate corresponding powers of λ:

$$\frac{d}{dx}\left(T\frac{du^{(0)}}{dx}\right) + \rho_0\omega^{(0)2}u^{(0)} = 0 \qquad (1.218)$$

$$\frac{d}{dx}\left(T\frac{du^{(1)}}{dx}\right) + \rho_0\omega^{(0)2}u^{(1)} + \rho_0\omega^{(1)2}u^{(0)} + \rho_1\omega^{(0)2}u^{(0)} = 0 \qquad (1.219)$$

etc. Since the usefulness of the perturbation method depends on getting reasonably good results with a small number of terms of the series, we will not go beyond Eq. (1.219).

Let the eigenvalues of (1.218) be $\omega_n{}^2$ and the eigenfunctions y_n. We consider the perturbation of a particular solution, let us say y_s; that is, we take $u^{(0)} = y_s$, $\omega^{(0)2} = \omega_s{}^2$.

The last two terms of (1.219) may be taken to the right-hand side; the equation is then an inhomogeneous equation for $u^{(1)}$. However, we cannot directly apply the results of the previous section, because one of the denominators in G would be zero (that corresponding to $n = s$). Nevertheless, we may proceed as before and expand

$$u^{(1)} = \sum c_n y_n \tag{1.220}$$

In addition, let us write

$$\rho_1 u^{(0)} = \rho_1 y_s = \rho_0 \sum b_{sn} y_n \tag{1.221}$$

b_{sn} is then determined by the equation

$$b_{sn} = \int \rho_1 y_s y_n dx \tag{1.222}$$

where we omit the limits of the integral throughout for convenience.

Problem 1-20:   By the variational method find corrections to ω^2 up to the third order.

If each term in (1.219) is now expanded in terms of the functions y_n, the coefficient of each y_n on the left-hand side will be zero. It is necessary, however, to treat the $n = s$ term differently from the others. In that case the first two terms cancel and we have simply

$$\omega^{(1)2} = -b_{ss}\omega_s{}^2$$
$$= -\omega_s{}^2 \int \rho_1 y_s{}^2 dx \tag{1.223}$$

This shows, in particular, that a positive perturbation decreases the frequency, while a negative one increases it. This result might have been surmised, of course, from the dependence of frequency on density in the case of uniform density.

Going now to the coefficients in (1.219) of y_n $(n \neq s)$,

$$(\omega_s{}^2 - \omega_n{}^2)c_n = -b_{sn}\omega_s{}^2 \tag{1.224}$$

the third term in this case having contributed nothing. The coefficients in (1.220) are then determined, and to this approximation the perturbed eigenfunction is

$$y = y_s - \omega_s{}^2 \sum_{n \neq s} \frac{y_n(x)}{\omega_s{}^2 - \omega_n{}^2} \int \lambda \rho_1 y_s y_n dx' \tag{1.225}$$

This is, of course, not normalized as it stands.

It is interesting to note that if we were to use the variational principle to calculate the eigenvalue, and use as "trial function" simply the unperturbed eigenfunction,

$$u^{(0)} = y_s$$

$$\omega^2 = \frac{\int T(dy_s/dx)^2 dx}{\int (\rho_0 + \lambda\rho_1)y_s^2 dx}$$

$$= \frac{\omega_s^2}{1 + \int \lambda\rho_1 y_s^2 dx} \tag{1.226}$$

Problem 1-21: Solve the case in which the perturbation is a small δ-function mass $\epsilon\delta(x-x_1)$.

---

which corresponds to first powers in λ with the result of the perturbation calculation. This is consistent with the general features of the variational method, according to which an error of order λ in the trial eigenfunction leads only to an error of order λ^2 in the eigenvalue. The eigenvalue is thus given correctly to *first* order in λ without using the correction terms in (1.225). If (1.225) is used in the variational integral, the eigenvalue will, by the same argument, be in error by terms of order λ^4; that is, it will be correct to the third order.

19. The JWKB Method

A useful approximate method, usually associated with the names of Jeffreys, Wentzel, Kramers, and Brillouin, applies to the case of a slowly varying density. We start from the spatial equation (1.115) obtained from the method of separation of variables

$$\frac{d}{dx}\left(T\frac{df}{dx}\right) + \omega^2\rho f = 0$$

If T and ρ are constant, the solution has the form

$$f(x) = Ae^{i(\omega/c)x} \tag{1.227}$$

For slowly variable ρ, we try a solution of the form

$$f(x) = A(x)e^{iS(x)} \tag{1.228}$$

Substituting in the equation and cancelling e^{iS}, we derive the equation

$$A'' + 2iS'A' + iS''A - S'^2A + \frac{\omega^2}{u^2}A = 0 \tag{1.229}$$

where

$$u^2(x) = \frac{T}{\rho(x)} \tag{1.230}$$

and primes indicate derivatives with respect to x. The real and imaginary parts of the left-hand side must be separately zero, which leads to the equations

$$2S'A' + S''A = 0 \tag{1.231}$$

and

$$A'' + \frac{\omega^2}{u^2}A - S'^2A = 0 \tag{1.232}$$

If $u(x)$ is slowly varying, the derivatives of both S and A will be small compared with the functions themselves. In particular, in the second equation (1.232) $A'' \ll \omega^2/u^2A$ and will be neglected. This equation may then be solved to give

$$S = \int^x \frac{\omega}{u(x')}\, dx' \tag{1.233}$$

The lower limit may be chosen arbitrarily, but it is convenient to take it to be $x = 0$, which we assume to correspond with the left-hand end of the string.

Equation (1.231) may be integrated to give

$$A^2 = \frac{\alpha^2}{S'} = \alpha^2 \frac{u}{\omega} \tag{1.234}$$

where α^2 is a constant. The approximate solution of (1.228) is then

$$f(x) = \text{constant} \cdot \sqrt{u}\exp\left[i\int_0^x \frac{\omega}{u(x')}\, dx'\right] \tag{1.235}$$

This is a formal (complex) solution. The *real* solution satisfying the boundary condition $f = 0$ at $x = 0$ is

$$f(x) = \text{constant} \cdot \sqrt{u}\sin\left[\int_0^x \frac{\omega}{u(x')}\, dx'\right] \tag{1.236}$$

If the other fixed end of the string is at $x = L$, the eigenvalue ω is determined by

$$\omega \int_0^L \frac{dx'}{u(x')} = n\pi \tag{1.237}$$

20. An Example

As an illustration of both the perturbation method and JWKB, consider the problem of a string of density

$$\rho = \rho_0 + \epsilon \sin\frac{\pi x}{L} \tag{1.238}$$

stretching from $x = 0$ to $x = L$. ρ_0 and ϵ are taken as constants. If ϵ is small, the conditions for the validity of each approximation are satisfied.

The modified frequency in the perturbation method, in which ϵ_1 $\sin (\pi x/L)$ is taken to be the perturbation, is given by (1.223):

$$\omega^2 = \omega_s^2 \left[1 - \int_0^L \rho_1 y_s^2 dx \right]$$

where ω_s is the *unperturbed* frequency. Evaluating the integral, it is found that

$$\omega^2 = \omega_s^2 \left[1 - \frac{\epsilon}{\rho_0 \pi} \left(1 + \frac{1}{4s^2 - 1} \right) \right] \tag{1.239}$$

To determine the answer by the JWKB method, it is necessary to evaluate the integral

$$\int_0^L \sqrt{\frac{\rho(x')}{T}} \, dx' = \frac{1}{c} \int_0^L \sqrt{1 + \frac{\epsilon}{\rho_0} \sin \frac{\pi x'}{L}} \, dx'$$

Now, in the notation of (1.232), $A'' \ll (\omega^2/u^2)A$ only if $\epsilon \ll \rho_0$; consequently a sufficiently good approximation is obtained on putting

$$\sqrt{1 + \frac{\epsilon}{\rho_0} \sin \frac{\pi x'}{L}} = 1 + \frac{1}{2} \frac{\epsilon}{\rho_0} \sin \frac{\pi x}{L}$$

the integral of which is $L(1 + (\epsilon/\rho_0 \pi))$. Thus

$$\omega = \omega_s \frac{1}{1 + \epsilon/\rho_0 \pi} \doteq \omega_s \left[1 - \frac{\epsilon}{\rho_0 \pi} \right] \tag{1.240}$$

Problem 1-22: Calculate the neglected term using the approximate solution obtained by neglecting it. Determine the condition that this is small, and thus the condition under which the approximation is justified.

Can you suggest how the approximation could be improved?

In contrast with (1.239), this predicts that all frequencies change in the same proportion. The perturbation method gives the more accurate result in this case, since it is *exact* to first powers in ϵ. The difference between the two results approaches zero as the frequency becomes large. The tendency of the JWKB method to approach the correct value for large eigenvalues is related to the "correspondence principle" of quantum mechanics.

Problem 1-23: Do exactly the case of density p^2/x^2 between $x = a$ and $x = b > a$. Solve by JWKB when $(b-a)/a \ll 1$. The normal modes are given by

$$y_n = c_n \sin \left[\frac{n\pi}{\ln (b/a)} \ln \left(\frac{x}{a}\right) \right]$$

and the normal frequencies are

$$\omega_n = \frac{n\pi\sqrt{T}}{p \ln (b/a)}$$

By JWKB the normal frequencies are given correctly, but the normal modes are determined approximately to be

$$y_n = c_n' x^{1/2} \sin \left[\frac{n\pi}{\ln (b/a)} \ln \left(\frac{x}{a}\right) \right]$$

21. Lagrangian and Hamiltonian Formulations of the Vibrating String Problem

Equations (1.31) and (1.34) give formulas for the kinetic and potential energies, respectively, of the vibrating string. From them we can formulate the Lagrangian of the system

$$L = K - V$$

$$= \frac{1}{2} \int \left(\frac{\partial y}{\partial t}\right)^2 dx - \frac{1}{2} \int T \left(\frac{\partial y}{\partial x}\right)^2 dx \qquad (1.241)$$

For systems with a finite number of degrees of freedom, the Lagrangian involves *sums* over quadratic forms in the generalized coordinates and their time derivatives. In the string problem, the system is continuous and the number of coordinates or degrees of freedom is effectively infinite. (This is not strictly true, of course; the string has a finite number of particles in it. However, to describe it on that basis we would need to use quantum, not classical mechanics. We have somewhat idealized the problem, by considering it as truly continuous — a reasonable assumption from the macroscopic viewpoint — and of negligible cross section. The continuity of the system, and of the coordinates that describe it, are characteristic of this *idealized* string, and represent a reasonable compromise with literal truth dictated by our interest only in the macroscopic features of the system.)

The *coordinates* of the problem are the displacements $y(x, t)$. We expect them to be functions of time, but what of the variation with x? We see that x simply serves to enumerate the coordinates; to each value of x there is a coordinate. Rather than *summing* over coordinates we must, when they form a continuum, *integrate* over them.

Suppose that we wish to write down Lagrange's equations of motion. They may be expressed formally as

$$\frac{d}{dt}\left(\frac{\partial L}{\partial \dot{y}(x, t)}\right) - \frac{\partial L}{\partial y(x, t)} = 0 \tag{1.242}$$

or, in this case

$$\frac{d}{dt}\frac{\partial K}{\partial \dot{y}} + \frac{\partial V}{\partial y} = 0 \tag{1.243}$$

We must first, then, differentiate the Lagrangian with respect to $\dot{y}(x, t) = \partial y/\partial t$. That is, we must calculate the rate of change of the kinetic energy K with respect to a change in $\dot{y}$.

The problem is complicated, however, by the fact that, whereas, in the case of discrete coordinates, the contribution from each coordinate is finite, in the continuous case it is not. As a formal step around this difficulty, it is useful to consider, not just the point x, but a small region $(x - \delta/2, x + \delta/2)$ around it; δ can ultimately be made as small as we like.

The contribution to K whose rate of change we wish to consider is, then,

$$\tfrac{1}{2}\rho(x)[\dot{y}(x, t)]^2\delta + \text{terms of order } \delta^2$$

The rate of change with respect to $\dot{y}$ is, to the same order,

$$\rho\dot{y}(x, t)\delta$$

The time rate of change of this quantity, which constitutes the first term in the Lagrangian equation of motion for the variable $y(x, t)$, is

$$\rho\delta\ddot{y}(x, t) = \rho\delta\frac{\partial^2 y}{\partial t^2}$$

Consider now the second term in the Lagrangian. The contribution to V from the range δ is

$$\frac{1}{2}\int_{x-\delta/2}^{x+\delta/2} T\left(\frac{\partial y}{\partial x'}\right)^2 dx'.$$

To calculate the derivative of this with respect to y, we must change y only within this interval; let the change be $\eta(x, t)$, which can be taken to go to zero at the ends of the interval. The change in V is then

$$\int_{x-\delta/2}^{x+\delta/2} T\frac{\partial y}{\partial x'}\frac{\partial \eta}{\partial x'} dx'$$

This may be integrated by parts to give

$$-\int_{x-\delta/2}^{x+\delta/2} \eta(x') \frac{\partial}{\partial x'}\left(T\frac{\partial y}{\partial x'}\right) dx'$$

where we have neglected second powers of η, since they make no contribution to the required rate of change. If, now, the interval δ is made vanishingly small, $(\partial/\partial x')(T(\partial y/\partial x'))$ may be treated as a constant in the evaluation to order δ, giving

$$-\frac{\partial}{\partial x}\left(T\frac{\partial y}{\partial x}\right) \int_{x-\delta/2}^{x+\delta/2} \eta(x')dx' = -\frac{\partial}{\partial x}\left(T\frac{\partial y}{\partial x}\right)\delta\bar{\eta}$$

where $\bar{\eta}$ is the *mean* value of the change in y in the interval. The rate of change of V with respect to this quantity is then

$$-\frac{\partial}{\partial x}\left(T\frac{\partial y}{\partial x}\right)\delta$$

Cancelling out the δ, we then arrive at the Lagrangian equation

$$\frac{\partial^2 y}{\partial t^2} = \frac{\partial}{\partial x}\left(T\frac{\partial y}{\partial x}\right)$$

which is, of course, the equation of motion previously derived.

The preceding derivation represents an attempt to follow literally and directly the usual procedures for deriving equations of motion from a Lagrangian. It is a procedure singularly lacking in elegance, whatever its virtues in directness. The job may, however, be done more painlessly if a different starting point is used. It is shown in texts on analytic mechanics [see, for example, Goldstein, *Classical Mechanics*, Chapter 2, p. 30.] that Hamilton's variational principle

$$\delta \int_{t_1}^{t_2} L\,dt = 0 \tag{1.244}$$

is precisely equivalent to the set of Lagrange's equations for conservative systems. Equation (1.244) envisages "paths" in coordinate space, prescribed by using time as a parameter, all paths going through the same end points. The "true" path, the one corresponding to the actual history of the physical system, is that which minimizes the integral in (1.244).

There is no special difficulty in formulating the dynamics of our continuous system in this way. For the string problem, the required variational principle is

$$\delta \int_{t_1}^{t_2} dt \int_{x_1}^{x_2} \left[\frac{1}{2}\left(\frac{\partial y}{\partial t}\right)^2 - \frac{1}{2}T\left(\frac{\partial y}{\partial x}\right)^2\right]dx = 0 \tag{1.245}$$

The variational calculation parallels rather closely that in the section, earlier in this chapter, on the Rayleigh–Ritz method; we replace y by $y + \epsilon\phi$ in the integral, and differentiate with respect to ϵ at $\epsilon = 0$. This leads to the equation

$$\int_{t_1}^{t_2} dt \int_{x_1}^{x_2} \left[\rho \frac{\partial y}{\partial t}\frac{\partial \phi}{\partial t} - T \frac{\partial y}{\partial x}\frac{\partial \phi}{\partial x} \right] dx = 0 \qquad (1.246)$$

Since (1.246) must be satisfied for *arbitrary* values of ϕ, we integrate by parts—the first term with respect to t, the second with respect to x—so as to obtain ϕ as a common factor of the integrand. The integrated term goes out in the t-integration by virtue of the formulation of Hamilton's Principle; that in the x-integration by virtue of the boundary conditions of the problem. Thus, (1.246) becomes

$$\int_{t_1}^{t_2} dt \int_{x_1}^{x_2} \phi \left[-\rho \frac{\partial^2 y}{\partial t^2} + \frac{\partial}{\partial x}\left(T \frac{\partial y}{\partial x} \right) \right] dx = 0 \qquad (1.247)$$

The expression in the square bracket must then vanish for all values of x, and this leads to the Lagrangian equations of motion.

Thus, the equation for vibrating string, in its usual form, is a formulation of the Lagrangian equations when the point-by-point displacements $y(x, t)$ are taken as generalized coordinates.

This calculation may be related to the previous one in the following way: If we choose for ϕ a δ function

$$\phi = \delta(x - x') \qquad (1.248)$$

then we pick out simply the equation of motion for the coordinate x'; we vary only this coordinate. This is precisely what we did, rather awkwardly, in our direct calculation.

Finally, let us consider a different set of generalized coordinates for the problem. The Sturm–Liouville equation associated with the string problem was given in Eq. (1.115):

$$\frac{d}{dx}\left(T \frac{dy}{dx} \right) + \lambda \rho y = 0$$

where $\lambda = \omega^2$, the square of the circular frequency. If the eigenvalues are ω_n^2 and the eigenfunctions $y_n(x)$, we have shown that any function satisfying the boundary conditions of the problem can be expanded in a series in the eigenfunctions

$$f(x) = \sum a_n y_n(x)$$

In particular, then, the displacement $y(x, t)$ can be so expanded; the coefficients (amplitudes) of the various modes being functions of t:

$$y(x, t) = \sum c_n(t) y_n(x) \qquad (1.249)$$

The quantities $c_n(t)$ also serve as generalized coordinates for the problem; they determine the motion completely. If the expansion (1.249) is substituted into the Lagrangian as given in (1.241), the Lagrangian may be expressed in terms of the c_n's. It is immediately evident, by virtue of the orthogonality of the y_n's, that is, the fact that

$$\int \rho y_n y_m dx = 0, \qquad n \neq m$$

that the *kinetic* energy may be written

$$K = \tfrac{1}{2} \sum_n \dot{c}_n{}^2 \tag{1.250}$$

This follows the customary pattern, that the kinetic energy is a bilinear combination of the generalized velocities.

Turning to the potential energy, the situation is at first sight more complicated. Suppose, however, that we integrate V by parts, and make use of the boundary condition that the displacement y vanishes at the ends of the interval. V may then be written

$$V = -\int_{x_1}^{x_2} y \frac{\partial}{\partial x}\left(T \frac{\partial y}{\partial x}\right) dx \tag{1.251}$$

x_1 and x_2 being, as usual, the coordinates of the ends of the string.

Problem 1-24: If, in the derivation of the string equation, we had kept the next term in the expansion of the tensional force in terms of $\partial y/\partial x$, we should have obtained the equation

$$\frac{\partial^2 y}{\partial x^2}\left[1 - \frac{3}{2}\left(\frac{\partial y}{\partial x}\right)^2\right] = \frac{1}{c^2}\frac{\partial^2 y}{\partial t^2}$$

for the case of constant density and tension. Show that, to this approximation, the potential energy is

$$\frac{1}{2} T \int_0^L \left(\frac{\partial y}{\partial x}\right)^2 \left[1 - \frac{1}{4}\left(\frac{\partial y}{\partial x}\right)^2\right] dx$$

when the string stretches from $x = 0$ to $x = L$. Show that, in terms of the amplitudes of the normal modes, the Lagrangian becomes

$$L = \tfrac{1}{2} \sum_n \left[\dot{c}_n{}^2 - \omega_n{}^2 c_n{}^2\right] - \tfrac{3}{8} \sum_{mnrs} A_{mnrs} c_m c_n c_r c_s$$

where

$$A_{mnrs} = \rho \int_0^L y_m y_n \frac{\partial y_r}{\partial x} \frac{\partial y_s}{\partial x} \, dx$$

$$= \rho \frac{rs\pi^2}{8L} [\delta(m-n+r+s)$$
$$+ \delta(m-n-r-s) + \delta(m-n+r-s) + \delta(m-n-r+s)$$
$$- \delta(m+n+r+s) - \delta(m+n-r-s) - \delta(m+n+r-s)$$
$$- \delta(m+n-r+s)]$$

the δ being 1 when its argument is zero and zero otherwise.

Let us now substitute in (1.251) the eigenfunction expansion of y. Making use of the eigenvalue equation

$$\frac{\partial}{\partial x}\left(T \frac{\partial y_n}{\partial x}\right) = -\omega_n^2 \rho y_n \qquad (1.252)$$

and once again using orthogonality, we find that V can be written in terms of the $c_n(t)$'s:

$$V = \tfrac{1}{2} \sum \omega_n^2 c_n^2 \qquad (1.253)$$

Combining with (1.250) gives

$$L = K - V$$
$$= \tfrac{1}{2} \sum [\dot{c}_n^2(t) - \omega_n^2 c_n^2(t)] \qquad (1.254)$$

Thus, using as dynamical coordinates the amplitudes of the normal modes of the motion, we obtain the Lagrangian in a particularly simple form in which there is no coupling between the coordinates in the equations of motion. The Lagrangian is, in fact, simply that of a system of an infinite number of isolated, noninteracting, harmonic oscillators.

The peculiar simplicity of the problem when expressed in terms of these coordinates is worth noting, and has formed the basis for the commonest method of dealing with the quantum mechanics of continuous systems. The quantum dynamics of a system depends on a knowledge of the Hamiltonian as a function of generalized coordinates and momenta. [In the string problem, the momenta are $p_n = \partial L/\partial \dot{c}_n = \dot{c}_n$, and the Hamiltonian is $H = K + V = \tfrac{1}{2} \sum (p_n^2 + \omega_n^2 c_n^2)$.] Since these coordinates are denumerable (even though infinite in number) the dynamics of the continuous system is mathematically equivalent to that of an infinite system of particles, in this case noninteracting. This is both a simple problem and one that is more or less familiar.

This technique, which we have applied to a particular one-dimensional continuous system, may be applied to more complicated systems; for example, by analyzing the normal modes of the electromagnetic field in a given geometry, and with a knowledge of the field energy we may reduce the problems of electromagnetic theory to a form similar to those of the dynamics of particles. This makes it evident, then, that the quantum mechanics of continuous fields is no more difficult than that of systems of particles.

Two further remarks complete our discussion of this problem. One concerns the apparent discrepancy according to which our system may be equally well described by a continuous system of coordinates $y(x, t)$, or a discrete one $c_n(t)$. One has the feeling that these cannot describe the same number of degrees of freedom (though in fact the number is in each case infinite). The situation seems less strange, however, when we consider that in the second (normal mode) description we have *used the boundary conditions in defining our coordinates*, whereas this was not the case when we used $y(x, t)$. Furthermore, if the coordinates $y(x, t)$ are taken *a priori* as completely independent of each other, we could *in principle* describe completely discontinuous motions of the string. The expansion in normal modes, on the other hand, requires that the function describing the point-by-point displacement be a reasonably well-behaved function.

The fact is, then, that the description in terms of normal modes is, in the mathematical sense, a more restrictive one. This is not, however, an issue in the physical problem itself, since the boundary conditions, the good mathematical behavior of the displacement, are essential requirements imposed on us by the *physics* of the system. The description in normal modes, then, does not, in fact, put any significant restrictions on the problem.

The second remark concerns the problem of the *forced* motion of the system. The problem is, to determine the *generalized force* associated with the coordinate c_n. This problem is one which is solved by a well-known prescription: The work done in a virtual displacement is calculated in terms of the virtual infinitesimal change in the generalized coordinates; the coefficients of these infinitesimal coordinate changes are the generalized forces which appear as inhomogeneous terms in the Lagrangian equations.

In terms of our original coordinates, we can write down this virtual work very simply. If $F(x, t)$ is the force per unit mass on the string at x, it is

$$\delta W = \int_{x_1}^{x_2} \rho(x) F(x, t) \delta y(x, t) dx \qquad (1.255)$$

From (1.249),

$$\delta y(x, t) = \sum \delta c_n(t) y_n(x) \qquad (1.256)$$

If we similarly expand $F(x, t)$,

$$F(x, t) = \sum F_n(t) y_n(x) \tag{1.257}$$

and use the orthogonality of the y_n's,

$$\delta W = \sum F_n(t) \delta c_n(t) \tag{1.258}$$

The generalized forces are then the coefficients of the expansion of the *force per unit mass* in terms of the eigenfunctions of the associated Sturm–Liouville problem. The equations of forced motion are then

$$\ddot{c}_n + \omega_n^2 c_n = F_n(t) \tag{1.259}$$

PRELUDE TO CHAPTER 2

In a sense, this chapter represents the theoretical foundation on which all the rest of the book is built. We sketch briefly the theory of linear vector spaces, using the notation and terminology introduced by Dirac in his treatment of quantum mechanics. In this theory, the eigenvalue problem is represented geometrically in a multidimensional space, and the concept of *orthogonality* of eigenstates (normal modes) is given a geometrical interpretation.

The theory is very general; its application to the various linear problems of mathematical physics depends in each case on the definition of a *scalar product* of arbitrarily chosen vectors in the space. These vectors may, for instance, be functions, and the scalar product defined in terms of these functions. Using the so-called "Schmidt orthogonalization procedure," which permits the construction of mutually orthogonal "vectors" out of arbitrary independent ones, orthogonal polynomials are constructed using different definitions of orthogonality. Legendre, Laguerre, Hermite, and Tschebycheff polynomials are introduced in this way.

It is shown that in general vectors may be represented by column (or row) matrices and operators by square matrices. Thus, linear differential operators may be put in one-to-one correspondence with matrix ones. The equivalence of Schödinger wave mechanics and Heisenberg matrix mechanics follows from such a correspondence.

All vectors may be expressed as linear combinations of a set of "basis vectors," a complete set of normalized orthogonal vectors. Transformations between one basis and another leave the magnitudes of vectors unchanged. Such transformations are said to be "unitary." The unitary transformation of operators is determined.

The general theory of Hermitian operators is introduced. Since Hermitian operators are operators with real eigenvalues, and the physical quantities represented by those eigenvalues must be real, these operators play a central role in physical applications.

In addition to having real eigenvalues, Hermitian operators are shown to have the property that their eigenvectors are mutually orthogonal (provided that they correspond to different eigenvalues).

Fourier series are interpreted in vector terms, the vectors of the basis being the periodic expansion functions. This is in turn a particular case of the Sturm–Liouville problem. The differential operator of the problem (along with boundary conditions) determines a Hermitian operator. The eigenfunctions correspond to mutually orthogonal (basis) states. Thus, all the Sturm–Liouville problems occurring throughout the book are eigenvalue problems of Hermitian operators in linear vector spaces.

A more mathematically rigorous and complete treatment of linear vector spaces may be found in books on pure mathematics. Particular attention is called to T. Kato's *Perturbation Theory for Linear Operators*, especially Secs. 3 and 5 of Chapter 1 and Secs. 1 and 3 of Chapter 5.

REFERENCES

Dirac, P. A. M., *Principles of Quantum Mechanics*, 4th ed. London: Oxford, Clarendon Press, 1958.

Goertzel G. and N. Tralli, *Some Mathematical Methods of Physics*. New York: McGraw-Hill, 1960.

Hochstadt, H., *Special Functions of Mathematical Physics*. New York: Holt, Rinehart and Winston, 1961.

Jackson, D., *Fourier Series and Orthogonal Polynomials*. Carus Mathematical Monograph No. 6. Oberlin, Ohio: Mathematical Association of America, 1941.

Jackson, J. D., *Mathematics for Quantum Mechanics*. New York: Benjamin, 1962.

Kato, T., *Perturbation Theory for Linear Operators*. Berlin: Springer, 1966.

Mackey, G. W., *Mathematical Foundations of Quantum Mechanics*. New York: Benjamin, 1963.

Messiah, A., *Mécanique Quantique*. Paris: Dunod, 1959, Chapter 7.

Sagan, H., *Boundary and Eigenvalue Problems in Mathematical Physics*. New York: Wiley, 1961.

Temple, G., *General Principles of Quantum Theory*. New York: Barnes and Noble, 1934.

2

LINEAR VECTOR
SPACES

"Don't let your simplicity be imposed on"

R. B. Sheridan, The Rivals

1. Introduction

In the previous chapter we have developed a theory of eigenvalue problems in ordinary linear differential equations of the second order. These differential equations are formed with linear differential operators, and, in conjunction with boundary conditions, determine eigenvalue problems. Linear combinations of the eigenfunctions serve to describe a general class of functions satisfying the boundary conditions. This class of functions constitutes a particular example of what is called a "linear vector space"; the differential operators (with their boundary conditions) are operators in this vector space.

The concept of a linear vector space is, however, much more general than this example. It is, in particular, at the root of the formulation of quantum mechanics. It is, in fact, a concept so broad that it is able to encompass a large part of mathematical physics. We try therefore to formulate it in as comprehensive a form as possible. The elaboration of our scheme proceeds in the following way. First, we label the entities out of which the vector space is to be constructed. We adapt, for this purpose, a notation introduced by Dirac in his book on quantum mechanics and commonly used in quantum mechanics. These entities, accordingly, are designated by the symbol $|\psi\rangle$, which we call, following Dirac, "ket" vectors, and, in one-to-one correspondence with these vectors, their "conjugates" $\langle\psi|$, which are called "bra" vectors.

We then proceed to define certain operations on these vectors; that is, to specify their algebra. Some of these operations are defined, in the first instance, for the "ket" vectors; in each case we indicate the corresponding operations for the "bra" vectors. For simplicity, we refer in future to the $|\psi\rangle$'s simply as "vectors," and to the $\langle\psi|$'s as their conjugate vectors.

The algebraic postulates which follow are given an abstract formulation. Any set of entities conforming to these postulates will then constitute a particular realization of a linear vector space. It may be useful to consider the set of functions expansible in a series of the eigenfunctions of a Sturm–Liouville problem as a particular example, though, as we see, formally quite different realizations exist.

(a) We first define an operation of summation, that is, we associate with two vectors $|\psi_1\rangle$ and $|\psi_2\rangle$ another vector which we call $(|\psi_1\rangle + |\psi_2\rangle)$. This summation is assumed commutative:

$$|\psi_1\rangle + |\psi_2\rangle = |\psi_2\rangle + |\psi_1\rangle \tag{2.1}$$

and associative:

$$|\psi_1\rangle + (|\psi_2\rangle + |\psi_3\rangle) = (|\psi_1\rangle + |\psi_2\rangle) + |\psi_3\rangle$$
$$= (|\psi_1\rangle + |\psi_3\rangle) + |\psi_2\rangle \tag{2.2}$$

Also, denoting the fact that $\langle\psi|$ is the conjugate of $|\psi\rangle$ thus:

$$\langle\psi| \leftrightarrow |\psi\rangle$$

we assume that

$$\langle\psi_1| + \langle\psi_2| \leftrightarrow |\psi_1\rangle + |\psi_2\rangle$$

The relations (2.1) and (2.2) then imply similar relations for the conjugate vectors.

(b) Before proceeding to the next postulate, we note that it would be natural to define $|\psi\rangle + |\psi\rangle$ as $2|\psi\rangle$. In fact we define a "ray" subspace of our vector space, each element of which is associated with a complex number a in association with the fixed vector $|\psi\rangle$. The vectors of this subspace are designated $a|\psi\rangle$. The conjugate of $a|\psi\rangle$ is designated $\langle\psi|a^*$, where a^* is the complex conjugate of a.

These vectors are assumed to have the following properties; corresponding relations exist for their conjugates.

$$(i) \quad a(|\psi_1\rangle + |\psi_2\rangle) = a|\psi_1\rangle + a|\psi_2\rangle \tag{2.3}$$

$$(ii) \quad (a_1 + a_2)|\psi\rangle = a_1|\psi\rangle + a_2|\psi\rangle \tag{2.4}$$

$$(iii) \quad a(b|\psi\rangle) = b(a|\psi\rangle) = ab|\psi\rangle \tag{2.5}$$

$$(iv) \quad a|\psi\rangle = |\psi\rangle \quad \text{when} \quad a = 1 \tag{2.6}$$

(v) $a|\psi\rangle$ will be a vector known as the "zero" vector, and will be independent of $|\psi\rangle$, when $a = 0$:

$$0|\psi\rangle = |0\rangle \tag{2.7}$$

When the zero vector appears in a vector equation, it will often simply be designated as 0.

We note that if we add to $|\psi\rangle$ the vector $(-1)|\psi\rangle$ we get the zero vector. Therefore, $(-1)|\psi\rangle = -|\psi\rangle$ is called the negative of $|\psi\rangle$. It also follows from (2.4) that the result of adding the zero vector to $|\psi\rangle$ is to give $|\psi\rangle$ itself.

2. Vector Spaces

Let us now say what we mean by a "vector space." Suppose we have a set of vectors $|\psi_i\rangle$, enumerated by real numbers i which may form either a discrete or a continuous set. (In the former case the i's will be the integers $1, 2, \ldots, N$, but with the possibility that $N \to \infty$.) Then the set of vectors

$$\sum_i a_i|\psi_i\rangle \tag{2.8}$$

where the a_i's are arbitrary complex numbers, define a "vector space"; that is to say, the space consists of the set of vectors.

In the continuous case, the vectors are designated by $|\psi_x\rangle$, specified by the continuous real variable $x(x_1 \leqslant x \leqslant x_2)$. Then the vectors of the vector space are

$$\int_{x_1}^{x_2} a(x)|\psi_x\rangle dx \tag{2.9}$$

where $a(x)$ is an arbitrary complex function.

3. Linear Independence, Dimensionality, and Bases

We say that n vectors $|\psi_i\rangle$ are "linearly independent" if there exists no set of nonzero numbers c_i such that

$$\sum_{i=1}^{n} c_i|\psi_i\rangle = 0 \tag{2.10}$$

It is evident that a set of *linearly independent* vectors is sufficient to define a vector space. If, in the set used to define the vector space in Sec. 2.2 above, there are relations of linear dependence such as (2.10), they may be used to eliminate some of the vectors by expressing them in terms of the others. Clearly, all the vectors (2.8) may then be expressed in terms of the smaller set.

The set of vectors used to define a vector space may be called a "basis." We know, then, that a basis need only consist of linearly independent vectors.

By following this line of argument, we can define the *dimensionality* of our vector space. This is the maximum number N of linearly independent vectors which may be chosen from among the vectors of the space. Thus, between *any* set of more than N vectors $|\psi_i\rangle$, $(i = 1, 2, \ldots, n > N)$ there will exist a relation of linear dependence, that is, a relation of the form of (2.10).

In the case of infinite-dimensional vector spaces, the space will be defined by what we call a complete set of basis vectors. This is a set of linearly independent vectors of the space such that, if any further vector of the space is added to the set, there will exist a relation of linear dependence connecting the vectors of the enhanced set.

4. Scalar Products

At this point, we introduce into our scheme the concept of the *scalar product* of vectors. It is in specifying the scalar products of the vectors that the properties of the space are truly defined, and the concept of multiplication of a vector by a number given meaning.[1]

We confine attention to vectors defined over the field of complex numbers; that is to say, to vector spaces in which the scalar product of any two vectors is a complex number.

It is convenient to define the scalar product of two vectors in terms of one (ket) vector $|\psi_1\rangle$ and the conjugate (bra) of another, $\langle\psi_2|$. This product is specified by the symbol $\langle\psi_2|\psi_1\rangle$ and is referred to as the product of $|\psi_1\rangle$ with $|\psi_2\rangle$. It need not, of course, be commutative, and in fact scalar multiplication will be defined in such a way that

$$\langle\psi_1|\psi_2\rangle = \langle\psi_2|\psi_1\rangle^* \qquad (2.11)$$

that is, so that if the vectors are interchanged in the multiplication, the product is transformed into its complex conjugate.

It is now necessary to specify some further properties of scalar products. It is assumed that

(a) the scalar product of $|\psi_2\rangle + |\psi_3\rangle$ with $|\psi_1\rangle$ is

$$\langle\psi_1|\psi_2\rangle + \langle\psi_1|\psi_3\rangle$$

(distributive law)

---

[1] It is true that in certain mathematical problems in particle physics use has been made of spaces in which only the norms (magnitudes squared) of vectors are defined (Banach spaces). Such problems are, however, beyond the scope of this book.

(b) the product of $a|\psi_2\rangle$ with $|\psi_1\rangle$ is

$$a\langle\psi_1|\psi_2\rangle$$

It follows from (2.11) that the product of $|\psi_1\rangle$ with $a|\psi_2\rangle$ is

$$a^*\langle\psi_2|\psi_1\rangle$$

and the product of $a_2|\psi_2\rangle$ with $a_1|\psi_1\rangle$ is

$$a_1^*a_2\langle\psi_1|\psi_2\rangle$$

The following deductions may then be made:

(c) By (2.11) the scalar product of a vector with itself is of necessity a *real* number; however, we assume further that it is *positive*, and refer to it as the magnitude squared of the vector. Vectors whose magnitude squared is unity are called "normalized" or "unit" vectors. Clearly, any vector may be normalized by dividing it by the root of its magnitude squared.

(d) By virtue of (2.7) the scalar product of the zero vector with any other vector is zero.

5. Schmidt Inequality and Orthogonalization

If the scalar product of two vectors $|\psi_1\rangle$ and $|\psi_2\rangle$ is written as an imaginary number z, it can be shown that $|z|^2 = \zeta^2$ is less than or equal to the product of the squares of the magnitudes of $|\psi_1\rangle$ and $|\psi_2\rangle$:

$$\zeta^2 = |\langle\psi_2|\psi_1\rangle|^2 \leq \langle\psi_1|\psi_1\rangle\langle\psi_2|\psi_2\rangle \qquad (2.12)$$

To prove this result, we observe that the magnitude squared of the vector

$$|\psi_1\rangle + \lambda|\psi_2\rangle$$

is greater than or equal to zero for arbitrary complex λ, i.e.,

$$\langle\psi_1|\psi_1\rangle + \lambda\langle\psi_1|\psi_2\rangle + \lambda^*\langle\psi_2|\psi_1\rangle + |\lambda|^2\langle\psi_2|\psi_2\rangle \geq 0$$

Putting

$$\langle\psi_2|\psi_1\rangle = z = \zeta e^{i\alpha} \qquad (2.13a)$$

$$\lambda = \mu e^{i\beta} \qquad (2.13b)$$

this becomes

$$\mu^2\langle\psi_2|\psi_2\rangle + 2\mu\zeta\cos(\alpha-\beta) + \langle\psi_1|\psi_1\rangle \geq 0$$

for all μ and β, or

$$\left[\mu\langle\psi_2|\psi_2\rangle + \frac{\zeta\cos(\alpha-\beta)}{\sqrt{\langle\psi_2|\psi_2\rangle}}\right]^2 + \left[\langle\psi_1|\psi_1\rangle - \frac{\zeta^2\cos^2(\alpha-\beta)}{\langle\psi_2|\psi_2\rangle}\right] \geq 0$$

under the same conditions. Choosing $\beta = \alpha$ and μ such that the first term is then equal to zero, it follows that

$$\langle \psi_1 | \psi_1 \rangle - \frac{\zeta^2}{\langle \psi_2 | \psi_2 \rangle} \geq 0$$

which is Eq. (2.12).

In the case of a *real* vector space, that is, one in which all scalar products are real, one has simply that

$$\langle \psi_2 | \psi_1 \rangle^2 \leq \langle \psi_1 | \psi_1 \rangle \langle \psi_2 | \psi_2 \rangle \tag{2.14}$$

This relation is known as the Schmidt inequality. It is then possible to define an *angle* between two vectors $|\psi_2 \rangle$ and $|\psi_1 \rangle$ by the equation

$$\cos \theta = \frac{\langle \psi_2 | \psi_1 \rangle}{\sqrt{\langle \psi_1 | \psi_1 \rangle \langle \psi_2 | \psi_2 \rangle}} \tag{2.15}$$

where (2.14) ensures that $\cos \theta \leq 1$.

If $\cos \theta = 0$, that is to say, if $\langle \psi_2 | \psi_1 \rangle = 0$, it is then natural to say that the vectors are *orthogonal*. This terminology is used also in the case of complex vector spaces, though the concept of "angle" is not then generally applicable.

It is easily seen that the angle between arbitrary vectors $x|a \rangle$ and $y|b \rangle$ of two different "rays" is independent of the particular vectors (that is, of x and y), since these coefficients cancel from numerator and denominator of (2.15).

We now define a method, known as the "Schmidt orthogonalization procedure," for constructing, out of any set of linearly independent vectors, a set whose members are normalized and mutually orthogonal. Let the vectors be

$$|\psi_1 \rangle, |\psi_2 \rangle, \ldots$$

The procedure may be described as follows: We may refer to the component of any vector $|\psi_j \rangle$ in the direction of a unit vector $|\varphi_i \rangle$ as $|\varphi_i \rangle \langle \varphi_i | \psi_j \rangle$. If this component is subtracted from $|\psi_j \rangle$, the resulting vector, $|\psi_j \rangle - |\varphi_i \rangle \langle \varphi_i | \psi_j \rangle$, is orthogonal to $|\varphi_i \rangle$ since its scalar product with $|\varphi_i \rangle$ is

$$\langle \varphi_i | \psi_j \rangle - \langle \varphi_i | \varphi_i \rangle \langle \varphi_i | \psi_j \rangle = 0$$

If, now, the first vector $|\psi_1 \rangle$ of the set considered is normalized by dividing it by its magnitude, we may subtract from the second vector its component in the direction of the first, to obtain a new vector which is orthogonal to the first. This in turn may be normalized. We may now take the third vector, and subtract from it its components in the directions of the first two, thus producing a vector orthogonal to each of the first two.

This procedure may be continued until a mutually orthogonal set is arrived at.

Problem 2-1: What happens to the Schmidt procedure if the original set of vectors are not all linearly independent?

The above remarks may be formalized by a mathematical induction. Let the normalized vector obtained from $|\psi_i\rangle$ by subtracting from it its components in the directions of $|\psi_1\rangle$, $|\psi_2\rangle$, . . ., $|\psi_{i-1}\rangle$ be designated as $|\varphi_i\rangle$. Then the vector

$$|\psi_{i+1}\rangle - \sum_{j=1}^{i} |\varphi_j\rangle\langle\varphi_j|\psi_{i+1}\rangle \qquad (2.16)$$

is orthogonal to the first i vectors. For if we take the scalar product with one of them, say the kth, by virtue of the orthogonality of all the vectors up to the ith, only the $j = k$ term will survive from the sum, leaving

$$\langle\varphi_k|\psi_{i+1}\rangle - \langle\varphi_k|\varphi_k\rangle\langle\varphi_k|\psi_{i+1}\rangle = 0$$

The vector (2.16) may now be normalized to give $|\varphi_{i+1}\rangle$, and the procedure repeated. In this way, orthogonal and normalized vectors may be generated until the original set is exhausted.

This method may be used to create, from any given basis for a vector space, a basis of normalized, mutually orthogonal vectors. Such a basis is called an "orthonormal" basis.

There is, of course, nothing unique in the Schmidt procedure. If one operates with the original set of vectors in a different order, a different orthonormal set will be generated. Still further orthonormal sets will be produced by using, in place of the given set, an arbitrary set of linearly independent combinations of them.

6. An Example of the Schmidt Procedure

As an illustration of a linear vector space, and of the generation of orthonormal bases, consider as the elements of our vector space all polynomials with real coefficients. These may be represented in terms of the basis

$$1, x, x^2, \ldots, x^n, \ldots$$

where the variable x has a range to be specified. Their scalar products are determined in terms of those of the pairs of the bases. The definition of scalar products is arbitrary except for the limitation of satisfying the postulates given above. Consider the definition

$$\langle x^n | x^m \rangle = \int_{-1}^{1} x^m x^n dx$$

$$= \frac{2}{m+n+1}, \qquad m+n \text{ even} \qquad (2.17)$$

$$= 0, \qquad\qquad m+n \text{ odd}$$

Thus, the scalar product of two polynomials $p_1(x), p_2(x)$ is

$$\int_{-1}^{1} p_1(x) p_2(x) \, dx.$$

With this definition of scalar product, it is now easy to verify that the postulates of a linear vector space are satisfied. Addition and multiplication by a constant are the operations of ordinary algebra. It is immediately evident that this vector space comprises two subspaces—those of even- and odd-ordered polynomials, respectively, all vectors of one of which are orthogonal to all vectors of the other. Thus, the Schmidt procedure can be carried out in each subspace separately.

Consider first the even subspace. Let the successive vectors of the orthonormal set be designated $|\varphi_0\rangle, |\varphi_2\rangle, |\varphi_4\rangle, \ldots$, while those of the odd subspace will be $|\varphi_1\rangle, |\varphi_3\rangle, |\varphi_5\rangle, \ldots$. The first vector of the even subspace is

$$|\varphi_0\rangle = \frac{1}{\sqrt{2}} \qquad (2.18)$$

The second is

$$N_2 \left(x^2 - \left\langle x^2 \left| \frac{1}{\sqrt{2}} \right\rangle \frac{1}{\sqrt{2}} \right) = N_2 \left(x^2 - \frac{1}{3} \right)$$

where N_2 is the normalizing factor. It is determined by the condition

$$N_2^2 (\langle x^2 | x^2 \rangle - \tfrac{2}{3}\langle x^2 | 1 \rangle + \tfrac{1}{9}\langle 1 | 1 \rangle) = 1$$

or

$$N_2 = \sqrt{\frac{45}{8}} = \frac{3}{2}\sqrt{\frac{5}{2}}$$

Therefore,

$$|\varphi_2\rangle = \sqrt{\frac{5}{2}} \frac{3x^2 - 1}{2} \qquad (2.19)$$

Problem 2-2: Show that

$$|\varphi_4\rangle = \sqrt{\frac{9}{2}} \frac{35x^4 - 30x^2 + 3}{8}$$

Similarly, the first odd vector is $|x\rangle$. Its magnitude squared is $\frac{2}{3}$ so that

$$|\varphi_1\rangle = \sqrt{\frac{3}{2}}x \tag{2.20}$$

Subtracting from x^3 the component in the direction of $|\varphi_1\rangle$ gives

$$x^3 - \frac{3}{2}\langle x^3|x\rangle x = x^3 - \frac{3}{5}x$$

$|\varphi_3\rangle$ is obtained by multiplying this by N_3 such that

$$N_3^2\langle x^3 - \tfrac{3}{5}x|x^3 - \tfrac{3}{5}x\rangle = 1$$

This yields the value $N_3 = \frac{5}{2}\sqrt{\frac{7}{2}}$. Therefore,

$$|\varphi_3\rangle = \sqrt{\frac{7}{2}}\frac{5x^3 - 3x}{2} \tag{2.21}$$

Problem 2-3: Show that

$$|\varphi_5\rangle = \sqrt{\frac{11}{2}}\frac{63x^5 - 70x^3 + 15x}{8}$$

The orthogonal polynomials which we have calculated here are the normalized Legendre polynomials. Other sets of orthogonal polynomials may be determined by choosing different definitions of scalar product. Some familiar cases are the following:

(a)   Laguerre polynomials

$$\langle x^m|x^n\rangle = \int_0^\infty e^{-x}x^m x^n dx = (m+n)! \tag{2.22}$$

(b)   Hermite polynomials

$$\langle x^m|x^n\rangle = \int_0^\infty e^{-x^2/2}x^m x^n dx = 0, \qquad m+n \text{ odd}$$

$$= 2^{(m+n-1)/2}\Gamma\left(\frac{m+n+1}{2}\right)$$

$$= \frac{(m+n)!}{\left(\frac{m+n}{2}\right)!2^{(m+n)/2}}\sqrt{\frac{\pi}{2}}, \qquad m+n \text{ even} \tag{2.23}$$

(c)   Tchebycheff polynomials

$$\langle x^m|x^n\rangle = \int_{-1}^1 \frac{x^m x^n}{\sqrt{1-x^2}}dx = 0, \qquad m+n \text{ odd}$$

$$= \frac{(n+m)!}{2^{n+m}\left[\left(\frac{n+m}{2}\right)!\right]^2}, \qquad m+n \text{ even} \tag{2.24}$$

These polynomials may be generated out of Sturm–Liouville equations as defined in the last chapter. Consider the four differential equations:

LEGENDRE EQUATION

$$\frac{d}{dx}\left[(1-x^2)\frac{dy}{dx}\right]+\lambda y = 0 \tag{2.25}$$

defined on the range $(-1, 1)$, and with boundary condition that y is to be finite at ± 1. We shall see in the next chapter that we must have $\lambda = l(l+1)$, where l is an integer, and that in this case the eigenfunctions are polynomials. Since the "weighting function" $r(x)$ [Eq. (1.122)] is unity, these polynomials will be mutually orthogonal in the sense of (2.17). Thus, the Legendre polynomials generated above are the solutions of this equation.

LAGUERRE EQUATION

$$x\frac{d^2y}{dx^2}+(1-x)\frac{dy}{dx}+\lambda y = 0 \tag{2.26}$$

on the range $(0 \le x \le \infty)$, with the boundary condition of finiteness at the limits. By (1.127) this is converted into a Sturm–Liouville equation by the "integrating factor"

$$g = \frac{1}{x}\exp\left[\int^x \frac{1-x'}{x'}\,dx'\right]$$

$$= e^{-x} \tag{2.27}$$

The equation becomes

$$\frac{d}{dx}\left[xe^{-x}\frac{dy}{dx}\right]+\lambda e^{-x}y = 0 \tag{2.28}$$

Thus, the weighting function defining orthogonality for the polynomial solutions of this equation (which occur for $\lambda = $ positive integers n) is e^{-x}, as in (2.22).

HERMITE EQUATION

$$\frac{d^2y}{dx^2}-x\frac{dy}{dx}+\lambda y = 0 \tag{2.29}$$

defined for $(-\infty \le x \le \infty)$ with the boundary condition of finiteness as $|x| \to \infty$. Polynomial solutions arise for $\lambda = $ positive integers n. The integrating factor required to make this a Sturm–Liouville equation is

$$g = \exp\int^x (-x')\,dx'$$

$$= e^{-x^2/2} \tag{2.30}$$

The equation then becomes

$$\frac{d}{dx}\left[e^{-x^2/2}\frac{dy}{dx}\right]+\lambda e^{-x^2/2}y=0 \tag{2.31}$$

The weighting function is therefore $e^{-x^2/2}$, as in (2.23).

TCHEBYCHEFF EQUATION

$$(1-x^2)\frac{d^2y}{dx^2}-x\frac{dy}{dx}+\lambda y=0 \tag{2.32}$$

on the range $(-1 \leqslant x \leqslant 1)$ with the boundary condition of finiteness at the limits. Polynomial solutions are found for $\lambda = n^2$, where n is an integer. The integrating factor (1.122) is

$$g(x) = \frac{1}{1-x^2}\exp\left[-\int^x \frac{x'}{1-x'^2}dx'\right]$$

$$= \frac{1}{\sqrt{1-x^2}} \tag{2.33}$$

The equation in Sturm–Liouville form is, then,

$$\frac{d}{dx}\left(\sqrt{1-x^2}\frac{dy}{dx}\right)+\frac{\lambda}{\sqrt{1-x^2}}y=0 \tag{2.34}$$

The weighting factor in the orthogonality relation, and therefore in the definition of scalar product (2.24) is thus $r(x) = 1/\sqrt{1-x^2}$.

Problem 2-4: Calculate the Laguerre, Hermite, and Tchebycheff polynomials up to $n = 3$.

Problem 2-5: Show that in the case of the associated Laguerre equation

$$x\frac{d^2y}{dx^2}+(\alpha+1-x)\frac{dy}{dx}+ny=0$$

the weighting function in the corresponding Sturm–Liouville equation is $x^\alpha e^{-x}$. Thus, determine the definition of the scalar product of polynomials, and calculate the polynomials up to the third order.

Problem 2-6: Show that the substitution $x = \cos z$ makes it possible to find solutions of (2.34) in the form

$$y = \cos(n\cos^{-1}x) \tag{2.35}$$

Thus, find in an alternative way the Tchebycheff polynomials up to $n = 4$.

7. Matrix Representation of Vectors and Transformation of Basis

We now show that it is possible to represent our vectors by column or row matrices. This representation, however, is relative to an orthonormal *basis*; if one changes the basis, the vector is represented by a different matrix.

Suppose that the vectors $|\varphi_i\rangle$ constitute an orthonormal basis for our vector space. Any vector in the space may then be written

$$|\psi\rangle = \sum_i a_i |\varphi_i\rangle \qquad (2.36)$$

Taking the scalar product with $|\varphi_j\rangle$, and using the orthogonality of the $|\varphi\rangle$'s, we get

$$a_j = \langle \varphi_j | \psi \rangle \qquad (2.37)$$

so that

$$|\psi\rangle = \sum_i |\varphi_i\rangle \langle \varphi_i | \psi \rangle \qquad (2.38)$$

The quantities a_j may be taken to be the elements of a column matrix

$$\begin{pmatrix} a_1 \\ a_2 \\ \cdot \\ \cdot \\ \cdot \\ a_i \\ \cdot \\ \cdot \\ \cdot \end{pmatrix}$$

which represents the vector $|\psi\rangle$ relative to the basis $|\varphi_i\rangle$.

If we consider two vectors $|\psi\rangle$ and $|\psi'\rangle$, where

$$|\psi\rangle = \sum a_i |\varphi_i\rangle \qquad (2.39)$$

and

$$|\psi'\rangle = \sum b_i |\varphi_i\rangle \qquad (2.40)$$

the scalar product of $|\psi\rangle$ with $|\psi'\rangle$ is

$$\langle \psi' | \psi \rangle = \sum_i b_i^* a_i \qquad (2.41)$$

If $\langle \psi' |$ is represented by the *row* matrix

$$(b_1^* b_2^* \ldots b_i^* \ldots)$$

than the scalar product $\langle \psi' | \psi \rangle$ can be represented by the *matrix* product

$$
(b_1^* b_2^* \ldots b_i^* \ldots)
\begin{pmatrix}
a_1 \\
a_2 \\
\cdot \\
\cdot \\
\cdot \\
a_i \\
\cdot \\
\cdot
\end{pmatrix}
$$

Conjugate vectors $\langle \psi |$ will then be represented by matrices which are the transpose conjugates (Hermitian conjugates) of those which represent the vectors $|\psi\rangle$. The matrix multiplication representing $\langle \psi | \psi' \rangle$ is derived from that representing $\langle \psi' | \psi \rangle$ by taking the transpose conjugate of the matrix product.

Consider a transformation from a basis $|\varphi_i\rangle$ to a basis $|\varphi_i'\rangle$. The vectors of the second basis may be expressed in terms of those of the first:

$$|\varphi_i'\rangle = \sum_j |\varphi_j\rangle \langle \varphi_j | \varphi_i' \rangle \tag{2.42}$$

Alternatively,

$$|\varphi_i\rangle = \sum_j |\varphi_j'\rangle \langle \varphi_j' | \varphi_i \rangle \tag{2.43}$$

Formally,

$$\sum_j |\varphi_j\rangle \langle \varphi_j | \text{ and } \sum_j |\varphi_j'\rangle \langle \varphi_j' |$$

may be considered to represent the unit operator. The vector $|\psi\rangle$ may be expressed relative to either basis:

$$|\psi\rangle = \sum a_i |\varphi_i\rangle$$

and

$$|\psi\rangle = \sum a_i' |\varphi_i'\rangle$$

Substituting in the first of these equations for $|\varphi_i\rangle$ from (2.43) we find that

$$\sum_{i,j} |\varphi_i'\rangle \langle \varphi_i' | \varphi_j \rangle a_j = \sum |\varphi_i'\rangle a_i'$$

from which it follows that

$$a_i' = \sum_j \langle \varphi_i' | \varphi_j \rangle a_j \tag{2.44}$$

This may be considered as a matrix equation, in which the column matrix (a_j) is transformed into the column matrix (a_i') by the square matrix S whose elements are

$$\langle \varphi_i' | \varphi_j \rangle$$

the index i designating the row and j designating the column; thus $\langle \varphi_i' | \varphi_j \rangle$ is the element of the ith row and the jth column of S.

Using the orthogonality of the bases, it can be shown that

$$S^+ S = 1 \tag{2.45}$$

where S^+ is the transpose conjugate of S, that is, the matrix with elements $\langle \varphi_j | \varphi_i' \rangle$. For if we take the scalar product of (2.43) with $|\varphi_k\rangle$ we find that

$$\delta_{ik} = \sum_j \langle \varphi_k | \varphi_j' \rangle \langle \varphi_j' | \varphi_i \rangle \tag{2.46}$$

δ_{ik} being the Kronecker delta, which is equal to unity when $i = k$ and zero when $i \neq k$. In matrix terms, this is simply Eq. (2.45).

If Eq. (2.42) is multiplied by $|\varphi_k'\rangle$ it follows similarly that

$$SS^+ = 1 \tag{2.47}$$

A basis transformation of the sort considered here is known as a *unitary* transformation. The transformation on the conjugate vectors is represented by the Hermitian conjugate matrix S^+, which is the inverse of the matrix of the original transformation.

8. Linear Operators and Their Matrix Representations

A linear operator L on a vector $|\psi\rangle$ is an operator which produces another vector $|\bar{\psi}\rangle$:

$$|\bar{\psi}\rangle = L|\psi\rangle \tag{2.48}$$

and is such that if

$$|\bar{\psi}_1\rangle = L|\psi_1\rangle$$

and

$$|\bar{\psi}_2\rangle = L|\psi_2\rangle$$

then

$$a_1|\bar{\psi}_1\rangle + a_2|\bar{\psi}_2\rangle = L(a_1|\psi_1\rangle + a_2|\psi_2\rangle) \tag{2.49}$$

Let us write the vectors in (2.48) in terms of a basis $|\varphi_i\rangle$:

$$|\psi\rangle = \sum a_i|\varphi_i\rangle$$

and

$$|\bar{\psi}\rangle = \sum \bar{a}_i|\varphi_i\rangle$$

Then (2.48) becomes

$$\sum_i \bar{a}_i|\varphi_i\rangle = \sum_i a_i L|\varphi_i\rangle$$

Taking the scalar product of both sides with $|\varphi_j\rangle$ yields the equation

$$\bar{a}_j = \sum_j \langle \varphi_j | (L|\varphi_i\rangle) a_i \tag{2.50}$$

Thus the matrix with elements $\langle\varphi_j|(L|\varphi_i\rangle)$, operating on the matrix representing $|\psi\rangle$, produces the matrix representative of $|\bar\psi\rangle$. For brevity, the elements of this matrix operator are sometimes designated simply L_{ji}. Once again, this matrix representative of L depends on the basis chosen.

This then suggests the question: What is the relation between the matrices representing the operator in two different bases? If the matrices representing $|\psi\rangle$, $|\bar\psi\rangle$, the basis transformation, and the linear operator L are designated by $A, \bar A, S$, and L, respectively, (2.48) may be written

$$\bar A = LA \tag{2.51}$$

Let the matrices representing $|\psi\rangle$ and $|\bar\psi\rangle$ relative to the *transformed* basis be A' and $\bar A'$. Then, by (2.44),

$$A' = SA \tag{2.52}$$

and

$$\bar A' = S\bar A \tag{2.53}$$

Since, from (2.45) and (2.47), S has the inverse $S^{-1} = S^+$, (2.52) may be written

$$A = S^{-1}A' \tag{2.54}$$

Multiplying (2.51) by S and substituting for A from (2.53) yields the equation

$$L'\bar A' = SLS^{-1}A' \tag{2.55}$$

It follows that the matrix representing the operator relative to the basis $|\varphi_i'\rangle$ is

$$L' = SLS^{-1} \tag{2.56}$$

This transformation of the linear operator L is often referred to as a "similarity transformation."

Problem 2-7:   A general rotation in three dimensions may be defined in terms of the "Euler angles" θ, φ, ψ, introduced in Goldstein's *Classical Mechanics*.[2] The rotation is described by the sequence of three rotations on an original orthogonal triad: (i) rotation through angle φ about the z axis, (ii) rotation through angle θ about the *new x* axis, and (iii) rotation through angle ψ about the direction of the z axis following the two preceding rotations. The matrix for this rotation is

$$A = \begin{pmatrix} \cos\psi\cos\varphi & \cos\psi\sin\varphi & \sin\psi\sin\theta \\ -\cos\theta\sin\varphi\sin\psi & +\cos\theta\cos\varphi\sin\psi & \\ -\sin\psi\cos\varphi & -\sin\psi\sin\varphi & \cos\psi\sin\theta \\ -\cos\theta\sin\varphi\cos\psi & +\cos\theta\cos\varphi\cos\psi & \\ \sin\theta\sin\varphi & -\sin\theta\cos\varphi & \cos\theta \end{pmatrix}$$

[2]See bibliography for Chapter 1.

(a) Show that it is unitary, that is, that $A^+A = AA^+ = 1$.

(b) Find the components of the final vectors of the triad relative to the original ones, and verify their orthogonality.

(c) Show directly that, if two vectors are orthogonal, they remain so under the transformation A.

(d) Determine the transformed form of the operator defining rotation through angle α about the original z axis.

---

9. The Eigenvalue Problem for Hermitian Operators

Suppose that an operator L, acting on a vector $|\psi\rangle$, produces a vector $|\bar\psi\rangle$. We denote the fact that it operates on a vector to its right, a ket vector, by an arrow pointing to the right:

$$|\bar\psi\rangle = \vec{L}|\psi\rangle \tag{2.57}$$

The corresponding operator which transforms the bra vector $\langle\psi|$ into $\langle\bar\psi|$ will be written as $\overleftarrow{L}$:

$$\langle\bar\psi| = \langle\psi|\overleftarrow{L} \tag{2.58}$$

Consider then the quantities $\langle\psi_2|\vec{L}|\psi_1\rangle$ and $\langle\psi_2|\overleftarrow{L}|\psi_1\rangle$.[3] They are not in general equal; one is the product of $L|\psi_1\rangle$ with $|\psi_2\rangle$ and the other the product of $|\psi_1\rangle$ with $L|\psi_2\rangle$. An operator for which these quantities *are* equal will be called a *Hermitian* operator.

Hermitian operators are of very great importance in all branches of mathematical physics. Their importance stems from the fact that their eigenvalues are necessarily real, and an operator that has only real eigenvalues is necessarily Hermitian. Since, as we shall see, many of the most important problems of mathematical physics are eigenvalue problems, in which the eigenvalues are physically observable quantities and must therefore be real, the theory of Hermitian operators is of central importance.

The condition for a Hermitian operator is, then, that for any two vectors $|\psi_1\rangle$ and $|\psi_2\rangle$

$$\langle\psi_2|\vec{L}|\psi_1\rangle = \langle\psi_2|\overleftarrow{L}|\psi_1\rangle \tag{2.59}$$

Thus, for Hermitian operators, it will not be necessary to put an arrow over the operator.[4] This condition may be written in another way. By virtue of Eq. (2.11), the right-hand side of (2.59) may be written as $\langle\psi_1|\vec{L}|\psi_2\rangle^*$, leading to a formula in which right-hand operators L occur on

---

[3]Note that, in the present notation, the operator L in (2.50) should be written $\vec{L}$.
[4]Which may equally well be taken to operate in either direction.

both sides of the equation:

$$\langle \psi_2 | \vec{L} | \psi_1 \rangle = \langle \psi_1 | \vec{L} | \psi_2 \rangle^* \qquad (2.60)$$

Two remarks should be made at this point concerning the matrix representations of Hermitian operators. First, from (2.50) and (2.60) it follows that Hermitian matrices (matrices of Hermitian operators) are their own transpose conjugates; that is, if one interchanges rows and columns and takes the complex conjugate of every element, they are unchanged. Second, from (2.59) it follows that the matrix which transforms the row matrices representing bra vectors is the same as that transforming the column matrices representing the corresponding ket vectors. In the first case, however, it multiplies the vector from the right, and in the second from the left.

The eigenvalue problem is the following: To find vectors $|\psi\rangle$ and numbers λ such that

$$L|\psi\rangle = \lambda|\psi\rangle \qquad (2.61)$$

that is, to find vectors which are transformed by the operator L into multiples of themselves, and the corresponding multiples.[5]

We shall now prove the following three theorems:

(A) The eigenvalues of Hermitian operators are real.

(B) If $\langle \psi | \vec{L} | \psi \rangle$ is real for all vectors $|\psi\rangle$, then L is Hermitian.

(C) The eigenvectors of a Hermitian operator corresponding to different eigenvalues are orthogonal.

THEOREM A

Let $|\psi\rangle$ be an eigenvector and λ its eigenvalue. Then

$$\vec{L}|\psi\rangle = \lambda|\psi\rangle$$

The conjugate equation is

$$\langle \psi | \overleftarrow{L} = \lambda^* \langle \psi |$$

Multiply the first of these equations on the left by $|\psi\rangle$, and the second on the right by $\langle \psi |$, and subtract. The left-hand sides cancel by virtue of Eq. (2.59). Therefore

$$(\lambda - \lambda^*)\langle \psi | \psi \rangle = 0$$

If $|\psi\rangle$ is not the zero vector, $\langle \psi | \psi \rangle > 0$, so $\lambda = \lambda^*$ and λ is real.

THEOREM B

If $\langle \psi | \vec{L} | \psi \rangle$ is always real, then by (2.11) it is equal to $\langle \psi | \overleftarrow{L} | \psi \rangle$, so we can omit the arrow over the L in quantities of this form. Now take

---

[5]The vectors satisfying (2.61) will be called eigenvectors, and the corresponding numbers λ the eigenvalues.

$|\psi\rangle = |\psi_1\rangle + |\psi_2\rangle$, where $|\psi_1\rangle$ and $|\psi_2\rangle$ are arbitrary vectors. Since $\langle\psi_1|L|\psi_1\rangle$ and $\langle\psi_2|L|\psi_2\rangle$ are real, it follows that

$$\langle\psi_2|\vec{L}|\psi_1\rangle + \langle\psi_1|\vec{L}|\psi_2\rangle$$

is real, or

$$\text{Im}\,\langle\psi_2|\vec{L}|\psi_1\rangle = -\,\text{Im}\,\langle\psi_1|\vec{L}|\psi_2\rangle \tag{2.62}$$

("Im" means "imaginary part of.")

Consider next $|\psi\rangle = |\psi_1\rangle - i|\psi_2\rangle$. Then $\langle\psi|L|\psi\rangle$ is equal to

$$\langle\psi_1|\vec{L}|\psi_1\rangle + \langle\psi_2|\vec{L}|\psi_2\rangle - i[\langle\psi_1|\vec{L}|\psi_2\rangle - \langle\psi_2|\vec{L}|\psi_1\rangle]$$

which must also be real. Thus,

$$\text{Re}\,\langle\psi_2|\vec{L}|\psi_1\rangle = \text{Re}\,\langle\psi_1|\vec{L}|\psi_2\rangle \tag{2.63}$$

where "Re" means "real part of."

Adding to Eq. (2.63) i times Eq. (2.62) gives Eq. (2.60)

$$\langle\psi_2|\vec{L}|\psi_1\rangle = \langle\psi_1|\vec{L}|\psi_2\rangle^*$$

which is the condition that L is Hermitian.

THEOREM C

Let $|\psi_1\rangle$ be an eigenvector of L with eigenvalue λ_1, and $|\psi_2\rangle$ an eigenvector with eigenvalue λ_2. These statements may be expressed by the equations:

$$L|\psi_1\rangle = \lambda_1|\psi_1\rangle \tag{2.64}$$

and

$$\langle\psi_2|L = \lambda_2\langle\psi_2| \tag{2.65}$$

In (2.65) we have made use of the reality of the λ's. If (2.64) is multiplied on the left by $\langle\psi_2|$ and (2.65) on the right by $|\psi_1\rangle$, and the equations are subtracted, we obtain

$$(\lambda_1 - \lambda_2)\langle\psi_2|\psi_1\rangle = 0$$

If λ_1 and λ_2 are different, the second factor must be zero:

$$\langle\psi_2|\psi_1\rangle = 0,$$

so that the vectors are orthogonal.

It may, of course, sometimes happen that several eigenvectors have the *same* eigenvalue. This is referred to as the "case of degeneracy"; a set of independent eigenvectors corresponding to a single eigenvalue is called a degenerate set. In this case they need not be orthogonal. However, if $|\psi^{(1)}\rangle$, $|\psi^{(2)}\rangle, \ldots, |\psi^{(\alpha)}\rangle$ all belong to the eigenvalue λ, so does any linear combination of them. Consequently, by means of the Schmidt orthogonalization procedure, it is possible to construct from the given degenerate set a set of mutually orthogonal vectors. If this is done, all the eigenvectors of L will be mutually orthogonal. We shall see that such a choice has great convenience, and so will commonly be adopted.

It is in general assumed that the eigenvectors of a Hermitian operator L form a basis for the vector space over which the operator is defined. In other words, any vector of the space will be expressible in terms of these eigenvectors $|\psi_i\rangle$:

$$|\psi\rangle = \sum_i C_i |\psi_i\rangle \qquad (2.66)$$

We do not discuss here the conditions of validity of this assumption; it is a purely mathematical point which the reader will find discussed elsewhere.

The orthogonality of the $|\psi_i\rangle$'s makes it possible to write (2.66) in the form

$$|\psi\rangle = \sum_i |\psi_i\rangle\langle\psi_i|\psi\rangle \qquad (2.67)$$

It is worth noting, incidentally, that $|\varphi\rangle\langle\chi|$, where $|\varphi\rangle$ and $\langle\chi|$ are arbitrary, is an operator, which may operate either to the right on a ket vector or to the left on a bra vector. From (2.42) or (2.67) it may be seen that

$$\sum_i |\psi_i\rangle\langle\psi_i|$$

is the *unit* operator if the $|\psi_i\rangle$ form a complete set of basis vectors. It also follows that

$$\sum_i \langle\psi|\psi_i\rangle\langle\psi_i|\psi\rangle = 1 \qquad (2.68)$$

if $|\psi\rangle$ is a normalized vector. We call this relation the "completeness relation."

10. Another Example

A further example of a vector space is one consisting of the totality of functions expansible in Fourier series on the interval $(-1, 1)$

$$f(x) = a_0 + a_1 \cos \pi x + a_1' \sin \pi x$$
$$+ a_2 \cos 2\pi x + a_2' \sin 2\pi x$$
$$\cdot$$
$$\cdot$$
$$\cdot$$
$$+ a_n \cos 2\pi n x + a_n' \sin 2\pi n x$$
$$\cdot$$
$$\cdot \quad \cdot$$
$$\cdot \qquad (2.69)$$

The set of coefficients may be taken to be the components of a vector or column matrix representing the function $f(x)$.

The vector space postulates are satisfied if the operation of addition is ordinary addition, and the scalar product of f_1 and f_2 is taken as

$$\langle f_1 | f_2 \rangle = \int_{-1}^{1} f_1(x) f_2(x) dx \tag{2.70}$$

If we consider only the space of those functions whose derivative also has a convergent Fourier expansion, then the operator of different-iation is a linear operator. It converts the vector

$$\begin{pmatrix} a_0 \\ a_1 \\ a_1' \\ a_2 \\ a_2' \\ \cdot \\ \cdot \\ \cdot \end{pmatrix} \quad \text{into} \quad \begin{pmatrix} 0 \\ \pi a_1' \\ -\pi a_1 \\ 2\pi a_2' \\ -2\pi a_2 \\ \cdot \\ \cdot \\ \cdot \end{pmatrix}$$

and is represented by the square matrix

$$\begin{pmatrix} 0 & 0 & 0 & 0 & 0 & \cdot & \cdot & \cdot \\ 0 & 0 & \pi & 0 & 0 & \cdot & \cdot & \cdot \\ 0 & -\pi & 0 & 0 & 0 & \cdot & \cdot & \cdot \\ 0 & 0 & 0 & 0 & 2\pi & \cdot & \cdot & \cdot \\ 0 & 0 & 0 & -2\pi & 0 & \cdot & \cdot & \cdot \\ \cdot & \cdot & \cdot & \cdot & \cdot & \cdot & & \\ \cdot & \cdot & \cdot & \cdot & \cdot & & & \\ \cdot & \cdot & \cdot & \cdot & \cdot & & & \end{pmatrix} \tag{2.71}$$

which is, of course, of infinite order.

The operator d/dx is not Hermitian, since

$$\int_{-1}^{1} f_2 \frac{d}{dx} f_1 dx = f_1 f_2 \Big|_{-1}^{1} - \int_{-1}^{1} f_1 \frac{d}{dx} f_2 dx$$

If we define the functions outside the interval $(-1, 1)$ in such a way that they are periodic over this interval, so that the integrated term vanishes, it is clear that the condition for a Hermitian operator is not satisfied. The operator $-i(d/dx)$ is, however, Hermitian. The determination of its eigenvectors and eigenvalues is left as a problem.

Problem 2-8:   For functions which, along with their first derivative, are expansible in a Fourier series on the interval $(-1, 1)$, show that $-i(d/dx)$ is Hermitian and find its eigenvalues and eigen-vectors.

It should be noted in passing that we have established in this particular case a correspondence between a differential operator $-i(d/dx)$ and a matrix operator $[-i$ times (2.71)]. This correspondence is of the sort which is involved in the connection between the Schrödinger and Heisenberg schemes of quantum mechanics (see Chapter 8).

11. Sturm–Liouville Problem and Linear Vector Spaces[6]

In this section we interpret the problem of the Sturm–Liouville equation, discussed in Chapter 1, as a problem in linear vector spaces. The linear vector space involved is, however, a slightly special one, in that it will be possible to define scalar products as *real* quantities rather than complex ones.

Consider the Sturm–Liouville equation

$$\left[-\frac{d}{dx}\left(p\,\frac{d}{dx}\right)+q\right]y = L(y) = \lambda r y \qquad (2.72)$$

defined on the range (a, b) and subject to boundary conditions such that, if y_m and y_n are two solutions

$$p(y_m y_n' - y_n y_m')\Big|_a^b = 0 \qquad (1.133)$$

Then, as we have shown, y_n and y_m are orthogonal, in the sense that

$$\int_a^b y_m(x) y_n(x) r(x)\,dx = 0, \qquad m \neq n \qquad (1.132)$$

There may be at most a twofold degeneracy (two y's for a given eigenvalue λ); two independent solutions may then be found, using the Schmidt procedure, which are mutually orthogonal. We may also *normalize* all the eigenfunctions y_n:

$$\int_a^b y_n^2(x) r(x)\,dx = 1 \qquad (1.137)$$

The vectors of our linear vector space will then be taken to be the set of functions which may be expanded in terms of the eigenfunctions of L:

$$|\psi\rangle \leftrightarrow \sum a_n y_n(x) = f(x) \qquad (2.73)$$

where the a_n's are taken to be real constants. The operations of addition and multiplication by a constant can obviously be defined, and the existence of the zero vector (all a_n's $= 0$) verified.

As for the conjugate vectors, we may in this case define them as

[6]For a more critical and rigorous discussion of this question the reader is referred to the book of Kato mentioned earlier, Chapter 5, Sec. 3.6.

identical with the vectors themselves (self-conjugate system). The scalar product of two vectors

$$|\psi_1\rangle \leftrightarrow \sum a_n^{(1)} y_n = f_1(x) \qquad (2.74)$$

and

$$|\psi_2\rangle \leftrightarrow a_n^{(2)} y_n = f_2(x) \qquad (2.75)$$

will be defined as

$$\langle \psi_1 | \psi_2 \rangle = \langle \psi_2 | \psi_1 \rangle = \int_a^b f_1(x) f_2(x) r(x) dx \qquad (2.76)$$

Here "$\leftrightarrow$" designates a correspondence. The vector $|\psi_1\rangle$ is the function f_1; we, however, do not use the equality symbol "$=$" because we preempt it to indicate relations of equality between elements designated in the same way [either in the vector notation $|\psi\rangle$ or the conventional functional notation $f(x)$].

It follows immediately from (2.76) that the y_n's constitute an orthonormal basis

$$|\varphi_n\rangle \leftrightarrow y_n \qquad (2.77)$$

Arbitrary vectors can be represented by column matrices whose elements are the expansion coefficients a_n where

$$a_n = \int_a^b f y_n r \, dx = \langle \varphi_n | \psi \rangle \qquad (2.78)$$

Similarly, the conjugate vectors may be represented by row matrices having the same set of components. The vectors $|\varphi_n\rangle$ correspond to matrices with all components zero except the nth, which is unity. The scalar product of $|\psi_1\rangle$ and $|\psi_2\rangle$, (f_1 and f_2), is

$$\langle \psi_1 | \psi_2 \rangle = \int_a^b f_1 f_2 r \, dx = \sum a_n^{(1)} a_n^{(2)} \qquad (2.79)$$

Various kinds of linear operators on the functions f can be defined; for example, any linear differential operator, or linear integral operator, of which an example could be

$$I(f) = \int_a^b f(x') g(x-x') dx' \qquad (2.80)$$

where g is a suitably chosen function.

It is easily seen that the operators associated with the Sturm–Liouville equation are Hermitian; this follows from our definition of the set of vectors as self-conjugate. We have

$$\vec{L} |\psi\rangle \leftrightarrow \frac{1}{r} L(f) \qquad (2.81)$$

The conjugate vector

$$\langle \psi | \overset{\leftarrow}{L}$$

also corresponds to the function $(1/r)L(f)$. Then

$$\langle\psi_2|\vec{L}|\psi_1\rangle = \int_a^b f_2 L(f_1)\,dx \tag{2.82}$$

and

$$\langle\psi_2|\overleftarrow{L}|\psi_1\rangle = \int_a^b L(f_2)f_1\,dx \tag{2.83}$$

Now, using the form (2.72) for $L(f)$,

$$\int_a^b f_2 L(f_1)\,dx - \int_a^b f_1 L(f_2)\,dx = p(f_1 f_2' - f_2 f_1')\Big|_a^b \tag{2.84}$$

Since the f's satisfy the same boundary conditions as the eigenfunctions y_n in terms of which they may be expanded, it follows from (1.133) that this is zero. We have therefore verified the Hermitian property (2.59).

It is useful, finally, to write the Rayleigh–Ritz variational principle (1.139) in terms of vector notation. From (2.82)

$$\begin{aligned}
\langle\psi|L|\psi\rangle &= \int_a^b fL(f)\,dx \\
&= \int_a^b f\left\{-\frac{d}{dx}\left[p(x)\frac{df}{dx}\right]+q(x)f\right\}dx \\
&= \int_a^b (pf'^2 + qf^2)\,dx
\end{aligned} \tag{2.85}$$

under the condition of validity of the Rayleigh–Ritz principle, that is, that

$$pff'\Big|_a^b = 0 \tag{1.149}$$

Thus, the numerator in (1.139) is $\langle\psi|L|\psi\rangle$. But the denominator is $\langle\psi|\psi\rangle$. Thus, (1.139) becomes

$$I(f) = 0 \tag{2.86}$$

where

$$I(f) = \frac{\langle\psi|L|\psi\rangle}{\langle\psi|\psi\rangle} \tag{2.87}$$

This is equivalent, by (1.151), to

$$\delta[\langle\psi|L|\psi\rangle - \lambda\langle\psi|\psi\rangle] = 0 \tag{2.88}$$

If the $|\psi\rangle$'s are expanded in terms of an orthonormal basis set $|n\rangle$ (not necessarily eigenstates of L),

$$|\psi\rangle = \sum |n\rangle\langle n|\psi\rangle = \sum |n\rangle a_n \tag{2.89}$$

where the a_n's are to be determined, (2.88) becomes

$$\delta\left[\sum_{n,m} a_n a_m(\langle m|L|n\rangle - \lambda\delta_{mn})\right] = 0 \tag{2.90}$$

Treating the a's as variational parameters, this leads to the equations

$$\sum_n (\langle m|L|n\rangle - \lambda\delta_{mn})a_n = 0 \tag{2.91}$$

or

$$\sum_n \langle m|L|n\rangle a_n = \lambda a_m \tag{2.92}$$

This is, of course, simply the matrix formulation for the eigenvalue equation for L in the representation $|n\rangle$.

In summary: The Rayleigh–Ritz variational principle for the eigenvalues (and eigenvectors) of L may be written in the vector form (2.87). If the vector is expressed in terms of an arbitrary basis, the Rayleigh–Ritz principle leads to the matrix eigenvalue problem in which the matrix representing L has the "matrix elements" $\langle m|L|n\rangle$.

If a *nonorthogonal* basis $|n'\rangle$ is used, the principle takes the form

$$\sum_{n'} (\langle m'|L|n'\rangle - \lambda\langle m'|n'\rangle)a_{n'} = 0 \tag{2.93}$$

Further Problems for Chapter 2

Problem 2-9:   Consider the vector space in which the vectors are the column matrices

$$|V\rangle = \begin{pmatrix} V_1 \\ V_2 \\ V_3 \end{pmatrix}$$

the V's being complex numbers, and the operator

$$L_n = i\begin{pmatrix} 0 & -n_3 & n_2 \\ n_3 & 0 & -n_1 \\ -n_2 & n_1 & 0 \end{pmatrix}$$

where $n_1^2 + n_2^2 + n_3^2 = 1$

(a) Show that L_n is Hermitian.

(b) Show that its eigenvalues are ± 1 and 0, and that the corresponding eigenvectors are

$$\frac{1}{\sqrt{2}}\begin{pmatrix} p_1 \pm iq_1 \\ p_2 \pm iq_2 \\ p_3 \pm iq_3 \end{pmatrix} \quad\text{and}\quad \begin{pmatrix} n_1 \\ n_2 \\ n_3 \end{pmatrix}, \quad\text{respectively,}$$

where

$$\begin{pmatrix} p_1 \\ p_2 \\ p_3 \end{pmatrix} \quad\text{and}\quad \begin{pmatrix} q_1 \\ q_2 \\ q_3 \end{pmatrix}$$

are unit vectors orthogonal to

$$\begin{pmatrix} n_1 \\ n_2 \\ n_3 \end{pmatrix}$$

and to each other.

(c) Show directly that $L_n{}^3 = L_n$, and that this permits an alternative derivation of the eigenvalues.

(d) Show that

$$S = \begin{pmatrix} p_1 & p_2 & p_3 \\ q_1 & q_2 & q_3 \\ n_1 & n_2 & n_3 \end{pmatrix}$$

is a unitary matrix. Determine the transform of the matrix L_n, namely, SL_nS^{-1}. By transformation, determine its eigenvectors.

(e) Find a unitary matrix which transforms L_n to diagonal form.

Problem 2-10:   Consider the vector space in which the vectors are the two-dimensional complex matrices $|s\rangle = \binom{s_1}{s_2}$, and the operator

$$\sigma_n = \begin{pmatrix} n_3 & n_1 - in_2 \\ n_1 + in_2 & -n_3 \end{pmatrix}$$

where $n_1{}^2 + n_2{}^2 + n_3{}^2 = 1$.

(a) Show that σ_n is Hermitian.

(b) Show that the eigenvalues are ± 1, and that if the n's are written in polar coordinates $n_1 = \sin\theta\cos\varphi$, $n_2 = \sin\theta\sin\varphi$, $n_3 = \cos\theta$, the eigenvectors are

$$\begin{pmatrix} \cos\theta/2 \, e^{-i\varphi/2} \\ \sin\theta/2 \, e^{i\varphi/2} \end{pmatrix} \quad \text{and} \quad \begin{pmatrix} \sin\theta/2 \, e^{-i\varphi/2} \\ -\cos\theta/2 \, e^{i\varphi/2} \end{pmatrix}$$

respectively.

(c) Determine the eigenvalues by verifying that $\sigma_n{}^2 = 1$.

(d) Show that

$$S = \begin{pmatrix} \cos\theta/2 \, e^{i\varphi/2} & \sin\theta/2 \, e^{-i\varphi/2} \\ \sin\theta/2 \, e^{i\varphi/2} & -\cos\theta/2 \, e^{-i\varphi/2} \end{pmatrix}$$

is a unitary matrix. Use it to find the transform $S\sigma_nS^{-1}$ of the matrix σ_n. Solve for the eigenvectors of the transformed matrix, and from them derive the eigenvectors of σ_n.

Problem   2-11:   Representing   the   fourth-degree polynomial

$$a_0 + a_1x + a_2x^2 + a_3x^3 + a_4x^4$$

by the column vector whose components are the coefficients, show that the operator

$$I|x^n\rangle = \int_{-\infty}^{x} x'^n e^{-(x-x')} dx'$$

is a linear operator in the vector space. Find its matrix form.

Problem 2-12:   Find the eigenvalues and normalized eigenvectors of the matrices

$$\begin{pmatrix} 0 & 0 & 0 & 1 \\ 0 & 0 & 1 & 0 \\ 0 & 1 & 0 & 0 \\ 1 & 0 & 0 & 0 \end{pmatrix}$$

and

$$\begin{pmatrix} b & a & 0 & 0 & 0 & a \\ a & b & a & 0 & 0 & 0 \\ 0 & a & b & a & 0 & 0 \\ 0 & 0 & a & b & a & 0 \\ 0 & 0 & 0 & a & b & a \\ a & 0 & 0 & 0 & a & b \end{pmatrix}$$

Problem 2-13:   Without calculating the eigenvalues, find for the matrices in the preceding problem
  (a) the sum of the eigenvalues
  (b) the sum of their squares.

PRELUDE TO CHAPTER 3

This chapter is devoted to potential problems, in which
the potential satisfies either the Laplacian equation
($\nabla^2\phi = 0$) or the Poisson (inhomogeneous) equation
$\nabla^2\phi = \rho$, where ρ represents a source. The problem to
which most attention is given is that of electrostatic
potentials. However, the same equation occurs in
magnetostatics, the theory of steady-state heat conduction, irrotational fluid dynamics, or steady current
flow in electrodynamics.

    We first show that irrotational vector fields can be
represented as the curl of a potential. This is then true
of the electrostatic field. In electrostatics, by virtue of
the field equation (Gauss' law) the potential satisfies in
general Poisson's equation.

    However, in the magnetostatic problem a similar
equation is arrived at in a different way. Since magnetic
fields are solenoidal ($\nabla \cdot B = 0$) it is shown that they
can be written as the curl of a vector potential. But
since the curl of a constant magnetic field is proportional to a steady current, it follows that this field also
satisfies an equation of Poisson type, the source term
now involving the current.

    Taking first the case of the homogeneous (Laplace)
equation which holds away from sources, we use the
method of separation of variables to solve the equation
in spherical polar coordinates. (It is assumed that the
reader knows the form of the vector operators and the
Laplacian in arbitrary orthogonal curvilinear coordinate systems; they are derived in all books on elementary vector analysis.) The angular functions are
the so-called "spherical harmonics," which in turn
depend on the polar angle θ through associated Legendre polynomials. These functions are determined from
their separated differential equations by the "factorization method" first introduced by Schrödinger and later
elaborated by L. Infeld and T. Hull.[1] This method is

---

[1]The Factorization Method, *Rev. Mod. Phys.* **23**, 21 (1951).

used subsequently in certain quantum problems in Chapter 9 (harmonic oscillator, hydrogen atom).

Making use of a physical argument involving the potential of a charge on the polar axis we find an alternative way to generate the axially symmetric Legendre polynomials, without solving a differential equation. What is obtained is in fact a generating function for these polynomials. It is used to derive some useful recursion formulas for them.

Returning to the same generating function and expressing it in terms of *Cartesian* coordinates, it is shown that the Legendre polynomials can be expressed in terms of derivatives of the potential of a point source at the origin, namely, $1/r$. The derivatives of various orders may be interpreted as the potentials of electrostatic *multipoles*. It follows that the expansion of any potential in spherical harmonics is a multipole expansion, the coefficients being essentially *multipole moments* of the distribution.

Non-axially-symmetric multipoles are shown to be generated by appropriate derivatives of $1/r$ with respect to all three Cartesian coordinates. By comparing the r and φ dependence of these solutions with those obtained by separation of variables, we are able to express associated Legendre functions also in terms of such derivatives. Again, certain useful properties of these functions may be derived. In particular, an essential addition theorem for spherical harmonics, which is needed in Chapter 5 in the development of radiation theory, is derived.

Electrostatic problems are of two types. In one, charge distributions are given, and the problem is to calculate the resulting potentials. These problems may be solved by a Green's function method, in which the potential is considered to be the sum (integral) of those due to the point sources comprising the distribution.

A particular case which is easy to solve is that in which there is axial symmetry. In this case, it is shown how the potential at any point may be determined if that on the symmetry axis is known.

The second type of problem involves boundary conditions (for example, on conducting surfaces); charge distributions will in general exist on the boundaries to enable these conditions to be satisfied, but are not known *a priori*. In certain cases (for example, plane surfaces, spherical surfaces), methods of "images" may be evolved; these are fictitious charges which may be seen by symmetry to ensure the satisfaction of the boundary conditions. It is shown how to calculate the charge distributions on these surfaces once the potentials are known.

We next derive relations, applicable to any linear problem (and therefore applicable also in wave, diffusion, heat conduction problems, and so forth) between the Green's functions in different geometries. These connect the potentials of unit point source, unit plane source, and unit spherical shell source.

The chapter ends with a number of worked problems. There is an appendix on a recursion relation for associated Legendre functions, and another on linear differential equations of the second order. The latter appendix shows how to derive the Green's function for any such equation, and thus, how to solve the inhomogeneous equation knowing how to solve the homogeneous one.

REFERENCES

Coulson, C. A., *Electricity*. Edinburgh: Oliver and Boyd, 1951.

Jackson, J. D., *Classical Electrodynamics*. New York: Wiley, 1962.

Mac Robert, T. M., *Spherical Harmonics*. New York: Dover, 1947.

Sagan, H., *Boundary and Eigenvalue Problems in Mathematical Physics*. New York: Wiley, 1961, Chapter 8.

Sommerfeld, A., *Partial Differential Equations in Physics*. New York: Academic, 1949, Chapter 4.

3

THE POTENTIAL EQUATION

"The diversity of physical arguments and opinions embraces all sorts of methods"

de Montaigne

1. Introduction; Electrostatic Potential

The potential equation makes its appearance in physics whenever one has fields $\mathbf{F}(\mathbf{r})$ which, in some portion of space at least, are both source-free ($\nabla \cdot \mathbf{F} = 0$) and irrotational ($\nabla \times \mathbf{F} = 0$) ($\mathbf{r}$ is the position vector of an arbitrary point). Any irrotational field can be expressed as the gradient of a potential

$$\mathbf{F}(\mathbf{r}) = -\nabla\phi(\mathbf{r}) \qquad (3.1)$$

where

$$\phi(\mathbf{r}) = \int_{\mathbf{r}_0}^{\mathbf{r}} \mathbf{F}\,ds \qquad (3.2)$$

the integral being a line integral along an arbitrary path from an arbitrary base point $\mathbf{r}_0$.

The irrotational feature ensures that ϕ depends only on the point $\mathbf{r}$, and not on the path of integration. For if c and c' are two contours between $\mathbf{r}_0$ and $\mathbf{r}$

$$\int_c \mathbf{F} \cdot d\mathbf{s} - \int_{c'} \mathbf{F} \cdot d\mathbf{s}$$

101

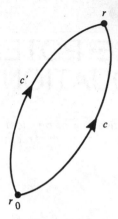

Figure 3.1

is the line integral of $\mathbf{F}$ around a closed contour. By Stokes' theorem, this is equal to the surface integral

$$\int_{s_0} \nabla \times \mathbf{F} \cdot \mathbf{n} dS$$

over any surface bounded by the contour. $\mathbf{n}$ is the unit normal vector to the surface. Since $\mathbf{F}$ is irrotational, this is zero, and therefore the two line integrals are equal.

If we substitute from (3.1) into the equation $\nabla \cdot \mathbf{F} = 0$ we obtain the potential equation

$$\nabla^2 \phi = 0 \tag{3.3}$$

This is the "Laplace equation."

It should be noted that, although we started with the problem of determining a vector field, it has now been reduced to one of solving for a *scalar* field.

It is, of course, possible to introduce the potential for any irrotational field, even when sources are present. Since the divergence of a vector is its efflux per unit volume, a source (of flux), which may be described by a density function $\rho(\mathbf{r})$, appears on the right-hand side of the divergence equation, so that one has

$$\nabla \cdot \mathbf{F} = \rho \tag{3.4}$$

in which case the potential satisfies the *inhomogeneous* equation

$$\nabla^2 \phi = -\rho \tag{3.5}$$

This is known as "Poisson's equation." It is generally true in physics that if one can solve the *homogeneous* (source-free) problem, one can construct from it the solution of the inhomogeneous one. We will see that this is the case with potential problems.

There is another type of problem, in which the Laplace or Poisson equation is satisfied by a *vector* field. In this problem the sources are sources of rotation (as in the problem of vortices in hydrodynamics), but the field is everywhere solenoidal ($\nabla \cdot \mathbf{F} = 0$). It is another standard theorem of vector calculus that a solenoidal field can always be expressed as the curl of another vector, which we call the vector potential:

$$\mathbf{F} = \nabla \times \mathbf{A} \tag{3.6}$$

Designating the density of the sources of circulation by the vector $\mathbf{J}$:

$$\nabla \times \mathbf{F} = \mathbf{J} \tag{3.7}$$

If (3.6) is substituted into (3.7), an equation is obtained for $\mathbf{A}$

$$\nabla \times (\nabla \times \mathbf{A}) = \mathbf{J} \tag{3.8}$$

Using a familiar vector identity on the left-hand side this becomes

$$\nabla(\nabla \cdot \mathbf{A}) - \nabla^2 \mathbf{A} = \mathbf{J} \tag{3.9}$$

Now $\mathbf{A}$ is not uniquely determined by (3.6) when $\mathbf{F}$ is known. In fact,

$$\mathbf{A}' = \mathbf{A} + \nabla \chi \tag{3.10}$$

represents the same field. But it is always possible to choose χ so that $\nabla \cdot \mathbf{A}' = 0$; it is a solution of the equation

$$\nabla^2 \chi = \nabla \cdot \mathbf{A} \tag{3.11}$$

If the vector potential is so chosen, (3.9) takes on the form of a Poisson equation:

$$\nabla^2 \mathbf{A} = -\mathbf{J} \tag{3.12}$$

If there are no sources, $\mathbf{J} = 0$ and we have a Laplace equation for $\mathbf{A}$

$$\nabla^2 \mathbf{A} = 0 \tag{3.13}$$

However, in this case it is clearly simpler to represent our field by a *scalar* potential.

The difference in the two cases which we have discussed is this:

Sources of flux are scalar, and so the field, too, is scalar. Sources of circulation, on the other hand, have a vectorial (directional) character, and hence produce a vector field.

Both of these situations are met in electromagnetic theory. To see this let us start from Maxwell's equations:

(I)

$$\nabla \cdot \mathbf{D} = \rho \tag{3.14}$$

which expresses Gauss' theorem, that the sources of electric flux are charges; ρ is the charge density.

(II)

$$\nabla \cdot \mathbf{B} = 0 \tag{3.15}$$

which says that there are no magnetic sources (magnetic poles).

(III)

$$\nabla \times \mathbf{E} = -\frac{\partial \mathbf{B}}{\partial t} \tag{3.16}$$

This is the vector expression of Faraday's law of induction; it states that a time-varying magnetic flux produces a circulation in the electric field.

(IV)

$$\nabla \times \mathbf{H} = \mathbf{J} + \frac{\partial \mathbf{D}}{\partial t} \tag{3.17}$$

This equation is a generalization of Ampère's law. In the steady state $\partial \mathbf{D}/\partial t = 0$, and the equation states that a current produces a circulation of magnetic field. The second term on the right-hand side expresses the discovery of Maxwell that a "displacement current" also produces a circulation in the magnetic field.[1]

The preceding equations must be supplemented by one which relates the displacement D to the electric field E:

$$\mathbf{D} = \epsilon \mathbf{E} \tag{3.18}$$

where ϵ is the "dielectric constant," and another which relates the magnetic induction $\mathbf{B}$ to the magnetic field $\mathbf{H}$:

$$\mathbf{B} = \mu \mathbf{H} \tag{3.19}$$

where μ is the "magnetic permeability." We consider only problems in which ϵ and μ are constant.

---

[1] See J. D. Jackson, *Classical Electrodynamics*, p. 178.

If we confine our attention to "steady-state" problems, the time derivatives on the right-hand sides of (3.16) and (3.17) are zero. Coupling between the electric and magnetic fields then ceases to exist, and the theory of the electrostatic field is to be found in Eqs. (3.14) and (3.16), while that of the magnetostatic field is given by (3.15) and (3.17).

Considering first the electrostatic case, we may write

$$\mathbf{E} = -\nabla\phi \tag{3.20}$$

where ϕ is the electrostatic potential, and

$$\nabla^2\phi = -\frac{\rho}{\epsilon} \tag{3.21}$$

It will be found convenient, to avoid writing awkward factors, to introduce

$$\Phi = 4\pi\epsilon\phi \tag{3.22}$$

from which it follows that

$$\nabla^2\Phi = -4\pi\rho \tag{3.23}$$

In terms of Φ, the electric field is

$$\mathbf{E} = -\frac{1}{4\pi\epsilon}\nabla\Phi \tag{3.24}$$

Turning to the magnetostatic case, the situation closely resembles that envisaged in Eqs. (3.6) and (3.7). From (3.15),

$$\mathbf{B} = \nabla \times \mathbf{A}$$

If $\mathbf{A}$ is chosen so that $\nabla \cdot \mathbf{A} = 0$, (3.17) yields for $\mathbf{A}$ the equation

$$\nabla^2\mathbf{A} = -\mu\mathbf{J} \tag{3.25}$$

There is some convenience in introducing

$$\mathfrak{A} = \frac{4\pi}{\mu}\mathbf{A} \tag{3.26}$$

so that

$$\nabla^2\mathfrak{A} = -4\pi\mathbf{J} \tag{3.27}$$

which is quite analogous to (3.23) except that it is a vector rather than a scalar equation.

2. Solution of Laplace's Equation in Spherical Coordinates

In spherical polar coordinates (r, θ, φ), the Laplacian equation takes the form

$$\frac{1}{r^2}\frac{\partial}{\partial r}\left(r^2\frac{\partial\Phi}{\partial r}\right) + \frac{1}{r^2\sin\theta}\frac{\partial}{\partial\theta}\left(\sin\theta\frac{\partial\Phi}{\partial\theta}\right) + \frac{1}{r^2\sin^2\theta}\frac{\partial^2\Phi}{\partial\varphi^2} = 0 \tag{3.28}$$

The equation may be solved by using the method of separation of variables in two stages. First, put

$$\Phi = R(r)S(\theta, \varphi) \qquad (3.29)$$

Substituting in the equation and dividing by Φ gives

$$\frac{1}{r^2 R}\frac{d}{dr}\left(r^2 \frac{dR}{dr}\right) + \frac{1}{S \sin^2 \theta}\left[\sin \theta \frac{\partial}{\partial \theta}\left(\sin \theta \frac{\partial S}{\partial \theta}\right) + \frac{\partial^2 S}{\partial \varphi^2}\right] = 0 \qquad (3.30)$$

The first term is a function of the radial coordinate only, and the second of the angular coordinates. For the sum to be zero, each must be a constant, that is,

$$\frac{d}{R dr}\left(r^2 \frac{dR}{dr}\right) = \lambda \qquad (3.31)$$

$$\frac{1}{S \sin^2 \theta}\left[\sin \theta \frac{\partial}{\partial \theta}\left(\sin \theta \frac{\partial S}{\partial \theta}\right) + \frac{\partial^2 S}{\partial \varphi^2}\right] = -\lambda \qquad (3.32)$$

We next multiply (3.32) by $S \sin^2 \theta$ and put

$$S = f(\theta)g(\varphi) \qquad (3.33)$$

to get

$$\left\{\frac{1}{f}\sin \theta \frac{d}{d\theta}\left(\sin \theta \frac{df}{d\theta}\right) + \lambda \sin^2 \theta\right\} + \frac{1}{g}\frac{d^2 g}{d\varphi^2} = 0 \qquad (3.34)$$

Again, the term in the curly bracket and the last term must be equal and opposite constants, which we write $m^2, -m^2$. Thus

$$\frac{1}{\sin \theta}\frac{d}{d\theta}\left(\sin \theta \frac{df}{d\theta}\right) + \left(\lambda - \frac{m^2}{\sin^2 \theta}\right)f = 0 \qquad (3.35)$$

and

$$\frac{d^2 g}{d\varphi^2} = -m^2 g \qquad (3.36)$$

Since the solution of (3.36) is

$$g = a \cos m\varphi + b \sin m\varphi \qquad (3.37)$$

and the solution must be single-valued (that is, must remain unaltered when φ is increased by 2π), m must be a real integer.

3.   The θ Equation and the Factorization Method

The next equation to be solved is (3.35), for the function $f(\theta)$. We see that this defines an eigenvalue problem, since f is finite for all angles θ only if λ has appropriately chosen values.

The method which we use for the solution of this problem is the "factorization method," which is discussed in detail for a wide range of problems in an article by L. Infeld and T. Hull.[2]

The Equation (3.35) for $f(\theta)$ may be written

$$\frac{d^2f}{d\theta^2} + \cot\theta\,\frac{df}{d\theta} - m^2\csc^2\theta f = -\lambda f \tag{3.38}$$

We try to "factorize" the left-hand side into two first-order operators:

$$\left(\frac{d}{d\theta} + \alpha\cot\theta\right)\left(\frac{d}{d\theta} + \beta\cot\theta\right)f$$

$$= \frac{d^2f}{d\theta^2} + (\alpha+\beta)\cot\theta\,\frac{df}{d\theta} - \beta\csc^2\theta f + \alpha\beta\cot^2\theta f$$

$$= \frac{d^2f}{d\theta^2} + (\alpha+\beta)\cot\theta\,\frac{df}{d\theta} - \beta(1-\alpha)\csc^2\theta f - \alpha\beta f \tag{3.39}$$

The first three terms of this expression can be made to correspond with the left-hand side of (3.38) by choosing α and β so that

$$\alpha+\beta = 1 \tag{3.40}$$

$$\beta(1-\alpha) = m^2 \tag{3.41}$$

These equations have two solutions:

(a) $\qquad\qquad\qquad \beta = m, \quad \alpha = -(m-1)$

(b) $\qquad\qquad\qquad \beta = -m, \quad \alpha = (m+1)$

Let us define two operators

$$M_m^{(+)} = \frac{d}{d\theta} - m\cot\theta \tag{3.42}$$

$$M_m^{(-)} = \frac{d}{d\theta} + (m+1)\cot\theta \tag{3.43}$$

The signs used to designate the M's, which seem at first sight illogical, will be justified as we proceed.

Using (3.39), (3.42), and (3.43), Eq. (3.38) may now be written in either of two ways:

$$M_{m-1}^{(+)}M_{m-1}^{(-)}f_m = -[\lambda - m(m-1)]f_m \tag{3.44}$$

or

$$M_m^{(-)}M_m^{(+)}f_m = -[\lambda - m(m+1)]f_m \tag{3.45}$$

[2] "The Factorization Method," *Rev. Mod. Phys.* **23**, 21 (1951).

Let us operate on the first of these equations with $M_{m-1}^{(-)}$:

$$M_{m-1}^{(-)}M_{m-1}^{(+)}[M_{m-1}^{(-)}f_m] = -[\lambda - m(m-1)][M_{m-1}^{(-)}f_m]$$

Comparing with (3.45) it may be seen that we may put

$$M_m^{(-)}f_{m+1} = p_m f_m \qquad (3.46)$$

where p_m is a constant.

Similarly, operating on (3.45) with $M_m^{(+)}$ gives

$$M_m^{(+)}M_m^{(-)}[M_m^{(+)}f_m] = -[\lambda - m(m+1)][M_m^{(+)}f_m]$$

which, on comparison with (3.44), permits us to conclude that

$$M_m^{(+)}f_m = q_m f_{m+1} \qquad (3.47)$$

Equations (3.46) and (3.47) show how, starting with the solution for a given m, we may get those for m increased or decreased by unity.

Let us next prove the following theorem: That

$$\int_0^\pi \bar{f} M_m^{(-)} f \sin \theta \, d\theta = -\int_0^\pi f M_m^{(+)} \bar{f} \sin \theta \, d\theta \qquad (3.48)$$

where $\bar{f}$ and f are arbitrary bounded functions of θ. The proof is quite direct. For

$$\int_0^\pi \sin \theta \bar{f} \left[\frac{d}{d\theta} + (m+1) \cot \theta \right] f \, d\theta$$

$$= \sin \theta \, \bar{f} f \Big|_0^\pi - \int_0^\pi f \left[\frac{d}{d\theta} (\sin \theta \bar{f}) - (m+1) \cos \theta \bar{f} \right] d\theta$$

$$= -\int_0^\pi \sin \theta f \left[\frac{d}{d\theta} - m \cot \theta \right] \bar{f} \, d\theta$$

We now apply this theorem to two different cases:

(i) $\qquad\qquad f = f_{m+1}, \qquad \bar{f} = M_m^{(-)} f_{m+1}$

and

(ii) $\qquad\qquad f = M_m^{(+)} f_m, \qquad \bar{f} = f_m$

In case (i) Eq. (3.48) yields

$$\int_0^\pi f_{m+1} M_m^{(+)} M_m^{(-)} f_{m+1} \sin \theta \, d\theta = -\int_0^\pi (M_m^{(-)} f_{m+1})(M_m^{(-)} f_{m+1}) \sin \theta \, d\theta$$

By (3.44) the left-hand side is

$$-[\lambda - m(m+1)] \int_0^\pi f_{m+1}^2 \sin \theta \, d\theta$$

whereas by (3.46) the right-hand side is

$$-p_m^2 \int_0^\pi f_m^2 \sin \theta \, d\theta$$

We may normalize our f_m functions so that the integrals are equal to unity; in that case

$$p_m = \sqrt{\lambda - m(m+1)} \qquad (3.49)$$

Proceeding in an exactly similar way in case (ii), it is also found that

$$q_m = \sqrt{\lambda - m(m+1)} \qquad (3.50)$$

Thus, with these values of p_m and q_m, (3.46) and (3.47) show that $M_m^{(-)}$ and $M_m^{(+)}$, respectively, convert normalized solutions of a given order into normalized solutions of order lower and higher, respectively, by one in m.

Since, however, (3.49) and (3.50) show that $m(m+1) \leqslant \lambda$, m can neither be raised nor lowered indefinitely. Letting m_2 be the maximum value of m, and m_1 its minimum, each must satisfy the equation

$$m(m+1) = \lambda \qquad (3.51)$$

since otherwise (3.46) or (3.47) would produce solutions with m lower than the minimum or higher than the maximum, respectively. Thus we must have

$$\lambda = l(l+1) \qquad (3.52)$$

where l is an integer. The maximum m is then seen to be l in (3.47), the minimum $-(l+1)$ in (3.46). Therefore, for a given l, the f's run from f_{-l} to f_l.

Equations (3.46) with $m = -(l+1)$ and (3.47) with $m = l$ give equations for f_{-l} and f_l, respectively, as follows:

$$\left(\frac{d}{d\theta} - l \cot \theta\right) f_{-l} = 0 \qquad (3.53)$$

and

$$\left(\frac{d}{d\theta} - l \cot \theta\right) f_l = 0 \qquad (3.54)$$

Aside from the question of normalization, the solution for each is $\sin^l \theta$. But normalization requires that it be multiplied by C_l where

$$C_l^2 \int_0^\pi \sin^{2l+1} \theta \, d\theta = 1$$

Evaluating the integral,

$$C_l^2 = \frac{(2l+1)!}{2^{2l+1}(l!)^2} \qquad (3.55)$$

Let the normalized solution for a given l and m be designated Θ_{lm}. Then what has just been shown is that

$$\Theta_{ll} = \sqrt{\frac{(2l+1)!}{2^{2l+1}(l!)^2}} \sin^l \theta \qquad (3.56)$$

The normalized solutions for the same l and other m's are then obtained, as may be seen from (3.46), (3.43), and (3.49), by the equation

$$\Theta_{l,m-1} = \frac{1}{\sqrt{l(l+1)-m(m-1)}} \left[\frac{d}{d\theta} + m \cot \theta \right] \Theta_{lm} \qquad (3.57)$$

The following are the explicit forms of these functions for $l=0$, $l=1$, and $l=2$.

(i) $l=0$ $\quad \Theta_{00} = \dfrac{1}{\sqrt{2}}$

(ii) $l=1$ $\quad \Theta_{11} = \dfrac{\sqrt{3}}{2} \sin \theta$

$\qquad\qquad\quad \Theta_{10} = \sqrt{\dfrac{3}{2}} \cos \theta$

$\qquad\qquad\quad \Theta_{1,-1} = -\dfrac{\sqrt{3}}{2} \sin \theta \qquad\qquad (3.58)$

(iii) $l=2$ $\quad \Theta_{22} = \dfrac{\sqrt{15}}{4} \sin^2 \theta$

$\qquad\qquad\quad \Theta_{21} = \dfrac{\sqrt{15}}{2} \sin \theta \cos \theta$

$\qquad\qquad\quad \Theta_{20} = \dfrac{\sqrt{5}}{2\sqrt{2}} (3 \cos^2 \theta - 1)$

$\qquad\qquad\quad \Theta_{2,-1} = -\dfrac{\sqrt{15}}{2} \sin \theta \cos \theta$

$\qquad\qquad\quad \Theta_{2,-2} = -\dfrac{\sqrt{15}}{4} \sin^2 \theta$

The minus signs appearing in the solutions for negative m are purely conventional; since we determined only $p_m{}^2$ and $q_m{}^2$ we might just as well have chosen them with negative signs for negative m's. Were that done, we would have found in each case that $\Theta_{l,-m} = \Theta_{l,m}$. This is, in fact, the more usual convention.

4. Spherical Harmonics

The normalized solutions to the angular equation (3.32) with $\lambda = l(l+1)$ are known as the "spherical harmonics" of order m, and are designated $Y_{lm}(\theta, \varphi)$. They are obtained by multiplying each of the functions Θ_{lm} by the normalized φ-function, namely,

$$\frac{1}{\sqrt{2\pi}} e^{im\varphi}$$

The θ-functions, with a somewhat different normalization which we discuss later, are the so-called "associated Legendre functions." Equation (3.35) with $\lambda = l(l+1)$ is the "associated Legendre equation."

5. Radial Solution and the General Solution of Laplace's Equation

If we put $\lambda = l(l+1)$ in the radial equation (3.31), it may be immediately solved (since it is an equation of the homogeneous type) to give

$$R = A_{lm}r^l + B_{lm}/r^{l+1} \tag{3.59}$$

The general separated solution then has the form

$$\Phi = \sum_{l=0}^{\infty} \sum_{m=-l}^{l} \left(A_{lm}r^l + \frac{B_{lm}}{r^{l+1}} \right) Y_{lm}(\theta, \varphi)$$

$$= \frac{1}{\sqrt{2\pi}} \sum_{l=0}^{\infty} \sum_{m=-l}^{l} \left(A_m r^l + \frac{B_m}{r^{l+1}} \right) \Theta_{lm}(\theta) e^{im\varphi} \tag{3.60}$$

This is a general form which must be satisfied by *all* potentials in free space.

The simplest solution for Φ, which represents the potential of a unit point charge, is

$$\Phi = \frac{1}{r} \tag{3.61}$$

where the origin is taken at the charge. This is, of course, a particular case of (3.60) with all A's equal to zero, as well as all B's except B_0, which is unity.

Potential problems are, in general, of two types: (1) those in which we know a charge distribution, and the aim is to find the resulting potential (these are problems in the solution of the Poisson equation); and (2) those in which a potential must be determined which satisfies given boundary conditions (e.g., Φ is a constant on conducting surfaces). The general solution of problems of the latter sort depends on the geometry of the surfaces on which the boundary conditions are given; to get analytic solutions, it is necessary to find a system of coordinates in which the equation "separates" (that is, in which a solution may be found which is a product of functions of the separate variables), and such that the boundary surfaces are parametric surfaces, that is, surfaces on which one of the coordinates is constant. Spherical polar coordinates are therefore suitable for problems involving spheres (r = constant), planes ($\theta = \pi/2$ or ϕ = constant), or cones (θ = constant). The subsequent sections are devoted to the exploration of problems which can be solved in these coordinates.

6. Legendre Polynomials

In all potential problems in which there is axial symmetry, we may choose the polar axis in the direction of the symmetry axis; the potentials will then be independent of the angle φ. The angular dependence of the solution for a given l [see (3.52)] is Θ_{l0}. These functions are polynomials in $\cos \theta$; with an appropriate normalization, they are the "Legendre polynomials." The preceding section does not provide a very convenient way of studying these polynomials. We therefore proceed in a different way, which permits a straightforward derivation of some of their properties.

Consider the potential of a unit charge located, not at the origin, but at a point S at a distance z_0 above the origin on the z axis. (See Fig. 3.2.) The potential at P is the inverse of the distance SP, which is

$$\Phi = \frac{1}{\sqrt{r^2 - 2rz_0 \cos \theta + z_0^2}} \tag{3.62}$$

We henceforth write

$$\mu = \cos \theta \tag{3.63}$$

If $r > z_0$, we write

$$\Phi = \frac{1}{r} \frac{1}{\sqrt{1 - 2(z_0/r)\mu + (z_0/r)^2}} \tag{3.64}$$

and expand the inverse square root in powers of z_0/r. The result is of the form

$$\Phi = \sum_{l=0}^{\infty} \frac{z_0^l}{r^{l+1}} P_l(\mu) \tag{3.65}$$

the P_l being a polynomial of degree l.

Equation (3.65) is, of course, an axially symmetric potential. Comparing it with (3.60) in the special case in which $A_{lm} = B_{lm} = 0$ for $m \neq 0$ we see that, to within a constant,

$$P_l(\mu) = \Theta_{l0}(\theta) \tag{3.66}$$

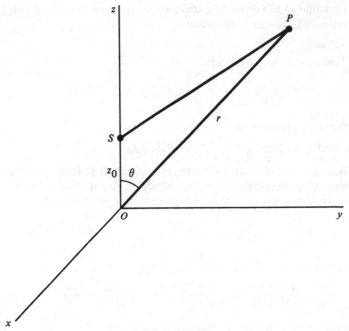

Figure 3.2

The expansion therefore gives us an alternative way of calculating Legendre polynomials. It simplifies matters a little to write the relation determining the Legendre polynomials in the form

$$\frac{1}{\sqrt{1-2t\mu+t^2}} = \sum_{l=0}^{\infty} t^l P_l(\mu) \qquad (3.67)$$

where $t < 1$.

It is easy to determine the Legendre polynomials of low order by the expansion:

$$\frac{1}{\sqrt{1-2t\mu+t^2}} = 1 + \tfrac{1}{2}(2t\mu - t^2) + \tfrac{3}{8}(2t\mu - t^2)^2$$
$$+ \tfrac{5}{16}(2t\mu - t^2)^3 + \tfrac{35}{128}(2t\mu - t^2)^4$$
$$+ \cdots$$

Picking out coefficients of the powers of t, we obtain the following formulas:

$$P_0(\mu) = 1, \qquad\qquad P_1(\mu) = \mu$$
$$P_2(\mu) = \tfrac{3}{2}\mu^2 - \tfrac{1}{2}, \qquad P_3(\mu) = \tfrac{5}{2}\mu^3 - \tfrac{3}{2}\mu \qquad (3.68)$$
$$P_4(\mu) = \tfrac{35}{8}\mu^4 - \tfrac{15}{4}\mu^2 + \tfrac{3}{8}, \quad \cdots$$

Equation (3.67) gives us a straightforward way to prove a number of properties of Legendre polynomials,

(I) $P_l(1) = 1$

Putting $\mu = 1$, (3.67) gives

$$\frac{1}{\sqrt{1 - 2t + t^2}} = \frac{1}{1 - t} = \sum t^l = \sum t^l P_l(1)$$

which proves the theorem.

(II) EVALUATION OF $\int_{-1}^{1} P_l(\mu) P_{l'}(\mu) d\mu$

Because the P_l's are solutions of a Sturm–Liouville problem they are mutually orthogonal; that is, the integral is zero if $l \neq l'$.

If we square Eq. (3.67) and integrate over μ, we find that

$$\int_{-1}^{1} \frac{d\mu}{1 - 2t\mu + t^2} = \frac{1}{t} \ln \frac{1 + t}{1 - t} = \sum_{l=0}^{\infty} t^{2l} \int_{-1}^{1} P_l^2(\mu) d\mu$$

Now the left-hand side may be expanded in powers of t, to give

$$\frac{1}{t} \ln \frac{1 + t}{1 - t} = 2 + \frac{2}{3} t^2 + \frac{2}{5} t^4 + \cdots + \frac{2}{2l + 1} t^{2l} + \cdots$$

Equating coefficients, we see that the integrals are

$$\int_{-1}^{1} P_l^2(\mu) d\mu = \frac{2}{2l + 1} \tag{3.69}$$

Since the Θ_{l0}'s are normalized to unity and the P_l's to $2/(2l + 1)$

$$P_l(\mu) = \sqrt{\frac{2}{2l + 1}} \Theta_{l0} \tag{3.70}$$

(III) EVALUATION OF $\int_{0}^{1} P_l(\mu) d\mu$

This integral will appear in Problem (3.15) below. It may be evaluated by integrating (3.67) as it stands. The integral of the left-hand side is

$$-\frac{1}{t} \sqrt{1 - 2t\mu + t^2} \Big|_{0}^{1} = \frac{1}{t} [\sqrt{1 + t^2} - (1 - t)]$$

which may be expanded in a power series to give

$$1 + \sum_{n=1}^{\infty} \binom{1/2}{n} t^{2n-1}$$

where $\binom{1/2}{n}$ is the coefficient of z^n in the expansion of $(1 + z)^{1/2}$.

Therefore

$$\int_0^1 P_l(\mu)\,d\mu = 1, \qquad l = 0$$
$$= 0, \qquad l \text{ even but} \neq 0 \qquad (3.71)$$
$$= \binom{1/2}{n} \qquad l = 2n - 1$$

(IV) A RECURSION RELATION

Differentiating with respect to t both sides of Eq. (3.67), we obtain the equation

$$\frac{\mu - t}{(1 - 2t\mu + t^2)^{3/2}} = \sum_{l=0}^{\infty} l t^{l-1} P_l(\mu) \qquad (3.72)$$

which may be written

$$\frac{\mu - t}{\sqrt{1 - 2t\mu + t^2}} = (1 - 2t\mu + t^2) \sum_{l=0}^{\infty} l t^{l-1} P_l(\mu)$$

or

$$(\mu - t) \sum_{l=0}^{\infty} t^l P_l(\mu) = (1 - 2t\mu + t^2) \sum_{l=0}^{\infty} l t^{l-1} P_l(\mu) \qquad (3.73)$$

If we now equate, on the two sides of this equation, the powers of t^l, we find that

$$-P_{l-1} + \mu P_l = (l+1)P_{l+1} - 2\mu l P_l + (l-1)P_{l-1}$$

which may be rearranged to give

$$(2l+1)\mu P_l = (l+1)P_{l+1} + l P_{l-1} \qquad (3.74)$$

This recursion relation, considered as an equation for P_{l+1} in terms of P_l and P_{l-1}, provides an alternative means of calculating the successive polynomials.

(V) A DERIVATIVE FORMULA

There are also many different relationships between Legendre polynomials and their derivatives. To obtain such relations we first differentiate (3.67):

$$\frac{t}{(1 - 2t\mu + t^2)^{3/2}} = \sum_{l=0}^{\infty} t^l P_l'(\mu)$$

or

$$t \sum t^l P_l = (1 - 2t\mu + t^2) \sum t^l P_l'$$

Equating coefficients of each power of t gives

$$P_l = P_{l+1}' - 2\mu P_l' + P_{l-1}' \qquad (3.75)$$

Differentiating (3.74) gives a second relation

$$P_l + \mu P_l' = \frac{l+1}{2l+1} P_{l+1}' + \frac{l}{2l+1} P_{l-1}' \qquad (3.76)$$

Solving for P'_{l+1}, P'_{l-1} in terms of P_l and P'_l:

$$P'_{l-1} = -lP_l + \mu P'_l \tag{3.77}$$

and

$$P'_{l+1} = (l+1)P_l + \mu P'_l \tag{3.78}$$

from which the relation

$$(2l+1)P_l = P'_{l+1} - P'_{l-1} \tag{3.79}$$

follows.

Another relation, which expresses P'_l in terms of Legendre functions, is obtained as follows: Multiply (3.77) by μ to get

$$\mu^2 P'_l = \mu P'_{l-1} + l\mu P_l$$

Substituting for $\mu P'_{l-1}$ from (3.78):

$$\mu^2 P'_l = P'_l - lP_{l-1} + l\mu P_l$$

Using (3.74),

$$(\mu^2 - 1)P'_l = \frac{l(l+1)}{2l+1}(P_{l+1} - P_{l-1}) \tag{3.80}$$

On differentiating (3.80) and using (3.79) we obtain directly the differential equation for P_l:

$$\frac{d}{d\mu}[(1-\mu^2)P'_l] + l(l+1)P_l = 0$$

By combining the relations already derived, numerous other equations involving the Legendre polynomials and their derivatives may be derived; however, (3.74), (3.79), and (3.80) are the most commonly used ones.

7. An Alternative Derivation of Legendre Polynomials: Multipoles

The potential at a point P of a unit point charge at a point on the z axis at a distance z_0 from the origin may be written in terms of the Cartesian coordinates (x, y, z) of P; it is

$$\Phi = \frac{1}{\sqrt{x^2 + y^2 + (z - z_0)^2}} \tag{3.81}$$

This may be expanded in a Taylor's series about the value $z_0 = 0$:

$$\Phi = \sum_{n=0}^{\infty} \frac{z_0^n}{n!}\left[\frac{\partial^n}{\partial z_0^n}\Phi\right]_{z_0=0} \tag{3.82}$$

A derivative with respect to z_0 of a function of $z - z_0$ is the negative of its derivative with respect to z. Therefore,

$$\Phi = \sum_{n=0}^{\infty} \frac{(-1)^n z_0^n}{n!} \frac{\partial^n}{\partial z^n} \left(\frac{1}{r}\right) \tag{3.83}$$

where

$$r^2 = x^2 + y^2 + z^2 \tag{3.84}$$

If this equation is compared with (3.65), it is seen that

$$\frac{1}{r^{l+1}} P_l(\mu) = \frac{(-1)^l}{l!} \frac{\partial^l}{\partial z^l} \left(\frac{1}{r}\right) \tag{3.85}$$

or

$$P_l(\mu) = \frac{(-1)^l}{l!} r^{l+1} \frac{\partial^l}{\partial z^l} \left(\frac{1}{r}\right) \tag{3.86}$$

Thus, the Legendre polynomials may be derived by successive differentiation with respect to z of $1/r$.

It is, of course, obvious that the derivatives of $1/r$ are solutions of Laplace's equation. It is possible to interpret the various terms as being the potentials of *multipoles* of successive orders.

The first derivative represents a *dipole* potential. If we take a charge $-e$ at $z = -\epsilon$ and a charge e at $z = \epsilon$, the potential is

$$\Phi_1 = \frac{-e}{\sqrt{x^2 + y^2 + (z+\epsilon)^2}} + \frac{e}{\sqrt{x^2 + y^2 + (z-\epsilon)^2}} \tag{3.87}$$

The "dipole" is obtained by letting $e \to \infty$ and $\epsilon \to 0$ in such a way that $2e\epsilon \to \mu_1$, the "dipole moment." The potential becomes, in this limit,

$$\lim \Phi_1 = -\mu_1 \frac{\partial}{\partial z} \left(\frac{1}{r}\right) \tag{3.88}$$

This dipole has, of course, a direction associated with it, the z axis, along which lies the line joining the two charges.

A pair of *dipoles* of opposite moment (μ_1 at $z = \epsilon$, $-\mu_1$ at $z = -\epsilon$) constitute a "quadrupole" when $\epsilon \to 0$, $\mu_1 \to \infty$ in such a way that $2\epsilon\mu_1 = \mu_2$, the "quadrupole moment." The quadrupole potential is

$$\Phi_2 = 2\epsilon\mu_1 \frac{\partial^2}{\partial z^2} \left(\frac{1}{r}\right)$$

$$= \mu_2 \frac{\partial^2}{\partial z^2} \left(\frac{1}{r}\right) \tag{3.89}$$

The procedure may be extended a step at a time, an "octupole" being the limit of two quadrupoles of opposite moment, and so forth. In fact, an inductive argument permits generalization to the statement that the potential of an n-pole is

$$\Phi_n = (-1)^n \mu_n \frac{\partial^n}{\partial z^n} \left(\frac{1}{r}\right) \tag{3.90}$$

Using (3.86) this may be written

$$\Phi_n = n! \mu_n \frac{1}{r^{n+1}} P_n(\cos \theta) \tag{3.91}$$

Problem 3-1:   Calculate the potential and multipole moments of a uniform linear charge distribution extending from $-l$ to l.

---

8. Multipoles without Axial Symmetry and Associated Legendre Functions

It is, of course, true that not only derivatives of $1/r$ with respect to z are solutions of Laplace's equation, but also derivatives with respect to x and y, or any combination of derivatives with respect to the three coordinates. If l derivatives are taken altogether, and the result expressed in polar coordinates, the r dependence will be $1/r^{l+1}$. Comparing with (3.60), it is evident that this solution is, then, some combination of spherical harmonics of order l. Conversely, each spherical harmonic must be some combination of derivatives of order l. To express the spherical harmonics in terms of derivatives, it is useful to introduce the operators

$$\partial_\pm = \frac{\partial}{\partial x} \pm i \frac{\partial}{\partial y} \tag{3.92}$$

and to consider as our three basic derivative operators not $\partial/\partial x$, $\partial/\partial y$, and $\partial/\partial z$ but ∂_+, ∂_-, and $\partial_z = \partial/\partial z$.

A general solution of order l is

$$\partial_+{}^p \partial_-{}^s \partial_z{}^{l-p-s} \left(\frac{1}{r}\right) \tag{3.93}$$

where $p + s \leqslant l$. If, however, use is made of the fact that $1/r$ satisfies Laplace's equation (except, of course, at $r = 0$), in calculating these derivatives we may use the relation

$$\partial_+ \partial_- = -\partial_z{}^2 \tag{3.94}$$

This may be used to eliminate entirely either the ∂_+ or ∂_- operators, depending on which occurs less frequently. Suppose, for instance, that $s < p$. If we put

$$p = s + m \tag{3.95}$$

we see that the solution is

$$f_{lm} = \partial_+{}^m \partial_z{}^{l-m} \left(\frac{1}{r}\right) \tag{3.96}$$

where we have omitted the signature $(-1)^s$ in front. If we had had $s = p + m > p$ we should have had, in place of (3.96), the solution

$$f_{l,-m} = \partial_-{}^m \partial_z{}^{l-m} \left(\frac{1}{r}\right) \tag{3.97}$$

It may now be shown that, aside from arbitrary multiplicative constants, f_{lm} is the same as Y_{lm}, and $f_{l,-m}$ as $Y_{l,-m}$. This may be most simply verified by showing that the dependence of the solutions on the polar angle φ is $e^{im\varphi}$ or $e^{-im\varphi}$, respectively.

It is not too difficult to show by induction that $\partial_\pm{}^m(1/r)$ may be written

$$\partial_\pm{}^m \left(\frac{1}{r}\right) = (x \pm iy)^m \rho_m(r) \tag{3.98}$$

where $\rho_m(r)$ is a function only of the radial coordinate, which will be determined later. For suppose it to be true for a given m. Let us then operate once more with $\partial_\pm$. Clearly

$$\partial_\pm (x \pm iy)^m = 0 \tag{3.99}$$

whereas

$$\partial_\pm \rho_m(r) = \frac{x \pm iy}{r} \rho_m'(r) \tag{3.100}$$

Therefore,

$$\partial_\pm{}^{m+1} \left(\frac{1}{r}\right) = (x \pm iy)^{m+1} \frac{1}{r} \rho_m'(r)$$
$$= (x \pm iy)^{m+1} \rho_{m+1}(r) \tag{3.101}$$

where

$$\rho_{m+1} = \frac{1}{r} \rho_m' \tag{3.102}$$

Since (3.98) is clearly true for $m = 0$, with $\rho_0 = 1/r$, the result is valid for all m. Moreover, using (3.102), it is easily shown that

$$\rho_m(r) = \left(-\frac{1}{2}\right)^m \frac{(2m)!}{m!} \frac{1}{r^{2m+1}} \tag{3.103}$$

Once (3.98) is proven, the dependence of (3.97) on φ follows immediately. For

$$\partial_\pm{}^m \left(\frac{1}{r}\right) = r^m \sin^m \theta \, e^{\pm im\varphi} \rho_m(r) \tag{3.104}$$

But if (3.98) is now differentiated $(l-m)$ times with respect to z, the derivatives pass through $(x \pm iy)^m$ and operate only on $\rho_m(r)$; this yields a function only of z and r, that is, of r and $\cos\theta$. Therefore, these further differentiations do not affect the φ dependence of the solution.

Looking now at $f_{l,m}$ from (3.96), we can conclude, because of its r and φ dependence, that

$$\partial_+{}^m \partial_z{}^{l-m} \left(\frac{1}{r}\right) = C_{lm} \frac{1}{r^{l+1}} \Theta_{lm} \, e^{im\varphi} \tag{3.105}$$

and

$$\partial_-{}^m \partial_z{}^{l-m} \left(\frac{1}{r}\right) = C_{lm} \frac{1}{r^{l+1}} \Theta_{lm} \, e^{-im\varphi} \tag{3.106}$$

That Θ is unchanged when m is changed into $-m$ in the φ-dependent term is clear from the fact that the equation satisfied by Θ_{lm} depends only on m^2.

These Θ_{lm}'s were defined to be normalized to unity. The "associated Legendre functions," on the other hand, are differently normalized. They are in fact defined by the relations

$$\partial_\pm{}^m \partial_z{}^{l-m} \left(\frac{1}{r}\right) = (-1)^l (l-|m|)! \frac{1}{r^{l+1}} P_l{}^{|m|} (\cos\theta) \, e^{\pm im\varphi} \tag{3.107}$$

It will be noted that, with this definition, $P_l{}^{|m|} (\cos\theta)$ reduces to the Legendre polynomial [Eq. (3.85)] for $m = 0$.

To determine the precise numerical relationship between Θ_{lm} and $P_l{}^{|m|}$ it is necessary to obtain the normalization of $P_l{}^{|m|}$. It is shown in Appendix 3A that the following recursion relation holds:

$$(2l+1)\mu P_l{}^m = (l-m+1)P_{l+1}^m + (l+m)P_{l-1}^m \tag{3.108}$$

This may be used to evaluate the normalization integral as follows: multiply (3.108) by P_{l-1}^m and integrate over $\mu = \cos\theta$ from -1 to 1:

$$(2l+1) \int_{-1}^{1} \mu P_l{}^m P_{l-1}^m \, d\mu = (l+m) \int_{-1}^{1} (P_{l-1}^m)^2 \, d\mu \tag{3.109}$$

We have used the fact that

$$\int_{-1}^{1} P_l{}^m P_{l'}{}^m \, d\mu = 0 \tag{3.110}$$

for $l \neq l'$, which follows from treating the differential equation (3.35) as a Sturm–Liouville problem, on the interval $(-1, 1)$.

If, now, on the left-hand side of (3.109) we substitute for P_{l-1}^m from (3.108), we get

$$(l-m)\frac{(2l+1)}{(2l-1)}\int_{-1}^{1}[P_l^m(\mu)]^2\,d\mu = (l+m)\int_{-1}^{1}[P_{l-1}^m(\mu)]^2\,d\mu \tag{3.111}$$

We may use this recursion relation repeatedly until l is reduced to the value m:

$$\int_{-1}^{1}[P_l^m(\mu)]^2\,d\mu = \frac{2m+1}{2l+1}\frac{(l+m)!}{(2m)!(l-m)!}\int_{-1}^{1}[P_m^m(\mu)]^2\,d\mu \tag{3.112}$$

Now, from (3.107) in the case $l=m$, using (3.103) and (3.104), we see that

$$(-1)^m\frac{1}{r^{m+1}}P_m^m(\mu)\,e^{im\varphi} = \sin^m\theta\,e^{im\varphi}\left(-\frac{1}{2}\right)^m\frac{(2m)!}{m!}\frac{1}{r^{m+1}}$$

Therefore,

$$P_m^m(\mu) = \frac{(2m)!}{m!2^m}\sin^m\theta \tag{3.113}$$

But

$$\int_0^\pi \sin^{2m+1}\theta\,d\theta = \frac{2^{2m+1}(m!)^2}{(2m+1)!} \tag{3.114}$$

We now have all that is required to evaluate our normalization integral:

$$\int_{-1}^{1}[P_l^m(\mu)]^2\,d\mu = \frac{2}{2l+1}\frac{(l+m)!}{(l-m)!} \tag{3.115}$$

It foilows, then, that

$$P_l^m(\mu) = \sqrt{\frac{2}{2l+1}\frac{(l+m)!}{(l-m)!}}\,\Theta_{lm} \tag{3.116}$$

Finally, the normalized spherical harmonics can be expressed in terms of the associated Legendre functions:

$$Y_l^m(\theta,\varphi) = \sqrt{\frac{2l+1}{4\pi}\frac{(l-m)!}{(l+m)!}}\,P_l^m(\cos\theta)\,e^{im\varphi} \tag{3.117}$$

It is easily seen that Y_l^m and $Y_{l'}^{m'}$ are orthogonal, in the sense that the integral over solid angle of the product of one with the complex conjugate of the other is zero. For

$$\int Y_l^m(\theta,\varphi) Y_{l'}^{m'*}(\theta,\varphi) \sin\theta \, d\theta \, d\varphi$$

$$= \left[\frac{(2l+1)}{4\pi} \frac{(l-m)!}{(l+m)!} \frac{(2l'+1)}{4\pi} \frac{(l'-m')!}{(l'+m')!} \right]^{1/2} \int_{-1}^{1} P_l^m(\mu) P_{l'}^{m'}(\mu) \, d\mu$$

$$\times \int_0^{2\pi} e^{i(m-m')\varphi} \, d\varphi$$

where $\mu = \cos\theta$. The φ integral gives $2\pi\delta_{mm'}$. The μ integral becomes

$$\int_{-1}^{1} P_l^m(\mu) P_{l'}^m(\mu) \, d\mu$$

which is zero unless $l = l'$ because the P_l^m's are solutions of a Sturm–Liouville problem for a given m. But when $l = l'$ we can evaluate the integral form (3.115). Putting in the value of the integral given there we get altogether

$$\frac{(2l+1)}{4\pi} \frac{(l-m)!}{(l+m)!} 2\pi \frac{2}{2l+1} \frac{(l+m)!}{(l-m)!} = 1$$

It follows then that the Y_l^m's are orthonormal, that is, that

$$\int Y_l^m(\theta,\varphi) Y_{l'}^{m'*}(\theta,\varphi) \, d\Omega = \delta_{ll'}\delta_{mm'} \tag{3.118}$$

where the element of solid angle $d\Omega$ is equal to $\sin\theta \, d\theta \, d\varphi$.

9. An Addition Theorem for Spherical Harmonics

In Fig. 3.3 let **r** and **r**′ be the position vectors of points with polar angles (θ,φ) and (θ',φ'), respectively, and α the angle between these vectors.

Consider the expansion of a function of direction $F(\Omega) = F(\theta,\varphi)$ in terms of spherical harmonics:

$$F(\theta,\varphi) = \sum A_{lm} Y_l^m(\theta,\varphi) \tag{3.119}$$

By the orthonormality theorem (3.118), we can find A_{lm} by multiplying by one of the Y_l^{m*}'s and integrating:

$$A_{lm} = \int F(\theta',\varphi') Y_l^{m*}(\theta',\varphi') \, d\Omega' \tag{3.120}$$

Substituting in (3.119), we obtain the equation

$$F(\theta,\varphi) = \int F(\theta',\varphi') \sum_{l,m} Y_l^m(\theta,\varphi) Y_l^{m*}(\theta',\varphi') \, d\Omega' \tag{3.121}$$

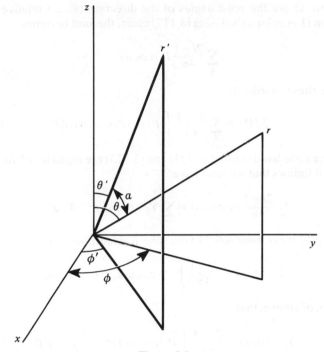

Figure 3.3

This shows that the sum

$$\sum_{l,m} Y_l^m(\theta,\varphi) Y_l^{m*}(\theta',\varphi')$$

is a δ-function, which we may designate $\delta(\Omega - \Omega')$, which, when integrated over solid angles, gives unity. Equation (3.121) becomes

$$F(\Omega) = \int F(\Omega')\delta(\Omega - \Omega')\, d\Omega' \qquad (3.122)$$

Consider this equation now in terms of a new set of coordinates, in which the polar axis is chosen along the direction (θ,φ). Then $Y_l^m(\theta,\varphi)$ becomes $Y_l^m(0,\varphi) = \sqrt{(2l+1)/4\pi}\, \delta_{m0}$ by virtue of (3.117). Then the sum

$$\sum_{l,m} Y_l^m(\theta,\varphi) Y_l^{m*}(\theta',\varphi')$$

becomes

$$\sum_l \sqrt{\frac{2l+1}{4\pi}} Y_l^{0*}(\alpha,\gamma)$$

where (α, γ) are the polar angles of the direction (θ', φ') relative to the direction Ω as polar axis. Using (3.117) again, the sum becomes

$$\sum_l \frac{2l+1}{4\pi} P_l(\cos \alpha).$$

Thus, in these coordinates,

$$F(\Omega) = \sum_l \frac{2l+1}{4\pi} \int F(\Omega') P_l(\cos \alpha) \, d\Omega' \qquad (3.123)$$

Since the right-hand sides of (3.121) and (3.123) are equal for *all* functions $F(\Omega')$, it follows that we must have

$$\sum_l \frac{2l+1}{4\pi} P_l(\cos \alpha) = \sum_{l,m} Y_l^m(\theta, \varphi) Y_l^{m*}(\theta', \varphi') \qquad (3.124)$$

By virtue of the orthogonality of the spherical harmonics, it follows that

$$Y_l^m(\theta, \varphi) = \frac{2l+1}{4\pi} \int P_l(\cos \alpha) Y_l^m(\theta', \varphi') \, d\Omega' \qquad (3.125)$$

and also, of course, that

$$Y_l^{m*}(\theta, \varphi) = \frac{2l+1}{4\pi} \int P_l(\cos \alpha) \, Y_l^{m*}(\theta', \varphi') \, d\Omega'$$

If we now make an expansion of $P_l(\cos \alpha)$ in the harmonics $Y_l^m(\theta', \varphi')$ with (θ, φ) fixed, the coefficients are

$$\int P(\cos \alpha) Y^{m*}(\theta', \varphi') \, d\Omega' = Y_l^{m*}(\theta, \varphi) \frac{4\pi}{2l+1}$$

Thus, we have the "addition theorem"

$$P_l(\cos \alpha) = \frac{4\pi}{2l+1} \sum_m Y_l^m(\theta', \varphi') Y_l^{m*}(\theta, \varphi) \qquad (3.126)$$

This merely says that in (3.124) the harmonics of each order may be equated.

If we substitute for the Y's in terms of associated Legendre functions, (3.126) becomes

$$P_l(\cos \alpha) = P_l(\cos \theta) P_l(\cos \theta') + 2 \sum_{m=1}^l \frac{(l-m)!}{(l+m)!} P_l^m(\cos \theta) P_l^m(\cos \theta')$$

$$\cos m(\varphi - \varphi') \qquad (3.127)$$

Problem 3-2:   A general change of axes can be described in terms of "Euler's angles," which define the transformation from old to new axes in three steps:

(a) rotation through angle Φ about the z axis
(b) rotation through angle Θ about the new x axis
(c) rotation through angle Ψ about the position of the z axis after (b).

(*i*) Show that the first rotation results in the transformation

$$\varphi' = \varphi - \Phi$$

with θ fixed, so that

$$Y_l^m(\theta, \varphi) = e^{im\Phi} Y_l^m(\theta, \varphi')$$

(*ii*) Show that the second rotation makes the transformation between derivatives in the primed axes and those in the new (double-primed) axes as follows:

$$\partial'_+ = \tfrac{1}{2}(1 + \cos\Theta)\partial''_+ + \tfrac{1}{2}(1 - \cos\Theta)\partial''_- - i\sin\Theta\,\partial''_z$$

$$\partial'_- = \tfrac{1}{2}(1 - \cos\Theta)\partial''_+ + \tfrac{1}{2}(1 + \cos\Theta)\partial''_- + i\sin\Theta\,\partial''_z$$

$$\partial'_z = (1/2i)\sin\Theta\,(\partial''_+ - \partial''_-) + \cos\Theta\,\partial''_z$$

(*iii*) Use the above results to prove the following transformation:

$$P_2(\cos\theta) = \frac{3\cos^2\theta - 1}{2}\,P_2(\cos\theta'') + \sin\theta\cos\theta\,P_2{}^1(\cos\theta'')\sin(\varphi'' + \Psi)$$

$$-\tfrac{1}{4}\sin^2\theta\,P_2{}^2(\cos\theta'')\cos 2(\varphi'' + \Psi)$$

where (θ'', φ'') are the polar coordinates relative to the final set of axes.

10.   Potential of a Given Charge Distribution

Consider an arbitrary charge distribution $\rho(\mathbf{r})$; let us try to calculate the potential of the electrostatic field which it produces. A formula may be written down for this potential making use of a physical argument. For the charge distribution may be considered to be made up of elementary parts, the potentials from which may be added to give the required potential. Since the distribution is continuous, the summation is a continuous one, that is to say, an integral. It is, in fact,

$$\Phi(r) = \int \frac{\rho(\mathbf{r}')}{|\mathbf{r} - \mathbf{r}'|}\,d^3\mathbf{r}' \tag{3.128}$$

We may obtain the expansion of this potential in spherical harmonics by first writing

$$\frac{1}{|\mathbf{r} - \mathbf{r}'|} = \sum_{l=0}^{\infty} \frac{r_<^l}{r_>^{l+1}}\,P_l(\cos\alpha)$$

where α is the angle between $\mathbf{r}$ and $\mathbf{r}'$, and then using the addition theorem (3.126):

$$\Phi = \sum_{l,m} \frac{4\pi}{2l+1} \int \rho(\mathbf{r}') \frac{r_<^l}{r_>^{l+1}} Y_l^{m*}(\theta', \varphi') r'^2 dr' d\Omega' Y_l^m(\theta, \varphi)$$

$$= \sum_{l,m} \frac{4\pi}{2l+1} \left\{ \frac{1}{r^{l+1}} \int_{r'=0}^{r} \rho(\mathbf{r}') r'^l Y_l^{m*}(\theta', \varphi') r'^2 dr' d\Omega' \right.$$

$$\left. + r^l \int_{r'=r}^{\infty} \frac{\rho(\mathbf{r}')}{r'^{l+1}} Y_l^{m*}(\theta', \varphi') r'^2 dr' d\Omega' \right\} Y_l^m(\theta, \varphi) \qquad (3.129)$$

It should be observed that this has the general form (3.60).

If $\rho(\mathbf{r}') = 0$ for $|\mathbf{r}'| > R$, then for $r > R$ we have

$$\Phi = \sum_{l,m} \frac{4\pi}{2+1} \int \rho(\mathbf{r}') r'^l Y_l^{m*}(\theta', \varphi') d^3\mathbf{r}' \frac{Y^m(\theta, \varphi)}{r^{l+1}} \qquad (3.130)$$

The integrals are defined as the "multipole moments" Q_l^{m*} of the distribution:

$$\Phi = \sum_{l,m} \sqrt{\frac{4\pi}{2l+1}} Q_l^{m*} \frac{Y_l^m(\theta, \varphi)}{r^{l+1}} \qquad (3.131)$$

where

$$Q_l^m = \sqrt{\frac{4\pi}{2l+1}} \int \rho(\mathbf{r}') r'^l Y_l^m(\theta', \varphi') d^3\mathbf{r}' \qquad (3.132)$$

As we have seen earlier, the various terms in (3.131) are all potentials of multipole distributions.

Problem 3-3: Calculate the potential of a hydrogen atom in its ground state, which consists of a nucleus of charge $+\epsilon$ and an electron charge distribution of density

$$\rho(r) = -\epsilon \frac{1}{\pi a_0^3} e^{-2r/a_0}$$

Problem 3-4: Calculate the potential of a hydrogen atom in an excited $2p$ state, in which the electron distribution is

$$\rho(r) = -\epsilon \frac{1}{32\pi a_0^3} r^2 \cos^2 \theta \, e^{-r/a_0}$$

Problem 3-5: Calculate the multipole moments of a distribution of four charges of alternating sign $(-\epsilon, +\epsilon)$ at the corners of a square.

Problem 3-6:   The electrons in inert gas atoms have charge distributions

$$\rho = f_i(r)\,|Y_l^m(\theta,\varphi)|^2$$

there being two electrons for each (l,m). Verify that the total charge density is

$$2\sum f_i(r)$$

11.   Potential of an Axially Symmetric Charge Distribution

If the charge distribution has an axis of symmetry, and we choose the axis of symmetry as the polar axis, it will be possible to expand the charge distribution in Legendre polynomials:

$$\rho(\mathbf{r}') = \sum_{l'} \rho_{l'}(r')P_{l'}(\cos\theta')\sqrt{\frac{2l'+1}{4\pi}} \tag{3.133}$$

where

$$\rho_{l'}(r') = \sqrt{\frac{2l'+1}{4\pi}}\int \rho(\mathbf{r}')P_{l'}(\cos\theta')\,d\Omega' \tag{3.134}$$

Noting that $\sqrt{(2l+1)/4\pi}\,P_l(\cos\theta') = Y_l^0(\theta',\varphi')$, substituting (3.133) into (3.132), and using the orthogonality of the Y's, we see that

$$Q_l^m = 0, \qquad m \neq 0$$

In fact, (3.131) and (3.132) then become

$$\Phi = \sum_l Q_l^0 \sqrt{\frac{4\pi}{2l+1}}\frac{P_l(\cos\theta)}{r^{l+1}} \tag{3.135}$$

where

$$\sqrt{\frac{4\pi}{2l+1}}Q_l^0 = \int \rho(r')r'^l P_l(\cos\theta')\,d^3\mathbf{r} \tag{3.136}$$

Suppose now that by some direct means it is possible to calculate the potential on the axis of symmetry, that is, for $\cos\theta = 1$. We have seen earlier that $P_l(1) = 1$, so

$$\Phi_{\text{axis}} = \sum_l \sqrt{\frac{4\pi}{2l+1}}Q_l^0\frac{1}{r^{l+1}} \tag{3.137}$$

Thus, if the potential on the axis of symmetry is known, expansion in $1/r$ gives the multipole moments. Then, from (3.135), the potential is known everywhere; one has only to multiply each term in the expansion of Φ_{axis} in powers of $1/r$ by the appropriate Legendre polynomial.

If we designate the density of the axially symmetric distribution as $\rho(z', \xi)$, ξ being distance measured perpendicularly from the symmetry axis, it is clear from first principles that

$$\Phi = \int \frac{\rho(z', \xi) 2\pi \xi \, d\xi \, dz'}{\sqrt{(z-z')^2 + \xi^2}} \tag{3.138}$$

where, on the axis, $z = r$. (See Fig. 3.4.)

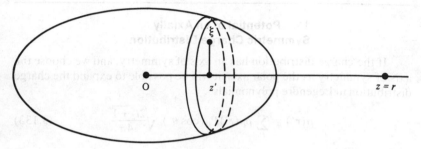

Figure 3.4

Consider a ring of radius a with a total charge ϵ; the charge per unit length is then $\epsilon/2\pi a$. If we take the origin at its center, its potential at a point distance z from the origin on the axis of symmetry is

$$\Phi = \frac{\epsilon}{\sqrt{z^2 + a^2}} \tag{3.139}$$

If $z > a$, we may expand

$$\Phi = \epsilon \sum_{n=0}^{\infty} \frac{a^{2n}}{z^{2n+1}} \binom{-1/2}{n} \tag{3.140}$$

where $\binom{-1/2}{n}$ are the binomial coefficients in the expansion of $(1+x)^{-1/2}$. The potential is then

$$\Phi = \epsilon \sum_{n=0}^{\infty} \binom{-1/2}{n} \frac{a^{2n}}{r^{2n+1}} P_{2n}(\cos \theta) \tag{3.141}$$

If, on the other hand, $z < a$, the potential becomes

$$\Phi = \epsilon \sum_{n=0}^{\infty} \binom{-1/2}{n} \frac{r^{2n}}{a^{2n+1}} P_{2n}(\cos \theta) \tag{3.142}$$

As another example, consider the potential of the charge distribution on a conducting disk of radius a and containing a charge ϵ. The charge per unit area on the surface is known to be

$$\rho(\xi) = \frac{\epsilon}{2\pi a \sqrt{a^2 - \xi^2}} \tag{3.143}$$

The potential on the axis is, if the origin is taken at the center of the disk,

$$\Phi = \frac{\epsilon}{2\pi a} \int_0^a \frac{1}{\sqrt{a^2 - \xi^2}} \frac{1}{\sqrt{\xi^2 + z^2}} 2\pi \xi \, d\xi \qquad (3.144)$$

Putting $a^2 - \xi^2 = u^2$ we get

$$\Phi = \frac{\epsilon}{a} \int_0^a \frac{1}{\sqrt{a^2 + z^2 - u^2}} \, du$$

$$= \frac{\epsilon}{a} \sin^{-1} \frac{a}{\sqrt{a^2 + z^2}}$$

$$= \frac{\epsilon}{a} \tan^{-1} \frac{a}{|z|}$$

For $z < a$

$$\Phi = \frac{\pi \epsilon}{2a} - \sum_{n=0}^{\infty} (-1)^{n+1} \frac{|z|^{2n+1}}{a^{2n+2}} \frac{1}{2n+1}$$

and for $z > a$

$$\Phi = \epsilon \sum_{n=0}^{\infty} (-1)^n \frac{a^{2n}}{|z|^{2n+1}} \frac{1}{2n+1}$$

Putting in the appropriate Legendre polynomials,

$$\Phi = \frac{\pi \epsilon}{2a} - \sum_{n=0}^{\infty} \frac{(-1)^{n+1}}{2n+1} \frac{r^{2n+1}}{a^{2n+2}} P_{2n+1} (\cos \theta) \qquad (3.145)$$

when $r < a$, and

$$\Phi = \epsilon \sum \frac{(-1)^n}{2n+1} \frac{a^{2n}}{r^{2n+1}} P_{2n} (\cos \theta) \qquad (3.146)$$

when $r > a$.

A word of caution about (3.145): although the potential is clearly symmetric on opposite sides of the disk, that is, symmetric in the transformation $\theta \rightarrow \pi - \theta$, this does not seem to be so for (3.145). Going back to the formula for the potential on the axis, however, we see that when $z \rightarrow -z$ all terms in the sum change sign. This, then, shows that, for $\theta > \pi/2$, the sign of the sum in (3.145) should be changed, so that the potential retains its symmetry.

Problem 3-7:   Calculate the *gravitational* potential of a thin disk of uniform mass σ per unit area.

Problem 3-8:   Calculate the gravitational potential of a homogeneous hemisphere of density (mass per unit volume) δ.

---

12. Potentials of Charge Distributions under Various Boundary Conditions — Green's Functions

In (3.128) the potential of a charge distribution of density $\rho(\mathbf{r})$ was given as

$$\Phi(\mathbf{r}) = \int \rho(\mathbf{r}')\frac{1}{|\mathbf{r}-\mathbf{r}'|}\,d^3\mathbf{r}'$$

This was valid in the absence of any boundary conditions; that is, for a charge distribution in an infinite empty space. The function

$$G(\mathbf{r}-\mathbf{r}') = \frac{1}{|\mathbf{r}-\mathbf{r}'|} \qquad (3.147)$$

is known as the Green's function for this problem. The Green's function is simply the potential of a point source under the boundary conditions (or lack of them) applying to the problem.

Let us consider now the potentials produced by charges when these potentials must satisfy various simple boundary conditions.

(I) PLANE BOUNDARY CONDITIONS

Suppose that we have an infinite plane conductor at $z = 0$. The potential is then subject to the boundary condition that $\Phi = 0$ on $z = 0$. Consider now a charge at $\mathbf{r}' = (x', y', z')$ where $z' > 0$. If there were an equal and opposite charge at $(x', y', -z')$, which we call the "image point," then the potential would in any case be zero on the plane $z = 0$; if, then, we took the conducting plane away, there would be no change of potential. Therefore, the potential is

$$\Phi = \frac{1}{|\mathbf{r}-\mathbf{r}'|} - \frac{1}{|\mathbf{r}-\mathbf{r}''|} \qquad (3.148)$$

where $\mathbf{r}''$ is the "image" of $\mathbf{r}'$ in $z = 0$. This is the "Green's function" for the boundary conditions of the problem.

If there is a charge distribution $\rho(\mathbf{r}')$ (lying entirely in the region $z' > 0$) its potential will then be

$$\Phi = \int \rho(\mathbf{r}')\left[\frac{1}{|\mathbf{r}-\mathbf{r}'|} - \frac{1}{|\mathbf{r}-\mathbf{r}''|}\right]d^3\mathbf{r}' \qquad (3.149)$$

If, in the second term, we introduce $-z'$ for z' as a variable, (3.149) becomes

$$\Phi = \int [\rho(\mathbf{r}') - \rho(\mathbf{r}'')]\frac{1}{|\mathbf{r}-\mathbf{r}'|}\,d^3\mathbf{r}' \qquad (3.150)$$

$\rho(\mathbf{r}'')$ is then an "image charge distribution."

Consider next the case in which the charge is placed between two parallel conducting planes, located, respectively, at $z = -a/2$ and $z = a/2$.

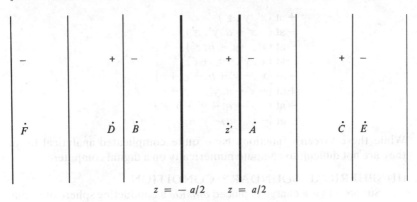

Figure 3.5

We may satisfy the boundary condition at $z = a/2$ by putting an image charge at A, where $z = a - z'$. But if we now wish to satisfy the boundary condition at $-a/2$ images *both* of the original charge and that at A must be introduced. This means a *negative* charge at $-a - z'$ (point B) and a positive one at D $(-2a + z')$. This then in turn requires further images in $z = a/2$, positive at C $(2a + z')$, and negative at E $(3a - z')$, etc. Continuing in this way, it is seen that the potential satisfying the boundary conditions is that which has positive image charges at the points

$$\mathbf{r}'_n = (x', y', z' + 2na), \qquad n = -\infty, \ldots, \infty$$

and negative ones at

$$\mathbf{r}''_n = [x', y', -z' + (2n+1)a]$$

The Green's function is then

$$G = \sum \frac{1}{|\mathbf{r} - \mathbf{r}'_n|} - \sum \frac{1}{|\mathbf{r} - \mathbf{r}''_n|} \qquad (3.151)$$

It should be noted, incidentally, that this potential shows a periodicity in z with period $2a$.

It is not essentially more difficult to deal with the case of a charge in a rectangle, with boundaries $x = \pm a/2$, $y = \pm b/2$, $z = \pm c/2$. The image system now forms a three-dimensional lattice. Taking the charge at the point (x', y', z') inside the cube, we can, by analogy with the one-dimensional case, take $2 \times 2 \times 2 = 8$ charges, four of each sign, in a cell of dimensions $2a \times 2b \times 2c$, and then repeat each cell by periodicity, the periods in the x, y, and z directions being $2a$, $2b$, and $2c$. The eight basic charges, with their signs, are

$$+ \text{at } (x', y', z')$$
$$- \text{at } (-x' + a, y', z')$$
$$- \text{at } (x', -y' + b, z')$$
$$- \text{at } (x', y', -z' + c)$$
$$+ \text{at } (x', -y' + b, -z' + c)$$
$$+ \text{at } (-x' + a, y', -z' + c)$$
$$+ \text{at } (-x' + a, -y' + b, z')$$
$$- \text{at } (-x' + a, -y' + b, -z' + c)$$

While these Green's functions have quite complicated analytical form, they are not difficult to compute numerically on a digital computer.

(II) SPHERICAL BOUNDARY CONDITION

Suppose that a charge is placed *outside* a conducting sphere of radius a. Let us first consider such a charge on the polar axis (or rather, *choose* the polar axis through the charge).

Going back to (3.65), the potential of a unit charge at r' ($> a$) at a point r ($< r'$ but $> a$) is

$$\Phi_1 = \sum \frac{r^l}{r'^{l+1}} P_l (\cos \theta) \tag{3.152}$$

Suppose that we take an image charge of strength $-\alpha$ at the point r'' ($< a$). Its potential is

$$\Phi_2 = -\alpha \sum \frac{r''^l}{r^{l+1}} P_l (\cos \theta) \tag{3.153}$$

These can give a net potential zero for $r = a$ if, for all l's,

$$\frac{a^l}{r'^{l+1}} = \alpha \frac{r''^l}{a^{l+1}} \tag{3.154}$$

This is clearly possible if $\alpha = a/r'$ and $r'' = a^2/r'$.

Problem 3-9:   Outside a grounded sphere of radius a, there is placed a line containing charge σ per unit length extending radially from $r = b$ to $r = c$. What is the net potential (a) outside the sphere and (b) inside the sphere?

Answer:

$$\sigma \left[\sinh^{-1} \frac{c - r \cos \theta}{r \sin \theta} - \sinh^{-1} \frac{b - r \cos \theta}{r \sin \theta} - \frac{a}{r} \sinh^{-1} \frac{cr - a^2 \cos \theta}{a^2 \sin \theta} \right.$$
$$\left. + \frac{a}{r} \sinh^{-1} \frac{br - a^2 \cos \theta}{a^2 \sin \theta} \right]$$

outside and zero inside. This may be written alternatively as

$$\Phi = \sigma \left[\ln \left| \frac{c - r \cos \theta + \sqrt{c^2 + r^2 - 2cr \cos \theta}}{b - r \cos \theta + \sqrt{b^2 + r^2 - 2br \cos \theta}} \right| \right.$$

$$\left. - \frac{a}{r} \ln \frac{cr - a^2 \cos \theta + \sqrt{c^2 r^2 + a^4 - 2cra^2 \cos \theta}}{br - a^2 \cos \theta + \sqrt{b^2 r^2 + a^4 - 2bra^2 \cos \theta}} \right]$$

Observing now the physical content of this result, we see that the boundary condition (zero potential on the sphere $r = a$) is satisfied if each charge e outside of the sphere at coordinates (r', θ') is matched by an image charge $-(a/r')e$ inside with coordinates $(a^2/r', \theta')$. If, on the other hand, the potential of the sphere is a constant Φ_0, we must add to the potential of the two charges a term $\Phi_0 a/r$.

It is then possible to write down the "Green's function" corresponding to these boundary conditions; it is

$$G_s(\mathbf{r}, \mathbf{r}') = \frac{1}{|\mathbf{r} - \mathbf{r}'|} - \frac{a}{r'} \frac{1}{|\mathbf{r} - (a^2/r'^2)\mathbf{r}'|} + \Phi_0 \frac{a}{r} \qquad (3.155)$$

The potential due to a charge distribution $\rho(\mathbf{r}')$ outside the sphere is then

$$\Phi(\mathbf{r}) = \int \rho(\mathbf{r}') G(\mathbf{r}, \mathbf{r}') \, d^3\mathbf{r}' \qquad (3.156)$$

when r is also outside the sphere.

From (3.128), we know that the potential due to a charge distribution $\rho(\mathbf{r})$ can be written

$$\Phi(\mathbf{r}) = \sum_{l=0}^{\infty} \int \frac{r_<^l}{r_>^{l+1}} \rho(\mathbf{r}') P_l(\cos \alpha) \, d^3\mathbf{r}' \qquad (3.157)$$

α being the angle between the vectors $\mathbf{r}$ and $\mathbf{r}'$, and $r_<, r_>$ are, respectively, the lesser and greater of r and r'. Let us make, in this formula, the substitutions

$$\mathbf{r}_1 = \frac{a^2}{r^2} \mathbf{r}, \qquad \mathbf{r}_1' = \frac{a^2}{r'^2} \mathbf{r}'; \qquad (3.158)$$

that is, let us write (3.157) in terms of the image points of $\mathbf{r}$ and $\mathbf{r}'$ in the sphere $r = a$. Since the angle between $\mathbf{r}_1, \mathbf{r}_1'$ is the same as that between $\mathbf{r}$ and $\mathbf{r}'$, $\cos \alpha$ can also relate to the former pair of vectors. The substitution (3.158) into (3.157) gives

$$\Phi \left(\frac{a^2}{r_1}, \Omega \right) = \sum_{l=0}^{\infty} \int \frac{1}{a^2} \frac{r_{1<}^{l+1}}{r_{1>}^l} \rho \left(\frac{a^2}{r_<'}, \Omega' \right) P_l(\cos \alpha) \frac{a^6}{r_1'^6} d^3\mathbf{r}_1' \qquad (3.159)$$

where Ω, Ω' stand for the polar angles of $\mathbf{r}$ and $\mathbf{r}'$ respectively. This may be written

$$\Phi \left(\frac{a^2}{r_1}, \Omega \right) = \frac{r_1}{a} \sum_{l=0}^{\infty} \int \frac{r_{1<}^l}{r_{1>}^{l+1}} \rho \left(\frac{a^2}{r_1'}, \Omega' \right) \left(\frac{a}{r_1'} \right)^5 P_l(\cos \alpha) \, d^3\mathbf{r}' \qquad (3.160)$$

From this it follows that

$$\frac{a}{r_1} \Phi\left(\frac{a^2}{r_1}, \Omega\right) = \Phi_1$$

is the potential generated by the charge distribution

$$\rho_1 = \left(\frac{a}{r}\right)^5 \rho\left(\frac{a^2}{r}, \Omega\right) \tag{3.161}$$

Thus, in general, if $\Phi(r)$ is a solution of the potential equation, so is

$$\Phi_1 = \frac{a}{r} \Phi\left(\frac{a^2}{r^2} \mathbf{r}\right) \tag{3.162}$$

Given any potential, inversion in a sphere may then be used to generate another potential; the relation between the corresponding charge distributions is given by (3.161).

Problem 3-10: Show that the Green's function for the region lying *outside* a hemisphere of radius a is the sum of infinite-medium Green's functions corresponding to sources at radius r_0 and at the three image points: (*i*) the image of r_0 in the full sphere and (*ii*) the negative images of the two preceding sources in the plane bounding the hemisphere.

(III) CHARGE ON CONDUCTING SURFACES

It is worth noting that, from a knowledge of the potential, it is easy to find the charge distribution on the surface. For let us apply Gauss' theorem

$$\int \nabla \cdot \mathbf{D} \, dV = \int \mathbf{D} \cdot \mathbf{n} \, dS \tag{3.163}$$

to a volume shaped like a "pillbox" bounding an area dS on the surface, and bounded by normals to the surface around the circumference of dS and by surfaces parallel to dS and a vanishingly small distance on each side of it (see Fig. 3.6). The theorem converts the integral of $\nabla \cdot \mathbf{D}$ over the enclosed volume, into an integral over its surface. The edge contributions can be made arbitrarily small. The surface integral is then

$$\langle \mathbf{D} \cdot \mathbf{n} \rangle_+ \, dS - \langle \mathbf{D} \cdot \mathbf{n} \rangle_- \, dS$$

where $\langle \quad \rangle$ designates an average over dS, and $+$ and $-$ denote the surfaces in the direction $\mathbf{n}$ and opposite to it, respectively. But by (3.14) the volume integral is $\langle \sigma \rangle \, dS$ where σ is the charge per unit surface area. Thus, letting $dS \to 0$,

$$\sigma = \langle \mathbf{D} \cdot \mathbf{n} \rangle_+ - \langle \mathbf{D} \cdot \mathbf{n} \rangle_- \tag{3.164}$$

Since, by (3.18) and (3.22)

$$\mathbf{D} = -\frac{1}{4\pi} \nabla \Phi \tag{3.165}$$

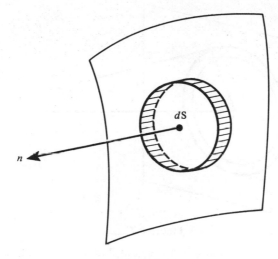

Figure 3.6

it follows that

$$\sigma = -\frac{1}{4\pi}\left[\left(\frac{\partial \Phi}{\partial n}\right)_+ - \left(\frac{\partial \Phi}{\partial n}\right)_-\right] \qquad (3.166)$$

where $\partial \Phi/\partial n$ is the normal derivative $\mathbf{n} \cdot \nabla \Phi$.

Problem 3-11:   A charge e is placed at a distance a from an infinite conducting plane kept at zero potential. Find the surface charge distribution on the plane.

Problem 3-12:   Using (3.153), show that, if a charge e is placed at a point with $r = b > a$ and $\theta = 0$ outside a sphere of radius a, the charge distribution on the surface is

$$\frac{e}{4\pi}\left\{\frac{1}{a^2} - \frac{(b^2 - a^2)\mu}{b(a^2 + b^2 - 2ab\mu)^{3/2}}\right\}$$

where μ is the cosine of the polar angle θ of the point on the sphere.

(IV) SOME RELATIONS BETWEEN GREEN'S FUNCTIONS

In this section we obtain relationships between the following Green's functions:

(1) That of a point source of unit strength.

(2) That of a uniform laminar source of unit strength per unit area.

(3) That of a spherically symmetric shell source of unit strength.

These relationships are not restricted to potential problems, but apply

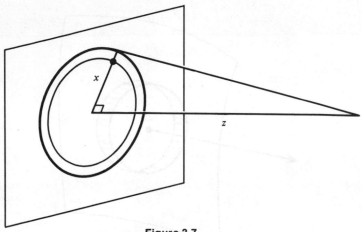

Figure 3.7

equally to all problems in which the contributions from different sources are *additive*, that is, all linear problems.

It is clear that if the solution for a point source is known, that for a uniform laminar source may be obtained by integration. Designating the point source solution by $\Phi_{pt}(r)$ and that of a laminar source by $\Phi_{pl}(z)$ (the source being at $z = 0$), Fig. 3.7 shows that

$$\Phi_{pl}(z) = \int_0^\infty \Phi_{pt}\left(\sqrt{x^2+z^2}\right) 2\pi x \, dx$$

Putting, in the integral, $r = \sqrt{x^2+z^2}$, the equation becomes

$$\Phi_{pl}(z) = \int_z^\infty \Phi_{pt}(r) 2\pi r \, dr \tag{3.167}$$

or

$$\Phi_{pt}(r) = -\left[\frac{1}{2\pi z}\frac{\partial \Phi_{pl}(z)}{\partial z}\right]_{z=r} \tag{3.168}$$

In the form (3.168) the theorem is particularly useful, since it shows that *the solution of a one-dimensional problem provides the basis for solving the most general three-dimensional problem.*

A trivial example is the potential problem, for there the one-dimensional homogeneous equation is

$$\frac{\partial^2 \Phi}{\partial z^2} = 0$$

which gives $\Phi_{pl} = A + Bz$. Equation (3.166) gives a boundary condition for unit surface charge:

$$1 = -\frac{2B}{4\pi}$$

so that $B = -2\pi$. Equation (3.168) then gives

$$\Phi_{pt}(r) = \frac{1}{r}$$

as expected.

It is further evident that the point source solution may be used to generate the shell source one. Simple geometry shows, on reference to Fig. 3.8, that

$$PQ = \sqrt{r^2 + r'^2 - 2rx}$$

where x is the distance OR. The area of the shaded strip is $2\pi r' dx$, and its source strength is $(1/4\pi r'^2)2\pi r' \, dx = (1/2r')dx$. Thus, the solution due to the shell source is

$$\Phi_{sh}(r) = \frac{1}{2r'} \int_{-r'}^{r'} \Phi_{pt}\left(\sqrt{r^2 + r'^2 - 2rx}\right) dx$$

Putting

$$r^2 + r'^2 - 2rx = \xi^2$$

this may be written as

$$\Phi_{sh}(r) = \frac{1}{2rr'} \int_{|r-r'|}^{r+r'} \Phi_{pt}(\xi)\xi \, d\xi \tag{3.169}$$

Reference to (3.167) shows that this can be put in the form

$$\Phi_{sh}(r) = \frac{1}{4\pi rr'}[-\Phi_{pl}(r+r') + \Phi_{pl}(|r-r'|)] \tag{3.170}$$

With this result, problems of spherical symmetry can also be obtained from the one-dimensional Green's function.

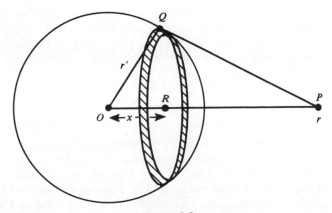

Figure 3.8

Problem 3-13:   The Yukawa equation is

$$\nabla^2 \psi - \mu^2 \psi = 0$$

If the source strength of a unit source at the origin is $-4\pi g \delta(\mathbf{r})$, show by first solving the one-dimensional equation that

$$\psi_{pl}(z) = \frac{2\pi g}{\mu} e^{-\mu z}$$

$$\psi_{pt}(r) = g \frac{e^{-\mu r}}{r}$$

$$\psi_{sh}(r) = \frac{g}{2\mu r r'} [e^{-\mu |r - r'|} - e^{-\mu(r + r')}]$$

Problem 3-14:   Show that the solution of the Yukawa potential for a uniform source of strength n per unit volume inside the sphere $r = a$ is

$$4\pi g n \frac{e^{-\mu r}}{\mu r} \left[\frac{a}{\mu} \cosh \mu a - \frac{1}{\mu^2} \sinh \mu a \right]$$

for $r > a$. Find the form for $r < a$.

---

13. Further Problems Involving the Potential Equation

In this section we propose several boundary value problems involving potentials or the potential equation, and indicate, without providing all the details of calculation, how they may be solved. The completion of the calculations will give the reader valuable practice in the use of the techniques of this chapter.

Problem 3-15:   Consider a conducting plate of large extent (so that edge effects can be neglected), from which is cut a circular hole of radius a. (See Fig. 3.9.) In this hole is attached a hemispherical conductor, also of radius a. The plane is charged so that, far from the sphere, the charge per unit area has a constant value σ_0. Find the potential, and the charge density on and near the sphere.

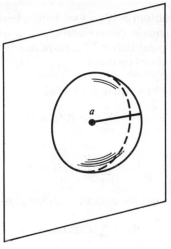

Figure 3.9

If we take the plate to lie in the plane $z = 0$, and the polar axis through the center of the sphere and perpendicular to it, the solution can be written, for $r > a$,

$$\Phi = \sum_{l=0}^{\infty} \left(A_l r^l + \frac{B_l}{r^{l+1}} \right) P_l (\cos \theta)$$

(a) If the condition is applied that the surface charge density has the *constant* value σ_0 far from the hole, it is seen that the only nonzero A coefficient is A_1, and that it must be

$$A_1 = -2\pi\sigma_0 \qquad (3.171)$$

(b) From the condition that $\Phi = 0$ for $r = a$ it follows that

$$B_1 = 2\pi\sigma_0 a^3, \qquad B_l = 0, \qquad l \neq 1 \qquad (3.172)$$

Thus,

$$\Phi = -2\pi\sigma_0 \left(r - \frac{a^3}{r^2} \right) \cos \theta$$

(c) The general form of the charge density on the plate is then

$$\sigma = \sigma_0 \left(1 - \frac{a^3}{r^3} \right) \qquad (3.173)$$

(d) The charge density on the sphere is

$$\sigma = \tfrac{3}{2}\sigma_0 \cos \theta \qquad (3.174)$$

Problem 3-16: Two hemispheres of radius a, separated around their common circumference by a negligibly thin strip of insulator, are kept at potentials $\pm\frac{1}{2}V_0$, respectively. Find their potentials and the charge distributions on them.

The potential may be written, for $r < a$,

$$\Phi_i = \sum_{l=0}^{\infty} A_l r^l P_l (\cos\theta) \tag{3.175}$$

and for $r > a$

$$\Phi_0 = \sum_{l=0}^{\infty} \frac{B_l}{r^{l+1}} P_l (\cos\theta) \tag{3.176}$$

where θ is measured from the axis of symmetry. At $r = a$ they reduce to the form

$$\Phi_a = \sum C_l P_l (\cos\theta) \tag{3.177}$$

where

$$A_l = \frac{C_l}{a^l}, \qquad B_l = C_l a^{l+1} \tag{3.178}$$

To find C_l, multiply (3.177) by $P_n (\cos\theta)$ and integrate from 0 to π.

(a) Using Eqs. (3.69) and (3.71) the nonzero C's are

$$C_{2s+1} = \frac{4s+3}{2} V_0 \binom{1/2}{s+1} \tag{3.179}$$

A few values are

$$C_1 = \tfrac{3}{4}V_0, \qquad C_3 = -\tfrac{7}{16}V_0, \qquad C_5 = \tfrac{11}{32}V_0, \qquad C_7 = -\tfrac{75}{256}V_0$$

(b) The charge density on the surface is

$$\sigma = -\frac{V_0}{2\pi a} \sum_{l=0}^{\infty} (l+1)(4l+3)\binom{1/2}{l+1} P_{2l+1}(\mu) \tag{3.180}$$

(c) The field at the center ($r = 0$) is

$$E = -\frac{1}{4\pi\epsilon} \frac{3V_0}{4a} \tag{3.181}$$

(d) The dipole moment of the system is

$$Q_1^{\,0} = \tfrac{3}{4}V_0 a^2 \tag{3.182}$$

Problem 3-17: In a conducting material there is a spherical hole of radius a. Far from the hole a uniform current J_0 is flowing in the z direction. Find the current at each point.

The current is

$$\mathbf{j} = \sigma \mathbf{E} = -\frac{\sigma}{4\pi\epsilon} \nabla \Phi \tag{3.183}$$

where σ is the conductivity of the material. The boundary condition of the problem is

$$\frac{\partial \Phi}{\partial r} = 0 \qquad \text{at } r = a \tag{3.184}$$

(a) At large distances

$$\Phi \rightarrow -\frac{4\pi\epsilon}{\sigma} J_0 r \cos \theta \tag{3.185}$$

The potential must therefore have the form

$$\Phi = -\frac{4\pi\epsilon}{\sigma} J_0 \left(r \cos \theta + \frac{\alpha \cos \theta}{r^2} \right) \tag{3.186}$$

Applying the boundary condition it is found that

$$\Phi = -\frac{4\pi\epsilon}{\sigma} J_0 \left(r + \frac{a^3}{2r^2} \right) \cos \theta \tag{3.187}$$

(b) The current in general has components as follows:
Radial

$$j_r = J_0 \cos \theta \left(1 - \frac{a^3}{r^3} \right) \tag{3.188}$$

Transverse

$$j_\theta = -J_0 \sin \theta \left(1 + \frac{a^3}{2r^3} \right) \tag{3.189}$$

(c) The electric field in the hole is constant and equal to

$$E = \frac{3}{2} \frac{J_0}{\sigma} \tag{3.190}$$

(d) A further problem which may be solved is to find the magnetic field **B** generated by the currents (3.188) and (3.189). Since the leading terms in these equations simply describe a constant current in the z direction, by calculating with the second terms only we can calculate the *perturbation* by the hole of the field of a constant current.

The equation determining the vector potential of the field is, by (2.25),

$$\nabla^2 \mathbf{A} = -\mu \mathbf{J}$$

This equation is readily solved for

$$J_r = -J_0 \cos \theta \frac{a^3}{r^3}, \qquad J_\theta = -J_0 \sin \theta \frac{a^3}{2r^3}, \qquad r > a$$

and $\mathbf{j} = 0$ for $r < a$.

The potentials come out to be: for $r < a$

$$A_r = -\tfrac{1}{6}\mu J_0 ar \cos \theta \qquad (3.191)$$

$$A_\theta = -\tfrac{1}{12}\mu J_0 ar \sin \theta \qquad (3.192)$$

and for $r > a$

$$A_r = -\mu J_0 \frac{a^3}{r^2} \cos \theta \left(\tfrac{1}{2}r - \tfrac{1}{3}a\right) \qquad (3.193)$$

$$A_\theta = -\mu J_0 \frac{a^3}{r^2} \sin \theta \left(\tfrac{1}{4}r - \tfrac{1}{6}a\right) \qquad (3.194)$$

From the equation

$$\mathbf{B} = \nabla \times \mathbf{A}$$

it is now possible to calculate B. The only nonzero component is B_φ, and it is found to be, for $r < a$

$$B_\phi = \tfrac{1}{3}\mu J_0 a \sin \theta \qquad (3.195)$$

and for $r > a$

$$B_\phi = \tfrac{1}{2}\mu J_0 \frac{a^3}{r^3} \left(r - \tfrac{1}{3}a\right) \sin \theta \qquad (3.196)$$

Problem 3-18: Calculate the potential of a charge distribution uniformly distributed over the interior of the spheroid

$$\frac{x^2 + y^2}{a^2} + \frac{z^2}{b^2} = 1 \qquad (3.197)$$

when the eccentricity is very small. To a good approximation, in this case

$$b^2 = a^2 (1 \pm \epsilon^2)$$

according as the ellipsoid is prolate or oblate.

The starting point is the formula

$$\Phi(\mathbf{r}) = \int \frac{\rho(\mathbf{r}')}{|\mathbf{r} - \mathbf{r}'|} d^3\mathbf{r}' = \rho_0 \int_{\text{inside}} \frac{1}{|\mathbf{r} - \mathbf{r}'|} d^3\mathbf{r}' \qquad (3.198)$$

It is useful to introduce the substitutions:

$$x' = R' a \sin \theta' \cos \phi'$$
$$y' = R' a \sin \theta' \sin \phi' \qquad (3.199)$$
$$z' = R' b \cos \theta'$$

and similar ones, dropping the primes, for r.

(a) It may be shown that, to the first order in ϵ^2,

$$\Phi = \rho_0 ab \int_{R' < 1} \left[\frac{1}{|\mathbf{R} - \mathbf{R}'|} \mp \frac{1}{2} \epsilon^2 \frac{(z - z')^2}{|\mathbf{R} - \mathbf{R}'|^3} \right] d^3\mathbf{R}' \qquad (3.200)$$

where

$$z = R \cos \theta, \qquad z' = R' \cos \theta' \tag{3.201}$$

From a straightforward calculation, the integral of the first term is found to be, for $R < 1$ (inside the spheroid)

$$2\pi(1 - \tfrac{1}{3}R^2) \tag{3.202}$$

and for $R > 1$ (outside the spheroid)

$$\frac{4\pi}{3R} \tag{3.203}$$

Consider next the second term; omitting the factor $\mp\tfrac{1}{2}\epsilon^2$ the integral is

$$\int_{R'<1} \frac{(z-z')^2}{|\mathbf{R}-\mathbf{R}'|^3} \, d^3\mathbf{R}' = \int_{R'<1} \left[\frac{\partial}{\partial z'} \frac{z-z'}{|\mathbf{R}-\mathbf{R}'|} + \frac{1}{|\mathbf{R}-\mathbf{R}'|} \right] d^3\mathbf{R}'$$

The second term in this integral combines with the one already calculated, so that it may be taken into account by multiplying the results above by $(1 \mp \tfrac{1}{2}\epsilon^2)$. The first term may be changed into a surface integral over $R' = 1$.

$$\int_{R'=1} \frac{z-z'}{|\mathbf{R}-\mathbf{R}'|} \frac{z'}{R'} \, dS' = \int \cos\theta' \, (R\cos\theta - \cos\theta')$$

$$\sum_{l,m} \frac{R_<^l}{R_>^{l+1}} \frac{4\pi}{2l+1} Y_l{}^m(\theta',\varphi') Y_l{}^{m*}(\theta,\varphi) \, d\Omega' \tag{3.204}$$

where the integral is over the solid angles of R'. The $m \neq 0$ terms drop out, and, from (3.117),

$$\frac{4\pi}{2l+1} Y_l{}^0(\theta',\varphi') Y_l{}^{0*}(\theta,\varphi) = P_l(\cos\theta) P_l(\cos\theta') \tag{3.205}$$

Using the orthogonality of Legendre polynomials it is now possible to evaluate (3.204). Only the $l = 0$, 1, and 2 terms in the sum contribute. After some calculation the results may be written:

(b) For $R < 1$

$$\frac{4\pi}{3} \left[\frac{4}{15} R^2 P_2(\cos\theta) + \frac{1}{3} R^2 - 1 \right] \tag{3.206}$$

and for R > 1

$$\frac{4\pi}{3R} \left[\left(\frac{2}{3} - \frac{2}{5R^2} \right) P_2(\cos\theta) - \frac{2}{3} \right] \tag{3.207}$$

(c) The various terms may now be combined. The total charge may be written.

$$q = \tfrac{4}{3}\pi\rho_0 a^2 b \tag{3.208}$$

The potentials come out to be, for $R < 1$

$$\Phi = \frac{q}{a}\left\{\frac{3}{2}\left(1-\frac{1}{3}R^2\right)\mp\epsilon^2\left[\frac{1}{2}-\frac{1}{6}R^2+\frac{2}{15}R^2 P_2\left(\cos\theta\right)\right]\right\} \quad (3.209)$$

and for $R > 1$

$$\Phi = \frac{q}{aR}\left\{1\mp\epsilon^2\left[\frac{1}{6}+\left(\frac{1}{3}-\frac{1}{5R^2}\right)P_2\left(\cos\theta\right)\right]\right\} \quad (3.210)$$

where

$$R^2 = \frac{x^2+y^2}{a^2}+\frac{z^2}{b^2} \quad (3.211)$$

In the terms of order ϵ^2, we may put $R = r/a$ without changing the result to order ϵ^2.

(d) The quadrupole moment of the distribution is

$$Q_2{}^0 = \pm\tfrac{1}{5}q\epsilon^2 a^2$$
$$= \tfrac{1}{5}q(b^2-a^2) \quad (3.212)$$

Problem 3-19: Find the potential of a uniformly charged circular disk of radius a and negligible thickness, carrying total charge Q. What are its multipole moments?

Problem 3-20:   Consider two identical rings of radius a and of negligible cross section separated by a distance b, the line joining their centers being perpendicular to the plane of each. Find the potentials and multipole moments when they carry equal charges q, and when they carry equal and opposite charges q.

Problem 3-21:   Find the potential of a charge distribution of quadrupole moment Q_2 distributed over the surface of a sphere of radius a, its angular variation being proportional to $P_2 \left(\cos\theta\right)$, θ being measured from the symmetry axis of the distribution.

Problem 3-22:   Consider two disks like those described in Problem 3-19, oppositely charged and at distance b, the line joining their centers being perpendicular to each. Find the potential as $q \to \infty$ and $b \to 0$ such that $\lim qb = \pi a^2\sigma$. (We call this a "dipole layer.") Find its multipole moments.

Problem 3-23:   Calculate the gravitational potential of a sphere of radius a with an eccentric spherical hole of radius b, whose center is at radius c. $(b+c < a)$.

APPENDIX 3A: A Recursion Relation for $P_l^m(\cos\theta)$

We start with the recursion relation (2.84) for the Legendre polynomial of order $(l-m)$:

$$(2l-2m+1)\mu P_{l-m} = (l-m+1)P_{l-m+1} + (l-m)P_{l-m-1}$$

Substituting the definitions of the P's as derivatives of $1/r$ with respect to z, and multiplying through by $(l-m)!(-1)^{l-m}/r^{l-m}$, we obtain the equation

$$-(2l-2m+1)z\partial_z^{l-m}\left(\frac{1}{r}\right) = r^2\partial_z^{l-m+1}\left(\frac{1}{r}\right) + (l-m)^2\partial_z^{l-m-1}\left(\frac{1}{r}\right)$$

We now operate on this with ∂_+^m to get

$$-(2l-2m+1)z\partial_+^m\partial_z^{l-m}\left(\frac{1}{r}\right) = (l-m)^2\partial_+^m\partial_z^{l-m-1}\left(\frac{1}{r}\right)$$

$$+r^2\partial_+^m\partial_z^{l-m+1}\left(\frac{1}{r}\right) + 2m(x+iy)\partial_+^{m-1}\partial_z^{l-m+1}\left(\frac{1}{r}\right)$$

Now $(x+iy)$ passes through ∂_+ operators, and

$$(x+iy)\partial_z\left(\frac{1}{r}\right) = z\partial_+\left(\frac{1}{r}\right)$$

so that the last term becomes

$$2m\partial_+^m\partial_z^{l-m}\left(\frac{z}{r}\right) = 2mz\partial_+^m\partial_z^{l-m}\left(\frac{1}{r}\right) + 2m(l-m)\partial_+^m\partial_z^{l-m-1}\left(\frac{1}{r}\right)$$

on using Leibnitz's rule for differentiation of the product (z times $1/r$).

Putting this back in the last term of the equation above, and using (3.107) to write the derivatives in terms of the associated Legendre functions, we get

$$(2l+1)\mu(-1)^l(l-m)!\frac{1}{r^l}P_l^m = r^2(-1)^{l+1}(l-m+1)!\frac{1}{r^{l+2}}P_{l+1}^m$$

$$+ (l^2-m^2)(-1)^{l-1}(l-m-1)!\frac{1}{r^l}P_{l-1}^m$$

Dividing through by $(-1)^l(l-m)!/r^l$, we arrive at the required recursion relation,

$$(2l+1)\mu P_l^m = (l-m+1)P_{l+1}^m + (l+m)P_{l-1}^m$$

APPENDIX 3B: Review of Theory of Linear Differential Equations of the Second Order

Consider the equation

$$p_0(x)y'' + p_1(x)y' + p_2(x)y = 0 \tag{B.1}$$

THEOREM 1

If we know one solution, $y_1(x)$, we can find a second, y_2.
If we have two solutions, y_1 and y_2, then

$$p_0 y_1'' + p_1 y_1' + p_2 y_1 = 0 \tag{B.2}$$

$$p_0 y_2'' + p_1 y_2' + p_2 y_2 = 0 \tag{B.3}$$

Let us multiply the first equation by y_2, the second by y_1, and subtract. We define the "Wronskian" W of two solutions as

$$\begin{vmatrix} y_1 & y_2 \\ y_1' & y_2' \end{vmatrix} = y_1 y_2' - y_2 y_1' \tag{B.4}$$

Then we obtain

$$p_0 W' + p_1 W = 0 \tag{B.5}$$

This may be solved to give

$$W = e^{-\int_{x_0}^{x} \frac{p_1(x')}{p_0(x')} dx'} \tag{B.6}$$

Once W is known, y_2 can be readily expressed in terms of y_1. For if we divide the equation

$$y_1 y_2' - y_2 y_1' = W$$

by y_1^2, we get

$$\frac{d}{dx}\left(\frac{y_2}{y_1}\right) = \frac{W}{y_1^2}$$

so that

$$y_2 = y_1 \int_{x_1}^{x} \frac{W(x')}{y_1^2(x')} dx' \tag{B.7}$$

x_1 being an arbitrary constant of integration.

THEOREM 2

If we know the solution of the homogeneous equation (B.1) we can find the solution of the inhomogeneous equation

$$p_0 y'' + p_1 y' + p_2 y = r(x) \tag{B.8}$$

First, we take this equation along with (B.1)

$$p_0 y_1'' + p_1 y_1' + p_2 y_1 = 0$$

We multiply each by the solution of the other, and subtract to get

$$p_0(y_1 y'' - yy_1'') + p_1(y_1 y' - yy_1') = r(x)y_1(x) \tag{B.9}$$

or

$$p_0 S' + p_1 S = ry \tag{B.9a}$$

where $S = y_1 y' - yy_1'$.

Next, we combine this equation with that for W, (B.5)

$$p_0 W' + p_1 W = 0$$

Again, we multiply each by the solution of the other and subtract:

$$p_0(WS' - SW') = ry_1 W \tag{B.10}$$

Dividing by W^2, this becomes

$$\frac{d}{dx}\left(\frac{S}{W}\right) = \frac{ry_1}{p_0 W} \tag{B.11}$$

On integration, this leads to

$$S = W \int^x \frac{r(x')y_1(x')}{p_0(x')W(x')} dx'$$

Substituting for S and dividing by y_1^2 gives

$$\frac{d}{dx}\left(\frac{y}{y_1}\right) = \frac{W}{y_1^2} \int^x \frac{r(x')y_1(x')}{p_0(x')W(x')} dx'$$

$$= \frac{d}{dx}\left(\frac{y_2}{y_1}\right) \int^x \frac{r(x')y_1(x')}{p_0(x')W(x')} dx' \tag{B.12}$$

We now integrate this equation, integrating the right-hand side by parts

$$\frac{y}{y_1} = \frac{y_2}{y_1} \int^x \frac{r(x')y_1(x')}{p_0(x')W(x')} dx' - \int^x \frac{y_2(x')r(x')}{p_0(x')W(x')} dx'$$

or

$$y = \int^x \frac{r(x')[y_2(x)y_1(x') - y_1(x)y_2(x')]}{p_0(x')W(x')} dx' \tag{B.13}$$

The choice of lower limit in the integrals is of no importance, since the contributions from the lower limit are simply multiples of y_2 and y_1, respectively. But we know, of course, that we may add to any particular solution of the inhomogeneous equation an arbitrary solution of the homogeneous equation; that is, an arbitrary linear combination of y_1 and y_2. The correct combination will be determined in a particular problem by the given boundary conditions.

Equation (B.13) may be written

$$y(x) = \int^x r(x')G(x, x') dx' \tag{B.14}$$

where

$$G(x, x') = \frac{y_2(x)y_1(x') - y_1(x)y_2(x')}{p_0(x')W(x')} \tag{B.15}$$

is the "Green's function" of the problem.

In general, the choice of constants in the solutions y_1 and y_2 enables us to find Green's functions satisfying any prescribed boundary conditions.

PRELUDE TO CHAPTER 4

This chapter is devoted to the very powerful methods of Fourier and Laplace transforms.

The Fourier transform in one dimension (Fourier integral) is first introduced as a limit of a Fourier series, and some simple properties and theorems derived (Parseval's theorem, the convolution theorem). As a simple example of the use of Fourier transforms, we solve the problem of the forced, damped, harmonic oscillator. It is shown that the solution is a convolution of the forcing function and a Green's function (corresponding to a unit impulse).

It is shown that, if there is no damping, permissible mathematical solutions are time-reversal invariant, and so may violate the physical condition of causality. This difficulty is removed by any irreversible process (for example, damping). For *causal* functions, relations are obtained between the real and imaginary parts of the solution (Kramers–Kronig dispersion relations). These relations have important applications in electrodynamics and elsewhere.

The next topic considered is that of correlation functions in time-evolving phenomena (both cross correlation, that is, the correlation between *different* quantities at different times, and autocorrelation, that is, correlation between values of a *given* quantity at different times). Defining the "power spectrum" of a time function as the magnitude squared of the Fourier transform of the function, it is shown that the power spectrum and the autocorrelation function are Fourier transforms one of the other. This will be used in Chapter 7 in connection with random signals, which are not well-defined and whose power spectrum can therefore not be calculated directly, but for which the autocorrelation function is directly attainable.

We next develop the theory of Fourier transforms in three dimensions. These will be used in a wide variety of problems in subsequent chapters. The basic properties needed for their use are derived.

As an application, it is shown how Fourier transforms may be used to solve the Poisson equation of the previous chapter, and to determine Green's functions. An interesting application of Fourier transforms is the proof of the Poisson summation formula, in both one and three dimensions. This formula makes possible a transformation from one infinite series involving a parameter to another. The relation between them is, that in the range of values of the parameter for which one series converges slowly, the other converges rapidly, and vice versa. In three dimensions the original series involves the periodicity of a lattice. The transformed form then involves the *inverse* lattice, exactly as defined in electron theory of solids. In fact, as in the solid-state problem what is involved is the Fourier transform of a function periodic in three dimensions.

As an application of the Poisson formula, we calculate approximately the energy of a simple model of an ionic crystal, consisting of a periodic array of charges of alternating sign. This involves the calculation of the so-called "Madelung constant."

After a brief note on three-dimensional δ functions, a further application is considered: the use of Fourier transforms to evaluate certain two- and three-center integrals occurring in atomic and molecular physics.

We next develop the basic theory of Laplace transforms, and fundamental theorems for such transforms—shifting theorem, convolution theorem. By calculating some simple transforms (of powers, exponentials, and trigonometric functions) we see that we may determine inverse transforms of all rational functions of the transform variable.

In order to calculate inverse transforms in general, we make the connection between Laplace and Fourier transforms. The Laplace inverse theorem may then be derived from its Fourier counterpart.

There follows a discussion of the Laplace transform of periodic functions, and the treatment of resonances in the response of second-order linear systems.

As a further exercise in the combined use of

Fourier and Laplace transforms, we use them to re-solve the vibrating string problem of Chapter 1.

The last part of the chapter is devoted to the discussion of special functions by transform methods. Basic properties of Γ and B functions are derived, and Stirling's approximate formula for the function of large argument, which will be used extensively in problems of probability in Chapter 7, is derived. This is done by the "method of steepest descents," the idea of which is first introduced here but which will be used again in Chapter 5 (Appendix).

It is then shown that Laplace transforms make possible the solution of the general linear differential equation with linear coefficients. The most important example of such an equation is the confluent hypergeometric equation, the theory of which is treated in some detail. Particular functions which can be expressed in terms of the confluent hypergeometric function are then discussed, for example, Laguerre functions and Hermite functions, which will arise in quantum problems in Chapter 9. A very general linear differential equation of second order, but with quadratic coefficients, is introduced, the solution of which can be obtained in terms of the confluent hypergeometric function by substitutions. The Bessel equation is seen to be a particular example of such an equation. By expressing the solutions in terms of the confluent hypergeometric function we obtain derivations of both integral and series forms for Bessel functions. The integral form is particularly useful for studying the general character of Bessel functions. The rest of the chapter is devoted to the theory of Bessel functions, including the following topics: the second (Neumann) solution of the Bessel equation; the distribution of zeros of the Bessel functions; the Hankel functions of first and second kind, which will be used in scattering theory in Chapters 5 and 9; the recursion and derivative formulas; the generating function and a second integral form related thereto.

REFERENCES

Churchill, R. V., *Fourier Series and Boundary Value Problems*. New York: McGraw-Hill, 1951.

Doetsch, G., *Theorie und Anwendung der Laplace-Transformation*. New York: Dover, 1937.

Franklin, P., *An Introduction to Fourier Methods and the Laplace Transformation*. New York: Dover, 1958.

Hochstadt, H., *Special Functions of Mathematical Physics*. New York: Holt, Rinehart and Winston, 1961.

Jaeger, J. C., *An Introduction to the Laplace Transformation*. London: Methuen, 1949.

Morse, P. M. and H. Feshbach, *Methods of Theoretical Physics*. New York: McGraw-Hill, 1953.

Papoulis, A., *The Fourier Integral and Its Applications*. New York: McGraw-Hill, 1962.

Sneddon, I. N., *Fourier Transforms*. New York: McGraw-Hill, 1951.

Sommerfeld, A., *Partial Differential Equations in Physics*. New York: Academic Press, 1949.

Tranter, C. J., *Integral Transforms of Mathematical Physics*. London: Methuen, 1956.

4

FOURIER AND LAPLACE TRANSFORMS AND THEIR APPLICATIONS

"Everything has two handles,—one by which it can be borne: another by which it cannot"

Epictetus

1. Introduction: Fourier Transform in One Dimension

The Sturm–Liouville problem

$$\frac{d^2y}{dx^2} + \lambda y = 0 \tag{4.1}$$

with the boundary condition of periodicity on the range $(-a/2 \leqslant x \leqslant a/2)$:

$$y\left(\frac{a}{2}\right) = y\left(-\frac{a}{2}\right) \tag{4.2}$$

produces the eigenvalues

$$\lambda = \frac{2\pi n}{a}, \qquad n = \text{integer} \tag{4.3}$$

and the corresponding eigenfunctions,

$$y = \sin\frac{2\pi nx}{a} \tag{4.4}$$

and

$$y = \cos\frac{2\pi nx}{a} \tag{4.5}$$

which form a degenerate pair. The expansion in these functions

$$y = f(x) = \sum_{n=0}^{\infty} \left(a_n \sin \frac{2\pi n x}{a} + b_n \cos \frac{2\pi n x}{a} \right) \tag{4.6}$$

is the well-known Fourier series expansion.

This expansion may be put into a form which involves complex exponentials:

$$y = \sum_{n=-\infty}^{\infty} c_n e^{2\pi i n x/a} \tag{4.7}$$

where

$$c_n = \tfrac{1}{2}(b_n - i a_n) \tag{4.8}$$

and

$$c_{-n} = c_n^* = \tfrac{1}{2}(b_n + i a_n) \tag{4.9}$$

The expansion functions in (4.7) are mutually orthogonal, in the sense that

$$\frac{1}{a} \int_{-a/2}^{a/2} e^{2\pi i n x/a} e^{-2\pi i m x/a} \, dx = \delta_{mn} \tag{4.10}$$

We note that, for complex functions, the orthogonality integral involves the product of one function with the complex conjugate of the other. The expansion coefficients in (4.7) are, then

$$c_n = \frac{1}{a} \int_{-a/2}^{a/2} f(x) e^{-2\pi i n x/a} \, dx \tag{4.11}$$

By going to the limit in which the interval of periodicity $a \rightarrow \infty$, we can get the Fourier integral theorem. The first step is to introduce

$$y_n = \frac{2\pi n}{a} \tag{4.12}$$

Then we can put

$$\frac{a}{2\pi} c_n = F(y_n) = \frac{1}{2\pi} \int_{-a/2}^{a/2} f(x) e^{-ixy_n} \, dx \tag{4.13}$$

At the same time, (4.7) becomes

$$f(x) = \frac{2\pi}{a} \sum_{n=-\infty}^{\infty} F(y_n) e^{ixy_n} \tag{4.14}$$

Now, let $a \rightarrow \infty$. We note that $2\pi/a$ is the interval between y's. Equation (4.14) becomes

$$f(x) = \lim_{\Delta y_n \rightarrow 0} \sum_{n=-\infty}^{\infty} F(y_n) e^{ixy_n} \Delta y_n$$

or

$$f(x) = \int_{-\infty}^{\infty} F(y) e^{ixy} \, dy \tag{4.15}$$

while (4.13) gives

$$F(y) = \frac{1}{2\pi} \int_{-\infty}^{\infty} f(x) e^{-ixy} \, dx \tag{4.16}$$

The two formulas (4.15) and (4.16), taken together, constitute the statement of the Fourier integral theorem. It is, of course, valid only when the integrals exist. General conditions of validity are discussed in various mathematical texts.[1] The function $F(y)$ is known as the Fourier transform of $f(x)$.

Let us now obtain a few valuable results involving Fourier transforms in one dimension (that is, Fourier transforms of functions of a single variable).

(a) If the transform of $f(x)$ is $F(y)$, the transform of $df/dx = f'(x)$, if it exists, is $iyF(y)$.

To show that this is so, we write down the transform of $f'(x)$:

$$\frac{1}{2\pi} \int_{-\infty}^{\infty} f'(x) \, e^{-ixy} dx = \frac{1}{2\pi} f(x) \, e^{-ixy} \Big|_{-\infty}^{\infty} + iy \frac{1}{2\pi} \int_{-\infty}^{\infty} f(x) \, e^{-ixy} dx \tag{4.17}$$

For the transform of $f(x)$ to exist, $f(x)$ must $\to 0$ at $\pm\infty$. Therefore the transform of $f'(x)$ is $iyF(y)$, as claimed.

By the same token, the transform of the nth derivative $f^{(n)}(x)$, if it exists, is $(iy)^n F(y)$.

(b) The second theorem of interest to us is Parseval's theorem:

$$\int_{-\infty}^{\infty} |f(x)|^2 dx = 2\pi \int_{-\infty}^{\infty} |F(y)|^2 dy \tag{4.18}$$

Consider the left-hand side of this equation with the limits of integration taken as $(-X, X)$; ultimately we will take $X \to \infty$. We substitute for $f(x), f^*(x)$ from (4.15):

$$\int_{-X}^{X} \left[\int_{-\infty}^{\infty} F(y) \, e^{ixy} dy \right] \left[\int_{-\infty}^{\infty} F^*(y') \, e^{-ixy'} dy' \right]$$

Let us now interchange the order of integration, doing the x integral first, to get

$$\int_{-\infty}^{\infty} dy \int_{-\infty}^{\infty} dy' F(y) F^*(y') 2 \, \frac{\sin X(y-y')}{y-y'} \tag{4.19}$$

Consider the properties of the function $[\sin X(y-y')]/[y-y']$ which occurs in this integral. When $y = y'$ it has the value X. It goes through the value zero whenever $y - y' = n\pi/X$. Its oscillations diminish rapidly with amplitude. Its integral over y from $-\infty$ to ∞ has the value π. The integral from $y' - \beta$ to $y' + \beta$ is

$$\int_{-\beta}^{\beta} \frac{\sin X\xi}{\xi} \, d\xi = \int_{-\beta X}^{\beta X} \frac{\sin Z}{Z} \, dZ$$

[1] I. I. Hirschman, Jr., *Infinite Series*. New York: Holt, Rinehart and Winston, 1962.

where we have put $Z = X\xi$. Thus, as $X \to \infty$, however small β may be, the contribution to the integral from $|y-y'| > \beta$ becomes negligibly small.

We see, therefore, that as $X \to \infty$, $2[\sin X(y-y')]/[y-y']$ becomes in effect 2π times the δ function $\delta(y-y')$, and the integral (4.19) becomes

$$2\pi \int_{-\infty}^{\infty} F(y')F^*(y')dy'$$

which proves the theorem (4.18).

It is interesting to consider this theorem when in place of x the variable is a time variable t, and the transform variable y is a circular frequency ω. Then $F(\omega)$ is the frequency amplitude of the time function $f(t)$. We call $2\pi|F(\omega)|^2$ the frequency spectrum; it gives the contribution to the "integrated intensity"

$$\int_{-\infty}^{\infty} |f(t)|^2 dt$$

from various frequency intervals. If the intensity is integrated over a *finite* range, it is clear that it does not break up into distinct contributions from each frequency, but there are "interference" terms. The longer the interval $T = 2X$ of integration, the narrower is the range of frequency overlap.

Problem 4-1:   Show that the transform of the function

$$f(t) = \sin \omega_0 t, \qquad |t| < \frac{2\pi N}{\omega_0}$$

$$= 0, \qquad |t| > \frac{2\pi N}{\omega_0}$$

is

$$F(\omega) = \frac{\omega_0}{\pi} \frac{1}{\omega^2 - \omega_0^2} \sin \frac{2\pi N\omega}{\omega_0}$$

Sketch its frequency spectrum as a function of ω. Show that the width between the first zeros on the two sides of its central peak is inversely proportional to N.

Problem 4-2:   Calculate the Fourier transform of $\cos x^2$.

Problem 4-3:   Calculate the Fourier transform of

$$-\frac{\partial}{\partial x}\frac{1}{e^x+1} = \frac{1}{(e^x+1)(e^{-x}+1)}$$

---

2. The Convolution Theorem

The Fourier transform of the convolution

$$C(x) = f_1(x) * f_2(x) = \int_{-\infty}^{\infty} f_1(x') f_2(x-x') \, dx' \tag{4.20}$$

is

$$\mathscr{C}(y) = 2\pi F_1(y) F_2(y) \tag{4.21}$$

To show that this is so, we write down the transform of $C(x)$:

$$\mathscr{C}(y) = \frac{1}{2\pi} \int_{-\infty}^{\infty} e^{-ixy} \left[\int_{-\infty}^{\infty} f_1(x') f_2(x-x') \, dx' \right] dx \tag{4.22}$$

Let us change the order of the two integrations, and then, in the x integral, introduce a new variable $x'' = x - x'$. Thus

$$\mathscr{C}(y) = \frac{1}{2\pi} \int_{-\infty}^{\infty} dx' f_1(x') \int_{-\infty}^{\infty} f_2(x'') e^{i(x'+x'')y} \, dx''$$

$$= 2\pi F_1(y) F_2(y)$$

as claimed.

Example

Consider the differential equation

$$\frac{d^2f}{dt^2} + k \frac{df}{dt} + \omega_0^2 f = \varphi(t) \tag{4.23}$$

Taking Fourier transforms of all terms of this equation, and letting the transforms of $f(t)$ and $\varphi(t)$ be $F(\omega)$ and $\Phi(\omega)$, respectively, we obtain the equation

$$[-\omega^2 + ik\omega + \omega_0^2]F = \Phi \tag{4.24}$$

or

$$F(\omega) = -\frac{\Phi(\omega)}{\omega^2 - \omega_0^2 - ik\omega} \tag{4.25}$$

Now if we designate the function whose transform is

$$-\frac{1}{2\pi} \frac{1}{\omega^2 - \omega_0^2 - ik\omega}$$

by $G(t)$, it follows from the convolution theorem that

$$f(t) = \int_{-\infty}^{\infty} \varphi(t') G(t-t') \, dt' \tag{4.26}$$

Consider, then

$$G(t) = -\frac{1}{2\pi} \int_{-\infty}^{\infty} \frac{e^{i\omega t} \, d\omega}{\omega^2 - \omega_0^2 - ik\omega} \tag{4.27}$$

This integral can be evaluated by contour integration (Fig. 4.1).

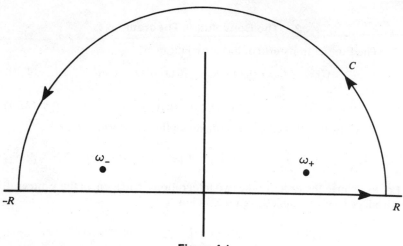

Figure 4.1

The zeros of the denominator are at the points

$$\omega_\pm = \tfrac{1}{2}[\pm\sqrt{4\omega_0{}^2 - k^2} + ik] \tag{4.28}$$

There are clearly two cases to consider: $k < 2\omega_0$ and $k > 2\omega_0$. In the second case, the two roots are both pure imaginary numbers. In the first case, they are complex, and are located as indicated in Fig. 4.1.

Instead of evaluating the integral from $-\infty$ to ∞, we first evaluate it from $-R$ to R, and then, at the end, let $R \to \infty$. Let us construct a contour by drawing the semicircle of radius R with center at the origin in the ω plane. The integral around the contour is

$$I = -\frac{1}{2\pi}\int_{-R}^{R}\frac{e^{i\omega t}\,d\omega}{(\omega - \omega_+)(\omega - \omega_-)} - \frac{1}{2\pi}\int_{C}\frac{e^{i\omega t}\,d\omega}{(\omega - \omega_+)(\omega - \omega_-)} \tag{4.29}$$

where C denotes the semicircular contour. The value of the integral is $2\pi i$ times the sum of the residues at the poles; thus

$$I = -i\left[\frac{e^{i\omega_+ t}}{\omega_+ - \omega_-} + \frac{e^{i\omega_- t}}{\omega_- - \omega_+}\right] = \frac{1}{\sqrt{\omega_0{}^2 - \tfrac{1}{4}k^2}}\,e^{-(1/2)kt}\sin\sqrt{\omega_0{}^2 - \tfrac{1}{4}k^2}\,t \tag{4.30}$$

It is easy to see that this represents $G(t)$ for $t > 0$, since in this case the integral on the semicircle C approaches zero as $R \to \infty$. For on the semicircle $\omega = Re^{i\theta} = R(\cos\theta + i\sin\theta)$ so

$$e^{i\omega t}d\omega = e^{iR\cos\theta t}e^{-R\sin\theta t}R\,d\theta$$

Since the rest of the integrand is bounded on the semicircle and $Re^{-R\sin\theta t} \to 0$ as $R \to \infty$, the integral on C must approach zero.

Thus, for $t > 0$

$$G(t) = \frac{1}{\sqrt{\omega_0^2 - \frac{1}{4}k^2}} \, e^{-(1/2)kt} \sin \sqrt{\omega_0^2 - \frac{1}{4}k^2} \, t \qquad (4.31)$$

On the other hand, for $t < 0$, it is not fruitful to take a contour around the upper half-plane, since the integral on this contour would approach infinity as $R \to \infty$. We therefore take, in this case, a semicircular contour around the *lower* half-plane. The integral on the semicircle again approaches zero as $R \to \infty$; however, in this case there are no poles inside the contour, and therefore

$$G(t) = 0, \qquad t < 0 \qquad (4.32)$$

This result is certainly what would be expected on physical grounds. For Eq. (4.26) may be interpreted in this way: that $G(t - t')$ is the "response" $f(t)$ at time t to the stimulus $\varphi(t')$ at time t'. This should be different from zero, then, only if $t > t'$. In the mathematics, however, we may note that this feature followed from the fact that $k > 0$. The $k(dy/dt)$ term in (4.23) is a *dissipative* term, causing a diminution in the response as t increases. It is the only term in the description of the system which is not time-reversible; thus, it is the only term which can distinguish a direction of time. Such a distinction in the direction of time is, of course, in turn, essential to the interpretation of causality.

In the light of such considerations, it is interesting to consider the case of no dissipation ($k = 0$). Now if there were no dissipative processes in nature, the direction of time would indeed be reversible (that is, if a given system were observed evolving in reverse time order, it would be observed to obey the familiar laws of physics). But we know that, in fact, dissipative processes are *always* present in physical systems, however small may be their effects; thus, one should really let $k \to 0$ from positive values rather than baldly putting it equal to zero; when ambiguities threaten to appear, they should be resolved by taking k to be *positive* but vanishingly small.

Now, as $k \to 0$ the poles in (4.27) approach the real axis. The first problem is, then, to interpret the integrals along the real axis, which now runs through these poles. Let us write (4.27) as

$$I = \frac{1}{2\pi(\omega_+ - \omega_-)} \left[\int \frac{e^{i\omega t} \, d\omega}{\omega - \omega_+} - \int \frac{e^{i\omega t} \, d\omega}{\omega - \omega_-} \right] \qquad (4.33)$$

We can consider only the second of the integrals in the square bracket: The other may then be evaluated by replacing ω_- by ω_+ in the result.

Let us define in this case a "principal part" integral; that is, we cut out from the range of integration an interval of length δ on both sides of $\omega_- = -\omega_0$. The principal part P is then obtained by taking the limit as $\delta \to 0$. A contour may then be constructed as shown in Fig. 4.2. This differs from that of Fig. 4.1 by virtue of the small semicircle of radius δ about $\omega = \omega_-$. We then have

$$P \int \frac{e^{i\omega t}\, d\omega}{\omega - \omega_-} + \int_\Gamma \frac{e^{i\omega t}}{\omega - \omega_-}\, d\omega = 2\pi i\,(\text{residue}) \tag{4.34}$$

$$= P \int \frac{e^{i\omega t}\, d\omega}{\omega - \omega_-} + \pi i \int e^{i\omega t}\, \delta(\omega - \omega_-)\, d\omega$$

Thus, when the pole comes on the real axis, we may write symbolically,

$$\frac{1}{\omega - \omega'} = P \frac{1}{\omega - \omega'} + \pi i \delta(\omega - \omega') \tag{4.35}$$

This is interpreted to mean, that an integral involving $1/(\omega - \omega')$ is to be understood as the principal-value integral $+ \pi i$ times the integral with the δ function.

If, instead of introducing the Green's function explicitly we had tried to calculate f directly by inverting (4.25), this could also have been done. The integral in

$$f(t) = - \int \frac{\Phi(\omega)}{\omega^2 - \omega_0^2 - ik\omega}\, d\omega \tag{4.36}$$

can usually also be evaluated by contour integration. To the contributions from the poles of the transform of the Green's function, we must add contributions from the poles of $\Phi(\omega)$. If $k = 0$, the contributions from $\pm \omega_0$ can be evaluated using (4.35); alternatively, the result may be evaluated for *finite k* and the limit $k \to 0$ taken later.

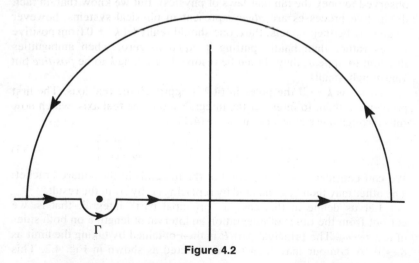

Figure 4.2

3. Causality and Dispersion Relations

We define a "causal" function as a function $g(t)$ which is zero for $t < t_0$: For simplicity, we choose $t_0 = 0$. Let us consider the question of the limitations on the Fourier transform of such a function.

Written in terms of its transform $G(\omega)$, the function $g(t)$ is [cf. (4.15)]

$$g(t) = \int_{-\infty}^{\infty} G(\omega) \, e^{i\omega t} \, d\omega \tag{4.15}$$

where

$$G(\omega) = \frac{1}{2\pi} \int_{-\infty}^{\infty} g(t) \, e^{-i\omega t} \, dt$$

$$= \frac{1}{2\pi} \int_{0}^{\infty} g(t) \, e^{-i\omega t} \, dt \tag{4.37}$$

This defines $G(\omega)$ for *real* ω; however, the definition can be extended into the *lower* complex ω plane, since the integral converges if ω has a negative imaginary part.

Consider now the integral

$$G(\omega) = \frac{1}{2\pi i} \int_{-\infty}^{\infty} \frac{G(\omega')}{\omega' - \omega} \, d\omega' \tag{4.38}$$

where ω is real. We evaluate this using the contour of Fig. (4.3). Let us suppose that $G(\omega)$ is such that the integral around the semicircle in the lower half-plane is zero as its radius approaches infinity. Taking now the

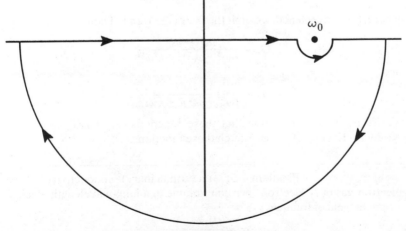

Figure 4.3

integral around the complete contour of Fig. 4.3, we find that its value, which is zero, is made up of two parts: A principal part integral and a contribution around the semicircular contour (δ-function contribution). Therefore it follows that

$$P\frac{1}{2\pi i}\int\frac{G(\omega')}{\omega'-\omega}\,d\omega'+\frac{1}{2}G(\omega)=0 \qquad (4.39)$$

If we now break up $G(\omega)$ into its real and imaginary parts

$$G(\omega)=G_1(\omega)+iG_2(\omega) \qquad (4.40)$$

and equate real and imaginary parts of (4.39), we get the relations (Kramers–Kronig relations):

$$G_1(\omega)=-\frac{1}{\pi}P\int\frac{G_2(\omega')}{\omega'-\omega}\,d\omega' \qquad (4.41)$$

and

$$G_2(\omega)=\frac{1}{\pi}P\int\frac{G_1(\omega')}{\omega'-\omega}\,d\omega' \qquad (4.42)$$

Equations (4.41) and (4.42) have been shown to be valid only if the integral on the semicircle of infinite radius around the lower half-plane vanishes.

Problem 4-4: Apply the foregoing to the transform of the Green's function (4.27) of the problem discussed in the last section. Take

$$G(\omega')=\frac{1}{\omega'^2-\omega_0^2-ik\omega'}$$

where, for convenience, we omit the factor $(-1/2\pi)$. Then

$$G_1=\frac{\omega'^2-\omega_0^2}{(\omega'^2-\omega_0^2)^2+k^2\omega'^2}$$

and

$$G_2=\frac{k\omega'}{(\omega'^2-\omega_0^2)^2+k^2\omega'^2}$$

Verify by direct integration that these functions G_1 and G_2 satisfy the Kramers–Kronig relations. Sketch these functions.

Problem 4-5: The (imaginary) conductivity of an electron gas of N electrons per unit volume in a long-wavelength electromagnetic field of frequency ω is

$$\sigma_i=\frac{iNe^2}{m}\frac{1}{\omega-\omega_c}$$

when there is a constant magnetic field **B** and the wave is right circularly polarized relative to this direction. $-e$ and m represent the electron charge and mass, respectively, and ω_c is the Larmor frequency eB/mc. Show that the real part of the conductivity, which determines the *absorption* of the wave, is

$$\sigma_r = -\frac{\pi N e^2}{m}\,\delta(\omega - \omega_c)$$

(The Kramers–Kronig relations obviously do not apply directly to "functions" as anomalous as the δ function. This difficulty may be overcome by attributing to ω a small imaginary part i/τ.)

Note the reason for choosing to deform the contour around the *lower* half-plane. From (4.37)

$$G(\omega_1 + i\omega_2) = \frac{1}{2\pi}\int_0^\infty g(t)\, e^{-i\omega_1 t}\, e^{\omega_2 t}\, dt \qquad (4.43)$$

Therefore, provided $g(t)$ remains bounded, the integral for $G(\omega)$ will converge so long as $\omega_2 < 0$.

The Kramers–Kronig relations (4.41) and (4.42) are also obviously only valid if the integrals on the right-hand side converge. This would not be the case, for instance, if G_1, G_2 approached a constant as $\omega \to \infty$ on the real axis. In that case, however, we could have obtained dispersion relations using in (4.38), instead of $G(\omega)$, the function $G(\omega)/(\omega - \omega_1)$, where ω_1 is arbitrary. Taking $\omega_1 = 0$, for instance, instead of (4.41) and (4.42) we would get the relations

$$G_1(\omega) = -\frac{\omega}{\pi} P \int \frac{G_2(\omega')}{\omega'(\omega' - \omega)}\, d\omega' \qquad (4.44)$$

and

$$G_2(\omega) = \frac{\omega}{\pi} P \int \frac{G_1(\omega')}{\omega'(\omega' - \omega)}\, d\omega' \qquad (4.45)$$

4. Linear Response Functions

The foregoing considerations are applicable to the following rather general problem. Take a physical system, subject to an external stimulus $\varphi(t)$, which responds *linearly* to this stimulus. This is expressed mathematically in the following way: if the function $f(t)$ measures the "response," then it may be written in the form

$$f(t) = \int_{-\infty}^t l(t - t')\varphi(t')\, dt' \qquad (4.46)$$

where $l(t)$ is "causal," that is to say

$$l(t) = 0, \qquad t < 0 \qquad (4.47)$$

Taking the Fourier transform of (4.46) and using the convolution theorem (4.21) gives

$$F(\omega) = 2\pi L(\omega)\Phi(\omega) \tag{4.48}$$

If the "stimulus" has a definite circular frequency, the response per unit stimulus at that frequency is $2\pi L(\omega)$.

It follows from (4.47) that this function must satisfy the Kramers–Kronig dispersion relations.

5. Cross Correlation and Auto-correlation Functions

Consider two time functions $f_1(t)$ and $f_2(t)$, whose square integral is finite. We may define a cross correlation between these functions for times separated by an interval τ

$$C_{12}(\tau) = \frac{\int_{-\infty}^{\infty} f_1(t) f_2(t+\tau)\, dt}{\left[\int_{-\infty}^{\infty} f_1{}^2(t)\, dt \int_{-\infty}^{\infty} f_2{}^2(t)\, dt \right]^{1/2}} \tag{4.49}$$

If $f_1 = f_2 = f$, this becomes the autocorrelation function

$$C(\tau) = \frac{\int_{-\infty}^{\infty} f_1(t) f_1(t+\tau)\, dt}{\int_{-\infty}^{\infty} f_1{}^2(t)\, dt} \tag{4.50}$$

This function has a number of interesting applications; probably the most interesting is its use in determining the frequency spectrum of certain stochastic functions, an application which we discuss in detail in Chapter 7.

Let us derive two important properties of this function.

The first is that $|C_{12}(\tau)| \leqslant 1$. This is seen by considering

$$N(\lambda) = \int_{-\infty}^{\infty} [f_1(t) + \lambda f_2(t+\tau)]^2\, dt \geqslant 0 \tag{4.51}$$

for arbitrary real values of λ. Putting

$$n_1{}^2 = \int_{-\infty}^{\infty} f_1{}^2(t)\, dt \tag{4.52}$$

and

$$n_2{}^2 = \int_{-\infty}^{\infty} f_2{}^2(t)\, dt \tag{4.53}$$

we can write (4.51) as

$$\lambda^2 n_2{}^2 + 2\lambda n_1 n_2 C_{12}(\tau) + n_1{}^2 \geqslant 0 \tag{4.54}$$

By completing the square, we find that

$$[\lambda n_2 + n_1 C_{12}(\tau)]^2 + n_1{}^2 [1 - C_{12}{}^2(\tau)] \geqslant 0$$

for all real λ's. It is therefore true for the value which makes the first (squared) term zero. Therefore

$$C_{12}{}^2(\tau) \leq 1 \qquad (4.55)$$

as stated above.

The second property may be expressed in the theorem that the Fourier transform of $C_{12}(\tau)$ is

$$\frac{2\pi}{n_1 n_2} F_1^*(\omega) F_2(\omega)$$

A special consequence is that the *autocorrelation* function is proportional to the frequency spectrum of the function $f(t)$.

To demonstrate the validity of this result, consider first the Fourier transform of the numerator of $C_{12}(\tau)$. It is

$$\frac{1}{2\pi} \int_{-\infty}^{\infty} e^{-i\omega\tau} \left[\int_{-\infty}^{\infty} f_1(t) f_2(t+\tau)\, dt \right] d\tau$$

If the order of the integrations is interchanged, this becomes

$$\frac{1}{2\pi} \int_{-\infty}^{\infty} f_1(t) \left[\int_{-\infty}^{\infty} f_2(t+\tau) e^{-i\omega\tau}\, d\tau \right] dt$$

For fixed t, in the inner integral it is possible to replace τ as the variable of integration by $t' = t + \tau$. Then we get

$$\frac{1}{2\pi} \int_{-\infty}^{\infty} f_1(t) e^{i\omega t}\, dt \int_{-\infty}^{\infty} f_2(t') e^{-i\omega t'}\, dt' = 2\pi F_1^*(\omega) F_2(\omega) \qquad (4.56)$$

Therefore the Fourier transform of $C_{12}(\tau)$, $\mathscr{F}(C_{12}(\tau))$, is

$$\mathscr{F}(C_{12}(\tau)) = \frac{2\pi F_1^*(\omega) F_2(\omega)}{n_1 n_2} \qquad (4.57)$$

The Fourier transform of the autocorrelation function is

$$\mathscr{F}(C(\tau)) = \frac{2\pi}{n^2} |F(\omega)|^2 \qquad (4.58)$$

From Parseval's theorem (4.18) the integral over frequency of this transform is 1. Thus, the Fourier transform of *the autocorrelation function is the frequency distribution function of* $f(t)$.

In practical terms, the usefulness of the result stems from the fact that, even if one has a function ("signal") for which no analytic form is known, but the autocorrelation function can be determined either analytically or numerically, the frequency distribution function of the signal can be obtained.

Consider the application of the above considerations to a system with a linear response. The correlation $C_{12}(\tau)$ between stimulus $f_1(t) =$

$\varphi(t)$ and response $f_2(t) = f(t)$ has a transform given by (4.57). Here $F_1^*(\omega)$ is $\Phi^*(\omega)$, whereas $F_2(\omega)$ is, from (4.48), $2\pi L(\omega)\Phi(\omega)$. Thus,

$$\mathscr{F}(C_{12}(\tau)) = \frac{4\pi^2|\Phi(\omega)|^2}{n_1 n_2} L(\omega) \tag{4.59}$$

Thus, from a knowledge of the correlation between stimulus and response as a function of time, and of the frequency spectrum of the stimulus, it is possible to determine $L(\omega)$, which determines the response of the system at each frequency.

The result (4.59) may be written in the simpler form

$$\mathscr{F}\left(\int_{-\infty}^{\infty} \varphi(t) g(t+\tau) dt\right) = 4\pi^2|\Phi(\omega)|^2 L(\omega) \tag{4.60}$$

6. Fourier Transform in Three Dimensions

A function of several variables may be transformed with respect to each of its variables; in taking the transform with respect to any one of these variables, the others are treated as constants. The commonest instance in physics is that of a function of the three Cartesian coordinates of a point in space. Just as, for a function of a single variable, the Fourier integral may be taken as the limit of a Fourier series as the interval of periodicity becomes infinitely large, so may the three-dimensional transform of a function of space coordinates be considered as the limit of a Fourier series in each of the three variables.

It is most convenient to represent a function of coordinates in terms of the position vector $\mathbf{r}$ with components x, y, and z. We define the Fourier transform $F(\mathbf{k}) = F(k_x, k_y, k_z)$ of the function $f(\mathbf{r}) = f(x, y, z)$ as

$$F(\mathbf{k}) = \left(\frac{1}{2\pi}\right)^3 \int f(\mathbf{r}) e^{-i\mathbf{k}\cdot\mathbf{r}} d^3\mathbf{r} \tag{4.61}$$

where $d^3\mathbf{r}$ represents the volume element $dx\,dy\,dz$ and the integral is over all space, that is, the range from $-\infty$ to ∞ in each coordinate. This follows from (4.16), each coordinate being transformed separately. The expression of the function in terms of its transform follows similarly from (4.15), and is

$$f(\mathbf{r}) = \int F(\mathbf{k}) e^{i\mathbf{k}\cdot\mathbf{r}} d^3\mathbf{k} \tag{4.62}$$

where the integral is over the whole of $\mathbf{k}$ space.

We now demonstrate a number of useful theorems. In all instances the validity of these theorems will, of course, be dependent on the existence of the integrals indicated. That is, the functions being transformed will be assumed to go to zero faster than $1/r^3$ as $r = |\mathbf{r}| \to \infty$. In all cases surface integrals over surfaces at infinity will be taken to be zero. Use will be made of Gauss' theorem (the divergence theorem) in vector form: Where $A(r)$ is a vector function of position

$$\int_{(v)} \nabla \cdot A(r)\, d^3r = \int_{(s)} A(r) \cdot n\, dS \tag{4.63}$$

where the integral on the left is taken over an arbitrary volume and that on the right is over the surface surrounding that volume. n is the unit normal vector drawn outward from the surface.

(a) The transform of the gradient ∇f is $ik F(k)$.

From (4.61) the transform of f is

$$\mathcal{F}(\nabla f) = \left(\frac{1}{2\pi}\right)^3 \int \nabla f \cdot e^{-ik \cdot r}\, d^3r$$

$$= \left(\frac{1}{2\pi}\right)^3 \left[\int \nabla (fe^{-ik \cdot r})\, d^3r + ik \int fe^{-ik \cdot r} d^3r \right]$$

The first integral may be transformed into the surface integral $\int fe^{-ik \cdot r} n\, dS$ over the surface at infinity and so is zero. Therefore,

$$\mathcal{F}(\nabla f) = ik F(k) \tag{4.64}$$

(b) The transform of the divergence of the vector function $a(r)$ is $ik \cdot A(k)$, where $A(k)$ is the transform of $a(r)$.

By definition the transform of $\nabla \cdot a$ is

$$\mathcal{F}(\nabla \cdot a) = \left(\frac{1}{2\pi}\right)^3 \int (\nabla \cdot a)e^{-ik \cdot r} d^3r$$

By the use of the identity

$$\nabla \cdot (ae^{-ik \cdot r}) = (\nabla \cdot a)e^{-ik \cdot r} - ik \cdot ae^{-ik \cdot r} \tag{4.65}$$

the integral may be written in two parts. The first, $\int \nabla \cdot (ae^{-ik \cdot r})\, d^3r$, transforms into the surface integral $\int e^{-ik \cdot r} a \cdot n\, dS$ which approaches 0; the second is $ik \cdot \int ae^{-ik \cdot r} d^3r$. Therefore,

$$\mathcal{F}(\nabla \cdot a) = ik \cdot A(k) \tag{4.66}$$

(c) The Fourier transform of the Laplacian of a function f, $\nabla^2 f$, is

$$\mathcal{F}(\nabla^2 f) = -k^2 F \tag{4.67}$$

This follows from the two previous results, if a is taken to be ∇f. Thus

$$\mathcal{F}(\nabla^2 f) = \mathcal{F}(\nabla \cdot \nabla f)$$

$$= ik \cdot \mathcal{F}(\nabla f)$$

$$= ik \cdot ikF$$

$$= -k^2 F$$

(d) The convolution of $f_1(\mathbf{r})$, $f_2(\mathbf{r})$ is

$$C(\mathbf{r}) = \int f_1(\mathbf{r}')f_2(\mathbf{r}-\mathbf{r}') \, d^3\mathbf{r}' = f_1(\mathbf{r}) * f_2(\mathbf{r})$$

The transform of the convolution is

$$\mathscr{F}(C(\mathbf{r})) = (2\pi)^3 F_1(\mathbf{k})F_2(\mathbf{k})$$

where F_1, F_2 are the transforms of $f_1(\mathbf{r})$ and $f_2(\mathbf{r})$, respectively.

By definition the transform of the convolution of $f_1(\mathbf{r})$ and $f_2(\mathbf{r})$ is

$$\left(\frac{1}{2\pi}\right)^3 \int e^{-i\mathbf{k}\cdot\mathbf{r}} \left[\int f_1(\mathbf{r}')f_2(\mathbf{r}-\mathbf{r}') \, d^3\mathbf{r}'\right]$$

Assuming that the order of the integrations may be changed, and replacing $\mathbf{r}$ as a variable of integration by $\mathbf{R} = \mathbf{r} - \mathbf{r}'$, this becomes

$$\left(\frac{1}{2\pi}\right)^3 \int f_1(\mathbf{r}')e^{-i\mathbf{k}\cdot\mathbf{r}} \left[\int f_2(\mathbf{R})e^{-i\mathbf{k}\cdot\mathbf{R}}d^3\mathbf{R}\right] d^3\mathbf{r}'$$

Therefore,

$$\mathscr{F}(C(\mathbf{r})) = (2\pi)^3 F_1(\mathbf{k})F_2(\mathbf{k}) \tag{4.68}$$

Conversely, it follows that

$$C(\mathbf{r}) = (2\pi)^3 \int F_1(\mathbf{k})F_2(\mathbf{k})e^{i\mathbf{k}\cdot\mathbf{r}} \, d^3\mathbf{k} \tag{4.69}$$

7. Solution of Poisson's Equation by Fourier Transforms

Using the preceding theorems, it is possible to solve most of the standard linear differential equations of mathematical physics. Not only does the Fourier transform method provide a very powerful tool for such problems, but it also provides considerable insight into the physical principles which they involve.

Consider, as a first illustration, Poisson's equation for the electrostatic potential arising from a charge distribution $\rho(\mathbf{r})$:

$$\nabla^2\Phi = -4\pi\rho \tag{3.23}$$

Taking the Fourier transforms of the two sides of this equation and designating the Laplace transforms of $\Phi(\mathbf{r})$ and $\rho(\mathbf{r})$ by $\Phi_0(\mathbf{k})$, $P(\mathbf{k})$, respectively, leads to the equation

$$\Phi_0(\mathbf{k}) = \frac{4\pi}{k^2} P(\mathbf{k}) \tag{4.70}$$

If we designate

$$G_0(\mathbf{k}) = \left(\frac{1}{2\pi}\right)^3 \frac{4\pi}{k^2} = \frac{1}{2\pi^2 k^2} \tag{4.71}$$

it follows from (4.68) that

$$\Phi_0(\mathbf{k}) = (2\pi)^3 G_0(\mathbf{k}) P(\mathbf{k}) \tag{4.72}$$

Therefore,

$$\Phi(\mathbf{r}) = (2\pi)^3 \int G_0(\mathbf{k}) P(\mathbf{k}) e^{i\mathbf{k} \cdot \mathbf{r}} \, d^3\mathbf{k} \tag{4.73}$$

which is, by (4.69),

$$\Phi(\mathbf{r}) = \int \rho(\mathbf{r}') G(\mathbf{r} - \mathbf{r}') \, d^3\mathbf{r}' \tag{4.74}$$

where $G(\mathbf{r})$ is the function whose transform is $G_0(\mathbf{k}) = 1/2\pi^2 k^2$. Again, we may refer to this as the "Green's function" of the problem — more specifically, in this case, the Green's function for the potential equation in free space.

We may determine directly the function $G(\mathbf{r})$ by the Fourier inverse theorem (4.62):

$$G(\mathbf{r}) = \frac{1}{2\pi^2} \int \frac{1}{k^2} e^{i\mathbf{k} \cdot \mathbf{r}} d^3\mathbf{k} \tag{4.75}$$

Spherical polar coordinates may be chosen for the variables in $\mathbf{k}$ space, using the direction of $\mathbf{r}$ as polar axis. If μ designates the cosine of the angle between $\mathbf{k}$ and $\mathbf{r}$, integration over polar angles gives

$$\begin{aligned}
G(\mathbf{r}) &= \frac{1}{\pi} \int_0^\infty k^2 dk \int_{-1}^1 d\mu \frac{1}{k^2} e^{ikr\mu} \\
&= \frac{2}{\pi} \int_0^\infty \frac{\sin kr}{kr} dk \\
&= \frac{1}{r} \tag{4.76}
\end{aligned}$$

The potential (4.74) then becomes

$$\Phi(\mathbf{r}) = \int \rho(\mathbf{r}') \frac{1}{|\mathbf{r} - \mathbf{r}'|} d^3\mathbf{r}' \tag{4.77}$$

a result which we wrote down earlier [Eq. (3.138)] on the basis of a physical argument.

8. Poisson's Summation Formula in One and Three Dimensions

In this section we derive a mathematical formula which is frequently used in mathematical physics in cases in which the solution to problems leads to an infinite series. We derive it first in its simplest one-dimensional form, and then give the generalization to three dimensions.

Consider the following function, defined in terms of a series:

$$S(x) = \sum_{n=-\infty}^{\infty} f(x + n x_0) \tag{4.78}$$

where f is an arbitrary function. It is clear that $S(x)$ is periodic, with period x_0, and so may be expressed as a Fourier series

$$S(x) = \sum_{l=-\infty}^{\infty} C_l\, e^{2\pi i l x/x_0} \tag{4.79}$$

The coefficients C_l are given by

$$C_l = \frac{1}{x_0} \int_{-(1/2)x_0}^{(1/2)x_0} S(x)\, e^{-2\pi i l x/x_0}\, dx \tag{4.80}$$

Substituting for $S(x)$ from (4.78),

$$C_l = \frac{1}{x_0} \sum_{n=-\infty}^{\infty} \int_{-(1/2)x_0}^{(1/2)x_0} f(x+nx_0)\, e^{-2\pi i l x/x_0}\, dx$$

In a term corresponding to a given n, introduce $y = x + nx_0$ to get

$$C_l = \frac{1}{x_0} \sum_{n=-\infty}^{\infty} \int_{(n-(1/2))x_0}^{(n+(1/2))x_0} f(y)\, e^{-2\pi i l y/x_0}\, dy$$

$$= \frac{1}{x_0} \int_{-\infty}^{\infty} f(y)\, e^{-2\pi i l y/x_0}\, dy$$

on summing the terms. Therefore,

$$C_l = \frac{2\pi}{x_0} F\left(\frac{2\pi l}{x_0}\right) e^{2\pi i l x/x_0} \tag{4.81}$$

where $F(k)$ is the Fourier transform of $f(x)$. Thus $S(x)$ may be written as

$$S(x) = \frac{2\pi}{x_0} \sum_{l=-\infty}^{\infty} F\left(\frac{2\pi l}{x_0}\right) e^{2\pi i l x/x_0} \tag{4.82}$$

This is the Poisson summation formula.

Example

Evaluate

$$S_0 = \sum_{n=0}^{\infty} \frac{1}{a^2+n^2} \tag{4.83}$$

In the first place the series may be written

$$S_0 = \frac{1}{2a^2} + \frac{1}{2} \sum_{n=-\infty}^{\infty} \frac{1}{a^2+n^2}$$

This then becomes an application of the Poisson formula in which $f(u) = 1/(a^2+u^2)$, $x_0 = 1$, and $x = 0$. The transform of $f(u)$ is

$$F(v) = \frac{1}{2\pi} \int \frac{1}{a^2+u^2} e^{-iuv}\, du$$

For $v > 0$ this may be evaluated as a contour integral in which the contour is completed with a semicircle of radius $R \to \infty$ around the lower half-plane. By the residue theorem, the integral is

$$-i\,[\text{residue at } u = -ia] = -i\left(\frac{1}{-2ia}\right)e^{-av} = \frac{1}{2a}e^{-av}$$

For $v < 0$ the contour is deformed around the *upper* half-plane, and the result is, similarly,

$$\frac{1}{2a}e^{-a|v|}$$

Therefore,

$$S_0 = \frac{1}{2a^2} + \frac{2\pi}{4a}\sum_{l=-\infty}^{\infty} e^{-a|2\pi l|}$$

$$= \frac{1}{2a^2} - \frac{\pi}{2a} + \frac{\pi}{a}\sum_{l=0}^{\infty} e^{-2\pi la} \tag{4.84}$$

The last series is a geometric series and so may be evaluated giving

$$S_0 = \frac{1}{2a^2} - \frac{\pi}{2a} + \frac{\pi}{a}\frac{1}{1-e^{-2\pi a}} \tag{4.85}$$

Problem 4-6:   Show that

$$\sum_{n=1}^{\infty}\frac{1}{n^2} = \frac{\pi^2}{6}$$

Problem 4-7:   Show that

$$S_1 = \sum_{n=0}^{\infty}\frac{1}{a^2 + (2n+1)^2} = \frac{\pi}{4a}\frac{1-e^{-\pi a}}{1+e^{-\pi a}}$$

Problem 4-8:   Show that

$$S_2 = \sum_{n=0}^{\infty}\frac{(-1)^n}{a^2+n^2} = \frac{1}{2a^2} + \frac{\pi}{a}\frac{e^{-\pi a}}{1-e^{-2\pi a}}$$

Problem 4-9:   Evaluate the series

$$\sum_{n=1}^{\infty}\frac{(-1)^n}{n^4} = -\frac{7\pi^4}{720}$$

Although in this case the Poisson formula has permitted an explicit evaluation of the series, this will, of course, not generally be the case. The formula still has considerable value, however, as an aid to numerical evaluation of the solution. Its value lies in this: That whereas the original series (4.83) is most rapidly convergent for small a, the transformed series

(4.84) converges most rapidly for *large* values of a. This feature is not peculiar to the present example; quite generally, it is found that precisely under those conditions in which one series converges slowly, the other converges rapidly.

Consider now the three-dimensional case, in which the sum to be evaluated is

$$S(\mathbf{r}) = \sum f(\mathbf{r} + \mathbf{R}_n) \qquad (4.86)$$

where the $\mathbf{R}_n$'s are "lattice vectors"

$$\mathbf{R}_n = n_1\boldsymbol{\tau}_1 + n_2\boldsymbol{\tau}_2 + n_3\boldsymbol{\tau}_3 \qquad (4.87)$$

the n's taking on all integral values from $-\infty$ to ∞. Although the theorem in this case follows much the same pattern as in the one-dimensional case, the difference is sufficient to justify following the derivation through in detail.

The function $S(\mathbf{r})$ is periodic with respect to displacements through each of the vectors $\boldsymbol{\tau}_1$, $\boldsymbol{\tau}_2$, and $\boldsymbol{\tau}_3$. It can therefore be expanded in a suitable Fourier series. To determine this series, it is necessary to introduce the "inverse lattice vectors"

$$\mathbf{b}_1 = \frac{\boldsymbol{\tau}_2 \times \boldsymbol{\tau}_3}{\Omega}, \qquad \mathbf{b}_2 = \frac{\boldsymbol{\tau}_3 \times \boldsymbol{\tau}_1}{\Omega}, \qquad \mathbf{b}_3 = \frac{\boldsymbol{\tau}_1 \times \boldsymbol{\tau}_2}{\Omega} \qquad (4.88)$$

where

$$\Omega = \boldsymbol{\tau}_1 \cdot \boldsymbol{\tau}_2 \times \boldsymbol{\tau}_3 \qquad (4.89)$$

It follows immediately that

$$\mathbf{b}_i \cdot \boldsymbol{\tau}_j = \delta_{ij} \qquad (4.90)$$

The Fourier series can then be expressed in the form

$$S(\mathbf{r}) = \sum \sigma(\mathbf{K}_l) e^{2\pi i \mathbf{K}_l \cdot \mathbf{r}} \qquad (4.91)$$

where

$$\mathbf{K}_l = l_1\mathbf{b}_1 + l_2\mathbf{b}_2 + l_3\mathbf{b}_3 \qquad (4.92)$$

the l's being integers. The sum in (4.91) is over all positive and negative integral values of the l's.

To determine the coefficients $\sigma(K_l)$, it is necessary to prove the orthogonality theorem:

$$\int e^{2\pi i (\mathbf{K}_l - \mathbf{K}_{l'}) \cdot \mathbf{r}} d^3\mathbf{r} = \Omega \delta_{\mathbf{K}_l, \mathbf{K}_{l'}} \qquad (4.93)$$

The integral is taken over a parallelpiped constructed of the vectors $\boldsymbol{\tau}_1, \boldsymbol{\tau}_2$, and $\boldsymbol{\tau}_3$, and $\delta_{\mathbf{K}_l, \mathbf{K}_{l'}}$ is zero when $\mathbf{K}_l \neq \mathbf{K}_{l'}$ and unity when $\mathbf{K}_l = \mathbf{K}_{l'}$.

The theorem is easily proven by introducing

$$\mathbf{r} = z_1\boldsymbol{\tau}_1 + z_2\boldsymbol{\tau}_2 + z_3\boldsymbol{\tau}_3 \qquad (4.94)$$

The integral then becomes, by virtue of (4.90),

$$\int_0^1 \int_0^1 \int_0^1 \exp 2\pi i [(l_1 - l_1')z_1 + (l_2 - l_2')z_2 + (l_3 - l_3')z_3] J \, dz_1 \, dz_2 \, dz_3 \tag{4.95}$$

J is the Jacobian of the components (x_1, x_2, x_3) of $\mathbf{r}$ with respect to (z_1, z_2, z_3). Thus

$$J = \begin{vmatrix} \tau_{1x} & \tau_{1y} & \tau_{1z} \\ \tau_{2x} & \tau_{2y} & \tau_{2z} \\ \tau_{3x} & \tau_{3y} & \tau_{3z} \end{vmatrix} = \boldsymbol{\tau}_1 \cdot \boldsymbol{\tau}_2 \times \boldsymbol{\tau}_3 = \Omega \tag{4.96}$$

The integral is, in fact, the product of three similar integrals of the form

$$\int_0^1 e^{2\pi i (l_1 - l_1')z_1} \, dz_1 = 0, \qquad l_1 \neq l_1'$$

$$= 1, \qquad l_1 = l_1' \tag{4.97}$$

Theorem (4.93) follows.

If (4.91) is now multiplied by $e^{-2\pi i \mathbf{K}_l \cdot \mathbf{r}}$, and we integrate as in (4.93), we find that

$$\sigma(\mathbf{K}_l) = \frac{1}{\Omega} \int S(\mathbf{r}) e^{-2\pi i \mathbf{K}_l \cdot \mathbf{r}} \, d^3 \mathbf{r} \tag{4.98}$$

the integral being over the cell of volume Ω.

We are now in a position to complete the derivation of the summation formula. Substituting for $S(\mathbf{r})$ from (4.86) in (4.98) it is found that

$$\sigma(\mathbf{K}_l) = \frac{1}{\Omega} \sum_n \int_{\text{cell}} f(\mathbf{r} + \mathbf{R}_n) e^{-2\pi i \mathbf{K}_l \cdot \mathbf{r}} \, d^3 \mathbf{r} \tag{4.99}$$

In a given term, the substitution

$$\mathbf{r}' = \mathbf{r} + \mathbf{R}_n \tag{4.100}$$

leads to the result

$$\sigma(\mathbf{K}_l) = \frac{1}{\Omega} \sum_n \int_{\text{cell } n} f(\mathbf{r}') e^{-2\pi i \mathbf{K}_l \cdot \mathbf{r}'} \, d^3 \mathbf{r}' \tag{4.101}$$

The sum is over cells displaced from the original one by the "lattice vectors" $\mathbf{R}_n$. But the sum of integrals over all such "cells" is the sum over all space, so that

$$\sigma(\mathbf{K}_l) = \frac{1}{\Omega} \int f(\mathbf{r}') e^{-2\pi i \mathbf{K}_l \cdot \mathbf{r}'} \, d^3 \mathbf{r}' \tag{4.102}$$

$$= \frac{(2\pi)^3}{\Omega} F(2\pi \mathbf{K}_l) \tag{4.103}$$

where $F(\mathbf{k})$ is the Fourier transform of $f(\mathbf{r})$. We thus arrive at the Poisson summation formula in three dimensions:

$$S(\mathbf{r}) = \sum_n f(\mathbf{r} + \mathbf{R}_n) = \frac{(2\pi)^3}{\Omega} \sum_l F(2\pi \mathbf{K}_l) e^{2\pi i \mathbf{K}_l \cdot \mathbf{r}} \tag{4.104}$$

This result has a number of applications, as we shall see in subsequent chapters. Suppose, for instance, that one wishes to solve a linear problem (potential, heat conduction, scattering, etc.) with a point source, and boundary conditions which can be satisfied by a system of image sources at $(-\mathbf{R}_n)$. If the Green's function (point source solution) in an *infinite* region is $f(\mathbf{r})$ for a source at the origin, the solution will be of the form (4.86). Equations (4.104) then makes it possible to transform from the solution in terms of images to one in the form of a Fourier series. In general, where the one solution converges slowly, the other converges rapidly.

A second application is in the theory of solids, where one may desire to evaluate sums of functions centered at the various lattice sites $\mathbf{R}_n$ of the atoms of a crystal. For example, the potential at any point in a crystal might be considered as the sum of contributions from the charge distribution in its different "cells." Equation (4.104) would then give the Fourier series representation of the potential. In particular, in an ionic crystal, one can calculate the potential energy per ion as the sum of the contributions from all the other ions.

Let us carry out this calculation for a cubic crystal, the successive ions having total charges of ϵ and $-\epsilon$. The charge distributions on the ions are in general different, but for simplicity we assume them to be the same, namely, $\pm \epsilon \rho_0(\mathbf{r} - \mathbf{R}_n)$, where $\mathbf{R}_n$ is the location of the nucleus of the nth ion. The total charge density may then be taken to be

$$\rho(\mathbf{r}) = \epsilon \sum_{\mathbf{R}_n} \cos \pi\boldsymbol{\lambda} \cdot \mathbf{R}_n \cdot \rho_0(\mathbf{r} - \mathbf{R}_n) \qquad (4.105)$$

Here

$$\mathbf{R}_n = n_1 a\mathbf{i} + n_2 a\mathbf{j} + n_3 a\mathbf{k} \qquad (4.106)$$

and

$$\boldsymbol{\lambda} = \frac{1}{a}(\mathbf{i} + \mathbf{j} + \mathbf{k}) \qquad (4.107)$$

where $\mathbf{i}$, $\mathbf{j}$, and $\mathbf{k}$ are unit vectors along the coordinate directions and a is the lattice spacing. Since $\boldsymbol{\lambda} \cdot \mathbf{R}_n = n_1 + n_2 + n_3$, the factor $\cos \pi\boldsymbol{\lambda} \cdot \mathbf{R}_n$ ensures the alternation of sign of the charges.

Let us use theorem (4.104) to transform the density (4.105). It may be considered as a sum $\Sigma f(\mathbf{R}_n)$, where

$$f(\mathbf{r}') = \epsilon \cos \pi\boldsymbol{\lambda} \cdot \mathbf{r}' \rho_0(\mathbf{r} - \mathbf{r}') \qquad (4.108)$$

The Fourier transform of this function is

$$F(\mathbf{k}) = \frac{\epsilon}{(2\pi)^3} \int \cos \pi\boldsymbol{\lambda} \cdot \mathbf{r}' \rho_0(\mathbf{r} - \mathbf{r}') e^{-i\mathbf{k} \cdot \mathbf{r}'} d^3\mathbf{r}' \qquad (4.109)$$

Writing the cosine as a sum of complex exponentials and introducing the variable

$$\mathbf{R} = \mathbf{r} - \mathbf{r}' \tag{4.110}$$

this becomes

$$F(\mathbf{k}) = \frac{\epsilon}{2} \left[e^{-i(\mathbf{k}-\pi\lambda)\cdot\mathbf{r}} P_0(\mathbf{k}-\pi\lambda) + e^{-i(\mathbf{k}+\pi\lambda)\cdot\mathbf{r}} P_0(\mathbf{k}+\pi\lambda) \right] \tag{4.111}$$

where $P_0(\mathbf{k})$ is the Fourier transform of $\rho_0(\mathbf{r})$. The required charge density may then be written as

$$\rho(\mathbf{r}) = \left(\frac{2\pi}{a}\right)^3 \frac{\epsilon}{2} \sum_l \left[e^{-2\pi i(\mathbf{K}_l-(1/2)\lambda)\cdot\mathbf{r}} P_0(2\pi\mathbf{K}_l - \pi\lambda) \right.$$

$$\left. + e^{-2\pi i(\mathbf{K}_l-(1/2)\lambda)\cdot\mathbf{r}} P_0(2\pi\mathbf{K}_l + \pi\lambda) \right] \tag{4.112}$$

This may now be used to calculate the potential at the position of the ion at the origin. However, since (4.112) includes the charge of that ion, we must subtract $\rho_0(r)$ from (4.112).

The potential is, then,

$$\Phi(\mathbf{r}) = \int \frac{\rho(\mathbf{r}')}{|\mathbf{r}-\mathbf{r}'|} d^3\mathbf{r}' - \int \frac{\rho_0(\mathbf{r}')}{|\mathbf{r}-\mathbf{r}'|} d^3\mathbf{r}' \tag{4.113}$$

$$= \Phi_1 + \Phi_2$$

Making use of the fact that

$$\int \frac{e^{-2\pi i \mathbf{k}\cdot\mathbf{r}'}}{r'} d^3\mathbf{r}' = \frac{1}{2\pi^2 k^2} \tag{4.114}$$

it is possible to calculate Φ_1.

$$\Phi_1(0) = \frac{4\pi}{a^3} \frac{\epsilon}{2} \sum_l \left\{ \frac{1}{|\mathbf{K}_l - \frac{1}{2}\lambda|^2} P_0(2\pi\mathbf{K}_l - \pi\lambda) + \frac{1}{|\mathbf{K}_l + \frac{1}{2}\lambda|^2} P_0(2\pi\mathbf{K}_l + \pi\lambda) \right\} \tag{4.115}$$

On the other hand,

$$\Phi_2(0) = -\epsilon \int \frac{1}{r'} \rho_0(\mathbf{r}') d^3\mathbf{r}' \tag{4.116}$$

For purposes of evaluation of the lattice sum, the precise form of ρ_0 should not matter provided the ions do not overlap; that is, provided ρ_0 is negligible outside the cubical "cell" surrounding a given atom. Assuming that the total ionic charge is unity, it is convenient to choose

$$\rho_0(\mathbf{r}) = \epsilon \left(\frac{\alpha}{\pi}\right)^{3/2} \frac{1}{a^3} e^{-\alpha r^2/a^2} \tag{4.117}$$

The transform of $\rho_0(\mathbf{r})$ is

$$P_0(\mathbf{k}) = \frac{1}{8\pi^3} e^{-k^2 a^2/4\alpha} \tag{4.118}$$

leading to a specific form for (4.115). The integral in (4.116) is elementary and gives

$$\Phi_2(0) = -\frac{2\epsilon}{a} \left(\frac{\alpha}{\pi}\right)^{1/2} \tag{4.119}$$

The potential energy of a point charge at the origin is, then,

$$V_M = \epsilon[\Phi_1(0) + \Phi_2(0)] \tag{4.120}$$

If we now take $\alpha \gg 1$, we should get a good approximation to the potential energy of a given charge due to alternating *point* charges at the other lattice sites. Using (4.115), (4.118), and (4.119) we obtain the formula

$$V_M = -\frac{2\epsilon^2}{a} \left\{ \left(\frac{\alpha}{\pi}\right)^{1/2} - \frac{1}{2\pi} \sum_{l_1,l_2,l_3} \frac{1}{(l_1 - \frac{1}{2})^2 + (l_2 - \frac{1}{2})^2 + (l_3 - \frac{1}{2})^2} \right.$$

$$\left. \exp\left(-\frac{\pi^2}{\alpha}\right)[(l_1 - \frac{1}{2})^2 + (l_2 - \frac{1}{2})^2 + (l_3 - \frac{1}{2})^2] \right\} \tag{4.121}$$

The coefficient of ϵ^2/a, evaluated for sufficiently large α, is the "Madelung constant." A quite accurate approximation (with an error of $< 1\%$) to the correct value of 1.7476 can be obtained by taking $\alpha = 2.5\pi^2$, a choice which leads to quite simple calculations. Adequate convergence in this case is obtained by summing all terms up to $|l_i - \frac{1}{2}| \leq \frac{7}{2}$ ($i = 1, 2, 3$). To obtain higher accuracy, it would be necessary to take a larger value of α ($5\pi^2$ would be a reasonable choice). It would then be necessary to sum more terms; the labor, however, is not prohibitive and in this case gives four-figure accuracy.

9. A Note on δ Functions and Three-Dimensional Transforms

It is interesting to note that many of the major results of Fourier transform theory may be obtained formally by the use of the Dirac "δ function" in three dimensions. Defining it by the relation

$$\int f(\mathbf{r}')\delta(\mathbf{r} - \mathbf{r}') \, d^3r' = f(\mathbf{r}) \tag{4.122}$$

where $f(\mathbf{r})$ is a well-behaved function, its transform is, from (4.61),

$$F(\mathbf{k}) = \frac{1}{(2\pi)^3} e^{-i\mathbf{k} \cdot \mathbf{r}'} \tag{4.123}$$

Therefore, from (4.62), if the Fourier integral theorem were valid for this "function," it could be written

$$\delta(\mathbf{r}-\mathbf{r}') = \frac{1}{(2\pi)^3} \int e^{i\mathbf{k}\cdot(\mathbf{r}-\mathbf{r}')} d^3k \qquad (4.124)$$

Of course, the integral on the right does not really exist. Equation (4.124) takes on meaning, however, if both sides are integrated with another function and (4.122) is used. If each side is multiplied by $f(\mathbf{r})$ and integrated over coordinates the resulting equation is

$$f(\mathbf{r}) = \left(\frac{1}{2\pi}\right)^3 \int f(\mathbf{r}') \left[\int e^{i\mathbf{k}\cdot(\mathbf{r}-\mathbf{r}')} d^3k \right] d^3r'$$

Reversing the order of the integrations on the left, a step which leads to well-defined integrals if $f(\mathbf{r})$ has a transform, reproduces the Fourier integral theorem:

$$f(\mathbf{r}) = \int F(\mathbf{k}) e^{i\mathbf{k}\cdot\mathbf{r}} d^3k$$

$$= \int \left[\left(\frac{1}{2\pi}\right)^3 \int f(\mathbf{r}') e^{-i\mathbf{k}\cdot\mathbf{r}'} d^3r' \right] e^{i\mathbf{k}\cdot\mathbf{r}} d^3k \qquad (4.125)$$

Equation (4.124), therefore, can be considered as valid in this sense.

The way in which it is most commonly used is, that when one has an integral such as that in (4.125), the order of the integrations is reversed, and the "invalid" $\mathbf{k}$ integral is replaced by the δ function, which may then be evaluated. The use of the δ function and its transform in this way is then completely equivalent to the use of a valid Fourier transform theorem.

Using δ functions in this way, the convolution theorem may be recovered quickly. For, expressing the functions $f(\mathbf{r})$ and $g(\mathbf{r})$ in the convolution by their transforms F and G, respectively, the convolution may be written

$$C(\mathbf{r}) = \int f(\mathbf{r}')g(\mathbf{r}-\mathbf{r}') d^3r'$$

$$= \int \left[\int F(\mathbf{k}) e^{i\mathbf{k}\cdot\mathbf{r}'} d^3k \right] \left[\int G(\mathbf{k}') e^{i\mathbf{k}'\cdot(\mathbf{r}-\mathbf{r}')} d^3k' \right] d^3r'$$

Doing the $\mathbf{r}'$ integral first gives

$$C(\mathbf{r}) = \int F(\mathbf{k})G(\mathbf{k}') e^{i\mathbf{k}'\cdot\mathbf{r}} (2\pi)^3 \delta(\mathbf{k}-\mathbf{k}') d^3k' d^3k$$

On performing the $\mathbf{k}'$ integration, the expression (4.69) is obtained for the convolution in terms of the transforms of its component functions:

$$C(\mathbf{r}) = (2\pi)^3 \int F(\mathbf{k})G(\mathbf{k}) e^{i\mathbf{k}\cdot\mathbf{r}} d^3k$$

The inverted equation (4.68) then follows.

10. Two- and Three-Center Integrals

Three-dimensional Fourier transforms provide a useful technique for doing integrals over functions centered on several points, such as are found in electrostatics and in atomic and molecular calculations. We consider a number of such integrals. First, however, we list several useful transforms.

(a)
$$\left(\frac{1}{2\pi}\right)^3 \int \frac{e^{-\alpha r}}{r} e^{-i\mathbf{k}\cdot\mathbf{r}} d^3\mathbf{r}$$

Putting $\mu = \cos\theta$ as the cosine of the angle between $\mathbf{k}$ and $\mathbf{r}$, using $\mathbf{k}$ as polar axis, and doing the angular integrations, the integral becomes

$$2\pi \int_0^\infty \frac{e^{-\alpha r}}{r}\left[\int_{-1}^1 e^{-ikr\mu} d\mu\right] r^2\,dr = \frac{4\pi}{k}\int_0^\infty e^{-\alpha r}\sin kr\,dr$$

$$= \frac{4\pi}{k}\,\text{Im}\int_0^\infty e^{-\alpha r} e^{ikr}\,dr$$

$$= \frac{4\pi}{k}\,\text{Im}\,\frac{1}{\alpha - ik}$$

$$= \frac{4\pi}{\alpha^2 + k^2} \tag{4.126}$$

so that the transform is $(1/2\pi^2)1/(\alpha^2 + k^2)$.

(b) Differentiating with respect to α, another useful transform is obtained:

$$\left(\frac{1}{2\pi}\right)^3 \int e^{-\alpha r} e^{-i\mathbf{k}\cdot\mathbf{r}} d^3\mathbf{r} = \frac{1}{\pi^2}\frac{\alpha}{(\alpha^2 + k^2)^2} \tag{4.127}$$

Transforms of $e^{-\alpha r}$ multiplied by positive powers of r may be obtained by further differentiations.

Problem 4-10:   Evaluate the transforms of

(i) $\mathbf{r}e^{-\alpha r}$

(ii) $\dfrac{\mathbf{r}}{r}\,e^{-\alpha r}$

and consequently express these functions as Fourier integrals.

Let us now consider some two- and three-center integrals.

(I) THE "OVERLAP INTEGRAL"

$$I_1 = \int e^{-\alpha r} e^{-\beta|\mathbf{r}-\mathbf{R}|} d^3\mathbf{r} \tag{4.128}$$

Replace $e^{-\alpha r}$ by its expression as an integral transform:

$$e^{-\alpha r} = \frac{\alpha}{\pi^2} \int \frac{1}{(\alpha^2 + k^2)^2} e^{i\mathbf{k} \cdot \mathbf{r}} \, d^3\mathbf{k} \tag{4.129}$$

Thus

$$I_1 = \frac{\alpha}{\pi^2} \int \frac{d^3\mathbf{k}}{(\alpha^2 + k^2)^2} \int e^{-\beta|\mathbf{r} - \mathbf{R}|} e^{i\mathbf{k} \cdot \mathbf{r}} \, d^3\mathbf{r}$$

But by (4.127), this is

$$I_1 = \frac{8}{\pi} \alpha\beta \int \frac{e^{i\mathbf{k} \cdot \mathbf{R}}}{(\alpha^2 + k^2)^2 (\beta^2 + k^2)^2} \, d^3\mathbf{k} \tag{4.130}$$

This may be evaluated by observing that

$$\frac{1}{(\alpha^2 + k^2)(\beta^2 + k^2)} = \frac{1}{\beta^2 - \alpha^2} \left[\frac{1}{\alpha^2 + k^2} - \frac{1}{\beta^2 + k^2} \right] \tag{4.131}$$

Therefore,

$$\int \frac{e^{i\mathbf{k} \cdot \mathbf{R}} d^3\mathbf{k}}{(\alpha^2 + k^2)(\beta^2 + k^2)} = \frac{2\pi^2}{\beta^2 - \alpha^2} \frac{e^{-\alpha R} - e^{-\beta R}}{R} \tag{4.132}$$

where R is the magnitude of the vector $\mathbf{R}$.

Calculating successively derivatives of the two sides with respect to α^2 and β^2, it is found that

$$\int \frac{e^{i\mathbf{k} \cdot \mathbf{R}} d^3\mathbf{k}}{(\alpha^2 + k^2)^2(\beta^2 + k^2)} = \pi^2 \left[\frac{1}{(\beta^2 - \alpha^2)\alpha} e^{-\alpha R} - \frac{2}{(\beta^2 - \alpha^2)^2} \left(\frac{e^{-\alpha R} - e^{-\beta R}}{R} \right) \right] \tag{4.133}$$

and

$$I_1 = \frac{8}{\pi} \alpha\beta \int \frac{e^{i\mathbf{k} \cdot \mathbf{R}} d^3\mathbf{k}}{(\alpha^2 + k^2)^2 (\beta^2 + k^2)^2}$$

$$= \frac{8\pi}{(\beta^2 - \alpha^2)^3} \left[(\beta^2 - \alpha^2)(\beta e^{-\alpha R} + \alpha e^{-\beta R}) - \frac{4\alpha\beta}{R} (e^{-\alpha R} - e^{-\beta R}) \right] \tag{4.134}$$

This formula is not too convenient when $\beta = \alpha$. It is, of course, possible to take the limit as $\beta \to \alpha$. Alternatively, it is possible to evaluate (4.130) by starting from

$$\int \frac{e^{i\mathbf{k} \cdot \mathbf{R}}}{k^2 + \alpha^2} \, d^3\mathbf{k} = 2\pi^2 \frac{e^{-\alpha R}}{R} \tag{4.135}$$

and differentiating successively with respect to α^2:

$$\int \frac{e^{i\mathbf{k}\mathbf{r}}}{(\alpha^2 + k^2)^2} \, d^3\mathbf{k} = \frac{\pi^2}{\alpha} e^{-\alpha R} \tag{4.136}$$

$$\int \frac{e^{i\mathbf{k} \cdot \mathbf{R}}}{(\alpha^2 + k^2)^3} \, d^3\mathbf{k} = \frac{\pi^2}{4\alpha^3} (1 + \alpha R) e^{-\alpha R} \tag{4.137}$$

and finally

$$\int \frac{e^{i\mathbf{k}\cdot\mathbf{R}}}{(\alpha^2+k^2)^4}\,d^3k = \frac{\pi^2}{24\alpha^5}(3+3\alpha R+\alpha^2 R^2)e^{-\alpha R} \tag{4.138}$$

(II) COULOMB INTERACTION BETWEEN TWO HYDROGEN ATOMS

A hydrogen atom in its lowest energy state consists of a nuclear charge $+\epsilon$ and a negative charge distribution

$$\rho(\mathbf{r}) = -\epsilon\,\frac{\alpha^3}{8\pi}\,e^{-\alpha R} \tag{4.139}$$

Taking the origin at one atom, and calling the position vector of the other relative to this one $\mathbf{R}$, the total potential energy of the Coulomb interaction between them is

$$V = \frac{\epsilon^2}{R} - 2\epsilon^2\,\frac{\alpha^3}{8\pi}\int \frac{1}{r}e^{-\alpha|\mathbf{r}-\mathbf{R}|}d^3r + \epsilon^2\left(\frac{\alpha^3}{8\pi}\right)^2 \int d^3r\,d^3r'\,e^{-\alpha r'}\,\frac{1}{|\mathbf{r}-\mathbf{r}'|}\,e^{-\alpha|\mathbf{r}-\mathbf{R}|} \tag{4.140}$$

The first term represents the interaction between the two nuclei, the second the two (equal) interactions between the electron distributions and and the nuclei of the other atom, and the third the interaction between the two electron distributions.

The second term, which we call V_1, can be done by first introducing the Fourier transform (4.129) of the exponential:

$$V_1 = -2\epsilon^2\,\frac{\alpha^3}{8\pi}\int \frac{d^3\mathbf{r}}{r}\int \frac{\alpha}{\pi^2}\frac{1}{(\alpha^2+k^2)^2}\,e^{i\mathbf{k}\cdot(\mathbf{r}-\mathbf{R})}\,d^3k \tag{4.141}$$

The $\mathbf{r}$ integral may then be evaluated, giving

$$V_1 = -\epsilon\,\frac{\alpha^4}{\pi^2}\int \frac{1}{k^2(\alpha^2+k^2)^2}\,e^{-i\mathbf{k}\cdot\mathbf{R}}\,d^3k \tag{4.142}$$

The integral is now a special case of (4.133), with $\beta = 0$:

$$V_1 = -\epsilon^2\,\frac{\alpha^4}{\pi^2}\pi^2\left[\frac{2}{\alpha^4}\frac{1-e^{-\alpha R}}{R}-\frac{1}{\alpha^3}e^{-\alpha R}\right]$$

$$= -\epsilon^2\left\{\frac{2}{R}(1-e^{-\alpha R})-\alpha e^{-\alpha R}\right\} \tag{4.143}$$

The last integral, which we call V_2, is a three-center integral ($\mathbf{r}$, $\mathbf{r}'$, and $\mathbf{R}$), and is the most difficult. Introducing again the Fourier transform of $e^{-\alpha|\mathbf{r}-\mathbf{R}|}$ it becomes

$$V_2 = \epsilon^2\,\frac{\alpha^7}{64\pi^4}\int d^3\mathbf{r}\,d^3\mathbf{r}'\,\frac{e^{-\alpha r'}}{|\mathbf{r}-\mathbf{r}'|}\int \frac{1}{(k^2+\alpha^2)^2}\,e^{i\mathbf{k}\cdot(\mathbf{r}-\mathbf{R})}d^3k \tag{4.144}$$

The **r** integral has the value $(4\pi/k^2)e^{i\mathbf{k}\cdot\mathbf{r}}$; substituting this value gives

$$V_2 = \epsilon^2 \frac{a^7}{16\pi^3} \int d^3r' e^{-\alpha r'} \int e^{i\mathbf{k}\cdot(\mathbf{r}'-\mathbf{R})} \frac{1}{k^2(k^2+\alpha^2)^2} d^3k \quad (4.145)$$

Next, it is possible to do the **r'** integral using (4.127), to get

$$V_2 = \epsilon^2 \frac{\alpha^8}{2\pi^2} \int \frac{1}{k^2(k^2+\alpha^2)^4} e^{-i\mathbf{k}\cdot\mathbf{R}} d^3k \quad (4.146)$$

To evaluate this integral, start from one already evaluated:

$$\int \frac{1}{(k^2+\alpha^2)^2 k^2} e^{-i\mathbf{k}\cdot\mathbf{R}} d^3k = \frac{\pi^2}{\alpha^4 R}(2 - 2e^{-\alpha R} - \alpha Re^{-\alpha R})$$

and differentiate twice with respect to α^2. This gives six times the required integral on the left. On the right, the differentiations yield

$$\frac{\pi^2}{4\alpha^8 R}[48 - e^{\alpha R}(48 + 33\alpha R + 9\alpha^2 R^2 + \alpha^3 R^3)]$$

Therefore,

$$V_2 = \frac{\epsilon^2}{R}\left[1 - e^{-\alpha R}\left(1 + \frac{11}{16}\alpha R + \frac{3}{16}\alpha^2 R^2 + \frac{1}{48}\alpha^3 R^3\right)\right] \quad (4.147)$$

The total potential energy of interaction is then

$$V = \frac{\epsilon^2}{R} + V_1 + V_2$$

$$= \frac{\epsilon^2}{R} e^{-\alpha R}\left[1 + \frac{5}{16}\alpha R - \frac{3}{16}\alpha^2 R^2 - \frac{1}{48}\alpha^3 R^3\right] \quad (4.148)$$

Another result which may be derived from the above calculation is that of the potential energy of interaction of two charge distributions both of which are centered on the same point. This result, which is needed for one of the standard calculations of the energy of the helium atom, is the limit of V_2 as $R \to 0$. It is, therefore,

$$V_2' = \tfrac{5}{16}\epsilon^2\alpha \quad (4.149)$$

Problem 4-11: Calculate the Coulomb interaction energy between two p-state hydrogen atoms, for which the electron charge distributions are, relative to their respective nuclei,

$$\frac{\beta^5}{32\pi} z^2 e^{-\beta r}$$

Consider the two cases:

(*i*) in which the z direction is along the line joining the two nuclei;

(*ii*) in which the z direction is perpendicular to the line joining the two nuclei.

Problem 4-12:   Calculate the integrals

$$\int \frac{\varphi_a(\mathbf{r}_1)\varphi_b(\mathbf{r}_1)\varphi_a(\mathbf{r}_2)\varphi_b(\mathbf{r}_2)}{r_{a2}} \, d^3\mathbf{r}_1 \, d^3\mathbf{r}_2$$

and

$$\int \frac{\varphi_a(\mathbf{r}_1)\varphi_b(\mathbf{r}_1)\varphi_a(\mathbf{r}_2)\varphi_b(\mathbf{r}_2)}{r_{12}} \, d^3\mathbf{r}_1 \, d^3\mathbf{r}_2$$

where

$$\varphi_{a,b}(\mathbf{r}) = \sqrt{\frac{\alpha^3}{\pi}} \, e^{-\alpha|\mathbf{r}-\mathbf{r}_{a,b}|}$$

and

$$\mathbf{R} = \mathbf{r}_b - \mathbf{r}_a$$

(These integrals enter into the calculation of the energy of a hydrogen molecule by the Heitler–London method.)

---

11. Laplace Transforms

The Laplace transform of the function $f(t)$ is

$$\mathscr{L}(f(t)) = \int_0^\infty f(t) \, e^{-st} \, dt \tag{4.150}$$

where $f(t)$ is assumed to be zero when the variable $t < 0$. This exists only for values of s such that the integral exists, say for Re $s > a$, where "Re" means "real part of." Thus, for any polynomial it exists for Re $s > 0$, but for $f(s) = e^{bt}$ it exists only for Re $s > b$.

Let us first calculate some theorems concerning Laplace transforms.

12. Transforms of Derivatives

The transform of $f'(t)$ is determined by integration by parts:

$$\mathscr{L}[f'(t)] = \int_0^\infty f'(t) \, e^{-st} \, dt$$

$$= [f(t) \, e^{-st}]_0^\infty + sF(s)$$

$F(s)$ being the transform of $f(t)$. Therefore,

$$\mathscr{L}[f'(t)] = -f(0) + sF(s) \tag{4.151}$$

Thus, the Laplace transform, unlike the Fourier transform, contains the "initial condition" on the function $f(t)$.

The derivative theorem may be applied repeatedly; for example,

$$\mathscr{L}[f''(t)] = -f'(0) + s\mathscr{L}[f'(t)]$$
$$= -f'(0) - sf(0) + s^2F(s) \tag{4.152}$$

and in general

$$\mathscr{L}[f^{(n)}(t)] = -f^{(n-1)}(0) - sf^{(n-2)}(0) \cdots$$
$$- s^{(n-1)}f(0) + s^nF(s) \tag{4.153}$$

Problem 4-13:   Prove (4.153) by induction.

13. "Shifting" Theorem

This theorem concerns the transform of the function $f(t-t_0)$, $(t_0 > 0)$. It is easily seen to be

$$\mathscr{L}[f(t-t_0)] = \int_0^\infty f(t-t_0) e^{-st} dt$$
$$= \int_{t_0}^\infty f(t-t_0) e^{-st} dt$$
$$= \int_0^\infty f(t') e^{-st'} dt' e^{-st_0}$$
$$= e^{-st_0}\mathscr{L}[f(t)] \tag{4.154}$$

14. Convolution Theorem

As in the case of Fourier transforms, the transform of a convolution is easily calculable in terms of the transforms of the functions from which it is formed. By "convolution" in this case we mean

$$C(t) = \int_0^t f_1(t')f_2(t-t') dt' \tag{4.155}$$

The upper limit of $t' = t$ arises from the fact that the functions are equal to zero for negative values of their argument.

The transform of the convolution is, by definition,

$$\mathscr{L}[C(t)] = \int_0^\infty dt\, e^{-st}\left[\int_0^t f_1(t')f_2(t-t') dt' \right]$$

This is a double integral whose range of integration in the (t, t') plane is as shown in Fig. 4.4; it is the cross-hatched region above the diagonal.

Reversing the order of integration gives

$$\mathscr{L}[C(t)] = \int_0^\infty f_1(t')\left[\int_{t'}^\infty e^{-st}f_2(t-t') dt \right] dt'$$

In the inner integral, substitute

$$t-t' = t'' \tag{4.156}$$

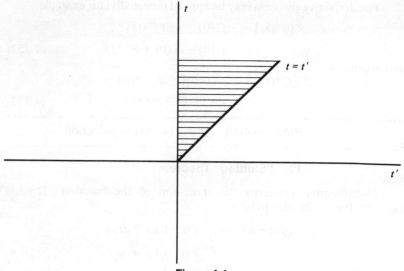

Figure 4.4

This leads to the result that

$$\mathcal{L}[C(t)] = \int_0^\infty f_1(t') \left[\int_0^\infty e^{-s(t'+t'')} f_2(t'') \, dt'' \right] dt'$$
$$= F_1(s) F_2(s) \tag{4.157}$$

That is, the transform of the convolution is the product of the transforms of the functions from which it is formed.

15. Some Simple Transforms

In this section we evaluate a few simple Laplace transforms which are frequently met in physical applications.

(a) $$f(t) = \theta(t) = 1, \quad t > 0$$
$$= 0, \quad t < 0$$

$$\mathcal{L}(\theta(t)) = \frac{1}{s} \tag{4.158}$$

(b) $$f(t) = \theta(t) t^n$$

$$\mathcal{L}(\theta(t) t^n) = \int_0^\infty t^n e^{-st} \, dt$$

$$= \frac{n!}{s^{n+1}} \tag{4.159}$$

(c) $$f(t) = \cos at\,\theta(t)$$

(d) $$f(t) = \sin at\,\theta(t)$$

These can be derived from

$$\mathscr{L}(e^{iat}) = \int_0^\infty e^{-st}\,e^{iat}\,dt$$

$$= \frac{1}{s - ia}$$

$$= \frac{s + ia}{s^2 + a^2}$$

Therefore,

$$\mathscr{L}(\cos at) = \frac{s}{s^2 + a^2} \tag{4.160}$$

and

$$\mathscr{L}(\sin at) = \frac{a}{s^2 + a^2} \tag{4.161}$$

(e) $$f(t) = e^{-bt}\theta(t)$$

In this case

$$\mathscr{L}(e^{-bt}) = \frac{1}{s + b} \tag{4.162}$$

(f) $$f(t) = e^{-bt}\cos at\,\theta(t)$$

This one can be done by replacing s by $(s + b)$ in (4.160); we find that

$$\mathscr{L}(e^{-bt}\cos at) = \frac{a}{(s + b)^2 + a^2} \tag{4.163}$$

(g) $$f(t) = e^{-bt}\sin at\,\theta(t)$$

By the same procedure as in (f), we obtain

$$\mathscr{L}(e^{-bt}\sin at) = \frac{s + b}{(s + b)^2 + a^2} \tag{4.164}$$

The above transforms form the basis of a table which can be read either to give the transforms from the functions or the functions from their transforms. Using it in the latter way, it may be shown that they permit us to calculate the function whose transform is $P(s)/Q(s)$, where $P(s)$ is a polynomial with real coefficients in s and $Q(s)$ is a similar polynomial of higher degree.

To prove that this is the case, let us suppose that $Q(s)$ is a polynomial of degree n. It then has n zeros, so that it is possible to write it as

$$Q(s) = a_0(s - s_1)(s - s_2)\cdots(s - s_n) \tag{4.165}$$

Because $Q(s)$ is real, any complex roots will occur in pairs; that is to say, if

$$s_i = a_i + ib_i \tag{4.166}$$

is a root, so is $a_i - ib_i$. The two linear factors can then be combined into a quadratic one:

$$(s - a_i)^2 + b_i^2$$

Let us first suppose that there are no repeated roots. Then we may expand

$$F(s) = \frac{P(s)}{Q(s)} \tag{4.167}$$

in partial fractions:

$$F(s) = \sum_j \frac{A_j}{s - s_j} + \sum_k \frac{B_k + C_k s}{(s - a_k)^2 + b_k^2} \tag{4.168}$$

where s_j are the real roots of $Q(s)$ and $a_k \pm ib_k$ are the complex ones. The coefficients A_j, B_k, and C_k may be obtained in terms of $P(s)$. If we first multiply across by $s - s_j$, and then put $s = s_j$, we get the constant A_j:

$$A_j = \lim_{s \to s_j} \frac{(s - s_j) P(s_j)}{Q(s)} \tag{4.169}$$

Both numerator and denominator of $(s - s_j)/Q(s)$ approach zero as $s \to s_j$, so we may use L'Hospital's rule to get

$$A_j = \frac{P(s_j)}{Q'(s_j)} \tag{4.170}$$

Equation (4.170) may in turn be used to get the terms in the partial fraction expression from two roots s_k and s_k^* which are complex conjugates of each other; together, they are

$$\frac{P(s_k)}{Q'(s_k)} \frac{1}{s - s_k} + \frac{P(s_k^*)}{Q'(s_k^*)} \frac{1}{s - s_k^*}$$

$$= \frac{[P(s_k)Q'(s_k^*) + P(s_k^*)Q(s_k)]s - P(s_k^*)Q'(s_k)s_k - P(s_k)Q'(s_k^*)s_k^*}{Q'(s_k)Q'(s_k^*)[(s - a_k)^2 + b_k^2]} \tag{4.171}$$

where

$$s_k = a_k + ib_k.$$

Comparing this with (4.168), it is found that

$$B_k = -\left[\frac{P(s_k^*)}{Q'(s_k^*)} s_k + \frac{P(s_k)}{Q'(s_k)} s_k^*\right] \tag{4.172}$$

and

$$C_k = \frac{P(s_k)}{Q'(s_k)} + \frac{P(s_k^*)}{Q'(s_k^*)} \tag{4.173}$$

This may be written in the alternative form

$$B_k = -2a_k \operatorname{Re}\left(\frac{P(s_k)}{Q'(s_k)}\right) - 2b_k \operatorname{Im}\left(\frac{P(s_k)}{Q'(s_k)}\right) \tag{4.174}$$

and

$$C_k = 2\operatorname{Re}\frac{P(s_k)}{Q'(s_k)} \tag{4.175}$$

To obtain the inverse of (4.168), that is, the function of which $F(s)$ is the transform, we need only write the numerators in the second sum in the form

$$B_k + C_k s = B_k + C_k a_k + C_k(s - a_k) \tag{4.176}$$

It is then evident that

$$f(s) = \sum_j A_j e^{s_j t} + \sum_k C_k e^{a_k t} \sin b_k t + \sum_k \frac{1}{b_k}(B_k + C_k a_k)e^{a_k t} \cos b_k t \tag{4.177}$$

Although this approach can also be extended to deal with the case of repeated roots, the problem may be treated more naturally by the inverse theorem, which we now discuss.

It is possible to see how to invert a Laplace transform by first starting from the Fourier integral theorem:

$$f(t) = \int_{-\infty}^{\infty} \phi(\omega)e^{i\omega t}\,d\omega \tag{4.178}$$

where

$$\phi(\omega) = \frac{1}{2\pi}\int_{-\infty}^{\infty} f(t)e^{-i\omega t}\,dt \tag{4.179}$$

Consider a function $g(t)$ which does not have a transform by virtue of the transform integral diverging at $+\infty$. Suppose, however, that $g(t)e^{-\sigma t}$ has a transform for large enough σ. Then consider the transform of $2\pi g(t)e^{-\sigma t}\theta(t)$ [or, alternatively, of $2\pi g(t)e^{-\sigma t}$, where $g(t) = 0$ for $t < 0$]:

$$\phi(\sigma + i\omega) = \int_0^{\infty} g(t)e^{-(\sigma + i\omega)t}\,dt \tag{4.180}$$

We write $s = \sigma + i\omega$, and call it the Laplace transform variable. The theorem expressing the function in terms of its transform is, then

$$2\pi g(t)e^{-\sigma t}\theta(t) = \int_{-\infty}^{\infty} \phi(\sigma + i\omega)e^{i\omega t}\,d\omega \tag{4.181}$$

The integral on the right-hand side may be considered as an integral over s along a contour parallel to the imaginary axis:

$$g(t)\theta(t) = \frac{1}{2\pi i}\int_{\sigma - i\infty}^{\sigma + i\infty} \phi(s)e^{st}\,ds \tag{4.182}$$

The integral on the right must, of course, vanish for $t < 0$. Now, in this

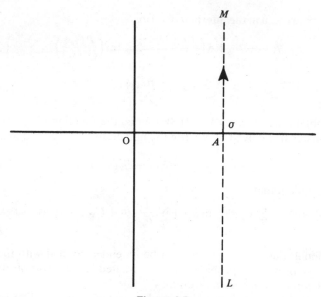

Figure 4.5

case it may be evaluated as a complex integral by using as a contour a semicircle with center at A and a radius which approaches ∞, and extending to the *right* of the dotted line LM representing the range of integration (see Fig. 4.5). The contribution to the integral from the semicircular contour is zero by virtue of the factor $e^{-R(\cos\theta + i\sin\theta)t}$, where $R = |s - \sigma| \rightarrow \infty$. For the integral to be zero, it is then necessary that there be no poles in $\phi(s)$ within this contour; that is, to the right of LM. The contour of integration may then be shifted to the right without affecting the value of the integral. It may be shifted to the left, however, only so long as it remains to the right of all singularities of $\phi(s)$. This consideration must then determine the choice of contour when a transform is known and the function from which it is derived is to be found.

With this understanding, (4.182) is the expression of the Laplace Inverse Theorem.

The inverse theorem of course provides an alternative means of evaluating the functions whose transforms are quotients $P(s)/Q(s)$ of polynomials. However, it can be used for a wider class of transforms $\phi(s)$. If $\phi(s)$ has no branch points, the integral (4.182) is $2\pi i$ times the sum of the residues at the poles. If there is an nth-order singularity at s_i, the residue is obtainable first by writing

$$\phi(s) = \frac{1}{(s - s_i)^n} \left[(s - s_i)^n \phi(s) \right] \tag{4.183}$$

and then picking out the coefficient of $(s-s_i)^{n-1}$ in the Taylor's series expansion of $(s-s_i)^n\phi(s)e^{st}$. The residue is then

$$\alpha_i(t) = \frac{1}{(n-1)!}\left[\frac{d^{n-1}}{ds^{n-1}}\{(s-s_i)^n\phi(s)e^{st}\}\right]_{s=s_i} \qquad (4.184)$$

The Laplace inverse is, then,

$$f(t) = \sum \alpha_i(t) \qquad (4.185)$$

Of course, for each pole s_i the order n of the singularity has its distinct value.

The derivative in (4.184) may be evaluated by Leibnitz's rule, giving

$$\alpha_i(t) = \sum_{m=0}^{n-1}\frac{1}{m!(n-1-m)!}\left[\frac{d^m}{ds^m}\{(s-s_i)^n\phi(s)\}\right]_{s=s_i} t^{n-1-m}e^{s_it} \qquad (4.186)$$

the factors $(n-1)!$ having cancelled in numerator and denominator. Equation (4.186) applies, of course, whether the poles s_i are real or imaginary. The imaginary ones may be combined in complex conjugate pairs to give the real function $f(t)$.

16.   Laplace Transform of a Periodic Function

Consider a function $\phi(t)$ defined on the interval $0 \leqslant t \leqslant T$. Construct the function $f(t)$ which is equal to $\phi(t)$ on that interval and defined by periodicity for other positive values of t. Its Laplace transform is

$$F(s) = \sum_{n=0}^{\infty}\int_{nT}^{(n+1)T}\phi(t-nT)e^{-st}\,dt \qquad (4.187)$$

In the nth term, we may substitute $t' = t - nT$ to get

$$F(s) = \sum_{n=0}^{\infty}e^{-snT}\int_0^T\phi(t)e^{-st}\,dt \qquad (4.188)$$

Designating the integral as $\Phi(s)$, this is

$$F(s) = \Phi(s)\sum_{n=0}^{\infty}e^{-snT}$$

$$= \Phi(s)\frac{1}{1-e^{-sT}} \qquad (4.189)$$

In inverting (4.189), there are contributions from all of the poles of the second factor, $(1-e^{-sT})^{-1}$. These poles are at

$$s = s_n = \frac{2\pi in}{T} \qquad (4.190)$$

where n is any arbitrary positive or negative integer. To find the contribution to the inverse from these poles, we must calculate the residues at the values s_n. Putting

$$s = s_n + \epsilon \qquad (4.191)$$

we see that

$$1 - e^{-sT} = 1 - e^{-(s_n+\epsilon)T}$$
$$= 1 - e^{-\epsilon T}$$
$$\simeq \epsilon T$$
$$= (s - s_n)T \qquad (4.192)$$

for ϵ small. The residue is, therefore,

$$a_n = \frac{\Phi(s_n)}{T} \qquad (4.193)$$

The total contribution from these poles is then

$$f_1(t) = \sum_{n=-\infty}^{\infty} a_n e^{s_n t}$$
$$= \frac{1}{T} \sum_{n=-\infty}^{\infty} \Phi(s_n) e^{s_n t}$$
$$= \frac{1}{T} \sum_{n=-\infty}^{\infty} \Phi\left(\frac{2\pi n}{T} i\right) e^{2\pi i n t/T} \qquad (4.194)$$

The coefficient of $e^{2\pi i n t/T}$ is

$$\frac{1}{T} \Phi\left(\frac{2\pi n}{T} i\right) = \frac{1}{T} \int_0^T \phi(t) e^{-2\pi i n t/T} \, dt \qquad (4.195)$$

We see, therefore, by comparison with (4.7) and (4.11), that (4.194) is the Fourier series expansion of the function $f(t)$. It follows that, for positive t, there are no other contributing poles to the inverse transform

$$f(t) = \frac{1}{2\pi i} \int_{\sigma-i\infty}^{\sigma+i\infty} \Phi(s) \frac{e^{st}}{1 - e^{-sT}} \, ds \qquad (4.196)$$

where, in this case, σ may be taken to be any value greater than zero. That there are no such poles is evident in any case, since if there were, they would contribute a negative experimental behavior to $\phi(t)$ when we invert $\Phi(s)$; we know, however, that $\phi(t)$ is precisely zero for $t > T$.

17. Resonance

Consider the second-order linear equation describing the reaction of a damped oscillator to an external stimulus $f(t)$:

$$\frac{d^2x}{dt^2} + \lambda \frac{dx}{dt} + \omega^2 x = f(t) \tag{4.197}$$

This equation may be solved by taking its Laplace transform. Initial conditions will be assumed

$$x(0) = x_0 \tag{4.198}$$

$$\dot{x}(0) = v_0 \tag{4.199}$$

The transform of $x(t)$ will be designated $X(s)$; that of $f(t)$, $F(s)$. The transform is, then,

$$-v_0 - sx_0 + s^2 X - \lambda x_0 + \lambda s X + \omega^2 X = F(s) \tag{4.200}$$

so that

$$X(s) = \frac{v_0 + (s + \lambda)x_0}{s^2 + \lambda s + \omega^2} + \frac{F(s)}{s^2 + \lambda s + \omega^2} \tag{4.201}$$

The first term represents a response to initial conditions, and thus has a transient character. Its only poles are

$$s = -\frac{\lambda}{2} \pm i \sqrt{\omega^2 - \frac{\lambda^2}{4}} \tag{4.202}$$

where we have assumed $\omega^2 > \lambda^2/4$. These roots are called s_1 (plus sign) and s_2 (minus sign). The contribution from these poles to the inverse of the first term of (4.201) is

$$x_1(t) = \frac{v_0 + (s_1 + \lambda)x_0}{s_1 - s_2} e^{s_1 t} + \frac{v_0 + (s_2 + \lambda)x_0}{s_2 - s_1} e^{s_2 t}$$

$$= (v_0 + \lambda x_0) \frac{e^{s_2 t} - e^{s_1 t}}{s_2 - s_1} + x_0 \frac{s_2 e^{s_2 t} - s_1 e^{s_1 t}}{s_2 - s_1} \tag{4.203}$$

Putting in the values of s_1 and s_2, the transient terms in the solution are seen to be

$$x_1(t) = e^{-\lambda t/2} \left\{ \left(v_0 + \frac{\lambda x_0}{2} \right) \frac{\sin \omega_0 t}{\omega_0} + x_0 \cos \omega_0 t \right\} \tag{4.204}$$

where

$$\omega_0 = \sqrt{\omega^2 - \frac{\lambda^2}{4}} \tag{4.205}$$

So far as the contributions to $x(t)$ from the second (forcing) term of (4.201) are concerned, they came from the poles of $F(s)$ as well as from s_1 and s_2. Let the poles of $F(s)$ be s_i, s_i^* ($i = 3, 4, \ldots$), and the residues be b_i, b_i^*, respectively. The contribution from such poles is

$$x_2(t) = \sum_i \left[\frac{b_i}{s_i^2 + \lambda s_i + \omega^2} e^{s_i t} + \frac{b_i^*}{s_i^{*2} + \lambda s_i^* + \omega^2} e^{s_i^* t} \right] \tag{4.206}$$

The contributions from s_1 and $s_2 = s_1^*$ are, on the other hand,

$$x_3(t) = \frac{F(s_1)e^{s_1 t}}{s_1 - s_2} + \frac{F(s_2)e^{s_2 t}}{s_2 - s_1} \qquad (4.207)$$

In view of the fact that

$$s^2 + \lambda s + \omega^2 = (s - s_1)(s - s_2)$$
$$= (s - s_1)(s - s_1^*) \qquad (4.208)$$

the denominators in (4.206) are

$$D_1 = (s_i - s_1)(s_i - s_1^*)$$

and

$$D_2 = (s_i^* - s_1)(s_i^* - s_1^*) = D_1^* \qquad (4.209)$$

respectively. While the terms (4.207) are again transient, if s_i is purely imaginary [that is, if $f(t)$ has a periodic component], (4.206) gives a non-transient contribution. Putting

$$b_i = \beta_i e^{i\delta_i} \qquad (4.210)$$

this contribution is

$$x_2(t) = \frac{\beta_i e^{i\delta_i}}{(i\omega_i - s_1)(i\omega_i - s_1^*)} e^{i\omega_i t} + \frac{\beta_i e^{-i\delta_i}}{(-i\omega_i - s_1)(-i\omega_i - s_1^*)} e^{-i\omega_i t}$$

$$\qquad (4.211)$$

where

$$s_i = i\omega_i \qquad (4.212)$$

With some simplification, this is

$$x_2(t) = \beta_i \left[\frac{1}{\omega_0^2 + \frac{1}{4}\lambda^2 - \omega_i^2 + i\lambda\omega_i} e^{i(\omega_i t + \delta_i)} + \frac{1}{\omega_0^2 + \frac{1}{4}\lambda^2 - \omega_i^2 - i\lambda\omega_i} e^{-i(\omega_i t + \delta_i)} \right]$$

$$= \frac{2\beta_i}{(\omega^2 - \omega_i^2)^2 + \lambda^2\omega_i^2} \left[(\omega^2 - \omega_i^2)\cos(\omega_i t + \delta_i) + \lambda\omega_i \sin(\omega_i t + \delta_i) \right]$$

$$\qquad (4.213)$$

As an example, consider the case in which

$$f(t) = A \cos \omega_1 t \qquad (4.214)$$

Then

$$F(s) = \frac{As}{s^2 + \omega_1^2} \qquad (4.215)$$

The poles are

$$s_1 = i\omega_1 \quad \text{and} \quad s_1^* = -i\omega_1$$

and the residues are

$$b_1 = \frac{A}{2} = b_1^*$$

The solution for $x_2(t)$ is, therefore,

$$x_2(t) = \frac{A}{(\omega^2 - \omega_1{}^2)^2 + \lambda^2\omega_1{}^2} [(\omega^2 - \omega_1{}^2) \cos \omega_1 t + \lambda\omega_1 \sin \omega_1 t] \qquad (4.216)$$

Putting

$$\tan \gamma_1 = \frac{\lambda\omega_1}{\omega^2 - \omega_1{}^2}$$

this may be put in the form

$$x_2(t) = \frac{A}{\sqrt{(\omega^2 - \omega_1{}^2)^2 + \lambda^2\omega_1{}^2}} \cos (\omega_1 t - \gamma_1) \qquad (4.217)$$

The resonant behavior at $\omega_1 = \omega$ is readily seen. At this resonance $\gamma_1 = \pi/2$, so that

$$x_2(t) = \frac{A}{\lambda\omega_1} \sin \omega_1 t \qquad (4.218)$$

Both the "height" and the "width" of the resonance are determined by the damping constant λ. The smaller the value of λ, the higher and narrower is the resonance.

The key to the analysis of the resonant behavior is in Eq. (4.211), where it is seen that the poles at $\pm i\omega_i$ pass near to the "natural" poles at $-(\lambda/2) \pm i\omega_0$, when ω_i is in the neighborhood of ω_0. The denominators then become small, leading to a peak in the magnitude of $x_2(t)$.

18.   Use of Laplace and Fourier Transforms to Solve the Vibrating String Problem

Consider the vibrating string equation

$$\frac{\partial^2 y}{\partial x^2} = \frac{1}{c^2} \frac{\partial^2 y}{\partial t^2} \qquad (4.219)$$

Subject to the initial conditions

$$y = y_0(x), \qquad t = 0 \qquad (4.220)$$

$$\frac{\partial y}{\partial t} = v_0(x), \qquad t = 0 \qquad (4.221)$$

If we take $Y(x, s)$ to be the Laplace transform of $y(x, t)$ with respect to the time variable, we may take the transform of (4.219) to obtain the equation

$$\frac{\partial^2 Y}{\partial x^2} = \frac{1}{c^2}(-sy_0 - v_0 + s^2 Y) \qquad (4.222)$$

If the string is infinitely long, we may now take the Fourier transform of this equation. Writing

$$\xi(k, s) = \mathscr{F}(Y(x, s)) \qquad (4.223)$$

the equation for the transform becomes

$$-k^2\xi = -\frac{s}{c^2}\bar{y}_0(k) - \frac{1}{c^2}\bar{v}_0(k) + \frac{s^2}{c^2}\xi \qquad (4.224)$$

where $\bar{y}_0(k)$, $v_0(k)$ are, respectively, the Fourier transforms of $y_0(x)$, $v_0(x)$.

Thus

$$\xi(k, s) = \frac{s\bar{y}_0 + \bar{v}_0}{s^2 + c^2 k^2} \qquad (4.225)$$

Taking the inverse Laplace transform of this equation, and calling the inverse of $\xi(k, s)$, $\eta(k, t)$

$$\eta(k, t) = \bar{y}_0 \cos kct + \bar{v}_0 \frac{\sin kct}{kc} \qquad (4.226)$$

The Fourier inverse of the first term is

$$\tfrac{1}{2}\int \bar{y}_0(k)(e^{ikct} + e^{-ikct})e^{ikx}\, dk = \tfrac{1}{2}[y_0(x + ct) + y_0(x - ct)] \qquad (4.227)$$

To obtain the inverse transform of the second term, we note that $v_0(x)$ replaces $y_0(x)$, and that $\sin kct/kc$ is the integral from 0 to t of $\cos kct$. Thus the inverse of the second term is

$$\frac{1}{2c}\left[\int_0^{x+ct} v_0(x')\, dx' - \int_0^{x-ct} v_0(x')\, dx'\right] = \frac{1}{2c}\int_{x-ct}^{x+ct} v_0(x')\, dx' \qquad (4.228)$$

This leads, then, to the solution obtained in the first chapter, Eq. (1.17):

$$y(x, t) = \tfrac{1}{2}\left[y_0(x + ct) + y_0(x - ct) + \frac{1}{c}\int_{x-ct}^{x+ct} v_0(x')\, dx'\right] \qquad (4.229)$$

Problem 4-14: Verify that the same result may be obtained by taking first the inverse *Laplace* transform of (4.225), followed by the inverse Fourier transform.

If the string is finite and of length L, it is possible to proceed from Eq.

(4.222), taking the Fourier *series* decomposition of the equation rather than the Fourier transform. If the string is of length L, and the ends are fixed, we write

$$Y(x, s) = \sum_{n=1}^{\infty} \bar{c}_n(s) \sin \frac{2\pi n x}{L} \qquad (4.230)$$

where

$$\bar{c}_n(s) = \frac{2}{L} \int_0^L Y(x', s) \sin \frac{2\pi n x'}{L} \, dx' \qquad (4.231)$$

We also let the nth Fourier coefficients of y_0, v_0 be a_n, b_n, respectively. Then, substituting in (4.222) and equating coefficients, it is found that

$$-\left(\frac{2\pi n}{L}\right)^2 \bar{c}_n = -\frac{s}{c^2} a_n - \frac{1}{c^2} b_n + \frac{s^2}{c^2} \bar{c}_n \qquad (4.232)$$

This may be solved to give

$$\bar{c}_n(s) = \frac{s a_n + b_n}{s^2 + (2\pi n c / L)^2} \qquad (4.233)$$

Taking the inverse Laplace transform, which we designate $c_n(t)$, we obtain

$$c_n(t) = a_n \cos \frac{2\pi n c t}{L} + \frac{L}{2\pi n c} b_n \sin \frac{2\pi n c t}{L} \qquad (4.234)$$

Then

$$y(x, t) = \sum_{n=1}^{\infty} c_n(t) \sin \frac{2\pi n x}{L}$$

$$= \sum \left(a_n \cos \frac{2\pi n c t}{L} + \frac{L}{2\pi n c} b_n \sin \frac{2\pi n c t}{L} \right) \cos \frac{2\pi n x}{L} \qquad (4.235)$$

19. The Gamma Function

The gamma function is defined for positive argument as

$$\Gamma(\alpha) = \int_0^{\infty} t^{\alpha - 1} e^{-t} \, dt \qquad (4.236)$$

This is met in calculating the Laplace transform of a general positive power of t:

$$\int_0^{\infty} t^{\alpha - 1} e^{-st} \, dt = \frac{\Gamma(\alpha)}{s^\alpha} \qquad (4.237)$$

If α is an integer, for example, $\alpha = n$,

$$\Gamma(n) = (n - 1)! \qquad (4.238)$$

On integration by parts it is easily verified that

$$\Gamma(\alpha) = -\int_0^\infty t^{\alpha-1} d(e^{-t})$$

$$= -[t^{\alpha-1}e^{-t}]_0^\infty + (\alpha-1)\int_0^\infty t^{\alpha-2}e^{-t}\, dt$$

$$= (\alpha-1)\Gamma(\alpha-1) \tag{4.239}$$

so long as we are dealing only with functions of positive argument.

If we wish to extend the definition of the Γ function to negative arguments, we may do so by using (4.239). Thus, for instance, from this formula

$$\Gamma(-4.2) = \left(\frac{1}{-4.2}\right)\left(\frac{1}{-3.2}\right)\left(\frac{1}{-2.2}\right)\left(\frac{1}{-1.2}\right)\left(\frac{1}{-0.2}\right)\Gamma(0.8)$$

We see, then, that the Γ function is defined for *all* arguments by (4.236) with $0 \le \alpha \le 1$, along with (4.239). It is easily seen that $\Gamma(z)$ has simple poles at $z = 0$ and at all negative integral values.

Problem 4-15:   Prove the following formulas:

$$\int_0^\infty t^{x-1}\cos t\, dt = \Gamma(x)\cos\frac{\pi x}{2}$$

$$\int_0^\infty t^{x-1}\sin t\, dt = \Gamma(x)\sin\frac{\pi x}{2}$$

---

20.   Stirling's Formula

This is a very useful formula in statistical physics, which gives an asymptotic form (that is, a form for large argument) for the Γ function. It is of particular interest for the case of integral argument (factorial).

In order to find it, we use a technique known as "method of steepest descents." We first write

$$\Gamma(z+1) = z^{z+1}\int_0^\infty e^{-tz}t^z\, dt$$

$$= z^{z+1}\int_0^\infty e^{-z(t-\ln t)}\, dt \tag{4.240}$$

We consider, then, the function $\Gamma(z+1)/z^{z+1}$ defined by the integral.

Our aim will now be to deform the contour so as to simplify the evaluation of the integral for large $|z|$. We note that the integrand is particularly large where $\operatorname{Re} z(\ln t - t)$ is positive. But if $|z|$ is large, the *imaginary* part will then in general produce very rapid oscillations of the integrand which will make evaluation very difficult. This difficulty may be overcome if we can find a contour on which the imaginary part is a constant.

Now it is well know that for a function of a complex variable (in this case t) the contours on which the real and imaginary parts are constant are orthogonal at every point. But the contour on which the imaginary part is constant is then in the direction of the gradient of the real part. If the real part is a *maximum* at $t = t_0$, its value will then decrease at a maximum rate along this contour.

We see, then, that if we take such a contour through t_0, the largest part of the integral will come from the immediate neighborhood of t_0.

Let us now carry this out for the case in point. The value t_0 is determined by

$$\frac{d}{dt}(\ln t - t) = 0$$

or

$$t = t_0 = 1 \qquad (4.241)$$

In fact, then, in this case it is not necessary to deform the contour at all; we simply expand

$$\ln t - t = -1 + \frac{(t-1)^2}{2}\left[\frac{d^2}{dt^2}(\ln t - t)\right]_{t=1} + \cdots$$

$$\simeq -1 - \frac{(t-1)^2}{2} + \text{higher powers} \qquad (4.242)$$

Assuming for the moment that when the real part of z is large the integral comes predominantly from the region of the saddle point $t = 1$, it follows that it is approximately

$$\Gamma(z+1) \simeq z^{z+1} e^{-z} \int_0^\infty e^{-z(t-1)^2/2}\, dt \qquad (4.243)$$

Putting $(t-1)/\sqrt{2} = \tau$ in the integrand this becomes

$$\Gamma(z+1) = z^{z+1} e^{-z} \sqrt{2} \int_{-1/\sqrt{2}}^\infty e^{-z\tau^2}\, d\tau$$

$$\simeq z^{z+1} e^{-z} \sqrt{2} \int_{-\infty}^\infty e^{-z\tau^2}\, d\tau$$

$$= \sqrt{2\pi}\, z^{z+1/2} e^{-z} \qquad (4.244)$$

This is Stirling's formula. For $z = $ the integer n it is often written in logarithmic form:

$$\ln n! \simeq (n + \tfrac{1}{2}) \ln n - n + \tfrac{1}{2} \ln 2\pi \qquad (4.245)$$

That this is increasingly accurate the larger the value of z (or n) is easily seen from the following facts.

(a) The real part of the argument $z(\ln t - t)$ has a single maximum at $t = 1$; it is monotonically decreasing (that is, takes on increasingly large *negative* values) as t moves away from unity in either direction.

(b) The argument of the exponential in the integral is

$$z\left[-1-\frac{(t-1)^2}{2}+a_3(t-1)^3+a_4(t-1)^4+\cdots\right]$$

In evaluating the integral, one may put

$$\frac{z(t-1)^2}{2} = u^2 \tag{4.246}$$

In terms of this variable this argument is

$$-z-u^2+\sum_{s=3}^{\infty} a_s\left(\frac{2u^2}{z}\right)^{s/2} \tag{4.247}$$

(c) By virtue of the factor e^{-u^2}, the integrand is negligible for $u \geqslant u_0$ (for instance, for u_0 between 5 and 6, $e^{-u_0^2} \approx 10^{-4}$). The higher terms in (4.247) are, then, negligible provided

$$z \gg 2u_0^2 \tag{4.248}$$

21.   The β Function

The β function is a function of two variables:

$$B(m, n) = \int_0^1 s^{m-1}(1-s)^{n-1}\, ds$$

$$= 2\int_0^{\pi/2} \sin^{2m-1}\theta \cos^{2n-1}\theta\, d\theta \tag{4.249}$$

on putting $s = \sin^2\theta$. This may now be expressed in terms of Γ functions as follows: Consider the product

$$\Gamma(m)\Gamma(n) = \int_0^{\infty} t^{m-1}\, e^{-t}\, dt \int_0^{\infty} v^{n-1}\, e^{-v}\, dv \tag{4.250}$$

The variables in this double integral may be written in terms of polar coordinates:

$$t = r^2 \sin^2\theta \tag{4.251}$$

$$v = r^2 \cos^2\theta \tag{4.252}$$

which gives

$$\Gamma(m)\Gamma(n) = 4\int_0^{\infty} r\, dr \int_0^{\pi/2} d\theta\, r^{2m+2n-2} \sin^{2m-1}\theta \cos^{2n-1}\theta\, e^{-r^2} \tag{4.253}$$

Putting $r^2 = x$, $\sin^2 \theta = s$, and writing the right-hand side as the product of two integrals, this takes the form

$$\Gamma(m)\Gamma(n) = \int_0^\infty x^{m+n-1} e^{-x} dx \int_0^1 s^{m-1}(1-s)^{n-1} ds$$

Since the second integral is the β function, it can obviously be expressed in the form:

$$B(m, n) = \frac{\Gamma(m)\Gamma(n)}{\Gamma(m+n)} \tag{4.254}$$

It might be noted in passing that if m and n are put equal to $\frac{1}{2}$ in (4.253),

$$[\Gamma(\tfrac{1}{2})]^2 = 4 \int_0^\infty re^{-r^2} dr \int_0^{\pi/2} d\theta = \pi$$

so that

$$\Gamma(\tfrac{1}{2}) = \sqrt{\pi} \tag{4.255}$$

These formulas involving γ and β functions are used in the discussion of the Bessel function below.

22. Use of Transforms for Equations with Linear Coefficients

A linear differential equation with constant coefficients can always be solved by the use of transforms, since the transforms of derivatives of a function are proportional to the transform of the function itself. We shall see in this section that equations with coefficients linear in the independent variable can also be solved by means of transforms.

Consider a differential equation of the form

$$[xf(D) + g(D)]y = 0 \tag{4.256}$$

where $f(D)$ and $g(D)$ are linear differential operators, D being synonymous with the derivative operator d/dx. If y is expressed in the form

$$y = \int_{s_1}^{s_2} e^{sx} F(s) \, ds \tag{4.257}$$

$$f(D)y = \int_{s_1}^{s_2} e^{sx} f(s) F(s) \, ds \tag{4.258}$$

and

$$xf(D)y = \int_{s_1}^{s_2} f(s) F(s) \frac{d}{ds} e^{sx} \, ds$$

$$= [f(s)F(s) e^{sx}]_{s_1}^{s_2} - \int_{s_1}^{s_2} e^{sx} \frac{d}{ds} \{f(s)F(s)\} \, ds \tag{4.259}$$

Equation (4.256) is then satisfied by choosing the function $F(s)$ and the limits s_1 and s_2 so that

$$[f(s)F(s)\,e^{sx}]_{s_1}^{s_2} - \int_{s_1}^{s_2} e^{sx}\left[\frac{d}{ds}\{f(s)F(s)\} - g(s)F(s)\right]ds = 0 \quad (4.260)$$

The procedure is now to choose $F(s)$ so as to make the quantity in the square bracket zero; once $F(s)$ is known, the limits may be chosen to make the integrated term vanish. Putting

$$f(s)F(s) = \phi(s) \quad (4.261)$$

we must first solve

$$\frac{d\phi}{ds} = \frac{g(s)}{f(s)}\phi(s) \quad (4.262)$$

The solution is

$$\phi(s) = \exp\int_{s_0}^{s} \frac{g(s')}{f(s')}\,ds' \quad (4.263)$$

Thus

$$F(s) = \frac{1}{f(s)}\exp\int_{s_0}^{s} \frac{g(s')}{f(s')}\,ds' \quad (4.264)$$

The integrated term is

$$\left[\exp\left(sx + \int_{s_0}^{s}\frac{g(s')}{f(s')}\,ds'\right)\right]_{s_1}^{s_2} \quad (4.265)$$

The choice of s_1 and s_2 to make this zero will, of course, depend on the functions $f(s)$ and $g(s)$, and is in general not unique.

Before turning to particular applications, we make two observations: first, that y in (4.257) has the form of a Laplace transform (that is, it is a continuous linear combination of exponentials). Taking s purely imaginary ($s = it$) it would, however, have the form of a Fourier transform. Since the choice of contour and of limits of integration is determined by the calculation, it cannot be said in advance which form will emerge. Secondly, if we tried a similar technique for an equation with quadratic coefficients, or ones with still higher powers, the method would in general not be a feasible one to employ. For if there were a term of the form $x^2h(D)$, we would have

$$x^2h(D)f = x^2\int_{s_1}^{s_2} e^{sx}h(s)F(s)\,ds$$

$$= \int_{s_1}^{s_2} h(s)F(s)\frac{d^2}{ds^2}e^{sx}\,ds \quad (4.266)$$

It would then be necessary to integrate by parts *twice*. This would lead to *two* integrated terms — which would create no fundamental problems — and to the integral term

$$\int_{s_1}^{s_2} e^{sx}\frac{d^2}{ds^2}\{h(s)F(s)\}\,ds \quad (4.267)$$

Adding such a term to (4.260), we see that $F(s)$ is determined now by the solution of a rather complicated *second-order* differential equation. In general, there is no reason to believe that this will be easier to solve than the original equation.

Though the method seems, therefore, to be applicable only to equations with linear coefficients, it should be borne in mind that in fact it applies to all equations which can be deduced from these by a transformation of either dependent or independent variable, or both.

23. The Confluent Hypergeometric Function

The confluent hypergeometric equation is

$$xy'' + (\gamma - x)y' - \beta y = 0 \tag{4.268}$$

This is a particular case of (4.256), where

$$f(s) = s^2 - s \tag{4.269}$$

and

$$g(s) = \gamma s - \beta \tag{4.270}$$

The exponent in (4.264) is then

$$\int_{s_0}^{s} \frac{\gamma s - \beta}{s^2 - s} \, ds = \int_{s_0}^{s} \left[\frac{\beta - \gamma}{1 - s} + \frac{\beta}{s} \right] ds$$

$$= (\gamma - \beta) \ln(1 - s) + \beta \ln s + \text{constant} \tag{4.271}$$

It follows that

$$F(s) = (1 - s)^{\gamma - \beta - 1} s^{\beta - 1} \tag{4.272}$$

The integrated term in (4.260) is

$$[(1 - s)^{\gamma - \beta} s^{\beta} e^{sx}]_{s_1}^{s_2} \tag{4.273}$$

If s_1 and s_2 are chosen to make this zero, the solution of the differential equation is

$$y_1 = C \int_{s_1}^{s_2} (1 - s)^{\gamma - \beta - 1} s^{\beta - 1} e^{sx} \, ds \tag{4.274}$$

where C is a constant.

Consider the case in which $\gamma > \beta > 0$. Then it is possible to choose $s_1 = 0$, $s_2 = 1$. The solution then has an essential (exponential) singularity when $x \to +\infty$. A series solution may be obtained by expanding the exponential:

$$y_1 = C \sum_{n=0}^{\infty} \frac{x^n}{n!} \int_0^1 (1 - s)^{\gamma - \beta - 1} s^{\beta + n - 1} \, ds$$

$$= C \sum_{n=0}^{\infty} \frac{\Gamma(\gamma - \beta)\Gamma(\beta + n)}{\Gamma(\gamma + n)} \frac{x^n}{n!} \tag{4.275}$$

If we write

$$\Gamma(\beta+n) = \beta(\beta+1) \ldots (\beta+n-1)\Gamma(\beta)$$
$$= \beta_n\Gamma(\beta) \qquad (4.276)$$

where β_n is the product of n factors, the first of which is β and each subsequent one is one greater than the preceding, the solution may be written

$$y_1(x) = C\frac{\Gamma(\gamma-\beta)\Gamma(\beta)}{\Gamma(\gamma)} \sum_{n=0}^{\infty} \frac{\beta_n}{\gamma_n}\frac{x^n}{n!}$$

The constant C may be chosen so that the factor in front of the sum is unity. The solution resulting from this choice is the "confluent hypergeometric function"

$$y_1(x) = F(\beta,\gamma,x)$$
$$= \sum_{n=0}^{\infty} \frac{\beta_n}{\gamma_n}\frac{x^n}{n!} \qquad (4.277)$$

With the value of C so chosen, the integral representation of the confluent hypergeometric function is

$$F(\beta,\gamma,x) = \frac{\Gamma(\gamma)}{\Gamma(\beta)\Gamma(\gamma-\beta)} \int_0^1 (1-s)^{\gamma-\beta-1}s^{\beta-1}e^{sx}\,ds \qquad (4.278)$$

This form is valid except when either β or $\gamma-\beta$ is negative or zero.

Problem 4-16: Show that when β is a negative integer, $-r$, a solution may be found in the form of a contour integral around the origin

$$F(-r,\gamma,x) = a\oint \frac{(1-s)^{\gamma+r-1}}{s^{r+1}}e^{sx}\,ds$$

By evaluating the contour integral show that

$$F(-r,\gamma,x) = 2\pi ia(-1)^r \sum_{m=0}^{r} (-1)^m \binom{\gamma+r-1}{r-m}\frac{x^m}{m!}$$

where $\binom{p}{q}$ is the coefficient of x^q in the expansion of $(1+x)^p$.

Using the condition that $F(-r,\gamma,x) = 1$ when $x=0$ determine a, and hence show that

$$F(-r,\gamma,x) = \sum_{m=0}^{r} \binom{r}{m}\frac{(-1)^m}{\gamma_m}x^m$$
$$= \frac{1}{2\pi i}(-1)^r\frac{r!}{\gamma_r}\oint \frac{(1-s)^{\gamma+r-1}}{s^{r+1}}e^{sx}\,ds$$

If $\gamma-\beta$ is a negative integer but β is not, show that the substitution $s' = 1-s$ may be used to derive a contour integral representation.

While this solution has been obtained in the first instance on the assumption that $\gamma > \beta > 0$, the fact that (4.277) satisfies the differential equation cannot depend on the values of β and γ in general, so that it provides a solution in all instances except when γ is a negative integer.

When $\gamma = -m$, a negative integer, it is evident from the series form (4.275) that, due to singularities of the denominator, all the terms of the series are equal to zero up to the term for which $n = m+1$. Therefore, at the origin the solution behaves in this case like x^{m+1}.

Problem 4-17:   Prove that

$$\frac{d}{dx} F(\beta, \gamma, x) = \frac{\beta}{\gamma} F(\beta+1, \gamma+1, x)$$

Let us substitute

$$y = x^{m+1} u \tag{4.279}$$

A simple calculation shows that u in turn satisfies a confluent hypergeometic equation

$$xu'' + (m+2-x)u' - (m+1+\beta)u = 0 \tag{4.280}$$

The function $u(x)$ therefore has the form

$$u = BF(m+1+\beta, m+2, x) \tag{4.281}$$

Problem 4-18:   Prove that

$$F(\beta, \gamma, x) = e^x F(\gamma - \beta, \gamma, x)$$

The value of the constant B which relates this solution to (4.275) may be found as follows: Using the known value of C,

$$C = \frac{\Gamma(\gamma)}{\Gamma(\beta)\Gamma(\gamma - \beta)} \tag{4.282}$$

in Eq. (4.275), we find that

$$\frac{y_1}{\Gamma(\gamma)} = \sum_{n=0}^{\infty} \frac{\Gamma(\beta+n)}{\Gamma(\beta)} \frac{1}{\Gamma(\gamma+n)} \frac{x^n}{n!} \tag{4.283}$$

The coefficient of the leading term as $\gamma \to -m$, that is, the coefficient of x^{m+1}, is

$$\lim_{x \to 0} \lim_{\gamma \to -m} \frac{y_1}{\Gamma(\gamma) x^{m+1}} = \frac{\Gamma(\beta+m+1)}{\Gamma(\beta)(m+1)!} = \frac{\beta_{m+1}}{(m+1)!}$$

But the limit of y, given by (4.279) and (4.281), is simply equal to B. Therefore,

$$\lim_{\gamma \to -m} \frac{1}{\Gamma(\gamma)} F(\beta, \gamma, x) = \frac{\beta_{m+1}}{(m+1)!} x^{m+1} F(m+1+\beta, m+2, x) \tag{4.284}$$

Problem 4-19: From the integral representation (4.278), show that, for large positive x

$$F(\beta,\gamma,x) \approx \frac{e^x}{x^{\gamma-\beta}}\frac{\Gamma(\gamma)}{\Gamma(\beta)}$$

provided that γ and β are not negative integers.

We are, of course, able to conclude, from the fact that their leading terms are equal, that the functions on the right- and left-hand sides are equal for all x, since they satisfy the same differential equation.

We have now shown how one solution of the confluent hypergeometic is found, for arbitrary values of β and γ; (4.275) provides a solution for $\gamma \neq -m$ and (4.284) for $\gamma = -m$. Let us see next how to find a second independent solution.

If we attempt to find a solution of the confluent hypergeometic of the form

$$y = x^k w(x) \tag{4.285}$$

where $w(x)$ is finite and analytic at the origin, we find, on substituting in the differential equation

$$x[x^k w'' + 2kx^{k-1}w' + k(k-1)x^{k-2}w] + (\gamma-x)[x^k w' + kx^{k-1}w] - \beta x^k w = 0 \tag{4.286}$$

Equating those terms which are predominant near the origin, that is, those which behave like x^{k-1}, we find that

$$k(k-1) + k\gamma = 0$$

so that $k = 0$ or

$$k = 1 - \gamma \tag{4.287}$$

Since the confluent hypergeometric function corresponds to the case $k = 0$, consider the other case. Putting $k = 1 - \gamma$ in (4.286), and dividing through by x^k, we find for $w(x)$ the equation

$$xw'' + (2-\gamma-x)w' - (1+\beta-\gamma)w = 0 \tag{4.288}$$

This is again a confluent hypergeometric equation. Therefore, the second solution is

$$y_2(x) = x^{1-\gamma}F(1+\beta-\gamma, 2-\gamma, x) \tag{4.289}$$

The general solution, then, except in special cases to be noted, is

$$y = C_1 y_1(x) + C_2 y_2(x)$$

$$= C_1 F(\beta,\gamma,x) + C_2 x^{1-\gamma}F(1+\beta-\gamma, 2-\gamma, x) \tag{4.290}$$

It is interesting to note that, when $\gamma = -m$, $y_2(x)$ becomes $x^{m+1}F(m +1+\beta, m+2, x)$, which is, to within a constant, identical with (4.284).

The problem remains, then, how to find a second solution when γ is a negative integer. Such a solution may be written down formally using (B.7) of Appendix 3B. The Wronskian is

$$W = \exp \int \left(1 - \frac{\gamma}{x}\right) dx$$

$$= x^{-\gamma} e^{x} \tag{4.291}$$

Knowing y_2, it follows that a second solution is

$$y = y_2(x) \int^{x} \frac{W(x')}{y_2{}^2(x')}\, dx'$$

$$= y_2(x) \int^{x} \frac{e^{x'}\, dx'}{x'^{m+2}F^{2}(m+1+\beta, m+2, x')} \tag{4.292}$$

Since $e^{x'}/F^2(m+1+\beta, m+2, x')$ can be expanded in a series about the origin, we see that there is a logarithmic term in the integral. The other solution is, therefore, not analytic in x, but has the form

$$y(x) = C_0 y_2(x) \ln x + \text{power series}$$

C_0 being a constant.

A remaining case in which only one solution has been found is that in which $\gamma = 1$, since, as may easily be seen, for this value of γ, $y_2 = y_1$. In this case, too, the second solution may be found in terms of the first from (B.7) as above; it has the form

$$y = F(\beta, 1, x) \int^{x} \frac{e^{x'}\, dx'}{x'F^{2}(\beta, 1, x')} \tag{4.293}$$

and again has a logarithmic singularity at $x = 0$.

We now turn to some special cases of functions related to the confluent hypergeometric function.

24. Laguerre Functions

The Laguerre equation is

$$xy'' + (\alpha + 1 - x)y' + ny = 0 \tag{4.294}$$

where we assume α to be > -1. This is directly a confluent hypergeometric; one solution is then

$$y = L_n{}^{(\alpha)}(x) = \frac{(\alpha+1)_n}{n!} F(-n, \alpha+1, x) \tag{4.295}$$

which is designated the Laguerre function. The choice of the multiplying constant is conventional. If n is an integer, we see immediately that the function is a polynomial of degree n.

The second independent solution of the equation, which has a singularity at the origin, is

$$y_2(x) = x^{-\alpha}F(-n-\alpha, 1-\alpha, x) \qquad (4.296)$$

unless α is zero or a positive integer. If α is zero, the solution is obviously identical with the first one. If it is a positive integer, formula (4.284) with $\alpha = m+1$ again shows that the "second" solution is identical with the first. In either case, an independent second, non-analytic solution exists, as demonstrated in the last section.

An integral representation of the Laguerre polynomials is, following (4.274),

$$L_n^{(\alpha)}(x) = \text{(constant)} \oint \frac{(1-s)^{\alpha+n}e^{sx}}{s^{n+1}} \, ds \qquad (4.297)$$

where we have chosen as contour of integration a path enclosing the singularity at the origin. This may be transformed by

---

Problem 4-20:   Taking the constant in (4.297) to be $1/2\pi i$, show, by making the substitution $sx = x - \sigma$ and evaluating the contour integration in σ, that

$$L_n^{(\alpha)}(x) = \frac{(-1)^n}{n!} e^x \frac{d^n}{dx^n} \left(x^{n+\alpha} e^{-x} \right)$$

---

putting

$$\frac{1-s}{s} = \frac{1}{t} \qquad (4.298)$$

The integral then becomes

$$L_n^{(\alpha)}(x) = \text{(constant)} \oint t^{-(n+1)} \frac{1}{(1+t)^{\alpha+1}} e^{tx/(1+t)} \, dt \qquad (4.299)$$

around the transformed contour. Now since the origin in s transforms into the origin in t, and $s = \infty$ transforms into $t = -1$, a contour surrounding $s = 0$ transforms into one surrounding $t = 0$ but not enclosing $t = -1$. The residue at $t = 0$ is the coefficient of t^n in

$$g(t) = \frac{1}{(1+t)^{\alpha+1}} \exp \frac{tx}{1+t} \qquad (4.300)$$

$L_n^{(\alpha)}(x)$ is therefore $2\pi i$ (constant) times this coefficient; thus

$$g(t) = \frac{1}{(1+t)^{\alpha+1}} \exp \frac{tx}{1+t} = \frac{1}{2\pi i (\text{constant})} L_n^{(\alpha)}(x) t^n$$

The value of the constant may be determined by putting $x = 0$. From (4.277) and (4.295)

$$L_n^{(\alpha)}(0) = \frac{(\alpha+1)_n}{n!} \qquad (4.301)$$

But

$$\frac{1}{(1+t)^{\alpha+1}} = \sum_{n=0}^{\infty} (-1)^n \frac{(\alpha+1)_n}{n!} t^n$$

Therefore the constant must be $1/(2\pi i)$, and we have

$$\frac{1}{(1+t)^{\alpha+1}} \exp \frac{tx}{1+t} = \sum_{n=0}^{\infty} (-1)^n L_n^{(\alpha)}(x) t^n \qquad (4.302)$$

The function on the left is therefore known as the "generating function" for the Laguerre polynomials.

Problem 4-21: By expanding the exponential in (4.302) show that

$$L_n^{(\alpha)}(x) = \sum_{s=0}^{n} \frac{\Gamma(\alpha+n+1)}{\Gamma(\alpha+s+1)} \frac{1}{(n-s)!s!} (-x)^s$$

Calculate the Laguerre polynomials for $\alpha = 0$ and $n = 0, 1, 2, 3, 4$ and compare with the forms obtained in Chapter 2.

Problem 4-22: By differentiation of the generating function with respect to x, show that

$$L_n^{(\alpha)\prime}(x) = L_{n-1}^{(\alpha+1)}(x)$$

Problem 4-23: By differentiating the generating formula with respect to t, derive the recursion formula:

$$nL_n^{(\alpha)}(x) = (x-\alpha-2n+1)L_{n-1}^{(\alpha)}(x) - (\alpha+n-1)L_{n-2}^{(\alpha)}(x)$$

---

25. Hermite Functions

The Hermite equation, which arises in the treatment of the harmonic oscillator in quantum mechanics (Chapter 9), is

$$y'' - xy' + ny = 0 \qquad (4.303)$$

In the notation of (4.256),

$$f(s) = -s \qquad (4.304)$$

$$g(s) = s^2 + n \qquad (4.305)$$

Thus,

$$\phi(s) = \exp \int_{s_0}^{s} \frac{g(s')}{f(s')}\, ds'$$

$$= A \exp\left(-\tfrac{1}{2}s^2 - n \ln s\right)$$

$$= \frac{A}{s^n} e^{-(1/2)s^2} \tag{4.306}$$

A being a constant. A solution of the equation may then be written as

$$y = A \int_{s_1}^{s_2} \frac{1}{s^{n+1}} e^{-(1/2)s^2} e^{sx}\, ds$$

where s_1 and s_2 are values of the variable s for which

$$\exp\left[sx - \tfrac{1}{2}s^2\right]/s^n = 0 \tag{4.307}$$

If n is an integer, the obvious contour to take is a closed loop surrounding the origin. If n is not an integer, there is a branch point at the origin. Making a cut along the positive real axis, one may choose a contour coming in from infinity on one side of the cut, looping around the origin, and going out the other side. The integral is then that of the discontinuity along the cut.

In the case in which n is a positive integer, the contour integral can be evaluated; it is $-2\pi i A$ times the coefficient of s^n in $e^{sx}e^{-(1/2)s^2}$. Choosing

$$-2\pi i A = \frac{1}{n!} \tag{4.308}$$

we obtain the generating formula

$$e^{sx}e^{-(1/2)s^2} = \sum_{n=0}^{\infty} H_n(x) \frac{s^n}{n!} \tag{4.309}$$

where $H_n(x)$ is the standard solution of (4.303) known as the "Hermite polynomial," and is represented by the integral

$$H_n(x) = \frac{1}{2\pi i n!} \oint \frac{1}{s^{n+1}} e^{sx} e^{-(1/2)s^2}\, ds \tag{4.310}$$

Problem 4-24:   Differentiating (4.309) with respect to s, prove the recursion relation

$$xH_n = H_{n+1} + nH_{n-1}$$

for $n \geqslant 1$. What is the relation for xH_0?

Problem 4-25:   Differentiating (4.309) with respect to x, prove the recursion relation

$$H_n' = nH_{n-1}$$

Problem 4-26: From the recursion relation, write down the first five Hermite polynomials.

Problem 4-27: Multiplying $e^{sx}e^{-(1/2)s^2} = \Sigma H_n(x)(s^n/n!)$ and $e^{tx}e^{-(1/2)t^2} = \Sigma H_m(x)(t^m/m!)$ show that

$$\int_{-\infty}^{\infty} H_n(x)H_m(x)e^{-x^2/2}\,dx = 0$$

for $m \neq n$ and $\sqrt{2\pi}n!$ for $m = n$.

The Hermite polynomial may be written in another way by introducing $t = s - x$ as a variable. Equation (4.310) then becomes

$$H_n(x) = \frac{e^{x^2/2}}{2\pi i n!} \oint \frac{1}{(t+x)^{n+1}} e^{-(1/2)t^2}\,dt \qquad (4.311)$$

the contour now being taken around $t = -x$. The integral is $2\pi i$ times the residue of $e^{-(1/2)t^2}$ at $t = -x$, which is in turn $n!$ times the nth derivative of $e^{-(1/2)t^2}$ at $t = -x$. Therefore,

$$H_n(x) = (-1)^n e^{x^2/2} \frac{d^n}{dx^n} e^{-x^2/2} \qquad (4.312)$$

Substituting

$$\tfrac{1}{2}x^2 = z \qquad (4.313)$$

in (4.303) leads to the differential equation

$$xy'' + (\tfrac{1}{2} - z)y' + \frac{n}{2}y = 0 \qquad (4.314)$$

which is a confluent hypergeometric equation. Therefore one solution is

$$y_1 = F\left(-\frac{n}{2}, \frac{1}{2}, \frac{x^2}{2}\right) \qquad (4.315)$$

and a second is, from (4.289),

$$y_2 = \left(\frac{x^2}{2}\right)^{1/2} F\left(-\frac{n-1}{2}, \frac{3}{2}, \frac{x^2}{2}\right) \qquad (4.316)$$

In the case in which n is even, y_1 is a polynomial, since β is a negative integer. Therefore, (4.315) multiplied by some factor B_n is the Hermite polynomial. The coefficient can be obtained by putting $x = 0$:

$$H_n(0) = B_n F\left(-\frac{n}{2}, \frac{1}{2}, 0\right) \qquad (4.317)$$

From (4.309)

$$H_n(0) = (-\tfrac{1}{2})^{n/2} \frac{n!}{(n/2)!}$$

and since $F(\beta, \gamma, 0) = 1$, we have

$$H_n(x) = (-\tfrac{1}{2})^{n/2} \frac{n!}{(n/2)!} F\left(-\frac{n}{2}, \frac{1}{2}, \frac{x^2}{2}\right) \qquad (4.318)$$

for even n.

When n is an *odd* integer, on the other hand, it is the solution y_2 which is polynomial; we therefore identify

$$B_n' x F\left(-\frac{n-1}{2}, \frac{3}{2}, \frac{x^2}{2}\right)$$

with $H_n(x)$. B_n' may be determined by equating coefficients of x; it is the coefficient of x in $H_n(x)$. From (4.309) this is $(-\tfrac{1}{2})^{(n-1)/2}[n!/((n-1)/2)!]$. Therefore,

$$H_n(x) = \left(-\frac{1}{2}\right)^{(n-1)/2} \frac{n!}{((n-1)/2)!} F\left(-\frac{n-1}{2}, \frac{3}{2}, \frac{x^2}{2}\right) \qquad (4.319)$$

for odd n.

Let us prove, finally, an interesting addition formula for Hermite polynomials, which can be derived from the generating function (4.309). Consider the formulas

$$e^{a_i s x_i} e^{-(1/2) s^2 a_i^2} = \sum_{n_i=0}^{\infty} H_{n_i}(x_i) \frac{s^{n_i} a_i^{n_i}}{n_i!} \qquad (4.320)$$

where $i = 1, 2, \ldots, N$. Multiply together the right-hand sides and the left-hand sides, writing

$$\mathbf{a} \cdot \mathbf{x} = \sum_{i=1}^{N} a_i x_i \qquad (4.321)$$

$$a^2 = \sum_{i=1}^{N} a_i^2 = 1 \qquad (4.322)$$

Thus

$$e^{\mathbf{a} \cdot \mathbf{x} s} e^{-(1/2) s^2} = \sum_{n=0}^{\infty} H_n(\mathbf{a} \cdot \mathbf{x}) \frac{s^n}{n!}$$

$$= \prod_{i=1}^{N} \sum_{n_i=0}^{\infty} H_{n_i}(x_i) \frac{s^{n_i} a_i^{n_i}}{n_i!}$$

Equating coefficients of s^n on both sides

$$H_n(\mathbf{a} \cdot \mathbf{x}) = n! \sum_{n_1 + \cdots + n_N = n} \prod_{i=1}^{N} H_{n_i}(x_i) \frac{a_i^{n_i}}{n_i!} \qquad (4.323)$$

26. Equations Reducible to the Confluent Hypergeometric: The Bessel Equation

In the previous section, we have seen that the Hermite equation was reducible to a confluent hypergeometric by a change of independent variable. There are also equations which may be so reduced by a transformation of the *dependent* variable.

For example, let us find the equation for y such that

$$y = e^{-\alpha x} x^s f(x) \qquad (4.324)$$

where $f(x)$ satisfies the confluent hypergeometric. Substitution of $e^{\alpha x} x^{-s} y$ for f in the equation leads to the following differential equation for y:

$$x^2 y'' + [(2\alpha - 1)x^2 + (\gamma - 2s)x]y' + [\alpha(\alpha - 1)x^2 + (\gamma\alpha - \beta - 2s\alpha + s)x$$
$$- s(\gamma - s - 1)]y = 0 \qquad (4.325)$$

A slight further generalization is possible if we seek to have y in the form

$$y = e^{-\alpha x} x^s f(kx) \qquad (4.326)$$

This requires replacing x by kx and α by α/k in the previous equation, so that y satisfies

$$x^2 y'' + [(2\alpha - k)x^2 + (\gamma - 2s)x]y' + [\alpha(\alpha - k)x^2 + (\gamma\alpha - k\beta - 2s\alpha + ks)x$$
$$- s(\gamma - s - 1)]y = 0 \qquad (4.327)$$

The Bessel equation

$$y'' + \frac{1}{x}y' + \left(1 - \frac{n^2}{x^2}\right)y = 0 \qquad (4.328)$$

is a particular case of the above with the values

$$2\alpha - k = 0 \qquad (4.329a)$$
$$\gamma - 2s = 1 \qquad (4.329b)$$
$$\alpha(\alpha - k) = 1 \qquad (4.329c)$$
$$\gamma\alpha - k\beta - 2s\alpha + ks = 0 \qquad (4.329d)$$
$$s(\gamma - s - 1) = n^2 \qquad (4.329e)$$

The quantity n is assumed to be a positive number. From (4.329a) and (4.329c), $\alpha = \pm i$. From (4.329b) and (4.329e), $s = \pm n$ and $\gamma = 1 \pm 2n$, where either plus or minus sign is taken in each case. Finally, from (4.329d), $\beta = \frac{1}{2}\gamma = \frac{1}{2} \pm n$. Consequently, two solutions of the Bessel equation are

$$y = A_1 x^n e^{\mp ix} F(n + \tfrac{1}{2}, 2n + 1, \pm 2ix) \qquad (4.330)$$

and

$$y_2 = A_2 x^{-n} e^{\mp ix} F(-n+\tfrac{1}{2}, -2n+1, \pm 2ix) \tag{4.331}$$

It is not guaranteed that these solutions are really different. In fact, when n is an integer, it is shown in the section on the confluent hypergeometric function [Eq. (4.284)] that $F(-n+\tfrac{1}{2}, -2n+1, \pm 2ix)$ is proportional to $x^{2n} F(n+\tfrac{1}{2}, 2n+1, \pm 2ix)$. It follows that the "second" solution (4.331) is the same as the first to within a multiplying constant. This may also be seen by writing down the standard form of the second solution to the confluent hypergeometric as given in (4.290). As shown in that context, there is an independent second solution which has a logarithmic singularity at $x = 0$.

27. General Properties of Bessel Functions

To study in detail the properties of Bessel functions, it is desirable to write the solutions in a different way. First, let us substitute in (4.330) the integral representation (4.278) of the confluent hypergeometric:

$$y_1 = A_1 x^n e^{-ix} \frac{\Gamma(2n+1)}{[\Gamma(n+\tfrac{1}{2})]^2} \int_0^1 s^{n-(1/2)} (1-s)^{n-(1/2)} e^{2isx}\, ds \tag{4.332}$$

Making the substitution $2s - 1 = t$, this takes the form

$$y_1 = A_1 x^n \frac{\Gamma(2n+1)}{[\Gamma(n+\tfrac{1}{2})]^2} \frac{1}{2^{2n}} \int_{-1}^1 (1-t^2)^{n-(1/2)} \cos tx\, dt \tag{4.333}$$

Since the rest of the integrand is an even function, e^{itx} has been replaced by its even part, $\cos tx$.

With the appropriate choice of the multiplying constant A_1, this is the standard Bessel function $J_n(x)$. The constant is chosen in such a way that, in the series expansion, the coefficients have a simple form. The series expansion may be obtained by expanding $\cos tx$:

$$y_1 = A_1 x^n \frac{\Gamma(2n+1)}{[\Gamma(n+\tfrac{1}{2})]^2} \frac{1}{2^{2n}} \sum_{l=0}^\infty \frac{(-1)^l}{(2l)!} x^{2l} \int_{-1}^1 (1-t^2)^{n-(1/2)} t^{2l}\, dt \tag{4.334}$$

Putting $t^2 = u$, the integral becomes

$$\int_0^1 (1-u)^{n-(1/2)} u^{l-(1/2)}\, du = B(n+\tfrac{1}{2}, l+\tfrac{1}{2})$$

$$= \frac{\Gamma(n+\tfrac{1}{2})\Gamma(l+\tfrac{1}{2})}{\Gamma(n+l+1)}$$

by virtue of (4.249) and (4.254). Thus,

$$y_1 = A_1 \frac{\Gamma(2n+1)}{\Gamma(n+\tfrac{1}{2})} \frac{x^n}{2^{2n}} \sum_{l=0}^\infty \frac{(-1)^l}{(2l)!} \frac{\Gamma(l+\tfrac{1}{2})}{\Gamma(n+l+1)} x^{2l} \tag{4.335}$$

Now $\Gamma(l+\frac{1}{2})$ may be written in terms of factorials:

$$\Gamma(l+\tfrac{1}{2}) = \frac{(2l-1)(2l-3)\cdots 1}{2^l}\Gamma\left(\frac{1}{2}\right)$$

$$= \frac{(2l)!}{l!2^{2l}}\sqrt{\pi} \tag{4.336}$$

so that

$$y_1 = \sqrt{\pi}\,A_1\frac{\Gamma(2n+1)}{\Gamma(n+\frac{1}{2})}\frac{x^n}{2^{2n}}\sum_{l=0}^{\infty}\frac{(-1)^l}{l!}\frac{1}{\Gamma(n+l+1)}\left(\frac{x}{2}\right)^{2l} \tag{4.337}$$

The normalization of the standard Bessel function is now chosen so that all that remains multiplying the sum is $(x/2)^n$; that is

$$A_1 = \frac{2^n\Gamma(n+\frac{1}{2})}{\sqrt{\pi}\,\Gamma(2n+1)} \tag{4.338}$$

It follows that the series expansion of the Bessel function is

$$J_n(x) = \sum_{l=0}^{\infty}\frac{(-1)^l}{l!}\frac{1}{\Gamma(n+l+1)}\left(\frac{x}{2}\right)^{n+2l} \tag{4.339}$$

and that its integral representation

$$J_n(x) = \frac{2}{\sqrt{\pi}\,\Gamma(n+\frac{1}{2})}\left(\frac{x}{2}\right)^n\int_0^1 (1-t^2)^{n-(1/2)}\cos tx\,dt \tag{4.340}$$

The usual treatment of the Bessel function begins by solving the differential equation by the Frobenius method to obtain the series expansion (4.339). Unfortunately, the series does not give much insight into the properties or behavior of the Bessel function. The point is perhaps best illustrated by observing that the series expansion of a sine or cosine does not reveal in any obvious way the oscillations or the periodicity of those functions. The integral (4.340), on the other hand provides considerable insight into the character of the function, as we now show.

This insight derives from the fact that the integral represents simply a superposition of harmonic functions or waves; the lower limit represents waves of infinitely long wavelength and the upper one corresponds to a wavelength of 2π. The amplitudes of the waves in the superposition are given by

$$(1-t)^{n-(1/2)} = \left(1-\frac{2\pi}{\lambda}\right)^{n-(1/2)} \tag{4.341}$$

where λ is the wavelength (see Fig. 4.6). The following observations then follow:

(a) At $x = 0$ the waves are all in phase; that is, they add constructively. Thereafter the phases become more and more randomly distributed as x increases. Therefore, the integral has its maximum value at $x = 0$.

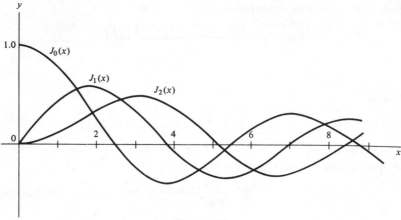

Figure 4.6

Furthermore, as $x \to \infty$, the randomness of the phases leads to the integral going to zero. In the case of $J_0(x)$, then the function has its maximum value, unity, at $x = 0$, and approaches zero for large x.

(b) Since all of the waves start with zero phase, none of them become negative before a quarter wavelength; for the shortest wave this is $\pi/2$. The integral, therefore, must be positive up to this value. After $\pi/2$, more and more of the periodic functions take on negative values, until a point is reached at which the function as a whole becomes negative. For the case of $n = 0$, this is the first zero of the function; for other n's, $J_n(0) = 0$ because of the factor x^n, but the next zero comes for $x > \pi/2$.

(c) For $n < \frac{1}{2}$, the amplitude $(1 - t^2)^{n-(1/2)}$ is greatest near $t = 1$, that is, for the shortest wavelengths. As n increases, the shorter wavelength components decrease relative to those of longer wavelength. After $n = \frac{1}{2}$, the amplitude starts at zero for $t = 1$ ($\lambda = 2\pi$). Consequently, as n increases, the first zero after $x = 0$ will come at larger and larger values of x.

The first zero of $J_0(x)$ is at $x = 2.405$, which is considerably bigger than $\pi/2 = 1.527$.

By the time that $x = 3\pi/2 = 4.712$, the waves with greatest amplitude have again become positive. Therefore, somewhere beyond this value, $J_0(x)$ should again become positive. The second zero is in fact found to be at $x = 5.520$.

(d) Though it does not follow rigorously, it seems reasonable that the Bessel functions would have an oscillatory character. This is a consequence of the fact that it is constructed from periodic functions, and that there are no components of very short wavelength.

28.   Second Solution of the Bessel Equation

Since the Bessel equation depends only on n^2, it should be possible to get a second solution from the first by replacing n by $-n$. This is certainly true for the series solution (4.339). The integral in (4.340), however, does not converge for negative n. The second solution is expressible in terms of a confluent hypergeometric by (4.289).

When n is an integer, it is found that the second solution is not independent of the first. Since the Γ function of a negative integer is infinite, the coefficient of the first n terms in $J_{-n}(x)$ becomes zero:

$$J_{-n}(x) = \sum_{l=0}^{\infty} \frac{(-1)^l}{l!} \frac{1}{\Gamma(-n+l+1)} \left(\frac{x}{2}\right)^{-n+2l}$$

$$= \sum_{m=0}^{\infty} \frac{(-1)^{n+m}}{(m+n)!} \frac{1}{\Gamma(m+1)} \left(\frac{x}{2}\right)^{n+2m}$$

where we have put $m = l - n$. Since $\Gamma(m+1) = m!$ and for integral n, $(m+n)! = \Gamma(m+n+1)$,

$$J_{-n}(x) = (-1)^n \sum_{m=0}^{\infty} \frac{(-1)^m}{m!} \frac{1}{\Gamma(n+m+1)} \left(\frac{x}{2}\right)^{n+2m}$$

$$= (-1)^n J_n(x) \qquad (4.342)$$

The other independent solution for n a negative integer may be derived by noting that (4.331) expresses it in terms of a confluent hypergeometric function with γ a negative integer. The situation therefore parallels that case, for which the essential results are given by (4.291) and (4.292). Since the Wronskian for the Bessel equation is $1/x$, the second solution may be written

$$y_2 = J_n(x) \int \frac{dx'}{x' J_n^2(x')} \qquad (4.343)$$

Subsequently we introduce the Hankel functions as two alternative independent but *complex* solutions of the Bessel equation. It is then possible to derive two real, independent solutions by taking real and imaginary parts.

29.   Zeros of Bessel Functions

It is possible to derive theorems for the zeros of Bessel functions of any sort, since these theorems follow only from a modified form of the differential equation itself.

We must first show that it is possible by a transformation to eliminate the first-derivative term in the differential equation. The transformation is

$$y = x^{-1/2} u \tag{4.344}$$

The function u satisfies the equation

$$u'' + \left(1 - \frac{n^2 + \frac{1}{4}}{x^2}\right) u = 0 \tag{4.345}$$

This function will have the same zeros as the solution of the Bessel equation except possibly for $x = 0$.

THEOREM 1

Between any two zeros of a Bessel function, there must always lie a zero of a sinusoidal function ($\sin x$ or $\cos x$ or some combination thereof).

This is proven in the following way. Starting from the equations

$$u'' + \left[1 - \frac{(n^2 + \frac{1}{4})}{x^2}\right] u = 0 \tag{4.345}$$

$$v'' + v = 0 \tag{4.346}$$

we multiply the first by v, the second by u, subtract and integrate between two successive zeros, x_1 and x_2, of u:

$$(vu' - uv')\Big|_{x_1}^{x_2} = (n^2 + \tfrac{1}{4}) \int_{x_1}^{x_2} \frac{1}{x^2} uv \, dx \tag{4.347}$$

Now, either $u'(x_1) > 0$, $u'(x_2) < 0$, and u is positive on the interval, or $u'(x_1) < 0$, $u'(x_2) > 0$, and u is negative on the interval. Consider the former case: the argument follows in exactly the same way in the latter. We have, then,

$$v(x_2)u'(x_2) - v(x_1)u'(x_1) = (n^2 + \tfrac{1}{4}) \int_{x_1}^{x_2} \frac{1}{x^2} uv \, dx \tag{4.348}$$

If v is positive on the whole interval (or if it is negative on the whole interval), the two sides of the equation have opposite signs. Therefore, v must change sign in the interval; that is, it must have a zero between successive zeros of the Bessel function.

It may further be shown that as $x \to \infty$, the zeros of the Bessel function approach π, for clearly, as $x \gg n$, Eq. (4.345) becomes more and more nearly identical to (4.346).

THEOREM 2

There is a zero of $J_n(x)$ between any two successive zeros of J_{n+1}. (in fact, the zeros of J_{n+1} are separated by more than those of J_n, though in both cases, the separations for large x approach π.)

The theorem is proven in much the same way as the one preceding. The equations for the functions are

$$u_n'' + \left(1 - \frac{n^2 + \frac{1}{4}}{x^2}\right) u_n = 0$$

$$u_{n+1}'' + \left(1 - \frac{(n+1)^2 + \frac{1}{4}}{x^2}\right) u_{n+1} = 0$$

Multiplying the first by u_{n+1}, the second by u_n, and integrating between x_1 and x_2 which are now two successive zeros of u_{n+1}, we obtain the equation

$$u_n(x_2)u_{n+1}'(x_2) - u_n(x_1)u_{n+1}'(x_1) = (2n+1) \int_{x_1}^{x_2} \frac{1}{x^2} u_n u_{n+1} \, dx \quad (4.349)$$

Let us take $u_{n+1}'(x_1) > 0$, $u_{n+1}'(x_2) < 0$, $u_{n+1}(x) > 0$ in the interval $x_1 \leqslant x \leqslant x_2$. Suppose then that u_n is positive on the whole interval; the left-hand side is negative and the right-hand side positive. If u_n were negative throughout, the sign of each side would be reversed. In either case we arrive at a contradiction. Therefore, u_n must change sign (that is, have a zero) between successive zeros of u_{n+1}.

Problem 4-28: Prove that between any two zeros of one solution of the Bessel equation, there is always a zero of the other.

---

30. Hankel Functions

Still a different pair of independent solutions of the Bessel equation may be obtained by returning to the equation (4.328) and making the substitution

$$y = x^n u \quad (4.350)$$

The function $u(x)$ satisfies the equation

$$xu'' + (2n+1)u' + xu = 0 \quad (4.351)$$

This is another particular case of (4.256), in which

$$f(s) = s^2 + 1 \quad (4.352)$$

and

$$g(s) = (2n+1)s \quad (4.353)$$

Solutions may then be found in the form

$$u = \int_{s_1}^{s_2} e^{sx} F(s) \, ds \quad (4.354)$$

where

$$F(s) = \frac{1}{s^2 + 1} \exp \int^s \frac{(2n+1)s'}{s'^2 + 1} \, ds' \quad (4.355)$$

and s_1 and s_2 are any two values for which

$$f(s)F(s)e^{sx} = 0 \tag{4.356}$$

Doing the integral in (4.355) we find that

$$F(s) = (s^2 + 1)^{n-(1/2)} \tag{4.357}$$

so that (4.356) becomes

$$(s^2 + 1)^{n+(1/2)}e^{sx} = 0 \tag{4.358}$$

There are various possible choices for s_1 and s_2. Taking $s_1 = -i$ and $s_2 = i$ the solution is, from (4.257),

$$y = C_n x^n \int_{-i}^{i} e^{sx}(s^2 + 1)^{n-(1/2)} \, ds \tag{4.359}$$

C_n being a constant. Putting $s = it$ and taking

$$C_n = -\frac{i}{2^n \sqrt{\pi}\,\Gamma(n + \tfrac{1}{2})} \tag{4.360}$$

this becomes the Bessel function $J_n(x)$.

Consider the integral around the contour shown in Fig. 4.7, including the factor $C_n x^n$. Since the integrand has no singularities inside the contour, the complete integral is zero. The contribution from the portion running from $i - R$ to $-i - R$ approaches zero as $R \to \infty$ when $x > 0$ by virtue of the factor e^{-Rx}. (When $x < 0$, the contour may be drawn to the *right* of the imaginary axis.)

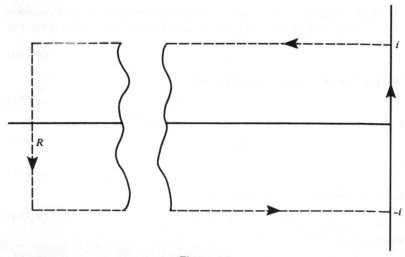

Figure 4.7

It follows, then, that the integral from $-i$ to i, which is the Bessel function $J_n(x)$, is equal to the sum of two other integrals:

(a) One running from $i-\infty$ to i, which we designate as $\tfrac{1}{2}H_n^{(1)}(x)$, $H_n^{(1)}$ being called the Hankel function of the first kind of order n, and

(b) One running from $-i$ to $-i-\infty$, to be designated $\tfrac{1}{2}H_n^{(2)}(x)$, $H_n^{(2)}$ being the Hankel function of the second kind of order n. Thus

$$J_n(x) = \tfrac{1}{2}[H_n^{(1)}(x) + H_n^{(2)}(x)] \qquad (4.361)$$

where

$$H_n^{(1)}(x) = -\frac{ix^n}{2^{n-1}\sqrt{\pi}\,\Gamma(n+\tfrac{1}{2})} \int_{i-\infty}^{i} e^{sx}(s^2+1)^{n-(1/2)}\,ds \qquad (4.362)$$

and

$$H_n^{(2)}(x) = -\frac{ix^n}{2^{n-1}\sqrt{\pi}\,\Gamma(n+\tfrac{1}{2})} \int_{-i}^{-i+\infty} e^{sx}(s^2+1)^{n-(1/2)}\,ds \qquad (4.363)$$

We note that $H_n^{(2)}(s)$ is the complex conjugate of $H_n^{(1)}(x)$, and consequently that $J_n(x)$ is the real part of both. It is also evident that $H_n^{(1)}$ and $H_n^{(2)}$ are independently solutions of the Bessel equation, since the condition (4.358) is satisfied at the limits in each case. Therefore, the *imaginary* parts of $H_n^{(1)}$ and $H_n^{(2)}$ are also solutions. The imaginary part of $H_n^{(1)}$ is known as the Neumann function $N_n(x)$, and is a second *real* solution independent of $J_n(x)$. The relations between the various solutions are

$$H_n^{(1)} = J_n + iN_n \qquad (4.364)$$

$$H_n^{(2)} = J_n - iN_n \qquad (4.365)$$

$$J_n = \tfrac{1}{2}[H_n^{(1)} + H_n^{(2)}] \qquad (4.366)$$

$$N_n = 1/(2i)[H_n^{(1)} - H_n^{(2)}] \qquad (4.367)$$

An alternative form of $H_n^{(1)}$ may be obtained by making use of the substitution

$$s = i - \sigma \qquad (4.368)$$

This leads to the form

$$H_n^{(1)}(x) = -\frac{ix^n e^{ix}}{2^{n-1}\sqrt{\pi}\,\Gamma(n+\tfrac{1}{2})} \int_0^\infty e^{-\sigma x}\sigma^{n-(1/2)}(\sigma - 2i)^{n-(1/2)}\,d\sigma \qquad (4.369)$$

Expanding in a binomial series the third term in the integrand, we obtain, for $(n-\tfrac{1}{2})$ not an integer,

$$H_n^{(1)}(x) = -\frac{ix^n e^{ix}(-2i)^{n-(1/2)}}{2^{n-1}\sqrt{\pi}\,\Gamma(n+\tfrac{1}{2})} \sum_{l=0}^{\infty} \binom{n-\tfrac{1}{2}}{l}\left(\frac{i}{2}\right)^l \int_0^\infty e^{-\sigma x}\sigma^{n+l-(1/2)}\,d\sigma$$

$$= \frac{(-i)^{n+(1/2)}}{\Gamma(n+\tfrac{1}{2})}\sqrt{\frac{2}{\pi x}}\,e^{ix} \sum_{l=0}^{\infty} \binom{n-\tfrac{1}{2}}{l}\Gamma(n+l+\tfrac{1}{2})\left(\frac{i}{2x}\right)^l \qquad (4.370)$$

This is an asymptotic expansion; for very large x one has the asymptotic forms

$$H_n^{(1)}(x) \approx \sqrt{\frac{2}{\pi x}} \exp i \left(x - \frac{n\pi}{2} - \frac{\pi}{4} \right) \qquad (4.371)$$

$$J_n(x) \approx \sqrt{\frac{2}{\pi x}} \cos \left(x - \frac{n\pi}{2} - \frac{\pi}{4} \right) \qquad (4.372)$$

$$N_n(x) \approx \sqrt{\frac{2}{\pi x}} \sin \left(x - \frac{n\pi}{2} - \frac{\pi}{4} \right) \qquad (4.373)$$

When n is an integer plus $\frac{1}{2}$, the sum in (4.370) terminates and we obtain closed forms for the functions as periodic functions times polynomials in inverse powers of x. In this case the functions of greatest interest are the so-called "spherical Bessel functions," which occur in the theory of spherical waves in the next chapter; they are designated

$$j_l(x) = \sqrt{\frac{\pi}{2x}} J_{l+(1/2)}(x) \qquad (4.374)$$

$$n_l(x) = \sqrt{\frac{\pi}{2x}} N_{l+(1/2)}(x) \qquad (4.375)$$

$$h_l^{(1,2)}(x) = \sqrt{\frac{\pi}{2x}} H_{l+(1/2)}^{(1,2)}(x) \qquad (4.376)$$

Again, it is convenient to deal with the Hankel function of the first kind. It is derivable from (4.370), and may be simplified by noting that the binomial coefficient in the sum becomes $\binom{n}{l}$, so that the sum terminates at $l = n$, and the Γ functions become simply factorials. Thus we find

$$h_n^{(1)}(x) = (-i)^{n+1} \frac{e^{ix}}{x} \sum_{l=0}^{n} \frac{(n+l)!}{l!(n-l)!} \left(\frac{i}{2x} \right)^l \qquad (4.377)$$

$h^{(2)}(x)$ is found by taking the complex conjugate equation, and $j_l(x)$ and $n_l(x)$ by picking out real and imaginary parts. For convenience the first three of these spherical Bessel functions are tabulated herewith:

$$j_0(x) = \frac{\sin x}{x}, \qquad\qquad\qquad n_0(x) = -\frac{\cos x}{x}$$

$$j_1(x) = \frac{\sin x}{x^2} - \frac{\cos x}{x}, \qquad\qquad n_1(x) = -\frac{\cos x}{x^2} - \frac{\sin x}{x} \qquad (4.378)$$

$$j_2(x) = \left(\frac{3}{x^2} - 1 \right) \frac{\sin x}{x} - 3 \frac{\cos x}{x^2}, \qquad n_2(x) = \left(1 - \frac{3}{x^2} \right) \frac{\cos x}{x} - 3 \frac{\sin x}{x^2}$$

31.   Further Formulas Involving Bessel Functions

Let us, for the moment, designate by Z_n an arbitrary solution of the Bessel equation. It is simple to verify that the equation can be written in the "factored" form

$$\left(\frac{d}{dx}+\frac{n+1}{x}\right)\left(\frac{d}{dx}-\frac{n}{x}\right)Z_n = -Z_n \tag{4.379}$$

Now let us define

$$\bar{Z}_n = \left(\frac{d}{dx}-\frac{n}{x}\right)Z_n \tag{4.380}$$

If we now operate on Eq. (4.379) with the operator $(d/dx - n/x)$, it becomes

$$\left(\frac{d}{dx}-\frac{n}{x}\right)\left(\frac{d}{dx}+\frac{n+1}{x}\right)\bar{Z}_n = -\bar{Z}_n \tag{4.381}$$

But, on expansion, this equation takes the form

$$\frac{d^2}{dx^2}\bar{Z}_n+\frac{1}{x}\frac{d}{dx}\bar{Z}_n+\left[1-\frac{(n+1)^2}{x^2}\right]\bar{Z}_n = 0 \tag{4.382}$$

which is the Bessel equation of order $(n+1)$. We may therefore put

$$\left(\frac{d}{dx}-\frac{n}{x}\right)Z_n = \bar{Z}_n = -Z_{n+1} \tag{4.383}$$

The minus sign is conventional; the reason for its choice will be evident as we proceed. In a similar way it is easily shown that

$$\left(\frac{d}{dx}+\frac{n}{x}\right)Z_n = Z_{n-1} \tag{4.384}$$

Problem 4-29:   Verify directly that if Z_n is $J_n(x)$, as given by the integral representation (4.340), the relations (4.383) and (4.384) are satisfied.

Problem 4-30:   Show that the relations (4.383) and (4.384) may be written

$$\frac{d}{dx}(x^{-n}Z_n) = -x^{-n}Z_{n+1}$$

$$\frac{d}{dx}(x^{n}Z_n) = x^{n}Z_{n-1}$$

The relations (4.383) and (4.384) may be combined to yield the following useful formulas:

$$2Z'_n = Z_{n-1} - Z_{n+1} \tag{4.385}$$

and

$$Z_{n-1} + Z_{n+1} = \frac{2n}{x} Z_n \tag{4.386}$$

Problem 4-31: Using the appropriate recursion relations, show that

$$\int \left[(a^2 - b^2)x - \frac{m^2 - n^2}{x} \right] Z_m(ax) Z_n(bx) \, dx$$
$$= x[bZ_m(ax)Z_{n-1}(bx) - aZ_{m-1}(ax)Z_n(bx)] - (m-n)Z_m(ax)Z_n(bx)$$

A commonly used representation of the Bessel functions of integral order can be obtained starting from the relation (4.385). We multiply both sides of the equation, which is of course valid for both positive and negative integral values of n, by ξ^n and sum over all n's. If we write

$$F(\xi, x) = \sum_{n=-\infty}^{\infty} J_n(x)\xi^n \tag{4.387}$$

we see that F satisfies the equation

$$2\frac{\partial F}{\partial x} = \left(\xi - \frac{1}{\xi}\right) F \tag{4.388}$$

This may be integrated to give

$$F(\xi, x) = F_0(\xi) \exp\frac{1}{2}\left(\xi - \frac{1}{\xi}\right) x$$

The value of $F_0(\xi)$ may be obtained by noting that $F(\xi, 0) = 1$; thus $F_0(\xi) = 1$ and

$$F(\xi, x) = \sum_{n=-\infty}^{\infty} J_n(x)\xi^n = \exp\frac{1}{2}\left(\xi - \frac{1}{\xi}\right) x \tag{4.389}$$

The function on the right is the "generating function" for the Bessel functions $J_n(x)$.

By the elementary theory of contour integration it is obvious that we may write

$$J_n(x) = \frac{1}{2\pi i} \oint \frac{1}{\xi^{n+1}} \exp\left[\frac{x}{2}\left(\xi - \frac{1}{\xi}\right)\right] d\xi \tag{4.390}$$

To obtain still another useful form for the Bessel function, let us put, in (4.389),

$$\xi = ie^{i\theta} \tag{4.391}$$

Equation (4.389) then takes the form

$$e^{ix \cos \theta} = \sum_{n=-\infty}^{\infty} i^n J_n(x) \, e^{in\theta} \tag{4.392}$$

The right-hand side is the Fourier series expansion of the periodic function $e^{ix \cos \theta}$. From the equation for Fourier coefficients (4.11),

$$J_n(x) = \frac{1}{2\pi i^n} \int_{-\pi}^{\pi} e^{ix \cos \theta} e^{-in\theta} \, d\theta \tag{4.393}$$

Because $\cos \theta$ is an *even* function, this may be written

$$J_n(x) = \frac{1}{\pi i^n} \int_{0}^{\pi} e^{ix \cos \theta} \cos n\theta \, d\theta \tag{4.394}$$

PRELUDE TO CHAPTER 5

In this chapter we develop first the theory of sound waves. The wave equation is derived, and shown to be satisfied by all the variables of a sound wave – velocity potential, velocity components, condensation, pressure. It is shown that all may be expressed in terms of a single scalar, that is, the velocity potential.

Kinetic and potential energy densities, as well as the Hamiltonian, are derived in general and for plane waves in particular. The Lagrangian is written down, and the equations of motion shown to be derivable from a Hamilton's principle.

In the next section the theory of guided waves in enclosures is developed for both rectangular and circular cross section. The "filtering out" of high-frequency waves by guides of small cross section is demonstrated.

From the wave equation with a source term, the Green's function and its transforms are derived, by carrying out a Fourier transform in spatial coordinates and a Laplace transform in time.

The spherical wave equation is solved by separation of variables. The angular functions are found to be spherical harmonics, and the radial ones spherical Bessel functions. It is shown that the two Hankel functions correspond to diverging and converging waves, respectively. An expression is obtained for the energy flux in the individual "partial waves."

At this point the basic result needed for the development of the theory of scattering of plane waves is derived, the expansion of a plane wave in spherical waves.

The radiation from an arbitrary source distribution is calculated, using the Green's function determined earlier. The expansion of the velocity potential in spherical harmonics (multipoles) is carried out. The energy radiated is shown to be a sum of that from the various multipoles. It is demonstrated that the intensities from successive multipoles decrease by factors approximately equal to the square of the ratio of the dimensions of the radiating system to the wavelength.

Expressing a general time-varying source in terms of its frequency components by Fourier transform, a general radiation formula is derived. In particular, the radiation from a moving source is evaluated, and the "bow wave" phenomenon demonstrated in the case in which the source has a velocity greater than the velocity of sound.

All elements of the preceding theory have analogs in the theory of electromagnetic waves, which is developed later in the chapter. The difference is simply that electromagnetic waves are *vector* waves, that is, what is propagated is a *vector* quantity.

We turn next to a more general discussion of Green's functions for waves. The previously developed Green's function was for an instantaneous source in an infinite medium. If one turns to the problem of sound waves determined by given initial conditions and subject to given boundary conditions, it is found that both the initial disturbance and the boundaries act as sources, and that the same Green's function describes the different kinds of sources.

The problem of waves in enclosures, discussed earlier in the chapter, is now looked at again from a different viewpoint, namely, that of the propagation of waves due to a particular disturbance. It is shown that *resonance* occurs as the stimulating frequency approaches a natural frequency in the enclosure. Rectangular, spherical, and cylindrical enclosure are considered.

The next major topic is the scattering of sound waves. There is, of course, a similar theory for electromagnetic waves. The theory developed here is, however, very close to that of particle waves in quantum mechanics, and results will be carried over to that problem when it is discussed in Chapter 9.

When the incident plane wave is expressed as a superposition of spherical waves, each is separately scattered, and each has a "partial cross section" for scattering. General formulas are derived for these cross sections. The case in which there is absorption of energy from the incident wave is also treated, and *absorption* and *total* (scattering + absorption) cross sections are calculated. The so-called "optical theorem," developed in the context of quantum scattering theory by Bohr, Peierls, and Placzek is proven, as are certain basic inequalities derived by them for the cross sections.

As a particular example, the problem of elastic scattering by a small sphere is solved, and the "Rayleigh sixth-power law" is proven.

We then turn to the propagation of electromagnetic waves. The theory of such waves is developed from Maxwell's equations. The problem is formulated both in terms of potentials and of the Hertz vector.

We next consider the problem of the intrinsic angular momentum of vector waves. A "spin operator" is derived, and its eigenstates are shown to correspond to right and left circularly polarized and longitudinally polarized waves.

By considering the interaction of electromagnetic waves with charged matter, we develop the expressions for their energy and momentum density and flux.

As a basis for the quantum theory of the electromagnetic field, we develop the Lagrangian and Hamiltonian formulation for the electromagnetic field. It is shown how to analyze the fields in terms of normal modes and normal coordinates. The fact that these normal modes satisfy oscillator dynamics provides the starting point for quantum electrodynamics. The Lagrangian of the interaction with charged particles is also derived.

The Coulomb or transverse gauge is used to discuss the propagation of plane waves, and their basic properties derived. Plane waves are shown to be suitable "normal modes" in an infinite medium.

Next, we return to the theory of multipole radiation, which is somewhat more complex for the vector electromagnetic field than for the scalar sound one. Dynamic multipole moments are derived for sources consisting of both charges and magnetic moments. Formulas for both "electric" and "magnetic" multipole radiation are derived, as is a simple relation between energy and magnetic moment of the radiation.

The theory of multipole radiation is of great importance in nuclear physics, where it is used to analyze the radiation emitted in nuclear transitions.

Another problem which originated in the liquid drop model of nuclei, that of the radiation from a uniformly charged vibrating liquid drop, is fully solved.

The next topic treated is that of Čerenkov radiation from particles moving through matter with a velocity greater than the velocity of light in the matter. This problem is very similar to that of supersonic sound waves dealt with earlier.

The final topic of the chapter is that of plasmas in a magnetic field, with special emphasis on metallic solid state plasmas. The problem is that of finding the normal modes of the coupled system of charges and electromagnetic field. This is solved by a "self-consistent" approximation, based on the following approach: The response of the charged particles (electrons) to an *arbitrary* electromagnetic field is calculated. This response comprises induced charged and current densities. But these charges and currents act as sources of electromagnetic fields. If these are then taken to be the same as the original fields, self-consistency is assured.

The dispersion relations of both longitudinal and circularly polarized transverse waves are obtained, and the normal modes calculated. Particular cases are plasmons (modified by the magnetic field) and helicons. The transmission of helicon waves through slabs is calculated.

There is an appendix on the method of steepest descents, with applications to asymptotic expansions of the gamma function and the Bessel and Hankel functions, and a short one on the transformation of units in electromagnetic theory.

REFERENCES

Abraham, M. and R. Becker, *Electricity and Magnetism.* London: Blackie, 1937.

Coulson, C. A., *Waves.* Edinburgh: Oliver and Boyd, 1948.

Jackson, J. D., *Electromagnetic Theory.* New York: Wiley, 1962.

Landau, L. D. and E. M. Lifshitz, *Electrodynamics of Continuous Media.* Reading, Mass.: Addison-Wesley, 1960.

Landau, L. D. and E. M. Lifshitz, *Fluid Mechanics.* New York: Wiley, 1959, Chapter 8.

Lord Rayleigh, *Theory of Sound.* New York: Dover, 1945, 2 vols.

Morse, P. M., *Vibration and Sound.* New York: McGraw-Hill, 1948.

Panofsky, W. K. H. and M. Phillips, *Classical Electricity and Magnetism.* Reading, Mass.: Addison-Wesley, 1955.

Sommerfeld, A., *Electrodynamics.* New York: Academic Press, 1952.

5

PROPAGATION AND SCATTERING OF WAVES

"Life is a wave, which in no two consecutive moments of its existence is composed of the same particles"

John Tyndall

1. Introduction

We consider in this chapter some problems of wave propagation in classical physics, in particular, sound waves in a gas and electromagnetic waves both in vacuo and in material media. Sound waves are considered first since they are, as will be shown, describable in terms of a single scalar quantity; we therefore refer to them as "scalar waves." (The scalar property is a consequence of the assumed isotropy of a gas. The situation is more complicated in the case of sound waves in a solid, where crystalline anisotropy may make possible the differentiation of waves of different polarizations.)

Electromagnetic waves, on the other hand, may be described by a vector quantity. The additional degrees of freedom correspond to different polarizations of the wave.

Other kinds of wavelike phenomena are possible, in the sense that the disturbance which is propagated may have another character. Electrons in wave mechanics are characterized by a spinor field; gravitational waves by a second-order tensor. We refer to the former in a later chapter, in which we discuss problems of quantum mechanics.

Many of the techniques to be developed here will be seen to be applicable in quantum mechanics, and mathematically similar problems will arise. There are also, however, important differences. In nonrelativistic quantum mechanics, the time dependence of the field is different from that of sound or electromagnetic waves and suggests a similarity to the problems of diffusion to be considered in the next chapter. Thus, there is a relation between the problems only in so far as the spatial dependence of the field is concerned. Also, the unique and varied physical situations met in quantum phenomena give rise to more complex and subtle problems, most of which have no classical analog.

Nevertheless, experience in the problems of classical wave propagation provides a valuable background of technique with which to approach these more difficult problems.

2. Sound Waves. Derivation of the Equations

Sound waves are small disturbances in the pressure and density of a gas. The fact that they are small permits the linearization of the equations.

Three equations are relevant. The first is the "equation of motion," which describes how a volume of gas reacts to the pressures exerted on it by the rest of the gas; the second is the "equation of continuity" which connects the density and velocity fields which enter into the equation of motion. The third and last is the "equation of state" relating pressure to density.

We use a field approach to describe the state of the gas in motion. The relevant fields are

(a) the density field, $\rho(\mathbf{r}, t)$

(b) the velocity field $\mathbf{v}(\mathbf{r}, t)$

(c) the pressure field $p(\mathbf{r}, t)$

Consider the gas which is instantaneously within an infinitesimal volume ΔV. We must first calculate mass times acceleration for this gas. The mass is, to the first order in ΔV, $\rho \Delta V$. The acceleration is the rate of change of velocity, viz.,

$$\lim_{\Delta t \to 0} \frac{1}{\Delta t} \{ \mathbf{v}(\mathbf{r} + \mathbf{v}\Delta t, t + \Delta t) - \mathbf{v}(\mathbf{r}, t) \}$$

the first term in the bracket being the velocity field at time $t + \Delta t$ at the point to which a gas molecule, originally at $\mathbf{r}$, is displaced in time Δt. This acceleration is, therefore,

$$\mathbf{f} = \frac{\partial \mathbf{v}}{\partial t} + (\mathbf{v} \cdot \boldsymbol{\nabla})\mathbf{v} \qquad (5.1)$$

So much for mass and acceleration; what of the force appearing on the other side of the Newtonian equation of motion? Assuming that there are no significant external forces, but only the force of pressure exerted by the rest of the gas, we must determine the net pressure force on our volume. If we assume that the gas may be considered inviscid, the pressure force is normal and is equal to

$$\mathbf{F}\Delta V = -\int_{\Delta V} p\mathbf{n}\, dS$$

(**F** is then the force field, or force per unit volume).

The integral is over the surface surrounding ΔV and $\mathbf{n}$ is the unit outward normal vector to the surface. But on applying Gauss' theorem to the vectors $p\mathbf{i}$, $p\mathbf{j}$, $p\mathbf{k}$, respectively, $\mathbf{i}$, $\mathbf{j}$, and $\mathbf{k}$ being unit vectors along the coordinate axes, we obtain

$$\int p n_x\, dS = \int \frac{\partial p}{\partial x}\, d^3\mathbf{r}$$

etc., the integrals on the right being volume integrals over the volume ΔV. Thus

$$\int_{\Delta V} p\mathbf{n}\, dS = \int \nabla p\, d^3\mathbf{r}$$

Since ΔV is small, again to the first order in ΔV this is equal to

$$\nabla p \Delta V$$

We are now in a position to write the equation of motion:

$$\rho \left(\frac{\partial \mathbf{v}}{\partial t} + (\mathbf{v} \cdot \nabla)\mathbf{v} \right) = -\nabla p \tag{5.2}$$

A continuity equation obviously connects ρ and $\mathbf{v}$, since the rate of change of density at a point is dependent on its rate of flow. For an arbitrary fixed volume, the rate of change of the mass of gas inside the volume is

$$\frac{\partial}{\partial t} \int \rho\, dV = \int \frac{\partial \rho}{\partial t}\, d^3\mathbf{r}$$

The efflux over the boundary is

$$\int \rho \mathbf{v} \cdot \mathbf{n}\, dS$$

which may be converted by Gauss' theorem into the volume integral

$$\int \nabla \cdot (\rho \mathbf{v})\, d^3\mathbf{r}$$

Since the rate of efflux is the negative of the rate of change of the mass inside, for *arbitrary* volume,

$$\frac{\partial \rho}{\partial t} + \nabla \cdot (\rho \mathbf{v}) = 0 \tag{5.3}$$

To define our problem completely, we must relate the pressure to the density through an equation of state

$$p = f(\rho) \tag{5.4}$$

We shall see that, because the range of variation of p in a sound wave is very small, the detailed form of the relation is not important. If conditions are adiabatic, the equation of state has the form

$$p = C\rho^\gamma \tag{5.5}$$

Equations (5.2)–(5.4) are quite complicated and in their general form would be very difficult to solve. Fortunately, sound waves are only rather minute disturbances of normal atmospheric conditions, and this fact makes possible the simplification of the equations. Let us look first at a static, silent atmosphere. Let the density be ρ_0 and the pressure p_0. The velocity field will be zero. Then we may write

$$\rho = \rho_0(1 + s) \tag{5.6}$$

where the quantity s, called the "condensation," is the fractional variation from equilibrium of the density. Then, from (5.4) and using the fact that s is small

$$\bar{p} = p - p_0 = \rho_0 s f'(\rho_0) \tag{5.7}$$

We note finally that (5.2) requires that $\mathbf{v} \to 0$ as $s \to 0$, so that $\mathbf{v}$ is also small. Keeping only terms of the first order in $\mathbf{v}$ and s, we may considerably simplify (5.2). The first term may be approximated by $\rho_0(\partial \mathbf{v}/\partial t)$, while the second is negligible. Therefore, Eq. (5.2) becomes

$$\rho_0 \frac{\partial \mathbf{v}}{\partial t} = -\rho_0 f'(\rho_0)\nabla s$$

or

$$\frac{\partial \mathbf{v}}{\partial t} = -f'(\rho_0)\nabla s \tag{5.8}$$

In the same approximation (5.3) becomes

$$\frac{\partial s}{\partial t} + \nabla \cdot \mathbf{v} = 0 \tag{5.9}$$

the ρ_0 cancelling out.

If we now take the divergence of (5.8) and the time derivative of (5.9) we may eliminate $\mathbf{v}$ to obtain the equation for s:

$$\frac{\partial^2 s}{\partial t^2} = f'(\rho_0)\nabla^2 s$$

We designate $f'(\rho_0)$ by c^2; we see that c is then the velocity of sound propagation. The resulting equation for s, viz.,

$$\nabla^2 s = \frac{1}{c^2} \frac{\partial^2 s}{\partial t^2} \qquad (5.10)$$

is the equation of wave propagation.

Since, from (5.7)

$$\bar{p} = \rho_0 c^2 s \qquad (5.11)$$

the pressure satisfies the same equation.

It is possible to show, finally, that if the flow is irrotational, that is $\nabla \times \mathbf{v} = 0$, $\mathbf{v}$ also satisfies the wave equation. For if we differentiate (5.8) with respect to time, and substitute for $\partial s/\partial t$ from (5.9), we get

$$\frac{\partial^2 \mathbf{v}}{\partial t^2} = c^2 \nabla(\nabla \cdot \mathbf{v})$$

Now, from a well-known vector identity, $\nabla \times (\nabla \times \mathbf{v}) = \nabla(\nabla \cdot \mathbf{v}) - \nabla^2 \mathbf{v}$. Thus, if $\nabla \times \mathbf{v} = 0$, the right-hand side of the equation for $\mathbf{v}$ becomes

$$c^2 \nabla^2 \mathbf{v}$$

Of course, it is also known that if $\mathbf{v}$ is irrotational, we can write it as the negative gradient of a potential ϕ, which is known as the "velocity potential"

$$\mathbf{v} = -\nabla\phi \qquad (5.12)$$

Equation (5.8) tells us that

$$\nabla\left(\frac{\partial\phi}{\partial t} - c^2 s\right) = 0$$

or $(\partial\phi/\partial t) - c^2 s =$ a constant in space. We may choose $\phi = 0$ in a quiescent atmosphere; since in this case $s = 0$ as well, we may write

$$s = \frac{1}{c^2} \frac{\partial\phi}{\partial t} \qquad (5.13)$$

ϕ, too, satisfies the wave equation, for if we differentiate (5.13) with respect to t and substitute from (5.9), at the same time expressing $\mathbf{v}$ in terms of ϕ by means of (5.12), we find that

$$\frac{1}{c^2} \frac{\partial^2 \phi}{\partial t^2} = \nabla^2 \phi \qquad (5.14)$$

Since we are dealing with a scalar field, we can choose any one of the above variables to describe the sound wave, and write the other quantities in terms of it. Thus, we may solve in terms of ϕ. Then, s is given by (5.13), p by (5.11), and the velocity components by (5.12).

As in the case of the vibrating string problem, the equations of sound propagation are only approximately linear. The approximations made are
(a) that the fractional density variations in a sound wave are small.

(b) that the second term in the equation of motion (5.2) is negligible compared with the first. To appreciate more fully what this latter approximation signifies, consider the identity

$$\mathbf{v} \times (\nabla \times \mathbf{v}) = \nabla(\tfrac{1}{2}v^2) - (\mathbf{v} \cdot \nabla)\mathbf{v}$$

If the flow is irrotational, $\nabla \times \mathbf{v} = 0$, so that the second term in the bracket on the left-hand side of (5.2) is $\nabla(\tfrac{1}{2}v^2)$. But the first term is $-\nabla(\partial\phi/\partial t)$. Therefore, the approximation is that

$$\tfrac{1}{2}v^2 \ll \frac{\partial\phi}{\partial t} = c^2 s$$

or

$$\frac{1}{2}\left(\frac{v}{c}\right)^2 \ll s \tag{5.15}$$

Consideration of some numerical magnitudes will now provide a test of the validity of our approximations. At the threshold of hearing for frequencies of 1000–5000 cycles/sec, s is of the order of 10^{-10}, while for sounds loud enough to cause pain it is 10^{-3} to 10^{-4}. Thus, over the whole range of normal "sound" the approximation of small s is indeed very good.

Problem 5-1:   A wind of velocity v blows across a Quonset hut in the shape of half a cylinder of radius a and length l. If the pressure inside is equal to that out, show that there is a vertical force on the hut of

$$\pi\rho v^2 a l$$

As for $\mathbf{v}$, let us note that a plane wave solution for the three-dimensional wave equation is

$$\phi = f(\mathbf{n} \cdot \mathbf{r} - ct) \tag{5.16}$$

where f is an arbitrary function, and $\mathbf{n}$ is the normal to the wave front. For if we define $u = \mathbf{n} \cdot \mathbf{r} - ct$

$$\nabla^2\phi = (n_x^2 + n_y^2 + n_z^2)\frac{d^2 f}{du^2} = \frac{d^2 f}{du^2}$$

and

$$\frac{1}{c^2}\frac{\partial^2\phi}{\partial t^2} = \frac{d^2 f}{du^2}$$

It follows that, for a plane wave

$$s = \frac{1}{c^2}\frac{\partial\phi}{\partial t} = -\frac{1}{c}f' \tag{5.17}$$

where f' is the derivative of f with respect to the argument u. But

$$\mathbf{v} \cdot \mathbf{n} = -\mathbf{n} \cdot \nabla f = -f' \tag{5.18}$$

Thus, $(\mathbf{v} \cdot \mathbf{n}/c)^2 = (1/c^2)f'^2 = s^2$. Consequently, the approximation involved in (5.15) is equivalent to the approximation that s is small.

3. Dynamics of Sound Waves

Let us first calculate the energy density (energy per unit volume) in a sound disturbance. The kinetic energy density is of course $\frac{1}{2}\rho v^2 = \frac{1}{2}\rho_0 (\nabla \phi)^2$ in an approximation. The potential energy may be calculated as follows: first we observe that the potential energy is equal to the work done by the excess pressure in locally compressing the gas. If we take an initial volume ΔV_0 this work is

$$-\int_{\Delta V_0}^{\Delta V} \bar{p}(\Delta V')\, d(\Delta V')$$

Now $\rho = A/\Delta V$, so $d(\Delta V') = d(A/\rho) = (-A/\rho^2)\rho_0\, ds$. Keeping only leading terms, the integral is

$$\frac{A}{\rho_0} \int_{\Delta V_0}^{\Delta V} \bar{p}\, ds$$

Using the fact that $\bar{p} = \rho_0 c^2 s$, and that $s = 0$ when the volume is ΔV_0, we obtain for the potential energy

$$\Delta V_0 \int_0^s \rho_0 c^2 s\, ds$$

so that the potential energy per unit volume is

$$\mathscr{V} = \tfrac{1}{2}\rho_0 c^2 s^2 \tag{5.19}$$

Using this along with the fact that the kinetic energy per unit volume is

$$\mathscr{T} = \tfrac{1}{2}\rho_0 (\nabla \phi)^2 \tag{5.20}$$

the energy density is then

$$\mathscr{H} = \tfrac{1}{2}\rho_0 [(\nabla \phi)^2 + c^2 s^2] \tag{5.21}$$

and the total energy is

$$H = \tfrac{1}{2}\rho_0 \int [(\nabla \phi)^2 + c^2 s^2]\, d^3\mathbf{r} \tag{5.22}$$

Using the expression (5.13) for s in terms of ϕ

$$H = \tfrac{1}{2}\rho_0 \int \left[(\nabla \phi)^2 + \frac{1}{c^2}\left(\frac{\partial \phi}{\partial t}\right)^2\right] d^3\mathbf{r} \tag{5.23}$$

An unusual feature of the use of the velocity potential ϕ as dynamical variable is that the potential energy ϕ involves time derivatives and the kinetic energy, spatial derivatives (in contrast, for instance, with the situation in the vibrating string problem).

We can use the expression (5.21) for the energy density to obtain one for the energy flux vector. Defining this latter quantity $\mathbf{P}$ such that $\mathbf{P} \cdot \mathbf{n}\, dS$ is the energy flow per unit time over a surface dS with unit normal vector $\mathbf{n}$, we can write down a continuity equation for the energy balance in a volume: rate of change of enclosed energy = rate of influx over the surface. We express this by the equation

$$\frac{\partial}{\partial t} \int \mathscr{H} d^3\mathbf{r} = -\int \mathbf{P} \cdot \mathbf{n}\, dS$$

Now

$$\frac{\partial \mathscr{H}}{\partial t} = \rho_0 \nabla \phi \cdot \nabla \left(\frac{\partial \phi}{\partial t}\right) + \frac{\rho_0}{c^2} \frac{\partial \phi}{\partial t} \frac{\partial^2 \phi}{\partial t^2}$$

$$= \rho_0 \nabla \phi \cdot \nabla \left(\frac{\partial \phi}{\partial t}\right) + \rho_0 \frac{\partial \phi}{\partial t} \nabla \cdot \nabla \phi$$

$$= \rho_0 \nabla \cdot \left(\frac{\partial \phi}{\partial t} \nabla \phi\right) \tag{5.24}$$

We may therefore identify $\mathbf{P}$ as

$$\mathbf{P} = -\rho_0 \frac{\partial \phi}{\partial t} \nabla \phi$$

$$= \bar{p}\mathbf{v} \tag{5.25}$$

It is evident that the energy flux over an element of surface is, then, the rate at which the excess pressure is doing work across this surface.

The momentum and angular momentum densities may also be written down on the basis of simple physical considerations:

$$\text{Momentum density} = \rho_0 s\mathbf{v} \tag{5.26}$$

$$\text{Angular momentum density} = \rho_0 s\mathbf{r} \times \mathbf{v} \tag{5.27}$$

It is interesting to see the relationships between these quantities for the plane disturbance (5.16). We find that

$$\mathscr{T} = \tfrac{1}{2}\rho_0 f'^2 \tag{5.28}$$

$$\mathscr{V} = \tfrac{1}{2}\rho_0 f'^2 \tag{5.29}$$

$$\mathscr{H} = \rho_0 f'^2 \tag{5.30}$$

$$\mathbf{P} = \rho_0 \mathbf{n} c f'^2 \tag{5.31}$$

(Angular momentum is not interesting for plane waves.) Note first that kinetic and potential energy densities are equal. The virial theorem of classical mechanics[1] says that this must be true of time averages in the case of quadratic forces, that is, the case of harmonic vibration. Here it is further true that these energies are equal at all times.

The second fact to observe is that the energy flux is c times the energy density. Since the loci of points of constant phase are the planes

$$\mathbf{n} \cdot \mathbf{r} - ct = \text{constant} \tag{5.32}$$

and the normal to these planes is $\mathbf{n}$, we see that energy is propagated perpendicular to the wave front with velocity c. It is this which justifies the statement that c is the velocity of sound. From (5.32) it appears that planes of constant phase also move in the direction of their normal with velocity c. Thus, the "phase velocity" and the "group velocity" for sound waves are equal.

The theory of Fourier transforms in three dimensions enables us to find quite general solutions. If we take the Fourier transform $\Phi(k, t)$ of the Eq. (5.14) we get

$$\frac{\partial^2 \Phi}{\partial t^2} = -k^2 c^2 \Phi$$

A simple solution of this equation is

$$\Phi(\mathbf{k}, t) = \Phi(\mathbf{k}, 0) e^{-ikct} \tag{5.33}$$

Thus

$$\phi(\mathbf{r}, t) = \int \Phi(\mathbf{k}, t) e^{i\mathbf{k} \cdot \mathbf{r}} d^3 \mathbf{k}$$
$$= \int \Phi(\mathbf{k}, 0) e^{i(\mathbf{k} \cdot \mathbf{r} - kct)} d^3 \mathbf{k} \tag{5.34}$$

This represents a general superposition of periodic plane waves. [The choice of the time dependence e^{ikct} in (5.33) would simply produce a wave with wave number vector $-\mathbf{k}$; since all $\mathbf{k}$'s are included in (5.34) this would give nothing new.]

4. Lagrangian and Hamiltonian Formulation

The Lagrangian of the system is, by definition,

$$L = \int (\mathcal{T} - \mathcal{V}) \, d^3 \mathbf{r} \tag{5.35}$$
$$= \tfrac{1}{2}\rho_0 \int \left[(\nabla\phi)^2 - \frac{1}{c^2}\left(\frac{\partial\phi}{\partial t}\right)^2 \right] d^3 \mathbf{r} \tag{5.36}$$

In view of the observation, made above, that in this problem velocities are represented by spatial derivatives and potential energy by time derivatives, there would at first sight seem to be some doubt as to how we should

[1]See H. Goldstein, *Classical Mechanics*. Reading, Mass.: Addison-Wesley, 1957, Chapter 3, p. 69.

define generalized momenta for the problem. Should we differentiate L with respect to the velocities $(-\nabla\phi)$ or the time derivatives $(\partial\phi/\partial t)$? The answer is in fact simply based on considerations of convenience. If we use (5.36) as the basis for an action principle,

$$\delta \int L \, dt = 0 \qquad (5.37)$$

it is easily verified that we get as the equations of motion, the wave equation (we use the plural term "equations of motion" rather than the singular because, in the sense of dynamics, there is an equation for each x). For

$$\delta \int L \, dt = \rho_0 \iint \left[\nabla\phi \cdot \delta\nabla\phi - \frac{1}{c^2} \frac{\partial\phi}{\partial t} \delta \frac{\partial\phi}{\partial t} \right] d^3\mathbf{r} \, dt$$

$$= \rho_0 \iint \left[\nabla\phi \cdot \nabla(\delta\phi) - \frac{1}{c^2} \frac{\partial\phi}{\partial t} \frac{\partial}{\partial t} \delta\phi \right] d^3\mathbf{r} \, dt$$

Using, in the first term,

$$\nabla \cdot (\delta\phi\nabla\phi) = \nabla\phi \cdot \nabla(\delta\phi) + \delta\phi\nabla^2\phi \qquad (5.38)$$

and noting that, on integration, the right-hand side transforms into a vanishing surface integral at infinity, that term becomes

$$-\rho_0 \iint \delta\phi\nabla^2\phi \, d^3\mathbf{r} \, dt$$

The other term can be integrated with respect to t, with the usual understanding that we must take $\delta\phi = 0$ at the limits of integration. It then becomes

$$-\frac{\rho_0}{c^2} \iint \delta\phi \frac{\partial^2\phi}{\partial t^2} d^3\mathbf{r} \, dt$$

The sum of these terms will then be zero for arbitrary $\delta\phi$ if and only if the wave equation

$$\nabla^2\phi - \frac{1}{c^2} \frac{\partial^2\phi}{\partial t^2} = 0 \qquad (5.39)$$

is satisfied.

Now, if we generalize this a bit, and take for the Lagrangian

$$L = \int \mathcal{L}\left(\phi, \nabla\phi, \frac{\partial\phi}{\partial t}\right) d^3\mathbf{r} \qquad (5.40)$$

the action principle gives

$$\delta \int L \, dt = 0 = \iint \left[\frac{\partial\mathcal{L}}{\partial\phi} \delta\phi + \frac{\partial\mathcal{L}}{\partial(\nabla\phi)} \cdot \nabla\delta\phi + \frac{\partial\mathcal{L}}{\partial(\partial\phi/\partial t)} \frac{\partial}{\partial t} \delta\phi \right] d^3\mathbf{r} \, dt$$

$$= \iint \delta\phi \left[\frac{\partial\mathcal{L}}{\partial\phi} - \nabla \cdot \frac{\partial\mathcal{L}}{\partial(\nabla\phi)} - \frac{\partial}{\partial t} \frac{\partial\mathcal{L}}{\partial(\partial\phi/\partial t)} \right] d^3\mathbf{r} \, dt$$

which leads to the equation of motion

$$\frac{\partial}{\partial t}\frac{\partial \mathscr{L}}{\partial (\partial \phi/\partial t)} + \nabla \cdot \frac{\partial \mathscr{L}}{\partial (\nabla \phi)} - \frac{\partial \mathscr{L}}{\partial \phi} = 0 \tag{5.41}$$

The negative of the quantity on the left-hand side of this equation is known as the variational derivative of L with respect to ϕ, and may be written $\delta L/\delta \phi$.

By analogy with the string equation (1.6) we may define the "generalized" momentum as

$$\pi(\mathbf{r}, t) = \frac{\partial \mathscr{L}}{\partial (\partial \phi/\partial t)} \tag{5.42}$$

If we do, we may follow the usual prescription for writing the Hamiltonian in terms of the Lagrangian, taking account of the fact that the usual sum $\sum p_i \dot{q}_i$ in

$$H = \sum p_i \dot{q}_i - L \tag{5.43}$$

must, because the coordinates are labeled by the continuous variables $\mathbf{r}$, be replaced by the integral $\int \pi (\partial \phi/\partial t) d^3\mathbf{r}$: Thus the Hamiltonian is

$$H = \int \left(\pi \frac{\partial \phi}{\partial t} - \mathscr{L} \right) d^3\mathbf{r}$$
$$= \int \mathscr{H} \, d^3\mathbf{r} \tag{5.44}$$

where $\mathscr{H}$ is known as the Hamiltonian density. In the case of our wave equation we have

$$\pi = -\frac{\rho_0}{c^2}\frac{\partial \phi}{\partial t}$$
$$= -\rho_0 s \tag{5.45}$$

by (5.13), so that

$$\mathscr{H} = -\frac{\rho_0}{c^2}\left(\frac{\partial \phi}{\partial t}\right)^2 - \left[\frac{1}{2}\rho_0(\nabla\phi)^2 - \frac{1}{2}\frac{\rho_0}{c^2}\left(\frac{\partial \phi}{\partial t}\right)^2\right]$$
$$= -\frac{1}{2}\rho_0(\nabla\phi)^2 - \frac{1}{2}\frac{\rho_0}{c^2}\left(\frac{\partial \phi}{\partial t}\right)^2 \tag{5.46}$$

This gives a Hamiltonian which is the negative of the total energy

$$H = \int \mathscr{H} \, d^3\mathbf{r} = -\int (\mathscr{T} + \mathscr{V}) \, d^3\mathbf{r} \tag{5.47}$$

The usual sign in this equation would have been reversed had we, at the beginning, changed the sign of $\mathscr{L}$; that is, if we had treated the quadratic term in $\partial \phi/\partial t$ as a kinetic energy term. Provided we are consistent, however, the change in sign has no consequences for the physics of the problem.

5. Guided Waves

Consider a sound wave of a definite frequency propagated down a long tube of rectangular cross section. The sides of the tube may be taken to be the directions of the x and y axes and have dimensions a cm $\times b$ cm. Let propagation be in the z direction. If we look for a solution of the wave equation with a circular frequency ω, it will be of the form

$$\phi = X(x)Y(y)Z(z)e^{i\omega t} \tag{5.48}$$

Substituting this in the wave equation, dividing through by XYZ, and cancelling out $e^{-i\omega t}$, we get the equation

$$\frac{1}{X}\frac{d^2X}{dx^2} + \frac{1}{Y}\frac{d^2Y}{dy^2} + \frac{1}{Z}\frac{d^2Z}{dz^2} = -\frac{\omega^2}{c^2} \tag{5.49}$$

Each of the three terms on the left must be a constant. The first two constants are determined by the boundary conditions, which are that $\partial\phi/\partial x = 0$ at $x = 0$ and $x = a$, and $\partial\phi/\partial y = 0$ at $y = 0$ and $y = b$. Thus we must have

$$X = \cos\frac{l\pi x}{a}, \qquad \frac{1}{X}\frac{d^2X}{dx^2} = -\frac{l^2\pi^2}{a^2} \tag{5.50}$$

$$Y = \cos\frac{m\pi y}{b}, \qquad \frac{1}{Y}\frac{d^2Y}{dy^2} = -\frac{m^2\pi^2}{b^2} \tag{5.51}$$

Consequently,

$$\frac{1}{Z}\frac{d^2Z}{dz^2} = \frac{l^2\pi^2}{a^2} + \frac{m^2\pi^2}{b^2} - \frac{\omega^2}{c^2} \tag{5.52}$$

Problem 5-2:   The end of a long cylindrical pipe of radius a is made to vibrate with velocity

$$V = f(r)\sin\omega t$$

Show that the velocity potential is given by

$$\phi = Im\,\frac{2e^{i\omega t}}{ia^2}\sum_{n=0}^{\infty}\frac{e^{-i\sqrt{k^2-\lambda_n^2/a^2}\,z}}{\sqrt{k^2-\lambda_n^2/a^2}}\frac{J_0(\lambda_n(r/a))}{J_1^2(\lambda_n)}$$

$$\times \int_0^a r'f(r')J_0\left(\frac{\lambda_n r'}{a}\right)dr'$$

the quantities λ_n being the roots of the zero-order Bessel function $J_0(x)$, and "Im" means "imaginary part of."

Now unless the constant on the right-hand side is negative, the disturbance will not be propagated but will die out exponentially with distance down the tube from the source. The case $l = m = 0$ represents propagation of a plane wave, and is always possible. But if ω is less than both $c\pi/a$ and $c\pi/b$, all other components will be attenuated. Thus, for either small enough ω or small enough tube dimensions, or both, the tube will "filter out" all but a plane wave.

The same phenomenon will hold for cylindrical guides. In this case, the equation is

$$\frac{1}{r}\frac{\partial}{\partial r}\left(r\frac{\partial\phi}{\partial r}\right) + \frac{1}{r^2}\frac{\partial^2\phi}{\partial\beta^2} + \frac{\partial^2\phi}{\partial z^2} = \frac{1}{c^2}\frac{\partial^2\phi}{\partial t^2} \qquad (5.53)$$

where r, β, and z are cylindrical coordinates. We can again use separation of variables, putting

$$\phi = R(r)e^{in\beta}Z(z)e^{-i\omega t} \qquad (5.54)$$

This leads to the equation

$$\frac{1}{R}\frac{d}{dr}\left(r\frac{dR}{dr}\right) + \frac{1}{Z}\frac{d^2Z}{dz^2} - \frac{n^2}{r^2} = -\frac{\omega^2}{c^2} \qquad (5.55)$$

The parts depending on r and z separately are constants, so that

$$\frac{1}{R}\frac{d}{dr}\left(r\frac{dR}{dr}\right) - \frac{n^2}{r^2} = -\alpha^2 \qquad (5.56)$$

and

$$\frac{1}{Z}\frac{d^2Z}{dz^2} = \alpha^2 - \frac{\omega^2}{c^2} \qquad (5.57)$$

Equation (5.56) is a Bessel equation for which the solution which is regular at the origin is

$$R = J_n(\alpha r) \qquad (5.58)$$

If α is real, this can be made to satisfy the condition $dR/dr = 0$ at $r = a$, a being the radius of the tube. This is so because $J_n(\alpha r)$ is oscillating, so that it has a zero derivative for an infinite number of values. But if α is imaginary, R will be monotonically increasing with r and its derivative will never be zero, so that the boundary conditions can never be satisfied.

The boundary conditions are in fact satisfied for

$$\alpha = \xi_{ns}/a \qquad (5.59)$$

ξ_{ns} being the sth point at which $J_n'(\alpha a)$ is zero. Equation (5.57) then becomes

$$\frac{d^2Z}{dz^2} + \left(\frac{\omega^2}{c^2} - \frac{\xi_{ns}^2}{a^2}\right)Z = 0 \qquad (5.60)$$

Again, therefore, unless $\xi_{ns} = 0$ (plane wave), propagation will only take place for sufficiently high ω or sufficiently large a.

6. Wave Equation with Sources

Let us describe an external disturbance by a "force density" $\mathbf{F}(\mathbf{r}, t)$: This signifies the force per unit volume acting on the gas at point $\mathbf{r}$. This may be incorporated into the equations by adding this quantity to the right-hand side of the Newtonian equation (5.2), so that that equation becomes

$$\rho\left[\frac{\partial \mathbf{v}}{\partial t} + (\mathbf{v} \cdot \nabla)\mathbf{v}\right] = -\nabla p + \mathbf{F} \tag{5.61}$$

Equation (5.8) is then replaced by

$$\frac{\partial \mathbf{v}}{\partial t} = -c^2 \nabla s + \frac{1}{\rho_0}\mathbf{F} \tag{5.62}$$

Taking the divergence of this equation and using (5.9) we see that

$$-\nabla^2 s + \frac{1}{c^2}\frac{\partial^2 s}{\partial t^2} = -\frac{1}{\rho_0 c^2}\,\nabla \cdot \mathbf{F} \tag{5.63}$$

We may alternatively obtain the equation for the velocity potential ϕ provided that $\mathbf{F}$ is conservative and may be written

$$\mathbf{F} = -\nabla \mathcal{U} \tag{5.64}$$

For, writing $\mathbf{v}$ in terms of ϕ, (5.62) then becomes

$$\nabla\left(\frac{\partial \phi}{\partial t} - c^2 s + \frac{1}{\rho_0}\mathcal{U}\right) = 0 \tag{5.65}$$

From this we may deduce

$$\frac{\partial \phi}{\partial t} - c^2 s + \frac{1}{\rho_0}\mathcal{U} = 0 \tag{5.66}$$

since in any case $\mathcal{U}$ is undetermined to within a constant. If we now differentiate with respect to t and substitute for $\partial s/\partial t$ from (5.9) we are led to the following equation:

$$-\nabla^2\phi + \frac{1}{c^2}\frac{\partial^2 \phi}{\partial t^2} = -\frac{1}{\rho_0 c^2}\frac{\partial \mathcal{U}}{\partial t} \tag{5.67}$$

Either (5.63) or (5.67) will serve as starting points for the theory of sound waves generated by forces, though (5.63) is more general since it involves no assumptions about $\mathbf{F}$.

Consider now the equation

$$-\nabla^2 f + \frac{1}{c^2}\frac{\partial^2 f}{\partial t^2} = h(\mathbf{r}, t) \tag{5.68}$$

of which (5.63) and (5.67) are special cases. Let us take Laplace transforms, where the transform of $f(r, t)$ is $F(r, \sigma)$ and that of $h(r, t)$ is $H(r, \sigma)$. The transformed equation is, then,

$$-\nabla^2 F + \frac{\sigma^2}{c^2} F = H \tag{5.69}$$

where f and $\partial t / \partial t$ are taken to be zero at $t = 0$. We now take Fourier transforms, $\bar{F}(\mathbf{k}, \sigma)$ and $\bar{H}(\mathbf{k}, \sigma)$ being, respectively, the transforms of F and H:

$$(k^2 + \sigma^2/c^2)\bar{F} = \bar{H}$$

so that

$$\bar{F} = \frac{1}{k^2 + \sigma^2/c^2} \bar{H} \tag{5.70}$$

We may next use the convolution theorem (4.21) and the fact that the inverse transform of $1/(k^2 + \sigma^2/c^2)$ is $(1/4\pi r) e^{-(\sigma/c)r}$ to obtain

$$F(\mathbf{r}, \sigma) = \frac{1}{4\pi} \int H(\mathbf{r}', \sigma) \frac{1}{|\mathbf{r} - \mathbf{r}'|} e^{-(\sigma/c)|\mathbf{r} - \mathbf{r}'|} \, d^3\mathbf{r}' \tag{5.71}$$

We may then take the inverse *Laplace* transform, making use of the "shifting theorem" (4.154) to get

$$f(\mathbf{r}, t) = \frac{1}{4\pi} \int \frac{1}{|\mathbf{r} - \mathbf{r}'|} h\left(\mathbf{r}', t - \frac{|\mathbf{r} - \mathbf{r}'|}{c}\right) d^3\mathbf{r}' \tag{5.72}$$

Having obtained this result, it is not difficult to see its physical interpretation. Let us look at it in the following way. If the velocity of propagation c were infinitely great, Eq. (5.68) would merely be a potential equation with a source $(1/4\pi)h$. The solution could then be obtained, as in Chapter 3 (3.128), by adding the contributions from the different parts of the source distribution, that is, we would have

$$f = \frac{1}{4\pi} \int h(\mathbf{r}', t) \frac{1}{|\mathbf{r} - \mathbf{r}'|} d^3\mathbf{r}' \tag{5.73}$$

The feature which the finite time of propagation introduces is, however, this: That the field at time t depends not on the source at time t, but rather on that at a time earlier than t by the time required to propagate the effect from $\mathbf{r}$ to $\mathbf{r}'$, that is, $(1/c)|\mathbf{r} - \mathbf{r}'|$. Consequently, the time t in (5.73) must be replaced by $t - (1/c)|\mathbf{r} - \mathbf{r}'|$, to give (5.72).

It should be noted, also, that we may deduce from (5.72) the "Green's function" for propagation in an infinite region. The Green's function has the property that the solution can be written

$$f(\mathbf{r}, t) = \int G(\mathbf{r}, \mathbf{r}', t, t') h(\mathbf{r}', t') d^3\mathbf{r}' \, dt' \tag{5.74}$$

The time integral is over all values $t' < t$. Thus the Green's function is

$$G(\mathbf{r}, \mathbf{r}'; t, t') = \frac{1}{4\pi|\mathbf{r}-\mathbf{r}'|} \delta\left(t - t' - \frac{|\mathbf{r}-\mathbf{r}'|}{c}\right) \tag{5.75}$$

since clearly substitution of this in (5.74) gives (5.72).

The Green's function for this problem is seen to depend only on *relative* space and time coordinates; that is, on $\mathbf{r}-\mathbf{r}'$ and $t-t'$. This is a consequence of the invariance of the physical system under the translations in space and time, so that neither the origin of space or of time coordinates can enter into the solution.

This translation invariance, along with the convolution theorems for Fourier and Laplace transforms, enable us to relate the *transforms* of the solution to those of the source. Using the subscripts f, l to relate to Fourier and Laplace transforms, respectively, it follows that

$$F_f(\mathbf{k}, t) = \int_0^t G_f(\mathbf{k}, t-t') h_f(\mathbf{k}, t') \, dt' \tag{5.76}$$

$$F_l(\mathbf{r}, \sigma) = \int G_l(\mathbf{r}-\mathbf{r}', \sigma) h_l(\mathbf{r}', \sigma) \, d^3r' \tag{5.77}$$

and

$$F_{fl}(\mathbf{k}, \sigma) = G_{fl}(\mathbf{k}, \sigma) h_{fl}(\mathbf{k}, \sigma) \tag{5.78}$$

the quantities in the last equation being transforms in both variables. It will be useful to record here for later reference the transforms of the Green's function:

$$G_f(\mathbf{k}, t) = \frac{1}{k} \sin kct \, \theta(t) \tag{5.79}$$

$$G_l(r, \sigma) = \frac{1}{4\pi r} e^{-\sigma r/c} \tag{5.80}$$

[cf. Eq. (5.71)] and

$$G_{fl}(\mathbf{k}, \sigma) = \frac{1}{k^2 + \sigma^2/c^2} \tag{5.81}$$

[cf. Eq. (5.70)]. $\theta(t)$ in (5.79) is 1 for $t > 0$ and 0 for $t < 0$.

It should be mentioned, finally, that the time variation can be analyzed by using a Fourier rather than a Laplace transform (see Chapter 4, Sec. 4). If we take Fourier transforms of the original equation (5.68) in both space and time variables, and let the Fourier transform in the time variable of $f(r, t)$ be

$$f_1(\mathbf{r}, \omega) = \frac{1}{2\pi} \int_{-\infty}^{\infty} f(\mathbf{r}, t) e^{i\omega t} \, dt \tag{5.82}$$

and the double transform

$$F_1(\mathbf{k}, \omega) = \frac{1}{(2\pi)^4} \int_{-\infty}^{\infty} dt \int d^3r f(\mathbf{r}, t) e^{-i(\mathbf{k}\cdot\mathbf{r}-\omega t)} \tag{5.83}$$

the transform of (5.63) is

$$\left(k^2 - \frac{\omega^2}{c^2}\right) F_1 = H_1 \tag{5.84}$$

f may be written in terms of its transform $F_1(\mathbf{k}, \omega)$ by the inverse theorem:

$$f(\mathbf{r}, t) = \int F_1(\mathbf{k}, \omega) e^{i(\mathbf{k} \cdot \mathbf{r} - \omega t)} d^3\mathbf{k} \, d\omega \tag{5.85}$$

It is therefore expressed as a superposition of plane waves [cf. (5.34)]. Since

$$F_1 = \frac{1}{k^2 - \omega^2/c^2} H_1 \tag{5.86}$$

the evaluation of the inverse transform is complicated by the existence of two poles on the real axis. Because by definition, f is the convolution (in both space and time coordinates), of $h(\mathbf{r}, t)$ and the Green's function $G(\mathbf{r} - \mathbf{r}', t - t')$, $F_1(\mathbf{k}, \omega)$ must be $(2\pi)^4$ times the product of the transforms of h and G. Therefore, the transform of G is

$$G_1(\mathbf{k}, \omega) = \left(\frac{1}{2\pi}\right)^4 \frac{1}{k^2 - \omega^2/c^2} \tag{5.87}$$

The problem of these singularities is then associated with the inversion of the transform of the Green's function. The resolution of the problem is related to the conditions of causality which G must satisfy, and this problem has already been discussed in the last chapter. It is instructive, however, to look at the problem from a different, and more physical, viewpoint. The wave equation involves only second derivatives with respect to t, and its solutions are, therefore, time-reversible. It cannot, therefore, by itself, pick out a direction of time such as is required for considerations of causality. Nature, however, contains irreversible features because of the existence of damping mechanisms. These could be incorporated in the wave equation by adding a term $\beta(\partial f/\partial t)$ on the left-hand side of (5.68). Such a term would appear, for instance, if one took account of the viscosity of the gas. Problems involving a linear time variation of this nature are fully studied in the next chapter.

In any case, the effect of such a term is to replace the equation (5.86) by

$$F_1 = \frac{1}{k^2 - \omega^2/c^2 - i\omega\beta} H_1 \tag{5.88}$$

However small β may be, the poles in ω have now been displaced to points below the real axis by an amount $\beta c^2/2$. Consequently, if the inverse is calculated with the poles in these positions, the requirements of causality will be met. After the evaluation, however, we may let β become as small as we choose.

It is possible now to invert (5.88) by using convolution theorems in both the space and the time variables. The inverse transform of

$$G_1(\mathbf{k}, \omega) = \frac{1}{(2\pi)^4} \frac{1}{k^2 - \omega^2/c^2 - i\omega\beta}$$

is, as $\beta \to 0$

$$G(\mathbf{r}, t) = \frac{1}{(2\pi)^4} \lim_{\beta \to 0} \int \frac{1}{k^2 - \omega^2/c^2 - i\omega\beta} e^{i(\mathbf{k} \cdot \mathbf{r} - \omega t)} \, d^3k \, d\omega \quad (5.89)$$

When $t < 0$, the ω integral can be evaluated by taking a contour which consists of the real axis and an infinite semicircle about the upper half-plane, on which the integrand goes exponentially to zero. Since there are no poles in this half-plane, $G = 0$ for $t < 0$, which is of course, the statement of causality. When $t > 0$, the contour has to be taken around the lower half-plane, and there are two poles. Evaluating the residues, one obtains

$$G(\mathbf{r}, t) = \frac{-ic}{16\pi^3} \int \frac{1}{k} e^{i\mathbf{k} \cdot \mathbf{r}} \, (e^{-ikct} - e^{ikct}) \, d^3k$$

$$= \frac{c}{8\pi^3} \int \frac{1}{k} e^{i\mathbf{k} \cdot \mathbf{r}} \, \sin kct \, d^3k \quad (5.90)$$

Taking $\mathbf{r}$ as polar axis for the $\mathbf{k}$ integration and evaluating the angular integrals gives

$$G(\mathbf{r}, t) = \frac{c}{2\pi^2 r} \int_0^\infty \sin kr \sin kct \, dk$$

$$= \frac{c}{8\pi^2 r} \int_{-\infty}^\infty [\cos k(r - ct) - \cos k(r + ct)] \, dk$$

$$= \frac{c}{4\pi r} [\delta(r - ct) - \delta(r + ct)] \quad (5.91)$$

The second δ function is, however, zero, since both $t \geq 0$ and $r \geq 0$, so its argument cannot be zero. Also

$$c\delta(r - ct) = c\delta(ct - r)$$

$$= \delta\left(t - \frac{r}{c}\right) \quad (5.92)$$

Thus

$$G(\mathbf{r}, t) = \frac{1}{4\pi r} \delta\left(t - \frac{r}{c}\right) \quad (5.93)$$

so that we have again obtained the result (5.75), but this time by using only Fourier transforms.

It should be noted that the Green's function (5.93) is a spherical wave; that is to say, a solution of the wave equation in spherical coordinates. It corresponds to the elementary solution $1/r$ in the case of the potential problem. Just as in that case we can, by separation of variables, get the most general such solution. Let us now solve that problem.

7. Spherical Waves

The wave equation in spherical polar coordinates is

$$\frac{1}{r^2}\frac{\partial}{\partial r}\left(r^2\frac{\partial f}{\partial r}\right)+\frac{1}{r^2}\left[\frac{1}{\sin\theta}\frac{\partial}{\partial\theta}\left(\sin\theta\frac{\partial f}{\partial\theta}\right)+\frac{1}{\sin^2\theta}\frac{\partial^2 f}{\partial\psi^2}\right]=\frac{1}{c^2}\frac{\partial^2 f}{\partial t^2} \quad (5.94)$$

Particular solutions may be found for which $f = e^{-i\omega t}S(\theta\cdot\varphi)R(r)$. For if we substitute this in (5.94) and divide by f, we get

$$\frac{1}{r^2 R}\frac{d}{dr}\left(r^2\frac{dR}{dr}\right)+\frac{1}{r^2 S}\Lambda(S)=-\frac{\omega^2}{c^2} \quad (5.94a)$$

where Λ is the differential operator which was studied extensively in the chapter on potential theory:

$$\Lambda=\frac{1}{\sin\theta}\frac{\partial}{\partial\theta}\left(\sin\theta\frac{\partial}{\partial\theta}\right)+\frac{1}{\sin^2\theta}\frac{\partial^2}{\partial\varphi^2} \quad (5.95)$$

Multiplying (5.94) by r^2, we see that it separates into a part which is a function only of r, and another which is a function only of the angular coordinates. Each must then be a constant. But we saw in the discussion of the potential problem that, for the solution to be well behaved as a function of angle,

$$\frac{1}{S}\Lambda(S)=-l(l+1) \quad (5.96)$$

where l is an integer. Normalized solutions of this equation were found to be

$$S = Y_l^m(\theta,\varphi) \quad (5.97)$$

as defined in Eq. (3.117).
The radial equation now takes the form

$$\frac{d^2 R}{dr^2}+\frac{2}{r}\frac{dR}{dr}+\left(\frac{\omega^2}{c^2}-\frac{l(l+1)}{r^2}\right)R=0 \quad (5.98)$$

As noted earlier, when the velocity of propagation $c\to\infty$, the equation becomes that of potential theory. It is homogeneous and has solutions r^l and $1/r^{l+1}$. For finite ω, however, it is more complicated.

For small r, however, the ω^2/c^2 term is negligible, so that in this range the solution has a behavior of this nature. Let us write

$$R(r)=r^l u_l(r) \quad (5.99)$$

and determine the equation for u_l. By straightforward calculation, we find it to be

$$r\frac{d^2u_l}{dr^2} + (2l+1)\frac{du_l}{dr} + k^2 r u_l = 0 \tag{5.100}$$

where

$$k = \omega/c \tag{5.101}$$

Since this equation has linear coefficients, we recognize it as one which can be solved by our generalized Laplace transform method. However, by a simple transformation, we can reduce it to a Bessel equation. If we put $kr = \xi$ and

$$R_l = \xi^{-1/2} v_l(\xi) \tag{5.102}$$

The equation for v_l is found to be

$$v_l'' + \frac{1}{\xi}v_l' + \left(1 - \frac{(l+\frac{1}{2})^2}{\xi^2}\right)v_l = 0 \tag{5.103}$$

The solution is, therefore,

$$R_l = \frac{\gamma_l}{\sqrt{kr}} Z_{l+(1/2)}(kr) \tag{5.104}$$

where Z is any linear combination of solutions of the Bessel equation of order l. If the solution must be regular at the origin, $Z_{l+(1/2)}$ must be the Bessel function $J_{l+(1/2)}$.

We may use the standard notation of "spherical Bessel functions" introduced in the last chapter [Eqs. (4.374) to (4.376)]. We shall extend the notation introduced there by defining

$$z_l(x) = \sqrt{\frac{\pi}{2x}} Z_{l+(1/2)}(x) \tag{5.105}$$

It should be noticed that $h_l^{(1)}(kr)$ defines a "diverging" spherical wave, with the property that at distances r such that $kr \gg 1$ the flow is away from the origin. $h_l^{(2)}(kr)$, on the other hand, represents a "converging" wave, with energy flowing *toward* the origin. To demonstrate these facts, we use the asymptotic forms of these functions for large argument, which may be derived from (4.377). The asymptotic forms are

$$h_l^{(1)}(kr) = e^{-i(l+1)(\pi/2)}\frac{e^{ikr}}{kr} \tag{5.106}$$

and

$$h_l^{(2)}(kr) = e^{i(l+1)(\pi/2)}\frac{e^{-ikr}}{kr} \tag{5.107}$$

At these distances, then, the complete solution is

$$\phi_{\text{out}} = A_1 \frac{e^{i[kr - \omega t - ((l+1)\pi/2)]}}{kr} Y_{lm}(\theta, \varphi) \qquad (5.108)$$

in the one case and

$$\phi_{\text{in}} = A_2 \frac{e^{-i[kr + \omega t - ((l+1)\pi/2)]}}{kr} Y_{lm}(\theta, \varphi) \qquad (5.109)$$

Let us calculate the energy flux in terms of these imaginary potentials. The radial energy flux has been shown to be [Eq. (5.25)]

$$P = -\rho_0 \int \frac{\partial \phi_r}{\partial t} \frac{\partial \phi_r}{\partial r} r^2 \, d\Omega \qquad (5.110)$$

where ϕ_r is the real part of the complex ϕ, and the integral is over solid angle at a given radius. P may be written

$$P = -\tfrac{1}{4}\rho_0 \int \left(\frac{\partial \phi}{\partial t} + \frac{\partial \phi^*}{\partial t} \right) \left(\frac{\partial \phi}{\partial r} + \frac{\partial \phi^*}{\partial r} \right) r^2 \, d\Omega \qquad (5.111)$$

Since $(\partial \phi/\partial t)(\partial \phi/\partial r)$ depends on the azimuthal angle φ as $e^{2im\varphi}$ and $(\partial \phi^*/\partial t)(\partial \phi^*/\partial r)$ depends on it as $e^{-2im\varphi}$, both of these terms integrate to zero. Consequently,

$$P = -\tfrac{1}{2}\rho_0 \mathcal{R}e \int \frac{\partial \phi}{\partial t} \frac{\partial \phi^*}{\partial r} r^2 \, d\Omega \qquad (5.112)$$

Now for the case of $\phi = \phi_{\text{out}}$

$$\frac{\partial \phi}{\partial t} = -i\omega A_1 \frac{e^{i[kr - \omega t - ((l+1)\pi/2)]}}{kr} Y_{lm} \qquad (5.113)$$

and

$$\frac{\partial \phi^*}{\partial r} = -ikA_1^* \frac{e^{-i[kr - \omega t - ((l+1)\pi/2)]}}{kr} Y_{lm}^* + O\left(\frac{1}{k^2 r^2} \right) \qquad (5.114)$$

By virtue of the normalization of the spherical harmonics, the integral may be evaluated immediately to give, for $kr \gg 1$,

$$P = \tfrac{1}{2}\rho_0 \frac{\omega}{k} |A_1|^2 \qquad (5.115)$$

$$= \tfrac{1}{2}\rho_0 c |A_1|^2$$

This is positive, corresponding to a flux in the positive direction, that is, in the direction of increasing r.

If we now do the same calculation for ϕ_{in}, only one thing is changed, that is, that k is replaced by $-k$. Therefore, in the light of (5.115), the flux in this case is radially inward.

A general spherical wave with a given frequency is

$$\phi = \sum_{l,m} A_{lm} z_l(kr) Y_{lm}(\theta, \varphi) e^{-i\omega t} \qquad (5.116)$$

For an arbitrary frequency spectrum, we have

$$\phi = \sum_{l,m} A_{lm}(\omega) z_l \left(\frac{\omega}{c} r\right) e^{-i\omega t} \, d\omega \, Y_{lm}(\theta, \varphi) \qquad (5.117)$$

In the case of axial symmetry (5.116) simplifies to

$$\phi = \sum B_l z_l(kr) P_l (\cos \theta) \, e^{-i\omega t} \qquad (5.118)$$

8. Expansion of a Plane Wave in Spherical Waves

A plane wave solution for arbitrary frequency is $e^{i(\mathbf{k}\cdot\mathbf{r}-\omega t)}$. Taking, for simplicity, the polar axis as the direction of the propagation vector $\mathbf{k}$, this becomes $e^{i(kz-\omega t)} = e^{ikr\cos\theta} e^{-i\omega t}$. Omitting the time-dependent term, $e^{ikr\cos\theta}$ is an axially symmetric solution of the wave equation. It must therefore be possible to put

$$e^{ikr\mu} = \sum B_l j_l(kr) P_l(\mu) \qquad (5.119)$$

Here we have put $\mu = \cos \theta$ and $z_l = j_l$ because $e^{ikr\mu}$ is finite at $r = 0$.

To determine B_l, multiply by $P_n(\mu)$ and integrate over μ between -1 and 1. We obtain

$$B_n j_n(kr) \frac{2}{2n+1} = \int_{-1}^{1} e^{ikr\mu} P_n(\mu) \, d\mu \qquad (5.120)$$

B_n may now be most simply obtained by equating coefficients of $(kr)^n$. Using (4.339) this coefficient in $j_n(kr)(2/2n+1)$ is

---

Problem 5-3:   A gas contained in a spherical resonator of radius R has an initial velocity potential $\phi(\mathbf{r}, t) = f_0(r)$ at time $t = 0$, and no condensation. Show that the velocity potential at time t is given by

$$\phi = \frac{2}{R} \sum_{n=1}^{\infty} \frac{\sin \alpha_n r/R}{r} \frac{\cos \alpha_n vt/R}{1 - (\sin 2\alpha_n/2\alpha_n)}$$

$$\times \int_0^R r' f_0(r') \sin \frac{\alpha_n r'}{R} \, dr'$$

the quantities α_n being the roots of $\tan x = x$. v is the velocity of propagation of sound in the gas.

Problem 5-4:   A conical horn has a flexible base of radius R and semivertical angle α. If the velocity of the base $r = R$ is

$$v(R, t) = v_0 \sin \omega t$$

show that the velocity potential is given by

$$\phi(\mathbf{r}, t) = 2v_0 R \sqrt{\frac{R}{r}} \frac{\sin\left[\omega t - k(r - R)\right] - 2ka\cos\left[\omega t - k(r - R)\right]}{1 + 4k^2 a^2}$$

where $k = \omega/c$ and c is the velocity of sound. Why does this not depend on α?

$$\sqrt{\frac{\pi}{2}}\frac{1}{\Gamma(n + \frac{3}{2})}\frac{2}{2^{n+1/2}}\frac{2}{2n+1} = \sqrt{\frac{\pi}{2}}\frac{(n+1)!2^{2n+2}}{(2n+2)!\sqrt{\pi}}\frac{1}{2^{n+1/2}}\frac{2}{2n+1}$$

$$= \frac{2^{n+2}(n+1)!}{(2n+2)!(2n+1)}$$

$$= \frac{n!2^{n+1}}{(2n+1)!(2n+1)}$$

where we have used (4.336) for $\Gamma(n + \frac{3}{2})$.

Consider now the right-hand side. The coefficient of $(kr)^n$ is

$$\frac{i^n}{n!}\int_{-1}^{1}\mu^n P_n(\mu)\, d\mu$$

This may be evaluated by using the recursion relation (3.84) for Legendre polynomials. Each factor μ yields polynomials of one higher and one lower order. But the integrals of all polynomials except $P_0 = 1$ are zero; therefore, the only terms which concern us are the polynomials of lower order in each instance. The n factors of μ make it just possible in this way to obtain a P_0 term. From (3.84) it is evident that the coefficient of P_0 in $\mu^n P_n(\mu)$ is

$$\frac{n}{2n+1}\frac{n-1}{2n-1}\cdots\frac{1}{1} = \frac{2^n(n!)^2}{(2n+1)!}$$

B_n is, therefore, determined by the equations

$$\frac{i^n 2^{n+1} n!}{(2n+1)!} = \frac{n!2^{n+1}}{(2n+1)!(2n+1)}B_n$$

or

$$B_n = i^n(2n+1) \qquad (5.121)$$

The expansion of a plane wave in spherical waves is then given by

$$e^{ikz} = \sum_{l=0}^{\infty} i^l(2l+1)j_l(kr)P_l(\cos\theta) \qquad (5.122)$$

An interesting corollary is the expression for the spherical Bessel function as an integral involving the Legendre polynomial of the same order. Multiplying (5.122) on the right by $P_n(\mu)$ gives

$$2i^n j_n(kr) = \int_{-1}^{1} e^{ikr\mu}P_n(\mu)\, d\mu$$

so that

$$j_n(kr) = \frac{(-i)^n}{2} \int_{-1}^{1} e^{ikr\mu} P_n(\mu) \, d\mu \qquad (5.123)$$

The expansion (5.122) plays an important role in the theory of scattering which will be developed subsequently.

9. Radiation from a Periodic Source

We have shown, in Eq. (5.72), that the radiation $\phi(\mathbf{r}, t)$ from a source $\sigma(\mathbf{r}, t)$ is given by

$$\phi(\mathbf{r}, t) = \frac{1}{4\pi} \int \frac{\sigma(\mathbf{r}', t - (|\mathbf{r} - \mathbf{r}'|/c))}{|\mathbf{r} - \mathbf{r}'|} \, d^3\mathbf{r}' \qquad (5.124)$$

If the time variation of the source is periodic with a frequency ω, that is if

$$\sigma(\mathbf{r}, t) = \sigma_0(\mathbf{r}) e^{-i\omega t} \qquad (5.125)$$

eq. (5.124) becomes

$$\phi(\mathbf{r}, t) = \frac{e^{-i\omega t}}{4\pi} \int \sigma_0(\mathbf{r}') \frac{1}{|\mathbf{r} - \mathbf{r}'|} \exp\left[i\omega \frac{|\mathbf{r} - \mathbf{r}'|}{c}\right] d^3\mathbf{r}' \qquad (5.126)$$

For regions outside the source, that is, for values of $\mathbf{r}$ such that $\sigma_0(\mathbf{r}) = 0$, this must have the form of (5.116). In order to obtain it in that form, however, we must expand

$$\frac{e^{i(\omega/c)|\mathbf{r} - \mathbf{r}'|}}{4\pi |\mathbf{r} - \mathbf{r}'|} \qquad (5.127)$$

in terms of spherical harmonics in the angular coordinates of the point $\mathbf{r}$.

This expansion may be conveniently carried out by writing the function $e^{i(\omega/c)R}/4\pi R$ in terms of its Fourier transform. For, inside the transform integral, the coordinates enter only through

$$e^{i\mathbf{k}\cdot\mathbf{R}} = e^{i\mathbf{k}\cdot\mathbf{r}} e^{-i\mathbf{k}\cdot\mathbf{r}'}$$

But, as we have seen in the previous section, the two terms of this product may be expanded in terms of the spherical harmonics of the coordinates of $\mathbf{r}$ and $\mathbf{r}'$, respectively, and we should therefore obtain the formula sought.

First then, what is the Fourier transform of $(1/4\pi)e^{i(\omega/c)R}/R$? Let us first calculate the transform of $(1/4\pi)e^{-\beta R}/R$, where β is some complex number with positive real part. This transform is

$$\left(\frac{1}{2\pi}\right)^3 \frac{1}{4\pi} \int \frac{e^{-\beta R}}{R} e^{i\mathbf{k}\cdot\mathbf{R}} d^3\mathbf{R}$$

Taking $\mathbf{k}$ to be the polar axis for the $\mathbf{R}$ integration, and integrating over the angular coordinates, we get

$$\left(\frac{1}{2\pi}\right)^3 \int \frac{e^{-\beta R}}{R} \frac{\sin kR}{kR} R^2 \, dR = \frac{1}{(2\pi)^3} \frac{1}{k^2 + \beta^2}$$

Therefore,

$$\frac{e^{-\beta R}}{4\pi R} = \int \frac{1}{k^2 + \beta^2} e^{i\mathbf{k}\cdot(\mathbf{r}-\mathbf{r}')} \, d^3\mathbf{k} \qquad (5.128)$$

Now

$$e^{i\mathbf{k}\cdot\mathbf{r}} = 4\pi \sum_{l,m} i^l j_l(kr) Y_l^m(\Omega) Y_l^{-m}(\Omega_k) \qquad (5.129)$$

and

$$e^{-i\mathbf{k}\cdot\mathbf{r}'} = 4\pi \sum (-i)^n j_n(kr') Y_n^{-m'}(\Omega') Y_n^{m'}(\Omega_k) \qquad (5.130)$$

Here we have made use of the formula (3.136) expressing the Legendre polynomial of the cosine of the angle between two vectors in terms of the polar angles of the vectors. Ω, Ω', Ω_k represent the polar angles of $\mathbf{r}$, $\mathbf{r}'$, and $\mathbf{k}$, respectively.

If these equations are substituted in (5.128), the integral over Ω_k may be done immediately to give, using the orthogonality of the $Y_l^m(\Omega_k)$'s,

$$\frac{e^{-\beta R}}{4\pi R} = \frac{2}{\pi} \sum_{l=0}^{\infty} \sum_{m=-l}^{l} Y_l^m(\Omega) Y_l^{-m}(\Omega') \int_0^{\infty} \frac{1}{k^2 + \beta^2} j_l(kr') j_l(kr) k^2 \, dk \qquad (5.131)$$

If, now, β is purely imaginary, the integral (5.131) becomes improper. It may, however, be evaluated for $\beta = -i(\omega/c) + \epsilon$, where ϵ is small and will ultimately be allowed to approach zero.

Let us now suppose that $r > r'$: If the reverse is the case, the appropriate formula may be obtained by interchanging $\mathbf{r}$ and $\mathbf{r}'$. The poles of the integrand in (5.131) are at

$$k = \pm i\beta = \pm\frac{\omega}{c} \pm i\epsilon$$

We know that

$$j_l(kr) = \tfrac{1}{2}[h_l^{(1)}(kr) + h_l^{(2)}(kr)] \qquad (5.132)$$

But $h_l^{(1)}(kr)$ approaches 0 on a semicircle whose radius approaches ∞ surrounding the upper half-plane, while $h_l^{(2)}(kr)$ behaves likewise in the *lower* half-plane. This will also be true for $h_l^{(1)}(kr) j_l(kr')$ and $h_l^{(2)}(kr) j_l(kr')$, by virtue of the fact that $r > r'$. Breaking up $j_l(kr)$ into its two parts, and evaluating the integrals involving them, respectively, by treating these integrals as parts of contours which are closed by semicircles of infinite radius encircling the appropriate half-plane, we find, as $\epsilon \to 0$, that

$$\int_0^{\infty} \frac{1}{k^2 + \beta^2} j_l(kr') j_l(kr) k^2 \, dk = \frac{1}{4} \int_{-\infty}^{\infty} \frac{1}{k^2 + \beta^2} j_l(kr') [h_l^{(1)}(kr)$$

$$+ h_l^{(2)}(kr)] k^2 \, dk = \frac{\pi i \omega}{4c} \left[j_l\left(\frac{\omega}{c} r'\right) h_l^{(1)}\left(\frac{\omega}{c} r\right)\right.$$

$$\left. + j_l\left(-\frac{\omega}{c} r'\right) h_l^{(2)}\left(-\frac{\omega}{c} r\right)\right] \qquad (5.133)$$

Now

$$j_l\left(-\frac{\omega}{c}r'\right) = (-1)^l j_l\left(\frac{\omega}{c}r'\right)$$

and

$$h_l^{(2)}\left(-\frac{\omega}{c}r'\right) = (-1)^l h_l^{(1)}\left(\frac{\omega}{c}r'\right)$$

The first follows from (4.377) and the second consequently from (5.132). Therefore,

$$\int_0^\infty \frac{1}{k^2+\beta^2}j_l(kr')j_l(kr)k^2\,dk = \frac{\pi i\omega}{2c}j_l\left(\frac{\omega}{c}r'\right)h_l^{(1)}\left(\frac{\omega}{c}r\right) \qquad (5.134)$$

Substituting this into (5.131) we obtain finally the formula we have been seeking:

$$\frac{e^{i(\omega/c)R}}{4\pi R} = \frac{i\omega}{c}\sum_{l=0}^\infty \sum_{m=-l}^l Y_l^m(\Omega)Y_l^{-m}(\Omega')j_l\left(\frac{\omega}{c}r'\right)h_l^{(1)}\left(\frac{\omega}{c}r\right) \qquad (5.135)$$

If this is now substituted in (5.126), we obtain the result

$$\phi(\mathbf{r},t) = \frac{i\omega}{c}\sum_{l=0}^\infty \sum_{m=-l}^l \int \sigma_0(\mathbf{r}')j_l\left(\frac{\omega}{c}r'\right)Y_l^{-m}(\Omega')\,d^3\mathbf{r}'$$

$$\times h_l^{(1)}\left(\frac{\omega}{c}r\right)Y_l^m(\Omega)e^{-i\omega t} \qquad (5.136)$$

This is now in the form (5.116), as required. The integral is a sort of radiation multipole moment; the lth term in the expansion is designated as multipole radiation of the lth order.

In the case in which the dimensions of the source are small compared with the wavelength of the radiation, i.e., in which $(\omega/c)r' \ll 1$ over the whole source, we may replace $j_l[(\omega/c)r']$ by the leading term in its series expansion:

$$j_l\left(\frac{\omega}{c}r'\right) \simeq \sqrt{\frac{\pi}{2(\omega/c)r'}}\frac{1}{\Gamma(l+\frac{3}{2})}\frac{1}{2^{l+(1/2)}}\left(\frac{\omega}{c}r'\right)^{l+(1/2)}$$

$$= \frac{2^{l+1}(l+1)!}{(2l+2)!}\left(\frac{\omega}{c}r'\right)^l$$

$$= \frac{2^l l!}{(2l+1)!}\left(\frac{\omega}{c}r'\right)^l$$

Then

$$\phi(\mathbf{r},t) \simeq \frac{i\omega}{c}\sum_{l=0}^\infty \sum_{m=-l}^l \frac{2^l l!}{(2l+1)!}\left(\frac{\omega}{c}\right)^l \int r'^l Y_l^{-m}(\Omega')\sigma_0(\mathbf{r}')\,d^3\mathbf{r}'$$

$$\times h_l^{(1)}\left(\frac{\omega}{c}r\right)Y_l^m(\Omega)\,e^{-i\omega t} \qquad (5.137)$$

This may then be written in terms of the electrostatic multipole moments defined in (3.142) (σ_0 here replaces ρ):

$$\phi(\mathbf{r}, t) = i \sum_{l=0}^{\infty} \frac{1}{\sqrt{4\pi(2l+1)}} \frac{2^l l!}{(2l)!} \left(\frac{\omega}{c}\right)^{l+1} Q_l^{-m} h_l^{(1)}\left(\frac{\omega}{c}r\right) Y_l^m(\Omega)\, e^{-i\omega t} \quad (5.138)$$

Let us calculate the intensity of the radiation at some distance from the source. The Hankel function can then be replaced by its asymptotic form. We also pick out the real (physical) part of ϕ to get

$$\phi = \sum_{l=0}^{\infty} \sum_{m=-l}^{l} \frac{1}{\sqrt{4\pi(2l+1)}} \frac{2^l l!}{(2l)!} \left(\frac{\omega}{c}\right)^{l+1} |Q_l^{-m}||Y_l^m(\Omega)|$$

$$\times \frac{c}{\omega r} \cos\left(\frac{\omega}{c}r - \frac{l\pi}{2} - \omega t + m\phi - \alpha_m\right) \quad (5.139)$$

where we have written

$$Q_l^{-m} = |Q_l^{-m}|e^{-i\alpha_m} \quad (5.140)$$

Now, by (5.25) the radial energy flux is

$$P_r = \bar{p} v_r$$

$$= -\rho_0 \frac{\partial \phi}{\partial t} \frac{\partial \phi}{\partial r} \quad (5.141)$$

where the subscript r indicates radial components. But at large distances $\partial \phi / \partial t$ decreases like $1/r$; so also does $\partial \phi / \partial r$, terms coming from the derivative of the r in the denominator of (5.139) being negligible. Thus

$$\frac{\partial \phi}{\partial t} = \sum_{l=0}^{\infty} \sum_{m=-l}^{l} \frac{1}{\sqrt{4\pi(2l+1)}} \frac{2^l l!}{(2l)!} \left(\frac{\omega}{c}\right)^{l} |Q_l^{-m}||Y_l^m(\Omega)|$$

$$\times \frac{\omega}{r} \sin\left(\frac{\omega}{c}r - \frac{l\pi}{2} - \omega t + m\phi - \alpha_m\right) \quad (5.142)$$

and

$$\frac{\partial \phi}{\partial r} = -\sum_{l'=0}^{\infty} \sum_{m'=-l'}^{l'} \frac{1}{\sqrt{4\pi(2l'+1)}} \frac{2^{l'} l'!}{(2l')!} \left(\frac{\omega}{c}\right)^{l'+1} |Q_{l'}^{-m'}||Y_{l'}^{m'}(\Omega)|$$

$$\times \frac{1}{r} \sin\left(\frac{\omega}{c}r - \frac{l'\pi}{2} - \omega t + m'\phi - \alpha_{m'}\right) \quad (5.143)$$

The *total* energy flowing from the source is

$$-\frac{\partial U}{\partial t} = \left\langle \int P_r r^2\, d\Omega \right\rangle \quad (5.144)$$

where $\langle\ \rangle$ designates a time average.

If we first perform the ϕ integral, we find that the only products which do not yield zero are those for which $m = m'$; when $m = m'$, the value of the ϕ integral is π. Next, the integral over θ may be done. By virtue of the fact that

$$\int |Y_l^m(\Omega)|\,|Y_{l'}^m(\Omega)|\,d\Omega = \frac{1}{2\pi}\,\delta_{ll'} \qquad (5.145)$$

as may be seen from (3.128), the rate of radiation of energy is

$$-\frac{\partial U}{\partial t} = \frac{\rho_0 c}{2} \sum_{l=0}^{\infty} \sum_{m=-l}^{l} \frac{1}{4\pi(2l+1)} \left[\frac{2^l l!}{(2l)!}\right]^2 \left(\frac{\omega}{c}\right)^{2l+2} |Q_l^{-m}|^2 \quad (5.146)$$

The terms of different l correspond to what is known as radiation of different multipole order; $l = 0$ corresponds to monopole radiation, $l = 1$ to dipole, $l = 2$ to quadrupole, etc. Since the lth term may be expressed in terms of a dimensionless quantity $(\omega a/c)^{2l+2}$, where a is the radius of the radiating source and $c/\omega = 2\pi\lambda$, λ being the wavelength of the radiation, the multipole expansion is in fact an expansion in powers of $(a/2\pi\lambda)^2$. If, therefore, the source is small compared with the wavelength, the intensity of the radiation decreases rapidly with the multipole order. A situation of this sort will be seen later to apply to electromagnetic radiation from atoms and from nuclei.

The formula (5.146) is valid even when $(\omega/c)a$ is not small, though the form of Q_l^{-m} is different in this case; it is, in fact,

$$Q_l^{-m} = \sqrt{4\pi(2l+1)}\,\frac{(2l)!}{2^l l!}\left(\frac{c}{\omega}\right)^l \int \sigma_0(\mathbf{r}')j_l\!\left(\frac{\omega}{c}r'\right)Y_l^{-m}(\Omega')\,d^3\mathbf{r}' \qquad (5.147)$$

Problem 5-5: The center of a sphere of radius a undergoes vibrations given by $Z = Z_0^{-i\omega t}$, where Z_0 is small. Show that the boundary condition at its surface is given approximately by

$$\left(\frac{\partial\phi}{\partial r}\right)_{r=a} = i\omega Z_0 \cos\theta\,e^{-i\omega t}$$

Problem 5-6: Show that the solution of the wave equation satisfying all the conditions of the problem is

$$\phi = \frac{icZ_0}{h_1^{(1)\prime}(ka)}\,h_1^{(1)}(kr)\cos\theta\,e^{-i\omega t}$$

where $k = \omega/c$.

Problem 5-7: Verify that the pressure is

$$\bar{p} = \frac{\rho_0 c\omega Z_0}{h_1^{(1)\prime}(ka)}\,h_1^{(1)}(kr)\cos\theta\,e^{-i\omega t}$$

and that the net force of reaction on the sphere is $-2\pi a^2 \int \bar{p}\cos\theta\,d\theta$ in the z direction, that is,

$$F_z = \frac{4}{3}\pi a^2 \frac{\rho_0 c \omega Z_0}{h_1^{(1)\prime}(ka)} h_1^{(1)}(ka) e^{-i\omega t}$$

Problem 5-8:   By calculating the physical (real) part of F_z show that this force has a component proportional to the velocity of the sphere and another proportional to its acceleration: They are, respectively,

$$\frac{4}{3}\pi a^2 \rho_0 c \omega Z_0 \frac{(ka)^4}{4+(ka)^4}\sin\omega t$$

and

$$\frac{4}{3}\pi a^2 \rho_0 c \omega Z_0 \frac{ka[2+(ka)^2]}{4+(ka)^4}\cos\omega t$$

whereas the corresponding components of velocity and acceleration of the center of the sphere are

$$-\omega Z_0 \sin\omega t \quad \text{and} \quad -\omega^2 Z_0 \cos\omega t$$

Problem 5-9:   Prove that the total radiated energy is

$$\frac{2}{3}\pi \frac{\rho_0 c^3 Z_0^2}{|h_1^{(1)\prime}(ka)|^2} = \frac{2}{3}\pi \frac{\rho_0 c^3 Z_0^2}{4+(ka)^4}(ka)^6$$

10.   Time-Varying Source

In the preceding section, we have dealt with the case of a simply periodic source. If it has an arbitrary time variation, the problem is no more difficult to solve. Consider, for example (5.63), where we designate the right-hand side as $S_0(\mathbf{r}, t)$. We may make a Fourier analysis

$$S_0(\mathbf{r}, t) = \int_{-\infty}^{\infty}\sigma_0(\mathbf{r},\omega)e^{i\omega t}\,d\omega \tag{5.148}$$

where

$$\sigma_0(\mathbf{r},\omega) = \frac{1}{2\pi}\int_{-\infty}^{\infty}S_0(\mathbf{r},t)e^{-i\omega t}\,dt \tag{5.149}$$

Then, the solution may be written

$$s(\mathbf{r},t) = \frac{1}{4\pi}\int\frac{S_0(\mathbf{r}', t-(|\mathbf{r}-\mathbf{r}'|/c))}{|\mathbf{r}-\mathbf{r}'|}\,d^3\mathbf{r}' \tag{5.150}$$

Introducing into this the Fourier transform (5.148), we find that

$$s(\mathbf{r},t) = \frac{1}{4\pi}\int_{-\infty}^{\infty}d\omega\int\frac{\sigma_0(\mathbf{r}',\omega)}{|\mathbf{r}-\mathbf{r}'|}e^{i\omega(t-|\mathbf{r}-\mathbf{r}'|/c)}\,d^3\mathbf{r}'$$

$$= \frac{1}{4\pi}\int_{-\infty}^{\infty}e^{i\omega t}\,d\omega\int\frac{\sigma_0(\mathbf{r}',\omega)}{|\mathbf{r}-\mathbf{r}'|}e^{-(i\omega/c)|\mathbf{r}-\mathbf{r}'|}\,d^3\mathbf{r}' \tag{5.151}$$

The inner integral is simply the solution for a simple periodic source, as calculated in the last section. The solution to the present problem, is, therefore, obtained by superimposing the solutions from such sources.

Problem 5-10: Prove that the total radiated intensity is made up of separate contributions from each frequency, and that interference terms vanish. (Note that this is only true if the radiated intensity is integrated over time.)

Problem 5-11: Determine the solution to the wave equation when the source is zero for negative t and has a time variation $e^{-\alpha t}$ for positive t.

---

11. Radiation from a Moving Source

Consider a point source of constant strength moving with a uniform velocity $\mathbf{u}$. This may be treated by the method of the previous section: In this case

$$S_0(\mathbf{r}, t) = S\delta(\mathbf{r} - \mathbf{u}t) \tag{5.152}$$

and, therefore,

$$\sigma_0(\mathbf{r}, \omega) = \frac{S}{2\pi} \int_{-\infty}^{\infty} \delta(\mathbf{r} - \mathbf{u}t) e^{-i\omega t}\, dt \tag{5.153}$$

It is convenient to take $\mathbf{u}$ in the z direction. Then

$$\sigma_0(\mathbf{r}, \omega) = \frac{S}{2\pi u} \delta(x)\delta(y) e^{-i\omega z/u} \tag{5.154}$$

Letting $\mathbf{z}'$ be a vector of magnitude z' in the z direction, and putting

$$s(\mathbf{r}, t) = \frac{S}{8\pi^2 u} \int_{-\infty}^{\infty} d\omega e^{i\omega t} \int \frac{\delta(x')\delta(y')}{|\mathbf{r} - \mathbf{r}'|} e^{i\omega(z'/u)} e^{-(i\omega/c)|\mathbf{r}-\mathbf{r}'|} \tag{5.155}$$

we may do the x', y', and t' integrals to get

$$s(\mathbf{r}, t) = \frac{S}{4\pi u} \int \frac{1}{|\mathbf{r} - \mathbf{z}'|} \delta\left(t - \frac{1}{c}|\mathbf{r} - \mathbf{z}'| - \frac{z'}{u}\right) dz' \tag{5.156}$$

In the first place, then, we note that s will be different from zero only if the argument of the δ function is zero, which requires that

$$ut - z' = \frac{u}{c}|\mathbf{r} - \mathbf{z}'| \tag{5.157}$$

This relation may be represented graphically, as in Fig. (5.1). For an arbitrary point z', points r satisfying (5.157) lie within a sphere of radius $(c/u)PQ$ of $P(z')$. We consider two cases. If $u < c$, that is, if the source is

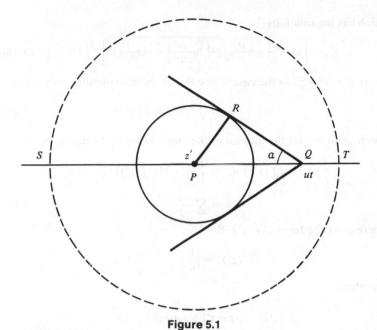

Figure 5.1

moving with a speed less than that of the velocity of propagation of sound, this sphere may have arbitrarily large radius as $z' \to -\infty$, and so $s = 0$ everywhere. The radiation at large distances will, however, originate from the source when it was at large negative z'. This situation is represented by the large (dotted) circle of diameter SPT.

In the case $u > c$ (source moving with supersonic velocity), (5.157) will be satisfied for a given z' for **r** inside the sphere represented by the radius $PR < PQ$. If we put

$$\frac{c}{u} = \frac{PR}{PQ} = \sin \alpha \qquad (5.158)$$

we see that $s = 0$ only inside the cone of semivertical angle α with vertex at $Q = ut$. This is the "bow wave" effect which is a familiar phenomenon (in a two-dimensional geometry) of water waves, where the velocity of wave propagation is very small. We discuss the analogous situation for electromagnetic waves in a dielectric medium (Čerenkov radiation) in the next section.

To evaluate (5.156), it is necessary to find the values of z' for which the argument of the δ function is zero. Squaring (5.157) we obtain

$$z'^2 \left(1 - \frac{u^2}{c^2}\right) - 2z' \left(ut - \frac{u^2}{c^2} z\right) + u^2 \left(t^2 - \frac{r^2}{c^2}\right) = 0 \qquad (5.159)$$

which has the solutions

$$z'_{1,2} = \left[ut - \frac{u^2}{c^2}z \pm \frac{u}{c}\sqrt{\xi^2\left(1-\frac{u^2}{c^2}\right)+(ut-z)^2}\right] \Big/ \left(1-\frac{u^2}{c^2}\right) \quad (5.160)$$

where $\xi^2 = r^2 - z^2$. In the case $u > c$ this gives real solutions only when

$$ut - z > \xi\sqrt{\frac{u^2}{c^2}-1} = \xi \cot \alpha \qquad (5.161)$$

which again defines the interior of the cone shown in the figure. Now

$$\int g(z')\delta(f(z'))\,dz' = \int g(z')\delta(|f(z')|)\,\frac{d|f(z')|}{|f'(z')|}$$

$$= \sum \frac{g(z_i)}{|f'(z_i)|} \qquad (5.162)$$

where z_i are the zeros of $f(z')$. But

$$|f'(z')| = \left| \frac{1}{u} + \frac{1}{c}\frac{z'-z}{|\mathbf{r}'-\mathbf{z}'|}\right|$$

Therefore,

$$\frac{g(z')}{|f'(z')|} = \frac{u}{||\mathbf{r}-\mathbf{z}'|+(u/c)(z'-z)|}$$

$$= \frac{u}{|ct-(c/u)z'+(u/c)(z'-z)|} \qquad (5.163)$$

when z' satisfies (5.157). If now z' is taken to be one of the values given by (5.160),

$$ct - \frac{u}{c}z - z'\left(\frac{c}{u}-\frac{u}{c}\right) = \frac{c}{u}\left[ut - \frac{u^2}{c^2}z - z'\left(1-\frac{u^2}{c^2}\right)\right]$$

$$= \mp\sqrt{\xi^2\left(1-\frac{u^2}{c^2}\right)+(ut-z)^2} \qquad (5.164)$$

We must now check to see whether one of the roots (5.160) might not be an extraneous one resulting from the squaring of (5.157). It is clear from (5.156) that for true roots $t - (z'/u)$ must be positive. Now from (5.160) it is easily seen that

$$t - \frac{z'}{u} = \frac{1}{u}\frac{(u^2/c^2)(z-ut)\mp\sqrt{\xi^2(1-(u^2/c^2))+(ut-z)^2}}{1-(u^2/c^2)}$$

If $u > c$ the denominator is negative. On the other hand, in the numerator the first term is greater than the square root. Therefore, the minus sign does correspond to an extraneous root. Likewise, if $u < c$, the denominator is positive and the square root is greater in magnitude than the first term, so that again we must reject the contribution from the term with the minus sign.

It follows, then, that only one root of (5.157) gives a contribution to (5.156); using (5.162), (5.163), and (5.164), we find that

$$s(\mathbf{r}, t) = \frac{S}{4\pi} \frac{1}{\sqrt{\xi^2(1 - (u^2/c^2)) + (ut - z)^2}} \tag{5.165}$$

This is then the form of the solution in general when $u < c$, and inside the cone mentioned earlier when $u > c$. It should be noted that in the latter case s becomes infinite on the surface of the cone, which violates the basic condition (namely, that of small s) on which our equations were derived. We cannot, therefore, believe the solution in detail in the neighborhood of the surface of the cone. No such limitation appears when we treat the corresponding electromagnetic problem, due to the presumably exact character of the wave functions in that case.

Problem 5-12: Show that the solution due to a point source of strength S_0 moving according to the general law

$$\mathbf{r}' = \boldsymbol{\xi}(t)$$

is

$$s(\mathbf{r}, t) = \frac{S_0}{4\pi} \frac{1}{|R(\tau) - (1/c)\mathbf{v}(\tau) \cdot \mathbf{R}(\tau)|}$$

where

$$\mathbf{R} = \mathbf{r} - \boldsymbol{\xi}$$
$$\mathbf{v} = d\boldsymbol{\xi}/dt$$

and $\tau(t)$ is the time satisfying

$$|\mathbf{r} - \boldsymbol{\xi}(\tau)| = c(t - \tau)$$

(this is the equivalent in sound wave theory of the Lienard–Wiechert potential of electromagnetic theory).

Use the above result to solve the problem of the source moving with uniform velocity.

Problem 5-13: Use the result of the previous problem to solve the problem of a sound wave generated by a point source moving periodically according to the law

$$z' = z_0 \cos \omega t$$

in the z direction.

---

12. Solution with Initial and Boundary Conditions

Let us consider the problem of solving the equation for the propagation of sound waves for given initial and boundary conditions. In particular, we suppose that the condensation $s(\mathbf{r}, t)$ is given to have the value $s_0(\mathbf{r}, t_0)$ at an initial time $t = t_0$, and that the normal velocity is given over given boundary surfaces which we designate as S. Thus we take

$$\frac{\partial \phi}{\partial t} = c^2 s_0 \quad \text{and} \quad v = 0 \quad \text{at} \quad t = t_0 \tag{5.166}$$

$$-\mathbf{n} \cdot \nabla \phi = -\frac{\partial \phi}{\partial n} = v_0 \quad \text{on} \quad S \tag{5.167}$$

Before showing that it is possible to express the solution in terms of Green's functions, we prove the "reciprocity theorem" for those Green's functions; that is,

$$G(\mathbf{r}', \mathbf{r}'', t' - t'') = G(\mathbf{r}'', \mathbf{r}', t' - t'') \tag{5.168}$$

This says that the solution at $\mathbf{r}''$, t' due to an instantaneous point source at $\mathbf{r}'$, t'' is the same as that at $\mathbf{r}'$, t' due to an instantaneous point source at $\mathbf{r}''$, t''.

This is a trivial result in an *infinite* medium, since, due to translation invariance, the Green's function can then only depend on $|\mathbf{r}' - \mathbf{r}''|$. However, we want to consider Green's functions satisfying boundary conditions of the form $(\partial G/\partial n) - \alpha G = 0$ on prescribed surfaces. (Particular cases are $\partial G/\partial n = 0$, corresponding to $\alpha = 0$, and $G = 0$, corresponding to $\alpha \to \infty$.)

The result (5.168) may be proven as follows: The Green's functions $G(\mathbf{r}, \mathbf{r}', t' - t)$ and $G(\mathbf{r}, \mathbf{r}'', t - t'')$ satisfy the equations

$$\frac{1}{c^2} \frac{\partial^2}{\partial t^2} G(\mathbf{r}, \mathbf{r}', t' - t) - \nabla^2 G(\mathbf{r}, \mathbf{r}', t' - t) = \delta(\mathbf{r} - \mathbf{r}')\delta(t' - t)$$

and

$$\frac{1}{c^2} \frac{\partial^2}{\partial t^2} G(\mathbf{r}, \mathbf{r}'', t - t'') - \nabla^2 G(\mathbf{r}, \mathbf{r}'', t - t'') = \delta(\mathbf{r} - \mathbf{r}'')\delta(t - t'')$$

If we multiply each of these equations by the solution of the other, subtract, and integrate, we obtain, on integration by parts,

$$\frac{1}{c^2} \int d^3\mathbf{r} \left[\frac{\partial}{\partial t} G(\mathbf{r}, \mathbf{r}', t' - t) \cdot G(\mathbf{r}, \mathbf{r}'', t - t'') \right.$$

$$\left. - G(\mathbf{r}, \mathbf{r}', t' - t) \frac{\partial}{\partial t} G(\mathbf{r}, \mathbf{r}'', t - t'') \right]_{t=t''_-}^{t'_+}$$

$$+ \iint \left[G(\mathbf{r},\mathbf{r}',t'-t)\frac{\partial}{\partial n} G(\mathbf{r},\mathbf{r}'',t-t'') \right.$$

$$\left. - G(\mathbf{r},\mathbf{r}'',t-t'')\frac{\partial}{\partial n} G(\mathbf{r},\mathbf{r}',t'-t) \right] dS\, dt$$

$$= G(\mathbf{r}',\mathbf{r}'',t'-t'') - G(\mathbf{r}'',\mathbf{r}',t'-t'') \qquad (5.169)$$

We have assumed that $t' > t''$. The time integral is taken from a time slightly less than t'' to one slightly greater than t'; the surface integral in the second term is over the bounding surfaces on which the boundary conditions are prescribed. The integral at infinity is assumed to vanish. The normal derivatives are drawn *outward* from the region in which we are solving, that is, *into* the bounding surfaces.

If the boundary conditions are of the form

$$\frac{\partial G}{\partial n} - \alpha G = 0 \qquad (5.170)$$

the second term on the left is zero. The *first* term is zero since the Green's function is zero for negative time argument (causality condition). Thus, the right-hand side is zero and the reciprocity theorem is proven.

We now show how the potential can be expressed in terms of Green's functions and the initial and boundary conditions.

When there is a source $S_0(\mathbf{r}, t)$ in the potential equation we may write that equation as

$$-\nabla'^2\phi(\mathbf{r}',t') + \frac{1}{c^2}\frac{\partial^2}{\partial t'^2}\phi(\mathbf{r}',t') = S_0(\mathbf{r}',t') \qquad (5.171)$$

We combine with this the Green's function equation

$$-\nabla'^2 G(\mathbf{r},\mathbf{r}',t-t') + \frac{1}{c^2}\frac{\partial^2}{\partial t'^2} G(\mathbf{r},\mathbf{r}',t-t') = \delta(\mathbf{r}-\mathbf{r}')\delta(t-t') \quad (5.172)$$

In the second equation we have used the reciprocity theorem.

We now multiply the first equation by G, the second by ϕ, and subtract. We then integrate as follows: with respect to $\mathbf{r}'$, over the whole space outside the bounding surfaces, and with respect to t' from t_0 to t_+ (a value slightly greater than t). We then use Green's theorem to transform the volume integrals of the Laplacian terms into surface integrals, to get

$$\int_{t_0}^{t_+} dt' \int dS' \left[\phi(\mathbf{r}',t')\frac{\partial}{\partial n'} G(\mathbf{r},\mathbf{r}',t-t') - G(\mathbf{r},\mathbf{r}',t-t')\frac{\partial}{\partial n'}\phi(\mathbf{r}',t') \right]$$

$$+ \frac{1}{c^2}\int d^3\mathbf{r}' \left[G(\mathbf{r},\mathbf{r}',t-t')\frac{\partial}{\partial t'}\phi(\mathbf{r}',t') - \phi(\mathbf{r}',t')\frac{\partial}{\partial t'} G(\mathbf{r},\mathbf{r}',t-t') \right]_{t_0}^{t_+}$$

$$= \int d^3\mathbf{r}'\, dt'\, S_0(\mathbf{r}',t') G(\mathbf{r},\mathbf{r}',t-t') - \phi(\mathbf{r},t) \qquad (5.173)$$

If we define the Green's function by the boundary condition $\partial G/\partial n = 0$ over the bounding surfaces, substitute the boundary condition (5.167) and the initial condition (5.166), and choose the potential ϕ to be zero at t_0 (since the potential is in any case only determined to within a constant), we obtain for $\phi(\mathbf{r}, t)$ the formula

$$\phi(\mathbf{r}, t) = \int S_0(\mathbf{r}', t') G(\mathbf{r}, \mathbf{r}', t - t') \, d^3\mathbf{r}' \, dt'$$

$$- \int_{t_0}^{t} dt' \int dS' v_0(\mathbf{r}', t') G(\mathbf{r}, \mathbf{r}', t - t')$$

$$+ \int G(\mathbf{r}, \mathbf{r}', t - t_0) s_0(\mathbf{r}', t_0) \, d^3\mathbf{r}' \qquad (5.174)$$

The first term is the expected source term, the second is a contribution from the boundaries, showing that they act like sources, and the third is the contribution from the initial disturbance. All contributions are determined in terms of the *same* Green's function.

If the medium is taken to be infinite and source-free but initially in motion, the corresponding terms in (5.173) may be neglected and the G may be taken to be that of Eq. (5.75), namely

$$G(\mathbf{r}, \mathbf{r}', t - t') = \frac{1}{4\pi|\mathbf{r} - \mathbf{r}'|} \delta\left(t - t' - \frac{|\mathbf{r} - \mathbf{r}'|}{c}\right)$$

This may be substituted into (5.173) to give

$$\phi(\mathbf{r}, t) = \frac{1}{4\pi} \int \frac{1}{|\mathbf{r} - \mathbf{r}'|} \delta\left(t - t_0 - \frac{|\mathbf{r} - \mathbf{r}'|}{c}\right) s(\mathbf{r}', t_0) \, d^3\mathbf{r}'$$

$$+ \frac{1}{4\pi c^2} \frac{\partial}{\partial t} \int \frac{1}{|\mathbf{r} - \mathbf{r}'|} \delta\left(t - t_0 - \frac{|\mathbf{r} - \mathbf{r}'|}{c}\right) \phi(\mathbf{r}', t_0) \, d^3\mathbf{r}' \quad (5.175)$$

Putting $\mathbf{r}' - \mathbf{r} = \mathbf{R}$, the first integral may be written

$$\frac{1}{4\pi} \int \frac{1}{R} \delta\left(t - t_0 - \frac{R}{c}\right) s(\mathbf{r} + \mathbf{R}, t_0) \, d^3\mathbf{R}$$

But

$$\frac{1}{4\pi} \int s(\mathbf{r} + \mathbf{R}, t_0) \, d\Omega_{\mathbf{R}}$$

where $d\Omega_{\mathbf{R}}$ is the element of solid angle subtended by the vector $\mathbf{R}$, is the average value of s over a sphere of radius R about the point $\mathbf{r}$; we call it $s_R(\mathbf{r}, t_0)$. Using this notation, the integral becomes

---

Problem 5-14:   Solve for the *velocity* in the sound wave subject to the conditions:

(a) normal components of velocity $\mathbf{V} \cdot \mathbf{n}_i = V_n(t)$ given on the ith surface;

(b) initial condensation $s_0(\mathbf{r})$ given;

(c) initial velocity $V_0(r)$ given.

In terms of the Green's function satisfying $G = 0$ on the bounding surfaces, show that the velocity is

$$\mathbf{V} = \int_0^t dt' \sum_i \int_{s_i} V_n(t') \mathbf{n}' \frac{\partial}{\partial n'} G(\mathbf{r}, \mathbf{r}', t - t') \, dS'$$

$$+ \int d^3\mathbf{r}' s_0(\mathbf{r}') \nabla' G(\mathbf{r}, \mathbf{r}', t) - \int d^3\mathbf{r}' \mathbf{V}_0(\mathbf{r}') \frac{\partial}{\partial t} G(\mathbf{r}, \mathbf{r}', t)$$

the initial time being $t = 0$.

$$\int \delta\left(t - t_0 - \frac{R}{c}\right) R s_R(\mathbf{r}, t_0) \, dR = c^2(t - t_0) s_{c(t-t_0)}(\mathbf{r}, t_0)$$

To evaluate the second term in (5.175), we note that

$$\frac{\partial}{\partial t} \delta\left(t - t_0 - \frac{R}{c}\right) = c \frac{\partial}{\partial R} \delta\left(t - t_0 - \frac{R}{c}\right)$$

This term then takes the form

$$\frac{1}{4\pi c} \int \frac{1}{R} \phi(\mathbf{r} + \mathbf{R}, t_0) \frac{\partial}{\partial R} \delta\left(t - t_0 - \frac{R}{c}\right) d^3\mathbf{R}$$

$$= \frac{1}{c} \int \phi_R(\mathbf{r}, t_0) R \frac{\partial}{\partial R} \delta\left(t - t_0 - \frac{R}{c}\right) dR$$

$$= -\frac{1}{c} \int \delta\left(t - t_0 - \frac{R}{c}\right) \frac{\partial}{\partial R} (R\phi_R(\mathbf{r}, t_0)) \, dR$$

$$= -\left[\frac{\partial}{\partial R} (R\phi_R(\mathbf{r}, t_0))\right]_{R=c(t-t_0)}$$

Substituting the values of the two integrals back into (5.175), we obtain the formula

$$\phi(\mathbf{r}, t) = \left[c R s_R(\mathbf{r}, t_0) - \frac{\partial}{\partial R} (R\phi_R(\mathbf{r}, t_0))\right]_{R=c(t-t_0)} \tag{5.176}$$

which expresses the velocity potential at a point in terms of the average of appropriate quantities (namely, ϕ and $s = 1/c^2 (\partial\phi/\partial t)$) over a sphere of radius R surrounding the point at a time earlier by an amount equal to the time required for a sound disturbance to travel from any point on this sphere to the point in question.

Problem 5-15: Calculate the infinite–medium Green's function for the one-dimensional ("plane") problem

$$\frac{\partial^2 \phi}{\partial z^2} - \frac{1}{c^2} \frac{\partial^2 \phi}{\partial t^2} = 0$$

Problem 5-16: Using the relation (3.178) relating "point source" to "plane source" solutions of a linear equation, obtain the formula (5.93) for the three-dimensional Green's function.

Problem 5-17: Using (3.180), obtain the "spherical shell" Green's function (that is, the solution for a unit spherical shell source).

---

13. Waves in Guides and Enclosures

Sound waves in gases inside an enclosed volume may only have certain definite frequencies. Consider, for example, a box of dimensions $a_1 \times a_2 \times a_3$ with perfectly rigid walls, so that the normal velocity must be zero over the surface. It is most convenient to take the boundaries to be defined by $x = 0, a_1; y = 0, a_2; z = 0, a_3$. We can solve the wave equation by separation of variables; if the time variation is $e^{-i\omega t}$, the equation for the velocity potential becomes

$$\nabla^2 \phi + \frac{\omega^2}{c^2}\phi = 0 \qquad (5.177)$$

Putting $\phi = X(x)\,Y(y)\,Z(z)$, substituting in the equation, and dividing by XYZ yields

$$\frac{1}{X}\frac{d^2X}{dx^2} + \frac{1}{Y}\frac{d^2Y}{dy^2} + \frac{1}{Z}\frac{d^2Z}{dz^2} + \frac{\omega^2}{c^2} = 0$$

Thus the first three terms must be constants, and their sum equal to $-(\omega^2/c^2)$. For the boundary conditions to be satisfied, these constants must be negative:

$$\frac{d^2X}{dx^2} = -\alpha^2 X \qquad (5.178a)$$

$$\frac{d^2Y}{dy^2} = -\beta^2 Y \qquad (5.178b)$$

$$\frac{d^2Z}{dz^2} = -\gamma^2 Z \qquad (5.178c)$$

$$\alpha^2 + \beta^2 + \gamma^2 = \omega^2/c^2 \qquad (5.179)$$

The solution to the first must be

$$X = A \cos \alpha x$$

where $\alpha a_1 = l\pi$, l being an integer. Similarly

$$Y = B \cos \beta y, \qquad \beta a_2 = m\pi$$

and

$$Z = C \cos \gamma z, \qquad \gamma a_3 = n\pi$$

m and n also being integers. Finally, substituting in (5.179) we find the possible circular frequencies ω:

$$\omega^2 = \pi^2 c^2 \left(\frac{l^2}{a_1^2} + \frac{m^2}{a_2^2} + \frac{n^2}{a_3^2}\right) = \omega_{lmn}^2 \qquad (5.180)$$

What happens if one attempts to create a disturbance with a different frequency ω_0 inside the enclosure, with a source of frequency $\omega_{0'}$, let us say? The equation now becomes

$$-\nabla^2\phi + \frac{1}{c^2}\frac{\partial^2\phi}{\partial t^2} = \delta(\mathbf{r} - \mathbf{r}_0)\, e^{-i\omega_0 t} \qquad (5.181)$$

Put $\phi = \phi_0(\mathbf{r})\, e^{-i\omega_0 t}$, so that

$$-\nabla^2\phi_0 - \frac{\omega_0^2}{c^2}\phi_0 = \delta(\mathbf{r} - \mathbf{r}_0) \qquad (5.182)$$

It is possible to expand ϕ_0 and the δ function in Fourier series of the form

$$\phi_0 = \sum A_{lmn} \cos\frac{l\pi x}{a_1} \cos\frac{m\pi y}{a_2} \cos\frac{n\pi z}{a_3} \qquad (5.183)$$

$$\delta(\mathbf{r} - \mathbf{r}_0) = \sum B_{lmn} \cos\frac{l\pi x}{a_1} \cos\frac{m\pi y}{a_2} \cos\frac{n\pi z}{a_3} \qquad (5.184)$$

In the latter, multiplying through by one of the expansion functions, integrating, and using orthogonality, it is found that

$$B_{lmn} = \frac{8}{a_1 a_2 a_3} \cos\frac{l\pi x_0}{a_1} \cos\frac{m\pi y_0}{a_2} \cos\frac{n\pi z_0}{a_3} \qquad (5.185)$$

where x_0, y_0, and z_0 are the components of $\mathbf{r}_0$.

Substituting in (5.182) and equating coefficients gives

$$A_{lmn} = \left[\frac{l^2\pi^2}{a_1^2} + \frac{m^2\pi^2}{a_2^2} + \frac{n^2\pi^2}{a_3^2} - \frac{\omega_0^2}{c^2}\right]^{-1} \frac{8}{a_1 a_2 a_3} \cos\frac{l\pi x_0}{a_1} \cos\frac{m\pi y_0}{a_2} \cos\frac{n\pi z_0}{a_3}$$

$$(5.186)$$

Therefore,

$$A_{lmn} = \frac{8}{a_1 a_2 a_3} \frac{c^2}{\omega_{lmn}^2 - \omega_0^2} \cos\frac{l\pi x_0}{a_1} \cos\frac{m\pi y_0}{a_2} \cos\frac{n\pi z_0}{a_3} \qquad (5.187)$$

Thus, in general, disturbances of any frequency can be sustained by a source of that frequency. There is a "resonance," however, when $\omega_0 = \omega_{lmn}$, that is, one of the "natural" frequences or normal modes. [The disturbance does not, of course, become infinite; the equations leading to (5.187) are only valid for $|s| = (\omega_0/c^2)|\phi|$ small.]

What then happens to this disturbance if the source is removed?

Clearly, the patterns corresponding to the various normal modes having been established, each will continue to vibrate with its appropriate frequency ω_{lmn}.

Problem 5-18: Find the frequencies of the normal modes of vibration of a gas in a cylindrical enclosure of radius b and length a. If $2b = 1$ ft and $a = 1$ ft, find the six lowest frequencies, and describe the patterns of condensation in each mode.

Problem 5-19: One end of a closed cylinder of radius b and length a undergoes small vibrations of amplitude $\epsilon f(r) \cos \omega_0 t$. Find the resulting sound wave inside the cylinder.

---

14. Spherical Enclosure

We have already seen [Eq. (5.116)] that the solutions of the wave equation in spherical coordinates are

$$\phi = z_l(kr) Y_{lm}(\theta, \varphi) e^{-i\omega t}$$

where $k = \omega/c$ and z_l is an arbitrary spherical Bessel function. Since the solution must be finite at $r = 0$, Z_l must be the Bessel function j_l. If the spherical enclosure is of radius a, the boundary condition that the radial velocity is zero at $r = a$, viz.,

$$V_r = -\left(\frac{\partial \phi}{\partial r}\right)_{r=a} = 0$$

yields for the frequencies the equation

$$j_l'\left(\frac{\omega}{c} a\right) = 0 \tag{5.189}$$

Thus, the frequency for a given l is independent of m. This is not surprising, since if we merely rotate coordinates, $Y_{lm}(\theta, \varphi)$ goes into a linear combination of $Y_{lm'}(\theta', \varphi')$ $(m' = -l, \ldots, l)$, and (θ', φ') are the new coordinates; but of course, such a rotation does not affect the frequency.

15. Propagation Down a Cylindrical Tube

In cylindrical coordinates $\xi = \sqrt{x^2 + y^2}$, z and $\varphi = \tan^{-1} y/x$, the spatial wave equation (5.177) becomes

$$\frac{1}{\xi}\frac{\partial}{\partial \xi}\left(\xi \frac{\partial \phi}{\partial \xi}\right) + \frac{1}{\xi^2}\frac{\partial^2 \phi}{\partial \varphi^2} + \frac{\partial^2 \phi}{\partial z^2} + \frac{\omega^2}{c^2} = 0 \tag{5.190}$$

Putting

$$\phi = F(\xi) G(\varphi) Z(z), \tag{5.191}$$

dividing through by ϕ, and imposing the condition that the resulting functions of ξ, φ, and z must be constants, we obtain

$$\frac{1}{\xi}\frac{d}{d\xi}\left(\xi\frac{dF}{d\xi}\right)-\frac{\beta^2}{\xi^2}F=-\alpha^2F \qquad (5.192a)$$

$$\frac{d^2G}{d\varphi^2}=-\beta^2G \qquad (5.192b)$$

$$\frac{d^2Z}{dz^2}=-\gamma^2Z \qquad (5.192c)$$

and

$$\frac{\omega^2}{c^2}=\alpha^2+\gamma^2 \qquad (5.193)$$

Since G must be single-valued, β must be an integer m, and

$$G=G_0\cos m(\varphi-\varphi_0) \qquad (5.194)$$

G_0 and φ_0 being constants. The solution of (5.192a) is then

$$F=F_0J_m(\alpha\xi); \qquad (5.195)$$

the second solution of the Bessel equation must be rejected because of the requirement of finiteness at $\xi=0$. To satisfy the boundary condition of zero radial velocity at the radius $\xi=a$ of the tube, we must have

$$J'_m(\alpha a)=0 \qquad (5.196)$$

which gives a set of values $\alpha=(\lambda_{ms}/a)$ which can be obtained from standard mathematical tables.[2]

If the frequency ω is fixed, γ is now determined by (5.193):

$$\gamma^2=\gamma^2_{ms}=\frac{\omega^2}{c^2}-\frac{\lambda^2_{ms}}{a^2} \qquad (5.197)$$

If the right-hand side is *positive*, the wave is propagated without attenuation down the tube, the solution being

$$\Phi=AJ_m\left(\lambda_{ms}\frac{\xi}{a}\right)\cos m(\varphi-\varphi_0)\,e^{i(\gamma_{ms}z-\omega t)} \qquad (5.198)$$

If, on the other hand, the right-hand side of (5.197) is *negative*,

$$\gamma^2=-\bar{\gamma}^2_{ms}=\frac{\lambda^2_{ms}}{a^2}-\frac{\omega^2}{c^2} \qquad (5.199)$$

and the solution has the form

$$\phi=AJ_m\left(\lambda_{ms}\frac{\xi}{a}\right)\cos m(\varphi-\varphi_0)\,e^{-\bar{\gamma}_{ms}z}\,e^{-i\omega t} \qquad (5.200)$$

In this case the wave is attenuated as it proceeds down the tube.

[2]See, for example, Morse and Feshbach, *Methods of Mathematical Physics.* New York: McGraw-Hill, 1953. p. 1576.

It may be seen from the foregoing that as a, the radius of the tube, decreases, more and more of the modes of propagation are attenuated. In fact, for sufficiently small radius, only one mode of propagation remains. For we note that, in the case $m = 0$, (5.196) is satisfied by $\alpha = 0$; in this case, the solution has neither radial nor angular dependence but is simply the plane wave

$$\phi = A \, e^{i\omega((z/c)-t)} \tag{5.201}$$

That this is a solution is quite evident from the fact that $\partial\phi/\partial\xi = 0$ *everywhere*, and is therefore satisfied on the sides of the tube. Thus, the propagation of a plane wave down the tube, with wave front parallel to the axis, is always possible.

The following are the first few zeros of the derivatives of the Bessel functions of low order:

$$\lambda_{01} = 3.83 \qquad \lambda_{02} = 7.02 \qquad \lambda_{03} = 10.17$$

$$\lambda_{11} = 1.84 \qquad \lambda_{12} = 5.33 \qquad \lambda_{13} = 8.54$$

$$\lambda_{21} = 3.05 \qquad \lambda_{22} = 6.71 \qquad \lambda_{23} = 9.97$$

$$\lambda_{31} = 4.20 \qquad \lambda_{32} = 8.02$$

Thus, if the circular frequency is lower than $\lambda_{11}c/a = 1.84c/a$ only a plane wave will fail to be attenuated; or alternatively, for given frequency, only a plane wave will be propagated if $a < (1.84c/\omega)$. Taking the velocity of sound in air at normal temperature and pressure to be $c = 1100$ ft/sec, and if the frequency is 500 cycles/sec, we find this limit on a to be about 7.73 in.; as the frequency changes, the limiting value of a changes in inverse proportion. Thus, for any frequency, a sufficiently small pipe "filters out" all but the plane wave component.

Problem 5-20: Calculate the limiting dimensions for propagation of a sound wave other than a plane wave, of circular frequency ω, down a rectangular pipe.

---

16. Scattering of Sound Waves

We consider the problem of a plane wave of a given circular frequency ω incident on a scattering object in the neighborhood of the origin of the coordinate system. The polar axis is taken to be in the direction of propagation of the incident wave. The velocity potential for this incident wave is then

$$\phi_{\text{inc}} = e^{i(kz-\omega t)} \tag{5.202}$$

The amplitude may be taken to be unity, since the amplitude of the scattered wave, and therefore of the complete disturbance, is proportional to that of the incident wave.

In order to understand the scattering problem, it is useful to express the plane wave in terms of spherical waves, using (5.122):

$$\phi_{\text{inc.}} = \sum_{l=0}^{\infty} i^l (2l+1) j_l(kr) P_l(\cos\theta) e^{-i\omega t} \qquad (5.203)$$

Furthermore, as shown in connection with Eqs. (5.106) and (5.107), if we write $j_l(kr)$ in terms of $h_l^{(1)}(kr)$ and $h_l^{(2)}(kr)$,

$$j_l(kr) = \tfrac{1}{2}[h_l^{(1)}(kr) + h_l^{(2)}(kr)] \qquad (5.204)$$

the two Hankel functions represent waves, respectively diverging from and converging on the scatterer.

If the complete velocity potential outside the region occupied by the scatterer is ϕ, we define the *scattered* wave as

$$\phi_{\text{sc}} = \phi - \phi_{\text{inc.}} \qquad (5.205)$$

The scattered wave will involve only partial waves *diverging* from the scatterer and the origin; it will, therefore, be of the form

$$\phi_{\text{sc}} = \sum C_{lm} h_l^{(1)}(kr) Y_{lm}(\theta,\varphi) e^{-i\omega t} \qquad (5.206)$$

If we consider only the special case for which the scatterer has axial symmetry about the polar axis, only the $m = 0$ terms appear. The scattered wave may then be written

$$\phi_{\text{sc}} = \tfrac{1}{2} \sum (W_l - 1) i^l (2l+1) h_l^{(1)}(kr) P_l(\cos\theta) e^{-i\omega t} \qquad (5.207)$$

The total velocity potential for the problem takes the form

$$\phi = \tfrac{1}{2} \sum i^l (2l+1) [W_l h_l^{(1)}(kr) + h_l^{(2)}(kr)] P_l(\cos\theta) e^{-i\omega t} \qquad (5.208)$$

The values of the constants W_l are determined by the fact that ϕ must satisfy boundary conditions at the "surface" of the scatterer. For sound waves, this "surface" is generally a physical boundary. The boundary conditions depend, as we shall see, on the physical nature of this surface. Problems quite similar mathematically to that of the scattering of sound waves occur in electromagnetic theory and in quantum mechanics. In electromagnetic theory, the situation is complicated by the fact that the wave has a vector character. Aside from that fact, the main feature distinguishing the various scattering problems is the difference in boundary conditions imposed by the physics of each system. In quantum mechanics, in particular, as we shall see later, we seldom have well-defined boundary surfaces, inside which the problem has a different character (and the same is true of electromagnetic theory). Nevertheless, the solution *inside* the scatterer can in each case be incorporated into a boundary condition at a "surface," the definition of which may, however, involve a certain arbitrariness.

For the moment, we shall be concerned with some results of a general character, for which the amplitudes W_l need not be specified. These results then apply equally to the scattering of sound waves, of electromagnetic waves, or of "quantum" (particle) waves.

These results are formulated in terms of "cross sections." The "scattering cross section" σ_{sc} is determined by the definition

$$\frac{\text{scattered flux}}{\text{incident flux per unit area}} \qquad (5.209)$$

The word "flux" here incorporates the idea of "per unit time." One must, however, specify flux of what? Here again, the situation differs according to the physics of the problem. For sound waves, we consider flux of sound energy, so that the flux over a surface is

$$\mathscr{P} = \left\langle -\rho_0 \int \frac{\partial \phi_r}{\partial t} \nabla \phi_r \cdot \mathbf{n} \, dS \right\rangle \qquad (5.210)$$

the integral being taken over the surface. ϕ_r means the physical, or real, part of ϕ. The brackets $\langle \quad \rangle$ indicate a time average over a sufficiently long period of time (for example, for a periodic wave, over many periods).

The cross section defined by (5.209) has the meaning that the rate of energy flux of the incident wave over σ_{sc} is equal to the total energy flux in the scattered wave; it is as though all the energy incident on the area σ_{sc} were scattered.

Let us calculate the flux per unit area in the incident wave (5.202). It is

$$\rho_0 \omega \langle \sin (kz - \omega t) k \sin (kz - \omega t) \rangle = \tfrac{1}{2} \rho_0 k^2 c \qquad (5.211)$$

since $\omega = kc$. On the other hand, the scattered flux at large distances from the origin may be calculated from the asymptotic form of ϕ_{sc}:

$$\phi_{sc}^{(as)} = \frac{1}{2} \sum (W_l - 1) i^l (2l+1) \frac{e^{i[kr-(l+1)(\pi/2)]}}{kr} P_l (\cos \theta) e^{-i\omega t} \qquad (5.212)$$

Putting

$$W_l - 1 = |W_l - 1| e^{i\beta_l} \qquad (5.213)$$

the real part of this is

$$\phi_{sc}^{(as)} = \frac{1}{2} \sum |W_l - 1| (2l+1) \frac{1}{kr} P_l (\cos \theta) \cos \left(\beta_l + kr - \frac{\pi}{2} - \omega t \right) \qquad (5.214)$$

Then the asymptotic value for the excess pressure $\rho_0 (\partial \phi / \partial t)$ is

$$p^{(as)} = \frac{1}{2} \frac{\rho_0 \omega}{kr} \sum |W_l - 1| (2l+1) P_l (\cos \theta) \sin \left(\beta_l + kr - \frac{\pi}{2} - \omega t \right) \qquad (5.215)$$

and for the radial velocity $-\partial\phi/\partial r$ is

$$-\left(\frac{\partial\phi}{\partial r}\right)^{(as)} = \frac{1}{2}\frac{1}{r}\sum |W_l - 1|(2l+1)P_l(\cos\theta)\sin\left(\beta_l + kr - \frac{\pi}{2} - \omega t\right) \quad (5.216)$$

These must now be multiplied together, and the time average of the integral of the product calculated over a sphere of sufficiently large radius. We may use the fact that the Legendre polynomials are mutually orthogonal, and that the time average of $\sin^2(\beta_l + kr - (\pi/2) - \omega t)$ is $\frac{1}{2}$. Therefore the scattered flux is

$$\mathscr{P}_{sc}^{(as)} = \tfrac{1}{8}\rho_0 c\, 4\pi \sum |W_l - 1|^2(2l+1) \quad (5.217)$$

Dividing (5.217) by (5.211) we find that

$$\sigma_{sc} = \frac{\pi}{k^2}\sum |W_l - 1|^2(2l+1) \quad (5.218)$$

We see that the scattering cross section is made up of distinct contributions from each order of spherical harmonic. The separate terms will be designated as the "partial cross sections" of different multipole orders. In the electromagnetic and quantum cases, the terms with various l are associated with different angular momenta about the origin. Such an identification is less apt in the case of sound waves.

From (5.212), we see that

$$\phi_{sc}^{(as)} = \frac{e^{i(kr-\omega t)}}{r}f(\theta) \quad (5.219)$$

where the "scattering amplitude" is

$$f(\theta) = \frac{1}{2ik}\sum (W_l - 1)(2l+1)P_l(\cos\theta) \quad (5.220)$$

The "forward scattering amplitude" is

$$f(0) = \frac{1}{2ik}\sum (2l+1)(W_l - 1) \quad (5.221)$$

since $P_l(1) = 1$ for all values of 1. This formula will be found useful shortly.

If we look at (5.208), we see that the total velocity potential can be split into a part converging on the origin

$$\phi_{conv} = \tfrac{1}{2}\sum i^l(2l+1)h_l^{(2)}(kr)P_l(\cos\theta)\,e^{-i\omega t} \quad (5.222)$$

and a part *diverging* from the origin

$$\phi_{div} = \tfrac{1}{2}\sum i^l(2l+1)W_l h_l^{(1)}(kr)P_l(\cos\theta)\,e^{-i\omega t} \quad (5.223)$$

If the scatterer is removed, $W_l = 1$. The energy flux in the *diverging* wave is easily obtained by observing that ϕ_{div} differs from ϕ_{sc} only in having W_l

in place of $(W_l - 1)$; thus, the flux is

$$\mathcal{P}_{\text{div}} = \frac{\pi}{2} \rho_0 c \sum |W_l|^2 (2l+1) \qquad (5.224)$$

On the other hand, the flux in the converging wave is in the opposite direction, [i.e., $k \rightarrow -k$ in (5.215) and (5.216)], but otherwise differs from (5.224) only in that W_l is replaced by unity:

$$\mathcal{P}_{\text{conv}} = \frac{\pi}{2} \rho_0 c \sum (2l+1) \qquad (5.225)$$

Summed over all l, this diverges; this is not really surprising, in that the plane wave, strictly speaking, contains an infinite amount of energy since it is of infinite extent perpendicular to the wave front. In practice, however, we are only interested in that part of the plane wave which is affected by the scatterer. Suppose, therefore, that the scatterer lies entirely within a sphere of radius R around the origin. We can then ignore those partial waves in the expansion of e^{ikz} which have negligible amplitude for $r < R$; that is, those for which $j_l(kR)$ is negligible. From the results of Appendix 5C, this is seen to be the case when l is substantially larger than kR. We therefore, cut off all of the series at a finite value $l = L > kR$.

If the "scatterer" cannot absorb energy from the wave, the flux in the diverging wave must equal that in the converging wave for each l; that is

$$|W_l| = 1 \qquad (5.226)$$

or

$$W_l = e^{2i\delta_l} \qquad (5.227)$$

(The factor 2 in the exponent is chosen for convenience, as will be seen as we proceed.) The angle δ_l is called the "phase shift for the lth partial wave." In terms of this phase shift, it is easily seen that the scattering cross section is

$$\sigma_{\text{sc}} = \frac{\pi}{k^2} \sum |e^{2i\delta_l} - 1|^2 (2l+1)$$

$$= \frac{4\pi}{k^2} \sum (2l+1) \sin^2 \delta_l \qquad (5.228)$$

From (5.213) and (5.214), the asymptotic form of the velocity potential is

$$\phi_{\text{sc}}^{(as)} = \frac{1}{kr} \sum (2l+1) \sin \delta_l P_l (\cos \theta) \cos \left(kr + \delta_l - \frac{\pi}{2} - \omega t \right) \qquad (5.229)$$

If the "scatterer" is also an absorber, the flux of the diverging wave is less than that in the converging wave, and the inequality

$$|W_l| < 1 \qquad (5.230)$$

holds. The absorbed energy will be the excess of the flux in the converging wave over that in the diverging wave:

$$\mathscr{P}_{ahs} = \frac{\pi}{2} \rho_0 c \sum (2l+1)(1-|W_l|^2) \tag{5.231}$$

We may then define the "absorption cross section"

$$\sigma_{abs} = \frac{\mathscr{P}_{abs}}{\text{incident flux per unit area}}$$

$$= \frac{1}{\frac{1}{2}\rho_0 k^2 c} \frac{\pi}{2} \rho_0 c \sum (2l+1)(1-|W_l|^2)$$

$$= \frac{\pi}{k^2} \sum (2l+1)(1-|W_l|^2) \tag{5.232}$$

This result, which relates the total cross section to the forward scattering difference between $|W_l|$ and 1 approaches zero very rapidly when $l > L$.

Finally, it is possible to define the "total cross section"

$$\sigma_{tot} = \sigma_{sc} + \sigma_{abs} \tag{5.233}$$

$$= \frac{\pi}{k^2} \sum (2l+1)\{|W_l-1|^2+1-|W_l|^2\}$$

$$= \frac{2\pi}{k^2} \text{Re} \sum (2l+1)(1-W_l) \tag{5.234}$$

"Re" signifying, as usual, "real part of." Comparing with (5.221), we may write

$$\sigma_{tot} = \frac{4\pi}{k} Im f(0) \tag{5.235}$$

This result, which relates the total cross section to the forward scattering amplitude, is known as the "optical theorem" of Bohr, Peierls, and Placzek.

17.   Inequalities Satisfied by the Cross Sections

Consider now the partial waves for each l, designating the partial scattering absorption, and total cross section, respectively, as

$$\sigma_{sc}^{(l)} = \frac{\pi}{k^2}(2l+1)|W_l-1|^2 = x_l \tag{5.236}$$

$$\sigma_{abs}^{(l)} = \frac{\pi}{k^2}(2l+1)(1-|W_l|^2) = y_l \tag{5.237}$$

and

$$\sigma_{\text{tot}}^{(l)} = \frac{2\pi}{k^2} (2l+1) \text{ Re } (1 - W_l) = z_l = x_l + y_l \qquad (5.238)$$

The fact that both the scattering and total cross sections can be expressed in terms of $(1 - W_l)$ leads to an interesting and useful inequality, namely,

$$\frac{[\sigma_{\text{tot}}^{(l)}]^2}{[(2\pi/k^2)(2l+1)]^2} \leqslant \frac{\sigma_{\text{sc}}^{(l)}}{(\pi/k^2)(2l+1)}$$

Writing

$$\frac{\pi}{k^2} (2l+1) = \sigma_l \qquad (5.239)$$

this may be written

$$[\sigma_{\text{tot}}^{(l)}]^2 \leqslant 4\sigma_l \sigma_{\text{sc}}^{(l)}$$

or

$$(x_l + y_l)^2 \leqslant 4\sigma_l x_l \qquad (5.240)$$

This inequality may be represented graphically by taking x_l and y_l as coordinates along orthogonal axes. The allowed values of the cross sections are then represented by the points lying between the x_l axis and the parabola

$$(x_l + y_l)^2 = 4\sigma_l x_l \qquad (5.241)$$

as illustrated in Fig. 5.2.

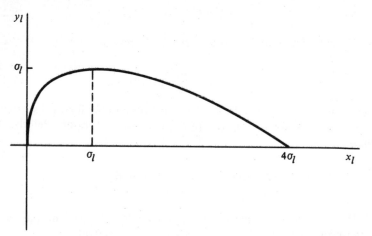

Figure 5.2

Thus, the maximum possible value of $\sigma_{\text{sc}}^{(l)}$ is $4\sigma_l = (4\pi/k^2)(2l+1)$; of $\sigma_{\text{abs}}^{(l)}$, $\sigma_l = (\pi/k^2)(2l+1)$; furthermore, if $\sigma_{\text{abs}}^{(l)}$ has its maximum value, $\sigma_{\text{sc}}^{(l)}$ must have the same value.

Somewhat less rigorously, we can obtain limits on the complete cross sections (the sums over all values of l).

Assuming that the scattering of all partial waves beyond $l = L$ can be neglected, we form L-dimensional vectors Z, X, and S with components $\sqrt{z_l}$, $\sqrt{x_l}$, and $\sqrt{\sigma_l}$, respectively. Since, by (5.241)

$$z_l \leqslant 2\sqrt{x_l}\sqrt{\sigma_l} \tag{5.242}$$

it follows on summing over l that

$$Z^2 \leqslant 2X \cdot S \tag{5.243}$$

where the right-hand side is a scalar product. But, therefore,

$$Z^2 \leqslant 2|X||S| \tag{5.244}$$

In terms of components this may be written as

$$\sigma_{\text{tot}} \leqslant 2\sqrt{\sigma_{\text{sc}}} \sqrt{\frac{\pi}{k^2} \sum_{l=1}^{L} (2l+1)} \tag{5.245}$$

Squaring both sides,

$$\sigma_{\text{tot}}^2 \leqslant \frac{4\pi}{k^2}(L+1)^2 \sigma_{\text{sc}} \tag{5.246}$$

This result is exact; the question is, however, what is the smallest value of L for which it is valid? As seen in Appendix 5A, the value should be taken somewhat larger than kR. However, since $l = kR$ divides the strongly scattered partial waves from the weakly scattered ones, and the additional contributions beyond $l = kR$ may be considered to be compensated by a diminished scattering as l approaches kR from below, the choice $L = kR$ is in fact not unreasonable. This leads to

$$\sigma_{\text{tot}} \leqslant 4\pi \left(R + \frac{1}{k}\right)^2 \sigma_{\text{sc}} \tag{5.247}$$

More detailed analysis shows, in fact, that this inequality is rather well satisfied for all values of R; even as $R \to 0$ it is correct, since then only the $l = 0$ wave is important. For $kR \gg 1$, $1/k$ may be neglected relative to R.

The inequality (5.247) may be represented graphically in the same manner as were the inequalities for the partial waves. It follows, then, that the maximum absorption cross section is, for scatterers whose dimensions are large compared with the wavelength, equal to πR^2, that is, the diameter of the scatterer. This seems a reasonable result; a completely absorbing body is one which absorbs all of the wave which impinges on it. But for maximum absorption, we have seen that the scattering cross section must be equal to the absorption cross section. This result is also reasonable, for we know that, for wavelength short compared to the dimensions of the scatterer, there is negligible diffraction into the region immediately behind the scatterer, which shadows this region. Thus, there must be a scattered wave which interferes *destructively* with the incident wave in this region. It must, therefore, have the same intensity as the incident wave would have in this region in the absence of the scatterer.

Thus, the total cross section for a perfect absorber is $2\pi R^2$.

18. Elastic Scattering by Small Sphere

Let us consider the problem of scattering by a hard sphere of radius "a" which does not absorb sound. By virtue of (5.208) and (5.227) the velocity potential of the problem has the form

$$\phi = \tfrac{1}{2} \sum i^l (2l+1) [e^{2i\delta_l} h_l^{(1)}(kr) + h_l^{(2)}(kr)] P_l (\cos\theta) e^{-i\omega t} \quad (5.248)$$

The phase shift of δ_l is determined by the boundary condition; once it is known, the cross section may be written down immediately. But at the surface of a hard sphere the radial component of velocity must be zero; we must therefore have

$$e^{2i\delta_l} h_l^{(1)\prime}(ka) + h_l^{(2)\prime}(ka) = 0 \quad (5.249)$$

If we write

$$h_l^{(1)\prime}(ka) = \sqrt{[j_l'(ka)]^2 + [n_l'(ka)]^2}\, e^{i\gamma_l} \quad (5.250)$$

$$h_l^{(2)\prime}(ka) = \sqrt{[j_l'(ka)]^2 + [n_l'(ka)]^2}\, e^{i\gamma_l} \quad (5.251)$$

where

$$\cos\gamma_l = \frac{j_l'(ka)}{\sqrt{[j_l'(ka)]^2 + [n_l'(ka)]^2}} \quad (5.252)$$

Eq. (5.249) takes the form

$$e^{2i\delta_l} + e^{2i\gamma_l} = 0$$

so that

$$\delta_l = \frac{\pi}{2} + \gamma_l \quad (5.253)$$

By (5.228) the scattering cross section is then

$$\sigma_{sc} = \frac{4\pi}{k^2} \sum (2l+1) \cos^2\gamma_l$$

$$= \frac{4\pi}{k^2} \sum \frac{(2l+1)[j_l'(ka)]^2}{[j_l'(ka)]^2 + [n_l'(ka)]^2} \quad (5.254)$$

When ka is small, the series converges rapidly. In order to find an approximate form, consider the behavior of $j_l'(x)$ and $n_l'(x)$ for small values of x. From (4.339)

$$j_l(x) \approx \frac{\sqrt{\pi}}{2^{l+1}\Gamma(l+\frac{3}{2})} x^l$$

and

$$j_l'(x) \approx \frac{\sqrt{\pi}\, l}{2^{l+1}\Gamma(l+\frac{3}{2})} x^{l-1} \qquad (5.255)$$

For the case $l = 0$, since the leading term in j_l is a constant we must go to the next term in its expansion. It is easily verified that

$$j_0'(x) \approx \tfrac{1}{3}x \qquad (5.256)$$

The n_l' terms in the denominator become very large compared with the j_l' ones. It is easily verified from (4.377) that for small x

$$n_l(x) \approx -\frac{(2l)!}{2^l l!}\frac{1}{x^{l+1}}$$

and

$$n_l'(x) \approx \frac{(2l)!(l+1)}{2^l l!}\frac{1}{x^{l+2}} \qquad (5.257)$$

It follows that

$$\cos\gamma_l = \text{constant } x^{2l+1} \qquad (5.258)$$

except for $l = 0$, for which

$$\cos\gamma_0 = -\tfrac{1}{3}x^3 \qquad (5.259)$$

The $l = 1$ term also behaves like x^3 for small x: Each further term is smaller by $(ka)^2$ and can, therefore, for small ka, be neglected. But

$$\cos\gamma_1 \approx \tfrac{1}{6}x^3 \qquad (5.260)$$

The approximate value of the cross section when the radius a of the scatterer is considerably smaller than the wavelength is

$$\sigma_{\text{sc}} = \frac{4\pi}{k^2}\left[\frac{1}{9}(ka)^6 + \frac{1}{12}(ka)^6\right]$$

$$= \frac{7}{9}\pi k^4 a^6 \qquad (5.261)$$

Thus, the cross section is proportional to the sixth power of the radius of the scatterer (Rayleigh's sixth-power law) and as the inverse fourth power of the wavelength of the incident wave.

19. Propagation of Electromagnetic Waves

We now develop the theory of electromagnetic wave propagation on the basis of Maxwell's equations. These equations have already been given in Chapter 3. They are

$$\nabla \cdot \mathbf{D} = \rho \tag{3.14}$$

$$\nabla \cdot \mathbf{B} = 0 \tag{3.15}$$

$$\nabla \times \mathbf{E} = -\frac{\partial \mathbf{B}}{\partial t} \tag{3.16}$$

$$\nabla \times \mathbf{H} = \mathbf{J} + \frac{\partial \mathbf{D}}{\partial t} \tag{3.17}$$

These are to be taken in conjunction with the "constitutive equations"

$$\mathbf{D} = \epsilon \mathbf{E} \tag{3.18}$$

$$\mathbf{B} = \mu \mathbf{H} \tag{3.19}$$

We treat only uniform media, for which ϵ and μ are constant. The form of the equations is that for rationalized MKS units; we do not concern ourselves here with various systems of units which are summarized in Appendix 5B. We note merely that it is possible to write

$$\epsilon = \epsilon' \epsilon_0 \tag{5.262}$$

$$\mu = \mu' \mu_0 \tag{5.263}$$

where ϵ_0, μ_0 are the dielectric constant and the magnetic permeability of a vacuum, and ϵ', μ' are called the *specific* dielectric constant and magnetic permeability, respectively.

The quantities ρ and $\mathbf{J}$ are electric charge and current density, respectively. A relation may be developed between them by making use of the fact that the rate of change of amount of charge within a volume V_0, namely,

$$\frac{\partial}{\partial t} \int_{V_0} \rho \, d^3\mathbf{r}$$

is equal to the rate at which it flows *in* over the boundary,

$$-\int \mathbf{J} \cdot \mathbf{n} \, dS$$

the integral being over the surface surrounding V_0 and $\mathbf{n}$ being the normal drawn *outward* from the surface. Using Gauss' theorem to change the surface integral into a volume integral, the "equation of continuity" has the form

$$\int_{V_0} \left(\frac{\partial \rho}{\partial t} + \nabla \cdot J \right) d^3\mathbf{r} = 0$$

where the integral is over an *arbitrary* volume; consequently

$$\frac{\partial \rho}{\partial t} + \nabla \cdot \mathbf{J} = 0 \tag{5.264}$$

Let us first consider ρ and **J** as *sources* of the electromagnetic field, so that (3.14) to (3.17) are the equations determining the field *produced* by a *given* system of charges and currents. This is the viewpoint of the theory of radiation, where ρ and **J** specify the properties of a radiating system — an "antenna." We might then call ρ and **J** the "external" charges and currents. There is, of course, an artificiality about this viewpoint, since charges are themselves in turn *acted upon* by fields; in the radiation problem, this reaction back upon the sources of field is known as "radiation reaction." Under appropriate circumstances, it may be ignored; the process here is the familiar one, employed in all areas of physics, of "isolating" a particular system from the whole of nature for study.

On the other hand, we may be interested in electromagnetic propagation in a *medium* containing charges and currents; in such a case, ρ and J describe the *medium*. In that case, in Maxwell's equations ϵ and μ should in the first instance be taken to have their vacuum values ϵ_0 and μ_0. The charge and current then not only determine the fields but are in turn determined by them, through the equations of motion of charged particles in a field. Thus, the problem is one of the behavior of a closed system under interaction between its parts.

In this case the electric current is normally determined by "Ohm's law"

$$\mathbf{J} = \sigma\mathbf{E} \tag{5.265}$$

In some circumstances (e.g., in an external magnetic field or in an anisotropic medium), σ may be a tensor; that is to say, Ohm's law may take the form

$$J_i = \sigma_{ij}E_j \tag{5.266}$$

$(i, j = 1, 2, 3)$. Once **J** is known, ρ can be determined from the continuity equation (5.264).

It should be noted that Eqs. (3.14) and (3.15) are not completely independent of (3.16) and (3.17). Taking the divergence of (3.16) leads to

$$-\frac{\partial}{\partial t}\mathbf{\nabla}\cdot\mathbf{B} = 0 \tag{5.267}$$

We may consider (3.15) simply as providing an initial condition for (3.16). Similarly, the divergence of (3.17) gives

$$\mathbf{\nabla}\cdot\mathbf{J} + \frac{\partial}{\partial t}\mathbf{\nabla}\cdot\mathbf{D} = 0$$

which, along with (5.264), gives

$$\frac{\partial}{\partial t}(\mathbf{\nabla}\cdot\mathbf{D} - \rho) = 0 \tag{5.268}$$

Equation (3.14), in turn, provides an initial condition for this equation. Provided that these initial conditions are satisfied, we need only use (3.16) and (3.17) to develop the theory of electromagnetic radiation.

Let us now determine the equations of wave propagation. Taking the curl of (3.16), and using the identity

$$\nabla \times (\nabla \times \mathbf{v}) = \nabla(\nabla \cdot \mathbf{v}) - \nabla^2\mathbf{v} \tag{5.269}$$

we obtain

$$\nabla(\nabla \cdot \mathbf{E}) - \nabla^2\mathbf{E} = -\frac{\partial}{\partial t}\nabla \times \mathbf{B}$$

Substituting from (3.14) and (3.17) this equation becomes

$$\nabla\left(\frac{\rho}{\epsilon}\right) - \nabla^2\mathbf{E} = -\mu\frac{\partial}{\partial t}\left(\mathbf{J} + \epsilon\frac{\partial\mathbf{E}}{\partial t}\right)$$

or

$$-\nabla^2\mathbf{E} + \epsilon\mu\frac{\partial^2\mathbf{E}}{\partial t^2} = -\mu\frac{\partial\mathbf{J}}{\partial t} - \frac{1}{\epsilon}\nabla\rho \tag{5.270}$$

Taking the curl of (3.17) and using the other equations, we find that

$$\nabla(\nabla \cdot \mathbf{H}) - \nabla^2\mathbf{H} = \nabla \times \mathbf{J} + \frac{\partial}{\partial t}\nabla \times \mathbf{D}$$

or

$$-\nabla^2\mathbf{H} = \nabla \times \mathbf{J} + \epsilon\frac{\partial}{\partial t}\left(-\mu\frac{\partial\mathbf{H}}{\partial t}\right)$$

so that

$$-\nabla^2\mathbf{H} + \epsilon\mu\frac{\partial^2\mathbf{H}}{\partial t^2} = \nabla \times \mathbf{J} \tag{5.271}$$

In a region free from external charges and currents, the right-hand sides of (5.270) and (5.271) are zero, so that both $\mathbf{E}$ and $\mathbf{H}$ satisfy wave equations of the form

$$-\nabla^2 f + \epsilon\mu\frac{\partial^2 f}{\partial t^2} = 0 \tag{5.272}$$

A general solution of this equation may be written [see (5.16)]

$$f = F(\mathbf{n} \cdot \mathbf{r} - ut) \tag{5.273}$$

where F is an arbitrary function and

$$u = \frac{1}{\sqrt{\epsilon\mu}}$$

is the wave velocity. From (5.262) and (5.263)

$$u = \frac{1}{\sqrt{\epsilon'\mu'}}\frac{1}{\sqrt{\epsilon_0\mu_0}} = \frac{c}{\sqrt{\epsilon'\mu'}} \tag{5.274}$$

where c is the velocity of light in vacuo, with a value very close to 3×10^{10} cm/sec. The quantity

$$n = \frac{c}{u} = \sqrt{\epsilon' \mu'} \qquad (5.275)$$

is the "index of refraction" of the medium.

Because of relations (3.16) and (3.17) which *relate* **E** and **H**, (5.270) and (5.271) may not be solved independently. But just as, in electrostatics, a vector quantity (the electric field) may be expressed in terms of a scalar potential ϕ, so in this case the components of the electric and magnetic field, six in number, can be expressed in terms of a smaller number of quantities. Since it is a standard result of vector analysis that a vector field whose divergence is zero can be expressed as the curl of another vector, it follows from (3.15) that **B** may be written as the curl of a vector **A**, known as the "vector potential":

$$\mathbf{B} = \nabla \times \mathbf{A} \qquad (5.276)$$

If this is now substituted in (3.16) that equation may be expressed in the form

$$\nabla \times \left(\mathbf{E} + \frac{\partial \mathbf{A}}{\partial t} \right) = 0 \qquad (5.277)$$

But since any *irrotational* vector can be expressed as the gradient of a scalar potential, we may put

$$\mathbf{E} + \frac{\partial \mathbf{A}}{\partial t} = -\nabla \phi$$

or

$$\mathbf{E} = -\nabla \phi - \frac{\partial \mathbf{A}}{\partial t} \qquad (5.278)$$

Thus, **E** and **B** can be expressed in terms of the vector potential **A** and the scalar potential ϕ.

Equations (5.276) and (5.278) are completely equivalent to the Maxwell equations (3.15) and (3.16).

The equations satisfied by **A** and ϕ must be determined by the *remaining* Maxwell equations (3.14) and (3.17). Substituting in them, we obtain

$$-\nabla^2 \phi - \frac{\partial}{\partial t} \nabla \cdot \mathbf{A} = \frac{\rho}{\epsilon} \qquad (5.279)$$

and

$$\frac{1}{\mu} \nabla \times (\nabla \times \mathbf{A}) + \epsilon \frac{\partial}{\partial t} \left(\nabla \phi + \frac{\partial \mathbf{A}}{\partial t} \right) = \mathbf{J} \qquad (5.280)$$

or

$$-\nabla^2 \mathbf{A} + \nabla \left(\nabla \cdot \mathbf{A} + \frac{1}{u^2} \frac{\partial \phi}{\partial t} \right) + \frac{1}{u^2} \frac{\partial^2 \mathbf{A}}{\partial t^2} = \mu \mathbf{J} \qquad (5.281)$$

In these equations $\mathbf{A}$ and ϕ are coupled. The situation may, however, be simplified by noting that (5.276) does not determine $\mathbf{A}$ uniquely. In fact, the vector $\mathbf{B}$ may be equally well represented by the vector potential $\mathbf{A}'$ where

$$\mathbf{A}' = \mathbf{A} + \nabla\chi \tag{5.282}$$

where χ is an arbitrary function. If we can take advantage of this arbitrariness to choose a vector potential such that the equation

$$\nabla \cdot \mathbf{A} + \frac{1}{u^2}\frac{\partial\phi}{\partial t} = 0 \tag{5.283}$$

(called the "Lorentz condition") is satisfied, (5.279) and (5.281) become

$$-\nabla^2\phi + \frac{1}{u^2}\frac{\partial^2\phi}{\partial t^2} = \frac{\rho}{\epsilon} \tag{5.284}$$

and

$$-\nabla^2\mathbf{A} + \frac{1}{u^2}\frac{\partial^2\mathbf{A}}{\partial t^2} = \mu\mathbf{J} \tag{5.285}$$

that is, inhomogeneous wave equations.

Is it then always possible to choose our potentials to satisfy (5.283)? We note that, although the transformation (5.282) leaves $\mathbf{B}$ invariant, it will not do so for $\mathbf{E}$ unless ϕ is also simultaneously changed. If, however, ϕ is transformed as follows:

$$\phi' = \phi - \frac{\partial\chi}{\partial t} \tag{5.286}$$

$\mathbf{E}$ and $\mathbf{B}$ will both be invariant under the simultaneous transformations (5.282) and (5.286). Such a transformation is called a "gauge transformation" on the potentials.

If, finally, the Lorentz condition (5.283) is to remain satisfied, that is, if

$$\nabla \cdot \mathbf{A}' + \frac{1}{u^2}\frac{\partial\phi'}{\partial t} = 0$$

is to follow from the corresponding equation in the *unprimed* quantities, χ will have to satisfy the equation

$$\nabla^2\chi - \frac{1}{u^2}\frac{\partial^2\chi}{\partial t^2} = 0 \tag{5.287}$$

that is, χ will also have to satisfy the wave equation. This condition defines the permissible gauge transformations which leave invariant the wave equation for the potentials.

Equation (5.283) implies, of course, that to specify the fields in terms of **A** and ϕ is still redundant, since ϕ and **A** are again related. If we introduce the "Hertz vector" **Π** such that

$$\mathbf{A} = \mu \frac{\partial \mathbf{\Pi}}{\partial t} \tag{5.288}$$

and

$$\phi = -\frac{1}{\epsilon} \nabla \cdot \mathbf{\Pi} \tag{5.289}$$

equation (5.283) is satisfied automatically. If at the same time we make use of the continuity equation (5.264) to express ρ and **J** in terms of a "polarization vector" **P**,

$$\rho = -\nabla \cdot \mathbf{P} \tag{5.290}$$

and

$$\mathbf{J} = \frac{\partial \mathbf{P}}{\partial t} \tag{5.291}$$

substituting (5.288)–(5.291) into the wave equations (5.284) and (2.285) gives

$$\nabla \cdot \left\{ \left(-\nabla^2 + \frac{1}{u^2} \frac{\partial^2}{\partial t^2} \right) \mathbf{\Pi} - \mathbf{P} \right\} = 0 \tag{5.292}$$

and

$$\frac{\partial}{\partial t} \left\{ \left(-\nabla^2 + \frac{1}{u^2} \frac{\partial^2}{\partial t^2} \right) \mathbf{\Pi} - \mathbf{P} \right\} = 0 \tag{5.293}$$

Thus, if we choose **Π** to satisfy the equation

$$-\nabla^2 \mathbf{\Pi} + \frac{1}{u^2} \frac{\partial^2 \mathbf{\Pi}}{\partial t^2} = \mathbf{P} \tag{5.294}$$

it is necessary only to satisfy this equation to ensure, not only the satisfaction of Maxwell's equations but also of the wave equations for ϕ and **A** and the Lorentz condition. It is evident, then, that a single vector is adequate to describe completely the electromagnetic field.

Finally, let us express the fields in terms of **Π**. The electric field is

$$\mathbf{E} = -\nabla \phi - \frac{\partial \mathbf{A}}{\partial t}$$

$$= \frac{1}{\epsilon} \nabla (\nabla \cdot \mathbf{\Pi}) - \mu \frac{\partial^2 \mathbf{\Pi}}{\partial t^2}$$

or

$$\mathbf{D} = \nabla \times (\nabla \times \mathbf{\Pi}) - \mathbf{P} \tag{5.295}$$

The magnetic field is

$$\mathbf{H} = \frac{1}{\mu} \nabla \times \mathbf{A} = \frac{\partial}{\partial t} \nabla \times \mathbf{\Pi} \tag{5.296}$$

20. "Spin" of Scalar and Vector Fields

In this section, we develop some results concerning vector fields in general, with a view to the analysis of the electromagnetic field in terms of potentials. The "spin" of such a field is defined, for the present heuristically, but in a manner convenient for use in connection with the quantum mechanics of fields, and it is related to the concept of "polarization".

Consider first a scalar field $f(\mathbf{r}) = f(x, y, z)$ expressed as a function of coordinates related to axes S. Let unit vectors along these axes be $\mathbf{i}$, $\mathbf{j}$, and $\mathbf{k}$. Consider now an infinitesimal rotation of the coordinates through an angle $\delta\theta$ about the direction of the vector $\mathbf{n}$, and designate the unit vectors along the new axes as $\mathbf{i}'$, $\mathbf{j}'$, $\mathbf{k}'$. Then we know that, to the first order in $\delta\theta$,

$$\mathbf{i}' - \mathbf{i} = \delta\theta \quad \mathbf{n} \times \mathbf{i} \tag{5.297}$$

$$\mathbf{j}' - \mathbf{j} = \delta\theta \quad \mathbf{n} \times \mathbf{j} \tag{5.298}$$

$$\mathbf{k}' - \mathbf{k} = \delta\theta \quad \mathbf{n} \times \mathbf{k} \tag{5.299}$$

Let us calculate next what happens to the coordinates of a fixed point under this transformation. If the new coordinates are x', y', z', it follows that

$$x'\mathbf{i}' + y'\mathbf{j}' + z'\mathbf{k}' = x\mathbf{i} + y\mathbf{j} + z\mathbf{k} \tag{5.300}$$

Writing the function f in terms of the transformed coordinates it becomes $\bar{f}(x', y', z')$. We now ask, how has the scalar field *function* changed under the rotation? The *rate* of change is

$$\frac{\partial f}{\partial \theta} = \lim_{\delta\theta = 0} \frac{\bar{f} - f}{\delta\theta} \tag{5.301}$$

where $\bar{f}$ and f are expressed in terms of the same coordinates. In other words, we ask: What is the change in the function describing the field, as a function of coordinates, when the coordinate axes are changed by a rotation?

The problem is most simply solved by stating (5.300) in the form

$$x(\theta)\mathbf{i}(\theta) + y(\theta)\mathbf{j}(\theta) + z(\theta)\mathbf{k}(\theta) = \text{constant} \tag{5.302}$$

The primed and unprimed quantities then refer to different values of θ. Differentiating with respect to θ and using (5.297)–(5.299) in the form

$$\frac{d\mathbf{i}}{d\theta} = \mathbf{n} \times \mathbf{i}, \quad \dots \tag{5.303}$$

we obtain

$$\frac{\partial \mathbf{r}}{\partial \theta} + \mathbf{n} \times \mathbf{r} = 0 \tag{5.304}$$

where
$$\mathbf{r}(\theta) = x(\theta)\mathbf{i} + y(\theta)\mathbf{j} + z(\theta)\mathbf{k} \qquad (5.305)$$
the axes being fixed.

But now, since the field at a fixed point is independent of the axes used,
$$\bar{f}(x', y', z') = f(x, y, z)$$
or
$$f_\theta(x(\theta), y(\theta), z(\theta)) = \text{constant} \qquad (5.306)$$

On differentiating with respect to θ we obtain
$$\frac{\partial f}{\partial \theta} + \frac{\partial \mathbf{r}}{\partial \theta} \cdot \boldsymbol{\nabla} f = 0 \qquad (5.307)$$
or
$$\frac{\partial f}{\partial \theta} = \mathbf{n} \times \mathbf{r} \cdot \boldsymbol{\nabla} f$$
$$= \mathbf{n} \cdot \mathbf{r} \times \boldsymbol{\nabla} f \qquad (5.308)$$

For reasons which will become evident later, we introduce the "rotation operator" $-i(\partial f/\partial \theta)$ on f, so that
$$-i\frac{\partial f}{\partial \theta} = \mathbf{n} \cdot \mathbf{L} f \qquad (5.309)$$
where
$$\mathbf{L} = -i\mathbf{r} \times \boldsymbol{\nabla} \qquad (5.310)$$
We shall see that this may be identified with the quantum "angular momentum" for a scalar field.

Let us now treat, in a similar way, the rotational properties of a *vector* field $\mathbf{A}(\mathbf{r})$. In this case, the rotation of coordinates not only changes $\mathbf{A}$ due to a change in *coordinates*, but also due to a change in its *components* under rotation. By analogy with (5.306) we have
$$\frac{d}{d\theta}\left[A_{\theta x}(\mathbf{r}(\theta))\mathbf{i}(\theta) + A_{\theta y}(\mathbf{r}(\theta))\mathbf{j}(\theta) + A_{\theta z}(\mathbf{r}(\theta))\mathbf{k}(\theta)\right] = 0 \qquad (5.311)$$

Differentiating out we obtain
$$\frac{\partial \mathbf{A}(\mathbf{r})}{\partial \theta} + \left(\frac{\partial \mathbf{r}}{\partial \theta} \cdot \boldsymbol{\nabla}\right)\mathbf{A} + \mathbf{n} \times \mathbf{A} = 0 \qquad (5.312)$$
Therefore,
$$-i\frac{\partial \mathbf{A}}{\partial \theta} = -i(\mathbf{n} \cdot \mathbf{L})\mathbf{A} + i\mathbf{n} \times \mathbf{A} \qquad (5.313)$$

The rotation operator now consists of two parts: $\mathbf{n} \cdot \mathbf{L}$, as in the case of scalar field, and the operator $i\mathbf{n}\times$, which is known as the "spin operator." If the components of $\mathbf{A}$ are represented as a column matrix, the spin operator for the direction **n** becomes

$$S_n = i \begin{pmatrix} 0 & -n_z & n_y \\ n_z & 0 & -n_x \\ -n_y & n_x & 0 \end{pmatrix} \qquad (5.314)$$

This may also be written

$$S_n = \mathbf{n} \cdot \mathbf{S} \qquad (5.315)$$

Where

$$S_x = \begin{pmatrix} 0 & 0 & 0 \\ 0 & 0 & -1 \\ 0 & 1 & 0 \end{pmatrix} i \qquad (5.316a)$$

$$S_y = \begin{pmatrix} 0 & 0 & 1 \\ 0 & 0 & 0 \\ -1 & 0 & 0 \end{pmatrix} i \qquad (5.316b)$$

and

$$S_z = \begin{pmatrix} 0 & -1 & 0 \\ 1 & 0 & 0 \\ 0 & 0 & 0 \end{pmatrix} i \qquad (5.316c)$$

Consider

$$S_n{}^3 \mathbf{A} = i^3 \mathbf{n} \times [\mathbf{n} \times (\mathbf{n} \times \mathbf{A})]$$

$$= -i^3 \mathbf{n} \times \mathbf{A}$$

$$= i \mathbf{n} \times \mathbf{A}$$

$$= S_n \mathbf{A} \qquad (5.317)$$

If we then look for eigenvectors of S_n, i.e., vectors such that

$$S_n \mathbf{A} = \lambda \mathbf{A} \qquad (5.318)$$

the eigenvalues must satisfy the equation

$$\lambda^3 = \lambda$$

so that

$$\lambda = 0, \pm 1 \qquad (5.319)$$

For the case $\lambda = 0$,

$$S_n \mathbf{A} = i \mathbf{n} \times \mathbf{A} = 0 \qquad (5.320)$$

so that the eigenvector is in the *direction* of **n**.

For $\lambda = \pm 1$, on the other hand

$$i \mathbf{n} \times \mathbf{A} = \pm \mathbf{A} \qquad (5.321)$$

In this case, it is clear that the vector must be complex, so we write it

$$\mathbf{A} = \mathbf{A}_1 + i \mathbf{A}_2 \qquad (5.322)$$

where A_1 and A_2 are real. Then, equating real and imaginary parts in (5.321),

$$\mathbf{n} \times \mathbf{A}_1 = \pm \mathbf{A}_2 \qquad (5.323)$$

$$-\mathbf{n} \times \mathbf{A}_2 = \pm \mathbf{A}_1 \qquad (5.324)$$

One of these equations is, of course, redundant, since they must be consistent. The eigenvectors corresponding to $\lambda = \pm 1$ are, respectively, of the form

$$\mathbf{A} = \mathbf{A}_1 \pm i\mathbf{n} \times \mathbf{A}_1 \qquad (5.325)$$

where $\mathbf{A}_1$ is an arbitrary vector perpendicular to $\mathbf{n}$.

If, for instance, $\mathbf{n}$ is taken in the z direction and $\mathbf{A}_1$ in the x direction, these eigenvectors become

$$\mathbf{A} = A_0(\mathbf{i} \pm i\mathbf{j})$$

where A_0 is a scalar.

It is, alternatively, possible to express these eigenvectors in terms of "polarization." We say that the vectors corresponding to $\lambda = \pm 1$ are *right* (+) and *left* (−) circularly polarized with respect to $\mathbf{n}$ and that that corresponding to $\lambda = 0$ is *longitudinally* polarized with respect to $\mathbf{n}$.

Problem 5-21: Show that Maxwell's equations in empty space may be written in the form

$$-iS_\beta \frac{\partial F}{\partial x_\beta} = \frac{i}{c} \frac{\partial F}{\partial t}$$

where F is the column matrix $\mathbf{E} + i\mathbf{H}$ and S_β are suitably defined matrices. (CGS units have been used.)

Verify that

$$S_1 S_2 - S_2 S_1 = iS_3$$

$$S_2 S_3 - S_3 S_2 = iS_1$$

$$S_3 S_1 - S_1 S_3 = iS_2$$

$$S_1{}^2 + S_2{}^2 + S_3{}^2 = 2$$

Use this to verify that the "spin" of the electromagnetic field is 1.

---

21. Interaction of Fields and Particles and the Dynamics of the Electromagnetic Field

The most direct physical approach to the dynamics of the electromagnetic field is through its interaction with charged particles. This approach rests on the assumption of conservation laws for energy and momentum (or, in the relativistic case, on the equivalent hypothesis of covariance). Thus, if a charged particle gains or loses energy, it is assumed to be acquired by the field. In this way, the dynamical variables of fields may be identified.

The fundamental empirical fact is that the force on a particle of charge e is

$$\mathbf{F} = e(\mathbf{E} + \mathbf{v} \times \mathbf{B}) \tag{5.326}$$

If one is concerned with a *continuous* (or quasicontinuous) distribution of charges, it is convenient to define instead a "force density"

$$\mathbf{F}(\mathbf{r}) = \rho\mathbf{E} + \mathbf{J} \times \mathbf{B} \tag{5.327}$$

whose integral over a volume gives the force on that volume.

We may now use this force equation, in conjunction with Maxwell's equations, to calculate the various dynamical variables of the field. To determine the energy density of the field, we use the relation

$$\frac{d}{dt}\int (\mathcal{U}_p + \mathcal{U}_f)\, d^3\mathbf{r} = -\int (\mathbf{P}_p + \mathbf{P}_f) \cdot \mathbf{n}\, dS \tag{5.328}$$

where $\mathcal{U}_p$, $\mathcal{U}_f$ are the energy densities, respectively, of the particles and the field and $\mathbf{P}_p$, $\mathbf{P}_f$ are their energy fluxes, and the integrals are taken over an arbitrary volume and its bounding surface. But

$$\frac{d}{dt}\int \mathcal{U}_p\, d^3\mathbf{r} = -\int \mathbf{P}_p \cdot \mathbf{n}\, dS + \int \mathbf{E} \cdot \mathbf{J}\, d^3\mathbf{r} \tag{5.329}$$

The last term represents the rate at which the particles gain energy from the field; the second term of (5.327) does not contribute because the force is perpendicular to $\mathbf{J}$. From (5.238) and (5.329) it follows that

$$\frac{d}{dt}\int \mathcal{U}_f d^3\mathbf{r} + \int \mathbf{P}_f \cdot \mathbf{n}\, dS = -\int \mathbf{E} \cdot \mathbf{J}\, d^3\mathbf{r} \tag{5.330}$$

We may substitute from (3.17) in the last integral:

$$\int \mathbf{E} \cdot \mathbf{J}\, d^3\mathbf{r} = \int \mathbf{E} \cdot \left(\nabla \times \mathbf{H} - \frac{\partial \mathbf{D}}{\partial t}\right) d^3\mathbf{r}$$

$$= -\int \nabla \cdot \mathbf{E} \times \mathbf{H}\, d^3\mathbf{r} + \int \mathbf{H} \cdot \nabla \times \mathbf{E}\, d^3\mathbf{r} - \frac{\partial}{\partial t}\frac{1}{2}\int \mathbf{D} \cdot \mathbf{E}\, d^3\mathbf{r}$$

where we have used the identity

$$\nabla \cdot \mathbf{E} \times \mathbf{H} = -\mathbf{E} \cdot \nabla \times \mathbf{H} + \mathbf{H} \cdot \nabla \times \mathbf{E}$$

Substituting for $\nabla \times \mathbf{E}$ from (3.16) and using Gauss' theorem to convert the first integral on the right to a surface integral, we find that

$$\int \mathbf{E} \cdot \mathbf{J} \, d^3\mathbf{r} = -\int \mathbf{E} \times \mathbf{H} \cdot \mathbf{n} \, dS - \frac{\partial}{\partial t} \frac{1}{2} \int (\mathbf{D} \cdot \mathbf{E} + \mathbf{B} \cdot \mathbf{H}) \, d^3\mathbf{r} \qquad (5.331)$$

Finally, therefore,

$$\frac{d}{dt} \int \mathscr{U}_f \, d^3\mathbf{r} - \int \mathbf{P}_f \cdot \mathbf{n} \, dS = \frac{\partial}{\partial t} \int \tfrac{1}{2}(\mathbf{D} \cdot \mathbf{E} + \mathbf{B} \cdot \mathbf{H}) \, d^3\mathbf{r} - \int \mathbf{E} \times \mathbf{H} \cdot \mathbf{n} \, dS \qquad (5.332)$$

It is, therefore, reasonable to interpret the energy density of the field as

$$\mathscr{U}_f = \tfrac{1}{2}(\mathbf{D} \cdot \mathbf{E} + \mathbf{B} \cdot \mathbf{H})$$

and the energy flux as

$$\mathbf{P}_f = \mathbf{E} \times \mathbf{H} \qquad (5.333)$$

The latter is known as the "Poynting vector."

Similar arguments may be used for momentum density and flux. Since momentum flux density is a tensor, it is convenient to use an index notation. We call $\mathbf{Q}^{(p)}$ and $\mathbf{Q}^{(f)}$ the momentum densities and $T_{ij}{}^{(p)}$, $T_{ij}{}^{(f)}$ the momentum flux tensors. Then, by conservation,

$$\frac{d}{dt} \int (Q_i{}^{(p)} + Q_i{}^{(f)}) \, d^3\mathbf{r} = -\int (T_{ij}^{(p)} + T_{ij}^{(f)}) n_j \, dS \qquad (5.334)$$

where we have used on the right-hand side the summation convention of tensor calculus, according to which we sum over the repeated index j.

Now, once again, we write down the momentum conservation law for the particles in the presence of the field. The rate at which the particles gain momentum from the field is given by the Lorentz force $\mathbf{F}$, so we have

$$\frac{d}{dt} \int Q_i{}^{(p)} \, d^3\mathbf{r} = -\int T_{ij}^{(p)} n_j \, dS + \int [\rho E_i + (\mathbf{J} \times \mathbf{B})_i] \, d^3\mathbf{r} \qquad (5.335)$$

Therefore,

$$\frac{d}{dt} \int Q_i{}^{(f)} \, d^3\mathbf{r} + \int T_{ij}^{(f)} n_j \, dS = -\int [\rho E_i + (\mathbf{J} \times \mathbf{B})_i] \, d^3\mathbf{r} \qquad (5.336)$$

We must now substitute for ρ and $\mathbf{J}$ and simplify the right-hand side:

$$\rho \mathbf{E} + \mathbf{J} \times \mathbf{B} = \mathbf{E}(\nabla \cdot \mathbf{D}) + (\nabla \times \mathbf{H}) \times \mathbf{B} - \frac{\partial \mathbf{D}}{\partial t} \times \mathbf{B}$$

$$= \mathbf{E}(\nabla \cdot \mathbf{D}) - \nabla(\tfrac{1}{2}\mathbf{B} \cdot \mathbf{H}) + (\mathbf{B} \cdot \nabla)\mathbf{H} - \frac{\partial \mathbf{D}}{\partial t} \times \mathbf{B}$$

Now

$$\mathbf{D} \times (\nabla \times \mathbf{E}) = \nabla(\tfrac{1}{2}\mathbf{D} \cdot \mathbf{E}) - (\mathbf{D} \cdot \nabla)\mathbf{E}$$

and in terms of indices

$$\partial_j(D_j E_i) = (\mathbf{D} \cdot \nabla)E_i + E_i(\nabla \cdot \mathbf{D})$$

Also
$$(\mathbf{B} \cdot \boldsymbol{\nabla})H_i = \partial_j(H_i B_j)$$
since
$$\boldsymbol{\nabla} \cdot \mathbf{B} = 0$$
Therefore,
$$[\rho\mathbf{E} + \mathbf{J} \times \mathbf{B}]_i = \partial_j(E_i D_j) - (\mathbf{D} \cdot \boldsymbol{\nabla})E_i$$
$$= \partial_j(E_i D_j) + [\mathbf{D} \times (\boldsymbol{\nabla} \times \mathbf{E})]_i - \partial_i(\tfrac{1}{2}\mathbf{D} \cdot \mathbf{E})$$
$$- \partial_i(\tfrac{1}{2}\mathbf{B} \cdot \mathbf{H}) + \partial_j(H_i B_j) - \left(\frac{\partial \mathbf{D}}{\partial t} \times \mathbf{B}\right)_i$$

Substituting $\boldsymbol{\nabla} \times \mathbf{E} = -(\partial\mathbf{B}/\partial t)$ in the second term and rearranging terms, we find

$$[\rho\mathbf{E} + \mathbf{J} \times \mathbf{B}]_i = \partial_j(E_i D_j + H_i B_j) - \partial_i(\tfrac{1}{2}\mathbf{D} \cdot \mathbf{E} + \tfrac{1}{2}\mathbf{B} \cdot \mathbf{H}) - \frac{\partial}{\partial t}(\mathbf{D} \times \mathbf{B})_i \tag{5.337}$$

Substituting in (5.336) leads to the equation

$$\frac{d}{dt}\int Q_i^{(f)} d^3\mathbf{r} + \int T_{ij}^{(f)} n_j \, dS = \frac{\partial}{\partial t}\int (\mathbf{D} \times \mathbf{B})_i \, d^3\mathbf{r} + \int \partial_j[\tfrac{1}{2}(\mathbf{D} \cdot \mathbf{E} + \mathbf{B} \cdot \mathbf{H})\delta_{ij}$$
$$- (E_i D_j + H_i B_j)] \, d^3\mathbf{r}$$

The last term may be converted into the surface integral

$$\int [\tfrac{1}{2}(\mathbf{D} \cdot \mathbf{E} + \mathbf{B} \cdot \mathbf{H})\delta_{ij} - E_i D_j - H_i B_j] n_j \, dS$$

We may, therefore, identify the momentum density and flux, respectively, as

$$\mathbf{Q}^{(f)} = \mathbf{D} \times \mathbf{B} \tag{5.338}$$
and
$$T_{ij}^{(f)} = \tfrac{1}{2}(\mathbf{D} \cdot \mathbf{E} + \mathbf{B} \cdot \mathbf{H})\delta_{ij} - E_i D_j - H_i B_j \tag{5.339}$$

It should be noted that

$$\mathbf{Q}^{(f)} = \epsilon\mu\mathbf{P}_f \tag{5.340}$$

The above results are valid only if ϵ and μ are constant. If they are variable, there are additional terms involving their variations.

Problem 5-22: Calculate the formulas for energy and momentum density and flux in the case of varying ϵ and μ.

Finally, an expression may be derived from (5.338) for the *angular momentum* of the field:

$$\mathbf{L}^{(f)} = \int \mathbf{r} \times \mathbf{Q}^{(f)} \, d^3\mathbf{r}$$
$$= \int \mathbf{r} \times (\mathbf{D} \times \mathbf{B}) \, d^3\mathbf{r} \tag{5.341}$$

Problem 5-23: Verify directly the constancy of the total *angular* momentum, that of the field plus that of the particles, in a given volume.

22. Lagrangian and Hamiltonian of the Field

We show first that the Lagrangian density of the free electromagnetic field may be written in the form

$$\mathscr{L} = \frac{1}{2}\left(\epsilon E^2 - \frac{1}{\mu} B^2\right) - \frac{1}{2\mu}\left(\boldsymbol{\nabla}\cdot\mathbf{A} + \frac{1}{u^2}\frac{\partial\phi}{\partial t}\right)^2 \tag{5.342}$$

This is to be considered as a function of the potentials and their space and time derivatives, which are treated for the present as independent. The Lagrangian is then the space integral of $\mathscr{L}$:

$$L = \int \mathscr{L}\left(\mathbf{A}, \varphi, \dot{\mathbf{A}}, \dot{\varphi}, \frac{\partial A_i}{\partial x_j}, \frac{\partial\phi}{\partial x_j}\right) d^3\mathbf{r} \tag{5.343}$$

and the action principle, from which the equations of motion may be derived, is

$$\delta \int_{t_1}^{t_2} L\,dt = \delta \int_{t_1}^{t_2} dt \int \mathscr{L}\,d^3\mathbf{r} \tag{5.344}$$

Proceeding now exactly as in the derivation of Eq. (5.41) for sound waves, but making independent variations of A_i and ϕ, it is found that the equations of motion are

$$\frac{\delta L}{\delta A_i} = 0 \tag{5.345}$$

and

$$\frac{\delta L}{\delta\phi} = 0 \tag{5.346}$$

Particular derivatives which are needed and may be readily verified are

$$\frac{\partial E_j}{\partial A_i} = -\delta_{ij}, \qquad \frac{\partial E_j}{\partial\phi_{,i}} = -\delta_{ij}$$

$$\frac{\partial B_j}{\partial A_{k,l}} = \epsilon_{jlk}, \qquad \frac{\partial}{\partial A_{k,l}}\boldsymbol{\nabla}\cdot\mathbf{A} = \delta_{kl} \tag{5.347}$$

where ϵ_{jlk} is the Kronecker permutation symbol whose value is zero unless j, l, and k are all different, 1 if they constitute an *even* permutation of 1, 2, 3, and -1 if they constitute an *odd* permutation, and "$,i$" signifies differentiation with respect to x_i.

Equations (5.345) now become

$$-\frac{\partial}{\partial t}(\epsilon E_i) + \frac{\partial}{\partial x_j}\left(\frac{1}{\mu}\epsilon_{ijl}B_l\right) - \frac{1}{\mu}\frac{\partial}{\partial x_i}\left(\mathbf{\nabla}\cdot\mathbf{A}+\frac{1}{u^2}\frac{\partial\phi}{\partial t}\right) = 0$$

or

$$-\epsilon\frac{\partial\mathbf{E}}{\partial t} + \frac{1}{\mu}\mathbf{\nabla}\times\mathbf{B} - \frac{1}{\mu}\mathbf{\nabla}(\mathbf{\nabla}\cdot\mathbf{A}) - \epsilon\frac{\partial}{\partial t}\mathbf{\nabla}\phi = 0 \qquad (5.348)$$

Writing $\mathbf{E}$ and $\mathbf{B}$ in terms of the potentials, this reduces to

$$-\nabla^2\mathbf{A} + \frac{1}{u^2}\frac{\partial^2\mathbf{A}}{\partial t^2} = 0 \qquad (5.349)$$

Similarly, (5.346) becomes

$$\epsilon\frac{\partial E_j}{\partial x_j} + \epsilon\frac{\partial}{\partial t}\left(\mathbf{\nabla}\cdot\mathbf{A} + \frac{1}{u^2}\frac{\partial\phi}{\partial t}\right) = 0$$

or

$$-\nabla^2\phi + \frac{1}{u^2}\frac{\partial^2\phi}{\partial t^2} = 0 \qquad (5.350)$$

If there are sources specified by charge and current densities ρ and $\mathbf{J}$, respectively, the Lagrangian density may be modified by adding to $\mathscr{L}$ the term

$$\mathscr{L}' = -\rho\phi - \mathbf{J}\cdot\mathbf{A} \qquad (5.351)$$

This adds a $\mu\mathbf{J}$ on the right-hand side of (5.349), and ρ/ϵ to the right-hand side of (5.350), as required.

We may now define generalized momenta in the usual way:

$$\pi_i = \frac{\partial\mathscr{L}}{\partial\dot{A}_i} = -\epsilon E_i \qquad (5.352)$$

$$\pi_0 = \frac{\partial\mathscr{L}}{\partial\dot{\phi}} = -\epsilon\left(\mathbf{\nabla}\cdot\mathbf{A} + \frac{1}{u^2}\frac{\partial\phi}{\partial t}\right) \qquad (5.353)$$

We see, then, that the Lorentz condition may be formulated as

$$\pi_0 = 0 \qquad (5.354)$$

From the generalized momenta, it is possible to calculate the Hamiltonian density:

$$\mathscr{H} = \pi_i\dot{A}_i + \pi_0\dot{\phi}_0 - \mathscr{L} \qquad (5.355)$$

Eliminating the time derivatives we obtain

$$\mathscr{H} = \frac{1}{2\epsilon}\boldsymbol{\pi}\cdot\boldsymbol{\pi} + \frac{1}{2\mu}(\mathbf{\nabla}\times\mathbf{A})^2 + \rho\phi + \mathbf{J}\cdot\mathbf{A}$$
$$- \boldsymbol{\pi}\cdot\mathbf{\nabla}\phi - u^2\pi_0\mathbf{\nabla}\cdot\mathbf{A} + u^2\pi_0^2/2\epsilon \qquad (5.356)$$

From this the equations of motion may be calculated, and are found to be

$$\dot{\phi} = \frac{\partial \mathcal{H}}{\partial \pi_0} = -u^2 \nabla \cdot \mathbf{A} + \frac{u^2}{\epsilon} \pi_0 \tag{5.357a}$$

$$\dot{\pi}_0 = -\frac{\partial \mathcal{H}}{\partial \phi} = \rho + \nabla \cdot \boldsymbol{\pi} \tag{5.357b}$$

$$\dot{\mathbf{A}} = \frac{\partial \mathcal{H}}{\partial \boldsymbol{\pi}} = \frac{1}{\epsilon} \boldsymbol{\pi} - \nabla \phi \tag{5.357c}$$

and

$$\dot{\boldsymbol{\pi}} = -\frac{\partial \mathcal{H}}{\partial \mathbf{A}} = \mathbf{J} - \frac{1}{\mu} \nabla \times (\nabla \times \mathbf{A}) + u^2 \nabla \pi_0 \tag{5.357d}$$

Equations (5.357a) and (5.357c) are *definitions* of π_0 and $\boldsymbol{\pi}$ respectively, while (5.357b) and (5.357d) are true dynamical equations.

There is an additional restriction upon these equations, arising from the fact that the continuity equation

$$\nabla \cdot \mathbf{J} + \frac{\partial \rho}{\partial t} = 0 \tag{5.358}$$

must be satisfied. From Eqs. (5.357b) and (5.357d) it follows that

$$\frac{\partial^2 \pi_0}{\partial t^2} - u^2 \nabla^2 \pi_0 = 0 \tag{5.359}$$

We note that Eqs. (5.357) lead to Maxwell's equations only if $\pi_0 = 0$. Thus, from all the solutions arising from our Lagrangian, physical reality is represented only by those which satisfy this condition. Since this is only one of the possible solutions of (5.359), the restriction is a real one. A sufficient hypothesis would be that at some initial time π_0 and $\partial \pi_0/\partial t$ were zero everywhere. It would then follow that they were zero at all times.

23. Normal Modes of the Electromagnetic Field

Consider the problem of expanding the electromagnetic field in normal modes, so as to introduce normal coordinates. If we consider first the *vector* field $\mathbf{A}$, we note that any vector field can be broken into two parts:

$$\mathbf{A} = \mathbf{A}_1 + \mathbf{A}_2 \tag{5.360}$$

one of which is solenoidal:

$$\nabla \cdot \mathbf{A}_1 = 0 \tag{5.361}$$

and the other irrotational:

$$\nabla \times \mathbf{A}_2 = 0 \tag{5.362}$$

[This is true for the source-free field. If there is a part of the field satisfying *both* (5.361) and (5.362) if can be written as ∇f, where f satisfies $\nabla^2 f = 0$. The only source-free solution of this equation is $f = 0$.]

Since an irrotational field can be expressed in terms of a potential, it is possible to write

$$A_2 = \nabla \psi \qquad (5.363)$$

If we define orthogonality of vector fields A_i and A_j by the condition

$$\int A_i^* \cdot A_j \, d^3 r = 0, \qquad i \neq j \qquad (5.364)$$

the integral being taken over the region of definition of the problem, then A_1 and A_2 as defined above are mutually orthogonal. For

$$\int A_2^* \cdot A_1 \, d^3 r = \int \nabla \psi^* \cdot A_1 \, d^3 r$$
$$= - \int \psi^* \nabla \cdot A_1 \, d^3 r$$
$$= 0 \qquad (5.365)$$

where we have used Gauss' theorem in the third line. We note that this orthogonality condition does not imply that the vectors A_1, A_2 are orthogonal at *each point*: only the *integral* of the scalar product is involved in the definition.

Since normal modes are source-free solutions of the wave equations for which the time dependence is $e^{-i\omega t}$, they satisfy the equations

$$-\nabla^2 \xi + \epsilon \mu \omega^2 \xi = 0 \qquad (5.366)$$

or

$$-\nabla^2 \phi + \epsilon \mu \omega^2 \phi = 0 \qquad (5.367)$$

For brevity we write

$$\epsilon \mu \omega^2 = k^2 \qquad (5.368)$$

Let the two independent sets of mutually orthogonal solenoidal vector solutions satisfying the boundary conditions of the problem be $\xi_{n\alpha}$, where $\alpha = 1, 2$ and n runs from 1 to ∞ [the third independent vector solution is of the form ∇v_n, where v_n are solutions of (5.367)]. By their definition

$$\nabla \cdot \xi_{n\alpha} = 0 \qquad (5.369)$$

The ξ's may also be taken to be real. The eigenvalues are taken to be k_{n0}^2 for v_n and $k_{n\alpha}^2$ for $\xi_{n\alpha}$, and both v_n and the $\xi_{n\alpha}$'s are assumed to be normalized.

We now write

$$A = \sum_{n,\alpha} a_{n\alpha}(t) \xi_{n\alpha}(r) + \sum_n b_n(t) \nabla v_n(r) \qquad (5.370)$$

$$\phi = \sum_n c_n(t) v_n(r) \qquad (5.371)$$

We then calculate in terms of these normal modes the Lagrangian

$$L = \int \mathscr{L} \, d^3\mathbf{r} \tag{5.372}$$

where $\mathscr{L}$ is given by (5.342). Now

$$\mathbf{E} = -\sum c_n \nabla v_n - \sum \dot{a}_{n\alpha}\boldsymbol{\xi}_{n\alpha} - \sum \dot{b}_n \nabla v_n$$

By orthogonality, and using (5.365),

$$\tfrac{1}{2}\epsilon \int \mathbf{E}^2 d^3\mathbf{r} = \tfrac{1}{2}\epsilon \left[\sum_{n,\alpha} \dot{a}_{n\alpha}^2 + \sum_n k_{n0}^2 (c_n + \dot{b}_n)^2 \right] \tag{5.373}$$

since

$$\int \nabla v_n \cdot \nabla v_m \, d^3\mathbf{r} = -\int v_n \nabla^2 v_m \, d^3\mathbf{r}$$
$$= k_{0m}^2 \int v_n v_m \, d^3\mathbf{r}$$
$$= 0 \quad \text{if } n \neq m$$

The integral of the magnetic term is

$$-\frac{1}{2\mu} \sum a_{m\alpha} a_{n\alpha'} \int (\nabla \times \boldsymbol{\xi}_{m\alpha}) \cdot (\nabla \times \boldsymbol{\xi}_{n\alpha'}) \, d^3\mathbf{r} \tag{5.374}$$

To evaluate this, consider first the vector identity

$$\nabla \cdot (\mathbf{a} \times \mathbf{b}) = -\mathbf{a} \cdot \nabla \times \mathbf{b} + \mathbf{b} \cdot \nabla \times \mathbf{a}$$

Putting $\mathbf{b} = \nabla \times \mathbf{a}'$, this becomes

$$\nabla \cdot [\mathbf{a} \times (\nabla \times \mathbf{a}')] = -\mathbf{a} \cdot \nabla \times (\nabla \times \mathbf{a}') + (\nabla \times \mathbf{a}) \cdot (\nabla \times \mathbf{a}')$$

Now put $\mathbf{a} = \boldsymbol{\xi}_{m\alpha}$, $\mathbf{a}' = \boldsymbol{\xi}_{n\alpha'}$, and integrate. The divergence of the left-hand side will be zero, since it may be transformed into a surface integral at infinity. Also

$$\nabla \times (\nabla \times \mathbf{a}') = \nabla \times (\nabla \times \boldsymbol{\xi}_{n\alpha'})$$
$$= \nabla (\nabla \cdot \boldsymbol{\xi}_{n\alpha'}) - \nabla^2 \boldsymbol{\xi}_{n\alpha'}$$
$$= k_{n\alpha'}^2 \boldsymbol{\xi}_{n\alpha'}$$

Therefore (5.374) is

$$-\frac{1}{2\mu} \int \mathbf{B}^2 d^3\mathbf{r} = -\frac{1}{2\mu} \sum a_{n\alpha}^2 k_{n\alpha}^2 \tag{5.375}$$

where we have made use of the normalization of the $\boldsymbol{\xi}$'s.

Finally, look at

$$\frac{1}{2\mu} \int \left(\nabla \cdot \mathbf{A} + \frac{1}{u^2}\frac{\partial \phi}{\partial t} \right)^2 d^3\mathbf{r} = \frac{1}{2\mu} \int \left[\nabla^2 \psi + \frac{1}{u^2}\left(\frac{\partial \phi}{\partial t}\right) \right]^2 d^3\mathbf{r}$$
$$= \frac{1}{2\mu} \int \left[-\sum k_{0n}^2 b_n v_n + \frac{1}{u^2}\sum \dot{c}_n v_n \right]^2 d^3\mathbf{r}$$
$$= \frac{1}{2\mu} \sum \left(\frac{1}{u^2}\dot{c}_n - k_{0n}^2 b_n \right)^2 \tag{5.376}$$

We thus obtain the Lagrangian in terms of the amplitudes of the normal modes and their derivatives:

$$L = \frac{1}{2}\epsilon \sum \dot{a}_{n\alpha}^2 - \frac{1}{2\mu} \sum k_{n\alpha}^2 a_{n\alpha}^2 + L_0 \tag{5.377}$$

where

$$L_0 = \frac{1}{2}\epsilon \sum k_{0n}^2 (c_n + b_n)^2 - \frac{1}{2\mu} \sum \left(\frac{1}{u^2}\dot{c}_n - k_{0n}^2 b_n\right)^2 \tag{5.378}$$

L_0 combines the scalar and longitudinal vector fields. The first two terms of (5.377) describe the "radiation" field. Generalized momenta can be defined; the problem of the dynamics of this field is that of a collection of oscillators, one for each normal mode.

If we use (5.378) to calculate the equations of motion for b_n and c_n we find

$$\ddot{b}_n + u^2 k_{0n}^2 b_n = 0 \tag{5.379}$$

and

$$\ddot{c}_n + u^2 k_{0n}^2 c_n = 0 \tag{5.380}$$

Since the only physically acceptable solutions are those which satisfy the Lorentz condition, however, these equations are not to be solved independently, but are related by the equation

$$c_n + \dot{b}_n = 0 \tag{5.381}$$

It is straightforward to generalize the foregoing formalism to take account of that part of the Lagrangian density describing the interaction of the field with particles [Eq. (5.351)]

$$\mathscr{L}' = -\rho\phi - \mathbf{J} \cdot \mathbf{A}$$

ρ and $\mathbf{J}$ must be expanded in the same way as ϕ and $\mathbf{A}$:

$$\rho = \sum_n \rho_n(t) v_n \tag{5.382}$$

and

$$\mathbf{J} = \sum_n J_{ln}\nabla v_n + \sum_{n,\alpha} J_{n\alpha}\boldsymbol{\xi}_{n\alpha} \tag{5.383}$$

where the subscript "l" designates the longitudinal current. The coefficients may be calculated by making use of the orthonormality of the v's and ξ's:

$$\rho_n(t) = \int \rho v_n \, d^3\mathbf{r} \tag{5.384}$$

$$J_{ln}(t) = -\frac{1}{k_{0n}^2} \int (\nabla \cdot \mathbf{J}) v_n \, d^3\mathbf{r}$$

$$= \frac{1}{k_{0n}^2} \int \frac{\partial \rho}{\partial t} v_n \, d^3\mathbf{r} \tag{5.385}$$

and

$$J_{n\alpha}(t) = \int \boldsymbol{\xi}_{n\alpha} \cdot \mathbf{J} \, d^3\mathbf{r} \tag{5.386}$$

Note that

$$\frac{\partial \rho_n}{\partial t} + k_{0n}^2 J_{ln} = 0 \tag{5.387}$$

The additional terms in the Lagrangian are then

$$L' = \int \mathscr{L}' \, d^3\mathbf{r}$$

$$= -\left[\sum_n \rho_n c_n + \sum_n k_{0n}^2 J_{ln} b_n + \sum_{n,\alpha} J_{n\alpha} a_{n\alpha} \right] \tag{5.388}$$

The additional terms in the Hamiltonian are given by the square bracket, without the minus sign. Equations (5.379) and (5.380) are now modified to

$$\ddot{b}_n + u^2 k_{0n}^2 b_n = -\frac{1}{\epsilon} J_{ln} = \frac{1}{\epsilon k_{0n}^2} \dot{\rho}_n \tag{5.389}$$

and

$$\ddot{c}_n + u^2 k_{0n}^2 c_n = u^2 \rho_n / \epsilon \tag{5.390}$$

Problem 5-24: Verify that the Lorentz condition $c_n + \dot{b}_n = 0$ is consistent with Eqs. (5.389) and (5.390).

---

24. Coulomb Gauge and Pure Radiation Field

The irrotational part of the vector potential, written in the last section as $\nabla \Psi$, can obviously be eliminated by the gauge transformation (5.282); one need only choose $\chi = -\Psi$. The resulting gauge, in which

$$\nabla \cdot \mathbf{A} = 0, \tag{5.391}$$

is known as the "Coulomb gauge." The resulting vector potential then describes the pure radiation field. Since, from (5.283), the scalar potential ϕ must then not depend on the time, it satisfies the equation

$$-\nabla^2 \phi = \rho / \epsilon \tag{5.392}$$

and describes the "longitudinal" part of the field. This is, therefore, a very convenient gauge for the treatment of radiation problems.

25. Plane Waves

It is easily verified that

$$\phi = f(\mathbf{n} \cdot \mathbf{r} - ut) \tag{5.393}$$

and

$$A = F(n \cdot r - ut) \tag{5.394}$$

are solutions of the source-free wave equation, n being an arbitrary unit vector and f and F arbitrary functions. These are plane wave solutions in the sense that the potentials are constant over the planes $n \cdot r - ut =$ constant; furthermore, the plane of constant phase moves perpendicular to n with a constant speed ("phase velocity") u.

If these solutions are to satisfy the Lorentz condition (5.283), the relation between f and F must be

$$f' = u n \cdot F' \tag{5.395}$$

where the primes designate derivatives of the functions with respect to their arguments. Since constants are of no importance in the potentials, we may take $f = u n \cdot F$.

It is now easily verified that the fields are given by

$$\begin{aligned} E &= -\nabla \phi - \frac{\partial A}{\partial t} \\ &= -u n \cdot (n \cdot F') + u F' \\ &= -u n \times (n \times F') \end{aligned} \tag{5.396}$$

and

$$\begin{aligned} B &= \nabla \times A \\ &= n \times F' \end{aligned} \tag{5.397}$$

The Poynting vector, describing the energy flux, is

$$\begin{aligned} P_f &= E \times H \\ &= -\frac{u}{\mu} [n \times (n \times F')] \times (n \times F') \\ &= \frac{u}{\mu} n(n \times F')^2 \end{aligned} \tag{5.398}$$

The energy *density* is given by

$$\begin{aligned} \mathscr{U} &= \frac{1}{2}(D \cdot E + B \cdot H) \\ &= \frac{1}{2} \epsilon u^2 (n \times F')^2 + \frac{1}{2\mu}(n \times F')^2 \\ &= \frac{1}{\mu}(n \times F')^2 \end{aligned} \tag{5.399}$$

From these results we make the following observations:

(a) The electric field $\mathbf{E}$ and the magnetic induction $\mathbf{B}$ are at right angles to each other and to the vector $\mathbf{n}$.

(b) The energy flux in the wave is in the direction of the vector $\mathbf{n}$, that is, perpendicular to the planes of constant phase.

(c) We may define a velocity of energy flow $\mathbf{u}$ by the equation

$$\mathbf{u}\mathcal{U} = \mathbf{P}_f \tag{5.400}$$

It follows from (5.389) and (5.390) that

$$\mathbf{u} = u\mathbf{n} \tag{5.401}$$

Thus, the velocity of propagation of energy is equal to the phase velocity.

(d) The energy density in the *electric* field and that in the *magnetic* field are equal.

Let us consider the question of the *polarization* of plane waves. It was shown above [Eq. (5.321)] that for waves polarized about the direction $\mathbf{n}$

$$i\mathbf{n} \times \mathbf{A} = \pm \mathbf{A}$$

the $+$ and $-$ signs corresponding to right and left circular polarization, respectively. If $\mathbf{F}_\pm$ represent polarized waves,

$$i\mathbf{n} \times \mathbf{F}_\pm = \pm \mathbf{F}_\pm \tag{5.402}$$

Substituting in (5.396) and (5.397) we obtain

$$\mathbf{E}_\pm = u\mathbf{F}'_\pm \tag{5.403}$$

$$\mathbf{B}_\pm = \mp i\mathbf{F}'_\pm \tag{5.404}$$

We must not, however, apply (5.398) and (5.399) to these solutions, since these latter equations are valid only for $\mathbf{F}$ real, whereas the polarized solutions are, by (5.326), complex.

Problem 5-25: Writing $\mathbf{F}'_\pm = (\mathbf{i} \pm i\mathbf{j})\lambda e^{i\theta}$, show that the real fields are

$$\mathbf{E}_\pm = u\lambda \, (\cos\theta \, \mathbf{i} \mp \sin\theta \, \mathbf{j})$$

$$\mathbf{B}_\pm = \lambda \, (\pm \sin\theta \, \mathbf{i} + \cos\theta \, \mathbf{j})$$

Hence calculate the energy flux and density.

Problem 5-26: Earlier in the chapter we discussed "guided" sound waves. Similar problems may be studied for electromagnetic waves. It is found that in guides of uniform cross section there exist transverse electric modes (no electric field along the guide) and transverse magnetic modes (no magnetic field along the guide).

(a) Calculate the electromagnetic fields in the transverse electric (TE) and transverse magnetic (TM) modes, in a rectangular guide of dimensions $a \times b$. (The walls of the guide are assumed perfectly conducting.)

(b) What is the maximum wavelength which can be propagated in each case?

Problem 5-27:   Solve the corresponding problem for a guide of circular cross section.

Problem 5-28:   *A problem on spherical waves.* Find the electromagnetic waves in a spherical resonator excited by an oscillating dipole of moment P and circular frequency ω located at its center.

---

26.   Plane Waves as Normal Modes

If the above functions of the form $f(\mathbf{n} \cdot \mathbf{r} - ut)$ are expressed as a Fourier integral:

$$f(\mathbf{n} \cdot \mathbf{r} - ut) = \int \bar{f}(k) e^{ik(\mathbf{n} \cdot \mathbf{r} - ut)} \, dk \qquad (5.405)$$

where

$$\bar{f}(k) = \frac{1}{2\pi} \int f(s) e^{-iks} \, ds \qquad (5.406)$$

we see that these solutions may be expressed in terms of the functions $e^{i(\mathbf{k} \cdot \mathbf{r} - \omega t)}$ where

$$\mathbf{k} = k\mathbf{n} \qquad (5.407)$$

and

$$\omega = ku \qquad (5.408)$$

In fact, the general solution may be so expressed since this is merely its expression as a three-dimensional Fourier transform. The normal modes are now characterized by the vector $\mathbf{k}$. The theory of the previous section may be applied if we make the identification

$$v_n(\mathbf{r}) \to \left(\frac{1}{2\pi}\right)^3 e^{i(\mathbf{k} \cdot \mathbf{r} - \omega t)} \qquad (5.409)$$

$$\mathbf{u}_{n\alpha}(\mathbf{r}) \to \left(\frac{1}{2\pi}\right)^3 \mathbf{e}_{k\alpha} e^{i(\mathbf{k} \cdot \mathbf{r} - \omega t)} \qquad (5.410)$$

The fact that the $\mathbf{u}_{n\alpha}$'s satisfy (5.369) implies that

$$\mathbf{k} \cdot \mathbf{e}_{k\alpha} = 0 \qquad (5.411)$$

so that the vectors $\mathbf{e}_{k\alpha}$ are transverse (orthogonal) to the propagation vector $\mathbf{k}$. The quantities $k_{0\alpha}^2$, $k_{n\alpha}^2$ are now simply k^2. The vectors $\mathbf{e}_{k\alpha}$ must also be mutually orthogonal, that is,

$$\mathbf{e}_{k1} \cdot \mathbf{e}_{k2} = 0 \tag{5.412}$$

27. Radiation from Given Sources: Multipole Fields

Equations (5.284) and (5.285) relate the fields ϕ and $\mathbf{A}$ to their sources, the charge density ρ, and current density $\mathbf{J}$. The solutions may be written down by analogy with that for sound waves [Eqs. (5.68) and (5.72)]:

$$\phi = \frac{1}{4\pi\epsilon} \int \frac{\rho(\mathbf{r}', t - R/c)}{R} d^3\mathbf{r}' \tag{5.413}$$

and

$$\mathbf{A} = \frac{\mu}{4\pi} \int \frac{\mathbf{J}(\mathbf{r}', t - R/c)}{R} d^3\mathbf{r}' \tag{5.414}$$

where

$$R = |\mathbf{r} - \mathbf{r}'|$$

It is possible, without difficulty, to generalize these equations to take account of a source of a different character, namely, a time-dependent external magnetization. The intrinsic magnetization is in general defined by

$$\mathbf{M} = \mathbf{B} - \mu_0 \mathbf{H} \tag{5.415}$$

so that

$$\mathbf{B} = \mu_0 \mathbf{H} + \mathbf{M}$$

This intrinsic magnetization is, for isotropic media, proportional to the magnetic field, the proportionality factor being $\mu_0\chi$, where χ is known as the magnetic susceptibility. Thus

$$\mathbf{B} = \mu_0(1 + \chi)\mathbf{H} \tag{5.416}$$

The external magnetization $\mathbf{M}_0$ must be added to the $\mathbf{M}$ above. Thus, the equation

$$\mathbf{B} = \nabla \times \mathbf{A}$$

must be replaced by

$$\mathbf{B} = \nabla \times \mathbf{A} - \mathbf{M}_0$$

If this is now substituted into the equation

$$\nabla \times \mathbf{B} = \mu\mathbf{J} + \mu\frac{\partial \mathbf{D}}{\partial t}$$

and $\mathbf{B}$, $\mathbf{E}$ are expressed in terms of potentials by the above and the equation (5.278), we are led to the equation

$$\nabla(\nabla \cdot \mathbf{A}) - \nabla^2\mathbf{A} + \frac{1}{u^2}\left(\nabla\frac{\partial\phi}{\partial t} + \frac{\partial\mathbf{A}}{\partial t}\right) = \mu\mathbf{J} + \nabla \times \mathbf{M}_0 \tag{5.417}$$

Therefore, the external magnetization may be taken into account throughout by adding to the current a "magnetization current" $(1/\mu)\mathbf{\nabla} \times \mathbf{M}_0$.

We treat, in particular, the problem of radiation in empty space from sources with a given frequency:

$$\rho(\mathbf{r}, t) = \rho_0(\mathbf{r})e^{-i\omega t} \tag{5.418}$$

$$\mathbf{J}(\mathbf{r}, t) = \mathbf{J}_0(\mathbf{r})e^{-i\omega t} \tag{5.419}$$

where, by virtue of the equation of continuity

$$\mathbf{\nabla} \cdot \mathbf{J}_0 - i\omega\rho_0 = 0 \tag{5.420}$$

The potentials then become

$$\phi = \frac{1}{4\pi\epsilon_0} \int \rho_0(\mathbf{r}')g(\mathbf{r} - \mathbf{r}') \, d^3r' e^{-i\omega t} \tag{5.421}$$

$$\mathbf{A} = \frac{\mu_0}{4\pi} \int \mathbf{J}_0(\mathbf{r}')g(\mathbf{r} - \mathbf{r}') \, d^3r' e^{-i\omega t} \tag{5.422}$$

where

$$g(R) = \frac{e^{i(\omega/c)R}}{R} \tag{5.423}$$

is the spherical "Green's function" of the radiation field. We note that

$$\nabla^2 g = -\frac{\omega^2}{c^2} g \tag{5.424}$$

We now show how to represent the fields in terms of a vector potential alone. To begin, we note the identity

$$(\mathbf{J}_0 \times \mathbf{\nabla}') \times \mathbf{\nabla}'g = -\mathbf{J}_0\nabla'^2 g + (\mathbf{J}_0 \cdot \mathbf{\nabla}')\mathbf{\nabla}'g$$

so that

$$\mathbf{J}_0 g = \frac{c^2}{\omega^2}[(\mathbf{J}_0 \times \mathbf{\nabla}') \times \mathbf{\nabla}'g - (\mathbf{J}_0 \cdot \mathbf{\nabla}')\mathbf{\nabla}'g] \tag{5.425}$$

Here $\mathbf{\nabla}'$ represents the gradient taken with respect to the coordinates $\mathbf{r}'$. Substituting this in $\mathbf{A}$ above, we get

$$\mathbf{A} = \frac{1}{\epsilon_0\omega^2} \int [(\mathbf{J}_0 + \mathbf{\nabla}') \times \mathbf{\nabla}'g - (\mathbf{J}_0 \cdot \mathbf{\nabla}')\mathbf{\nabla}'g] \, d^3r' e^{-i\omega t}$$

But

$$\mathbf{\nabla}'g = -\mathbf{\nabla}g \tag{5.426}$$

so it is possible to write

$$\mathbf{A} = (\mathbf{A}_1 + \mathbf{A}_2)e^{-i\omega t} \tag{5.427}$$

where

$$\mathbf{A}_1 = \frac{1}{\epsilon_0\omega^2} \mathbf{\nabla} \times \int (\mathbf{J}_0 \times \mathbf{\nabla}')g \, d^3r' \tag{5.428}$$

and

$$A_2 = \frac{1}{\epsilon_0 \omega^2} \nabla \int J_0 \cdot \nabla' g \, d^3r'$$

$$= -\frac{1}{\epsilon_0 \omega^2} \nabla \int (\nabla' \cdot J_0) g \, d^3r'$$

$$= -\frac{i}{\epsilon_0 \omega} \nabla \int \rho_0 g \, d^3r' \qquad (5.429)$$

Clearly, A_2 contributes nothing to the magnetic field. Also, A_2, along with ϕ, contribute nothing to **E**. For

$$-\nabla\phi - \dot{A}_2 = -\frac{1}{\epsilon_0} \nabla \int \rho_0 g \, d^3r' + \frac{1}{\epsilon_0} \nabla \int \rho_0 g \, d^3r' = 0$$

Thus, A_1 alone completely describes the electromagnetic field of the sources. By virtue of (5.426), it may equally well be written as

$$A_1 = \frac{1}{\epsilon_0 \omega^2} \nabla \times \left[\nabla \times \int J_0 g \, d^3r' \right] \qquad (5.430)$$

Having already shown that A_1 and A_2 are orthogonal, we now show that A_1 itself may be broken into two orthogonal parts. To do so, we first observe that if $\Phi'(\mathbf{R}, t)$ is a solution of the wave equation, so is

$$A_1' = L\Phi' = i\nabla \times r'\Phi' \qquad (5.431)$$

where **L** is the operator $-i\mathbf{r} \times \nabla$ defined in (5.310). In quantum mechanics, this is the angular momentum operator. Its most important property here is that it commutes with the Laplacian operator; that is,

$$\nabla^2 L\Phi' = L\nabla^2\Phi' \qquad (5.432)$$

From this result, which is simple to verify directly, it follows that

$$\nabla^2(L\Phi') = L \frac{1}{u^2} \frac{\partial^2 \Phi'}{\partial t^2} = \frac{1}{u^2} \frac{\partial^2}{\partial t^2} (L\Phi') \qquad (5.433)$$

If the time dependence $e^{-i\omega t}$ is substituted,

$$\Phi' = \Phi_0' e^{-i\omega t} \qquad (5.434)$$

the equation becomes

$$\nabla^2(L\Phi') + k^2 L\Phi' = 0 \qquad (5.435)$$

where $k^2 = (\omega^2/u^2) = \epsilon\mu\omega^2$ was defined in (5.368). Being a curl, A_1 is also orthogonal to A_2.

Finally, we show that

$$A_1'' = \frac{1}{k} \nabla \times L\Phi' \qquad (5.436)$$

is also a solution of the wave equation and is orthogonal to both $\mathbf{A}_1'$ and $\mathbf{A}_2$. Its orthogonality to $\mathbf{A}_2$ follows again from the fact that it is a curl. That it is a solution of the wave equation follows from the fact that ∇^2 commutes with both ∇ and $\mathbf{L}$; consequently

$$\nabla^2(\nabla \times \mathbf{L})\Phi_0 = (\nabla \times \mathbf{L})\nabla^2\Phi_0$$

The proof that $\mathbf{A}_1''$ is orthogonal to $\mathbf{A}_1'$ [in the sense of (5.364)] makes use of the fact that the operators $i\nabla$ and $\mathbf{L}$ are Hermitian, that is, satisfy the relations

$$\int f_2^* i\nabla f_1 \, d^3\mathbf{r} = \int f_1(i\nabla f_2)^* \, d^3\mathbf{r} \tag{5.437}$$

and

$$\int f_2^* \mathbf{L} f_1 \, d^3\mathbf{r} = \int f_1(\mathbf{L} f_2)^* \, d^3\mathbf{r} \tag{5.438}$$

The first results from a simple integration by parts (we assume, as always, that f_1 and f_2 vanish sufficiently rapidly at infinity that the surface integral $\int f_1 f_2^* \mathbf{n} \, dS$ is zero there). As for the second, it is sufficient to prove it for one component of $\mathbf{L}$; let us choose L_z. Now if we use spherical polar coordinates, we see that, since $x = r \sin \theta \cos \varphi$, $y = r \sin \theta \sin \varphi$, and $z = r \cos \theta$,

$$-i\frac{\partial f}{\partial \varphi} = -i\left[\frac{\partial f}{\partial x}\frac{\partial x}{\partial \varphi} + \frac{\partial f}{\partial y}\frac{\partial y}{\partial \varphi}\right]$$

$$= -i\left(x\frac{\partial}{\partial y} - y\frac{\partial}{\partial x}\right)$$

$$= L_z \tag{5.439}$$

Thus,

$$\int f_2^* L_z f_1 \, d^3\mathbf{r} = -i\int f_2^* \frac{\partial f_1}{\partial \varphi} \, d^3\mathbf{r}$$

$$= i\int f_1 \frac{\partial f_2^*}{\partial \varphi} \, d^3\mathbf{r}$$

$$= \int f_1(L_z f_2)^* \, d^3\mathbf{r} \tag{5.440}$$

The transition from the second to third line is accomplished by partial integration with respect to φ, and making use of the fact that f_1, f_2 are single-valued.

Having thus shown that the operators $i\nabla$ and $\mathbf{L}$ are Hermitian, we prove that $\mathbf{A}_1'$, $\mathbf{A}_1''$ are orthogonal as follows:

$$\int \mathbf{A}_1'^* \cdot \mathbf{A}_1'' \, d^3\mathbf{r} = \int (\mathbf{L}\Phi')^* \cdot \frac{1}{k}\nabla \times \mathbf{L}\Phi' \, d^3\mathbf{r}$$

$$= \frac{1}{k}\int \Phi'^* \mathbf{L} \cdot \nabla \times \mathbf{L}\Phi' \, d^3\mathbf{r} \tag{5.441}$$

But

$$\mathbf{L} \cdot \nabla \times \mathbf{L} = (\mathbf{L} \times \nabla) \cdot \mathbf{L}$$

$$= [(\mathbf{r} \times \nabla) \times \nabla] \cdot \mathbf{L}$$

$$= (\mathbf{r} \cdot \nabla)\nabla \cdot \mathbf{L} - \mathbf{r} \cdot \nabla^2 \mathbf{L}$$

$$= 0$$

since both $\nabla \cdot \mathbf{L}$ and $\mathbf{r} \cdot \mathbf{L} = 0$.

Thus, we have succeeded in proving that $\mathbf{A}'_1$ and $\mathbf{A}''_1$ are two solutions of the wave equation which are orthogonal to each other and to $\mathbf{A}_2$. Before proceeding with a multipole expansion, let us note the nature of the fields generated by $\mathbf{A}'_1$, $\mathbf{A}''_1$, respectively. The fields generated by $\mathbf{A}'_1$ are

$$\mathbf{E}' = i\omega\mathbf{L}\Phi' \tag{5.442}$$

$$\mathbf{H}' = \frac{1}{\mu_0}\nabla \times \mathbf{L}\Phi' \tag{5.443}$$

$\mathbf{E}'$ has no radial component, since $\mathbf{r} \cdot \mathbf{L} = 0$. The radial component of $\mathbf{H}'$ is on the other hand

$$\frac{1}{r}\mathbf{r} \cdot \mathbf{H}' = \frac{1}{\mu_0 r}\mathbf{r} \cdot \nabla \times \mathbf{L}\Phi' \tag{5.444}$$

$$= \frac{i}{\mu_0 r}\mathbf{L}^2\Phi'$$

Since only the magnetic field has a radial component, we refer to the radiation described by this potential as "magnetic" radiation. The energy flux is

$$\mathbf{P} = \mathrm{Re}\,\mathbf{E} \times \mathrm{Re}\,\mathbf{H} \tag{5.445}$$

since the real parts represent the physical fields. The real part of the product of two complex numbers z_1 and z_2 is

$$\tfrac{1}{2}(z_1 z_2^* + z_1^* z_2) = \mathrm{Re}\,(z_1 z_2^*)$$

Thus the energy flux contains both radial and transverse parts. The radial part, which describes the radiation of energy from the source, is

$$\frac{1}{r}\mathbf{r} \cdot \mathbf{P} = \frac{1}{\mu_0 r}\mathrm{Re}\,i\omega\mathbf{r} \cdot [(\mathbf{L}\Phi') \times (\nabla \times \mathbf{L}\Phi')^*]$$

$$= -\frac{\omega}{\mu_0 r}\mathrm{Re}\,i\mathbf{L}\Phi' \cdot \mathbf{r} \times (\nabla \times \mathbf{L}\Phi')^*$$

$$= \frac{\omega}{\mu_0 r}\mathrm{Re}\,i\mathbf{L}\Phi' \cdot (\mathbf{r} \cdot \nabla)\mathbf{L}\Phi'^*$$

$$= \frac{\omega}{\mu_0}\mathrm{Re}\,i\mathbf{L}\Phi' \cdot \mathbf{L}\frac{\partial}{\partial r}\Phi'^* \tag{5.446}$$

Finally, we can calculate the *angular* momentum of the radiation; this is

$$\mathbf{m} = \frac{1}{\mu_0 c^2}\, \mathbf{r} \times \mathbf{P} \tag{5.447}$$

$$= \frac{\omega}{\mu_0 c^2}\, \text{Re}\; i\mathbf{r} \times [\mathbf{L}\Phi' \times (\nabla \times \mathbf{L}\Phi')^*]$$

$$= \frac{\omega}{\mu_0 c^2}\, \text{Re}\; \mathbf{L}\Phi'(-i\mathbf{r} \cdot \nabla \times \mathbf{L}\Phi')^*$$

$$= \frac{\omega}{\mu_0 c^2}\, \text{Re}\; \mathbf{L}\Phi'(\mathbf{L}^2\Phi')^* \tag{5.448}$$

[The momentum density was shown in (5.338) and (5.340) to be $(1/c^2)\mathbf{P}$.]

Let us turn next to $\mathbf{A}_2''$. The corresponding fields are

$$\mathbf{E}'' = \frac{i\omega}{k}\, \nabla \times \mathbf{L}\Phi'' \tag{5.449}$$

and

$$\mathbf{H}'' = \frac{1}{\mu_0 k}\, \nabla \times (\nabla \times \mathbf{L})\Phi''$$

$$= -\frac{1}{\mu_0}\, k\mathbf{L}\Phi'' \tag{5.450}$$

It is now the magnetic field which has no radial component: The radiation is designated "electric" radiation.

The energy flux in the radial direction, as well as the angular momentum, are given by exactly the same formula as in the case of magnetic radiation.

We know that the potential (5.430) can be written as the sum of terms of the form (5.431) and (5.436). The problem remains, however, to find the functions Φ', Φ''. We do this in the following way: We expand the Φ's in spherical harmonics, and use the orthogonality of the terms in the expansion to determine the coefficients.

Let us first demonstrate this orthogonality. We write

$$\Phi' = \sum z_{lm}(kr) Y_{lm}(\theta, \varphi) \tag{5.451}$$

where Y_{lm} is a spherical harmonic. (It is convenient to write m as a subscript rather than a superscript from this point forward.) We now show that

(a) $$\int (\mathbf{L}Y_{l'm'})^* \cdot \mathbf{L}Y_{lm}\, d\Omega = l(l+1)\delta_{ll'}\delta_{mm'} \tag{5.452}$$

and that

(b) $$\int (i\nabla \times \mathbf{L}z_{l'm'}(kr)Y_{l'm'})^* \cdot i\nabla \times \mathbf{L}z_{lm}Y_{lm}\, d\Omega$$

$$= l(l+1)k^2 z_{lm}^2(kr)\delta_{ll'}\delta_{mm'} \tag{5.453}$$

The z's could be ignored in case (a) because $\mathbf{L}z_{lm}(kr)=0$. The integrals are over solid angle.

To prove (a), we first make use of the Hermitian property of the operator $\mathbf{L}$ to write the integral as

$$\int Y^*_{l'm'}\mathbf{L}^2 Y_{lm}\,d\Omega$$

Now

$$\mathbf{L}^2 = -(\mathbf{r}\times\nabla)\cdot(\mathbf{r}\times\nabla)$$

$$= \mathbf{r}\cdot\nabla\times(\mathbf{r}\times\nabla)$$

In index notation, we can write this

$$-L^2 = x_i\frac{\partial}{\partial x_j}x_i\frac{\partial}{\partial x_j}-x_i\frac{\partial}{\partial x_j}x_j\frac{\partial}{\partial x_i}$$

$$= [r^2\nabla^2+\mathbf{r}\cdot\nabla]-[3\mathbf{r}\cdot\nabla-(\mathbf{r}\cdot\nabla)^2+\mathbf{r}\cdot\nabla]$$

$$= r^2\nabla^2 - r\frac{\partial}{\partial r}\left(r\frac{\partial}{\partial r}\right)-r\frac{\partial}{\partial r}$$

$$= r^2\nabla^2 - r^2\frac{\partial^2}{\partial r^2}-2r\frac{\partial}{\partial r}$$

$$= \frac{1}{\sin\theta}\frac{\partial}{\partial\theta}\left(\sin\theta\frac{\partial}{\partial\theta}\right)+\frac{1}{\sin^2\theta}\frac{\partial^2}{\partial\varphi^2}$$

from (3.28), θ and φ being polar coordinates. But from (3.45) and (3.62) it follows that

$$\mathbf{L}^2 Y_{lm} = l(l+1)Y_{lm} \qquad (5.454)$$

Out integral is then equal to

$$l(l+1)\delta_{ll'}\delta_{mm'} \qquad (5.455)$$

as we wished to prove.

Turning to (b), and making use first of the Hermitian character of $i\nabla$ and then that of $\mathbf{L}$, we obtain

$$\int (\mathbf{L}z_{l'm'}(kr)Y_{l'm'})^*\cdot\nabla\times(\nabla\times\mathbf{L}z_{lm}Y_{lm})\,d\Omega$$

$$= -\int (\mathbf{L}z_{l'm'}Y_{l'm'})^*\cdot\nabla^2\mathbf{L}z_{lm}Y_{lm}\,d\Omega$$

$$= k^2 z_{l'm'}z_{lm}\int Y^*_{l'm'}\cdot\mathbf{L}^2 Y_{lm}\,d\Omega$$

$$= l(l+1)k^2 z_{l'm'}z_{lm}\delta_{ll'}\delta_{mm'}$$

as required.

We are now in a position to complete the evaluation of the potential in terms of an expansion in harmonics; that is, a multipole expansion. We start from the fact that

$$\mathbf{A}_1 = \frac{1}{\epsilon_0\omega^2}\nabla\times\left[\nabla\times\int \mathbf{J}_0 g\,d^3r'\right]=\mathbf{L}\sum z_{l'm'}Y_{l'm'}(\Omega)+\frac{1}{k}\nabla\times\mathbf{L}\sum\bar z_{l'm'}Y_{l'm'}(\Omega)$$

$$(5.456)$$

where $z_{l'm'}$, $\bar{z}_{l'm'}$ are to be determined. We first multiply on both sides of this equation by $(LY_{lm})^*$ and integrate over solid angle. This gives

$$\frac{1}{\epsilon_0\omega^2} \int (LY_{lm})^* \cdot \nabla \times \left[\nabla \times \int \mathbf{J}_0 g d^3\mathbf{r}'\right] d^3\mathbf{r} = l(l+1)z_{lm}(kr) \quad (5.457)$$

We now use the expansion for g given in (5.135),

$$g(\mathbf{r}-\mathbf{r}') = 4\pi i k \sum_l j_l(kr')h_l^{(1)}(kr)Y_{lm}(\Omega)Y_{lm}^*(\Omega') \quad (5.458)$$

In using this expansion, we use primes to designate coordinates within the source, and unprimed symbols to designate the coordinates of the (outside) point at which the field is being calculated. Thus $r' < r$. Equation (5.457) may now be written

$$\begin{aligned}
z_{lm}(kr) &= \frac{1}{l(l+1)} \int Y_{lm}^* \mathbf{L} \cdot \nabla \times \left[\nabla \times \int \mathbf{J}_0 g d^3\mathbf{r}'\right] d\Omega \frac{1}{\epsilon_0\omega^2} \\
&= \frac{k^2}{l(l+1)} \int Y_{lm}^* \mathbf{L} \cdot \left[\int \mathbf{J}_0 g d^3\mathbf{r}'\right] d\Omega \frac{1}{\epsilon_0\omega^2} \\
&= \frac{\mu_0}{l(l+1)} \int Y_{lm}^* \left[\int \mathbf{J}_0 \cdot \mathbf{L}' g d^3\mathbf{r}'\right] d\Omega \\
&= -\frac{i\mu_0}{l(l+1)} \int Y_{lm}^* \left[\int \nabla' \cdot (\mathbf{r}' \times \mathbf{J}_0) g d^3\mathbf{r}'\right] d\Omega \quad (5.459)
\end{aligned}$$

In going from the second to the third line, we have used the fact, which is easily verified, that

$$\mathbf{L}g = \mathbf{L}'g \quad (5.460)$$

Substituting for g from (5.458) we find that

$$z_{lm}(kr) = \frac{k\mu_0}{l(l+1)} h_l^{(1)}(kr) \int \nabla' \cdot \mathbf{r}' \times \mathbf{J}_0 j_l(kr')Y_{lm}^*(\Omega') d^3\mathbf{r}' \quad (5.461)$$

We may proceed similarly to find $\bar{z}_{lm}(kr)$. We multiply (5.456) by $(1/k)\nabla \times LY_{lm}^*$ and integrate over solid angle. Proceeding on the right-hand side as in the derivation of (5.453), we obtain the result that

$$\begin{aligned}
\bar{z}_{lm}(kr) &= \frac{1}{\epsilon_0\omega^2} \frac{1}{l(l+1)} \int \frac{1}{k}(\nabla \times LY_{lm})^* \cdot \nabla \times \left(\nabla \times \int \mathbf{J}_0 g d^3\mathbf{r}'\right) d\Omega \\
&= \frac{k}{\epsilon_0\omega^2} \frac{1}{l(l+1)} \int LY_{lm}^* \cdot \nabla \times \left(\int \mathbf{J}_0 g d^3\mathbf{r}'\right) d\Omega \\
&= \frac{\mu_0}{k} \frac{1}{l(l+1)} \int Y_{lm}^* \left[\int \mathbf{J}_0 \cdot \nabla' \times \mathbf{L}' g d^3\mathbf{r}'\right] d\Omega \quad (5.462)
\end{aligned}$$

The inner integral may be transformed into a more convenient form:

$$\int \mathbf{J}_0 \cdot \mathbf{\nabla}' \times \mathbf{L}' g d^3\mathbf{r}' = i \int \mathbf{J}_0 \cdot [k^2\mathbf{r}' + \mathbf{\nabla}'(1 + \mathbf{r}' \cdot \mathbf{\nabla}')] g d^3\mathbf{r}'$$

$$= ik^2 \int \mathbf{r}' \cdot \mathbf{J}_0 g d^3\mathbf{r}' + \omega \int \rho(1 + \mathbf{r}' \cdot \mathbf{\nabla}') g d^3\mathbf{r}' \tag{5.463}$$

To obtain the last term we have integrated by parts and used the continuity equation. Thus,

$$\bar{z}_{lm}(kr) = \frac{\mu_0}{l(l+1)} \int Y_{lm}^* \left[ik \int \mathbf{r}' \cdot \mathbf{J}_0 g \, d^3\mathbf{r}' \right.$$

$$\left. + \frac{\omega}{k} \int \rho(1 + \mathbf{r}' \cdot \mathbf{\nabla}') g d^3\mathbf{r}' \right] d\Omega \tag{5.464}$$

Substituting again for g this takes the form

$$\bar{z}_{lm}(kr) = \frac{i\mu_0}{l(l+1)} h_l^{(1)}(kr)\omega \int \left\{ \rho_0(1 + \mathbf{r}' \cdot \mathbf{\nabla}') \right.$$

$$\left. + \frac{ik}{c} \mathbf{r}' \cdot \mathbf{J}_0 \right\} j_l(kr') Y_{lm}(\Omega') \, d^3\mathbf{r}' \tag{5.465}$$

where c, the velocity of light, is equal to $1/\sqrt{\epsilon_0\mu_0}$ and $k = \omega/c$.

Let us now introduce

$$d'_{lm} = -\frac{ik\mu_0}{l(l+1)} \int \mathbf{\nabla}' \cdot \mathbf{r}' \times \mathbf{J}_0 j_l(kr') Y_{lm}(\Omega') \, d^3\mathbf{r}' \tag{5.466}$$

and

$$d''_{lm} = \frac{\omega\mu_0}{l(l+1)} \int \left\{ \rho_0(1 + \mathbf{r}' \cdot \mathbf{\nabla}') + \frac{ik}{c} \mathbf{r}' \cdot \mathbf{J}_0 \right\} j_l(kr') Y_{lm}(\Omega') \, d^3\mathbf{r}' \tag{5.467}$$

Problem 5-29: Show that, if there is also a source consisting of a magnetic moment, it contributes to d'_{lm} an additional term

$$-\frac{ik}{l(l+1)} \int \{\mathbf{\nabla}' \cdot \mathbf{M}_0(1 + \mathbf{r}' \cdot \mathbf{\nabla}') - k^2\mathbf{r}' \cdot \mathbf{M}_0\} j_l(kr') Y_{lm}(\Omega') \, d^3\mathbf{r}'$$

and to d''_{lm} a term

$$-\frac{ik^2}{l(l+1)} \int \mathbf{\nabla}' \cdot (\mathbf{r}' \times \mathbf{M}_0) j_l(kr') Y_{lm}(\Omega') \, d^3\mathbf{r}'$$

If we now use (5.431), (5.436), and (5.451), we can write the magnetic and electric radiation, respectively, in the following form:

Magnetic

$$\mathbf{A}'_1 = \mathbf{L}\Phi'$$

where

$$\Phi' = i \sum_{l,m} d'_{lm} h_l^{(1)}(kr) Y_{lm}(\Omega) \qquad (5.468)$$

Electric

$$\mathbf{A}''_1 = \frac{1}{k} \nabla \times \mathbf{L}\Phi''$$

$$\Phi'' = i \sum_{l,m} d''_{lm} h_l^{(1)}(kr) Y_{lm}(\Omega) \qquad (5.469)$$

The fields are of course given by (5.442) and (5.443) for the magnetic case and (5.449) and (5.450) for the electric. The terms of a given l in each case describe 2^l-pole radiation.

To calculate the intensity of radiation, we use the forms of these functions in the asymptotic region $kr \gg 1$. There

$$\Phi' \approx \frac{e^{ikr}}{kr} \sum (-i)^l d'_{lm} Y_{lm} \qquad (5.470)$$

and

$$\Phi'' \approx \frac{e^{ikr}}{kr} \sum (-i)^l d''_{lm} Y_{lm} \qquad (5.471)$$

Substituting in (5.446) and integrating over a sphere of large radius surrounding the source, we get the integrated radial energy flux

$$\mathscr{P}_r = \frac{\omega}{\mu_0 k} \sum_{l,m} l(l+1)|d_{lm}|^2 = \frac{c}{\mu_0} \sum_{l,m} l(l+1)|d_{lm}|^2 \qquad (5.472)$$

where the appropriate d_{lm} is used for each sort of radiation.

If the energy of the radiation per unit range of r is $\mathscr{U}$, then

$$\mathscr{U} = \frac{1}{c}\mathscr{P}_r = \frac{1}{\mu_0} \sum l(l+1)|d_{lm}|^2 \qquad (5.473)$$

From (5.447), it is also straightforward to calculate the angular momentum of the radiation of each multipole order. The z component of angular momentum in a shell of unit thickness dr at distance r in the asymptotic region is

$$m_z = \frac{1}{\mu_0 \omega} ml(l+1)|d_{lm}|^2 \qquad (5.474)$$

Therefore,

$$m_z = m \frac{\mathscr{U}}{\omega} \qquad (5.475)$$

In problems of radiation from atoms and nuclei, the size of the radiating system is generally small compared to the wavelength of the emitted radiation; that is, $ka \ll 1$, where a is the radius of the emitter. By virtue of this fact, in calculating the radiation of a given multipole order from (5.466) and (5.467) the spherical Bessel functions may be replaced to a good approximation by their value for small r':

$$j_l(kr') \doteq \frac{2^l l!}{(2l+1)!}(kr')^l \qquad (5.476)$$

In the electric multipole expansion, we note also that

$$\frac{(ik/c)\mathbf{r}' \cdot \mathbf{J}_0}{\rho} \approx (ka)^2$$

so that the second term in (5.467) can be neglected. Then

$$d''_{l\,m} \doteq \frac{\omega\mu_0}{l(l+1)} k^l \frac{2^l(l+1)!}{(2l+1)!} \int \rho_0(\mathbf{r}')r'^l Y_{lm}(\Omega')\, d^3\mathbf{r}' \qquad (5.477)$$

Recalling the definition of the electric multipole moment in Eq. (3.142) of Chapter 3,

$$Q_{lm} = \sqrt{\frac{4\pi}{2l+1}} \int \rho(\mathbf{r}')r'^l Y_{lm}(\Omega')\, d^3\mathbf{r}'$$

we may write

$$d''_{lm} = \frac{1}{\sqrt{4\pi(2l+1)}}\, \mu_0\, \frac{\omega^{l+1}}{c^l}\, \frac{2^l(l-1)!}{(2l)!}\, Q_{lm} \qquad (5.478)$$

The intensity of the corresponding electric 2^l-pole radiation is then

$$\mathscr{P}^{(eD)}_{r,lm} = \frac{c}{\epsilon_0}\left(\frac{\omega}{c}\right)^{2l+2} \frac{l(l+1)}{2l+1}\left[\frac{2^l(l-1)!}{(2l)!}\right]^2 |Q_{lm}|^2 \qquad (5.479)$$

Similarly, in the case of magnetic 2^l-pole radiation, we can define the *magnetic* multipole moment

$$\bar{Q}_{lm} = \sqrt{\frac{4\pi}{2l+1}} \int \boldsymbol{\nabla}' \cdot \mathbf{r}' \times \mathbf{J}_0 r'^l Y_{lm}(\Omega')\, d^3\mathbf{r}' \qquad (5.480)$$

in terms of which we may write

$$d'_{lm} = -i\, \frac{1}{\sqrt{4\pi(2l+1)}}\, \mu_0\left(\frac{\omega}{c}\right)^{l+1} \frac{1}{l(l+1)}\, \frac{2^l l!}{(2l)!}\, \bar{Q}_{lm} \qquad (5.481)$$

Therefore the intensity of magnetic 2^l-pole radiation is

$$\mathscr{P}^{(mag)}_{r,lm} = c\mu_0\left(\frac{\omega}{c}\right)^{2l+2} \frac{l}{(l+1)(2l+1)}\left[\frac{2^l(l-1)!}{(2l)!}\right]^2 |\bar{Q}_{lm}|^2 \qquad (5.482)$$

Problem 5-30: Calculate the contribution to the multipole moments of various orders from an oscillating magnetic moment $\mathbf{M} = \mathbf{M}_0 e^{-i\omega t}$.

---

28. Radiation from a Uniformly Charged Vibrating Liquid Drop

The problem of radiation from a vibrating liquid drop carrying a charge is of interest because the liquid drop provides under certain conditions a model of an atomic nucleus. The model is a suitable one when the currents which are the source of the radiation are primarily collective, that is, are carried by the whole charge of the nucleus, rather than being associated with specific particles. The model we treat is a completely *classical* one. The charge is assumed to be uniformly distributed. The undisturbed shape of the drop is taken to be spherical. The dynamics of the drop are that of an incompressible fluid acted upon by its internal Coulomb forces and by surface tension.

The Lagrangian therefore contains three terms: the kinetic energy, the Coulomb energy, and the surface energy. If ρ is the charge density and $\mathbf{J}$ the current, there is a continuity equation

$$\nabla \cdot \mathbf{J} + \frac{\partial \rho}{\partial t} = 0 \qquad (5.483)$$

Assuming the motion to be irrotational, $\mathbf{J}$ may be expressed in terms of a velocity potential

$$\mathbf{J} = -\nabla \phi \qquad (5.484)$$

Since the charge density is constant *inside* the drop,

$$\nabla^2 \phi = 0 \qquad (5.485)$$

If the surface of the drop is oscillating, the boundary conditions are time-dependent, so we can write

$$\phi = \sum a_{lm}(t) r^l Y_{lm}(\theta, \varphi), \qquad a_{l,-m} = a_{lm}^* \qquad (5.486)$$

We now assume the equilibrium radius to be R_0. The shape of the surface may in general be expressed as

$$r = r_0(t) = R_0 + \sum c_{lm}(t) Y_{lm}(\theta, \varphi) \qquad (5.487)$$

We assume that the distortion is small, so that the c_{lm}'s are small compared with R_0.

The boundary conditions now enable us to express the a_{lm}'s in terms of the c_{lm}'s. The kinetic energy can be calculated in terms of ϕ and therefore of the a's; the Coulomb and surface energies depend on the droplet shape and therefore on the c's.

The boundary condition is, to the first order in small quantities,

$$-\left(\frac{\partial \phi}{\partial r}\right)_{r=R_0} = r_0(t) \qquad (5.488)$$

On equating coefficients of each harmonic, this yields

$$\dot{c}_{lm} = - lR_0^{l-1}a_{lm} \tag{5.489}$$

It should be noted that, for the volume to remain constant, R_0 must itself be a function of t. If R is the radius of the drop at rest, so that the mass is

$$M = \tfrac{4}{3}\pi R^3 \eta \tag{5.490}$$

where η is the mass density, then

$$\tfrac{4}{3}\pi R^3 = \int d\Omega \int_0^{R_0 + \Sigma c_{lm}Y_{lm}} r^2 \, dr$$

$$= \int d\Omega \tfrac{1}{3} \left[R_0 + \sum c_{lm}Y_{lm} \right]^3$$

$$= \tfrac{4}{3}\pi R_0^3 + R_0 \sum |c_{lm}|^2 \tag{5.491}$$

if we omit as negligible terms cubic in the c's. Therefore,

$$R_0(t) = R\left[1 - \frac{1}{4\pi R^2} \sum |c_{lm}|^2 \right] \tag{5.492}$$

to the same approximation.

Let us now calculate the terms in the energy one at a time. The kinetic energy is

$$T = \tfrac{1}{2}\eta \int (\nabla\phi)^2 \, d^3\mathbf{r} \tag{5.493}$$

where η is the mass density. Since $\nabla^2\phi = 0$ inside the drop, T may be transformed to a surface integral

$$T = \tfrac{1}{2}\eta \int \phi \nabla\phi \cdot \mathbf{n} \, dS \tag{5.494}$$

where the integral is over the boundary surface of the drop, and $\mathbf{n}$ is the unit normal vector drawn outward. To a good approximation $\mathbf{n} \cdot \nabla\phi = \partial\phi/\partial r$, and

$$T = \tfrac{1}{2}\eta \sum_l lR_0^{2l+1} \sum_{m=-l}^{l} |a_{lm}|^2 \tag{5.495}$$

If we write

$$c_{lm} = \alpha_{lm} + i\beta_{lm} \tag{5.496}$$

where α_{lm}, β_{lm} are real and use (5.489), the kinetic energy becomes

$$T = \frac{1}{2}\eta \sum_{l=1}^{\infty} lR_0^{2l+1} \sum_{m=-l}^{l} [(\operatorname{Re} a_{lm})^2 + (\operatorname{Im} a_{lm})^2]$$

$$= \frac{1}{2}\eta R_0^3 \sum_{l,m} \frac{1}{l} (\dot{\alpha}_{lm}^2 + \dot{\beta}_{lm}^2)$$

$$= \frac{3M}{8\pi} \sum_{l,m} \frac{1}{l} (\dot{\alpha}_{lm}^2 + \dot{\beta}_{lm}^2) \tag{5.497}$$

where M is the total mass of the drop (replacing R_0 by R here involves only omitting *fourth* powers of the distortion parameters).

Next, consider the surface energy. The elements of solid angle and of surface area are connected by the relation

$$d\Omega = \frac{dS \cos \bar\theta}{r^2}$$

or

$$dS = \frac{r^2 \, d\Omega}{\cos \bar\theta} \tag{5.498}$$

where $\cos \bar\theta$ is the cosine of the angle between the radial direction and the normal to the surface. To calculate $\cos \bar\theta$, we note that the unit normal vector to the surface

$$r = R_0 + \sum c_{lm} Y_{lm} = f(\theta, \varphi)$$

has components

$$\frac{1}{\sqrt{1 + \frac{1}{r^2}\left[\left(\frac{\partial f}{\partial \theta}\right)^2 + \frac{1}{\sin^2 \theta}\left(\frac{\partial f}{\partial \varphi}\right)^2\right]}}\left[1, -\frac{1}{r}\frac{\partial f}{\partial \theta}, -\frac{1}{r \sin \theta}\frac{\partial f}{\partial \varphi}\right] \tag{5.499}$$

Thus, to the usual approximation,

$$\frac{1}{\cos \bar\theta} = \sqrt{1 + \frac{1}{r^2}\left[\left(\frac{\partial f}{\partial \theta}\right)^2 + \frac{1}{\sin^2 \theta}\left(\frac{\partial f}{\partial \varphi}\right)^2\right]}$$

$$\doteq 1 + \frac{1}{2r^2}\left[\left(\frac{\partial f}{\partial \theta}\right)^2 + \frac{1}{\sin^2 \theta}\left(\frac{\partial f}{\partial \varphi}\right)^2\right] \tag{5.500}$$

The surface area is therefore

$$A = \int d\Omega \left\{r^2 + \frac{1}{2}\left[\left(\frac{\partial f}{\partial \theta}\right)^2 + \frac{1}{\sin^2 \theta}\left(\frac{\partial f}{\partial \varphi}\right)^2\right]\right\}$$

$$= 4\pi R_0^2 + \sum |c_{lm}|^2 + \frac{1}{2}\int d\Omega \left[\left(\frac{\partial f}{\partial \theta}\right)^2 + \frac{1}{\sin^2 \theta}\left(\frac{\partial f}{\partial \varphi}\right)^2\right]$$

$$= 4\pi R^2 - \sum |c_{lm}|^2 - \frac{1}{2}\int f\left[\frac{1}{\sin \theta}\frac{\partial}{\partial \theta}\left(\sin \theta \frac{\partial f}{\partial \theta}\right) + \frac{1}{\sin^2 \theta}\frac{\partial^2 f}{\partial \varphi^2}\right] d\Omega \tag{5.501}$$

In obtaining the first two terms, use has been made of (5.492). The last follows on integrating by parts the angular integral. But the operator in the square bracket, operating on Y_{lm}, was shown in Chapter 3 to give $-l(l+1)Y_{lm}$. Thus, substituting for f and integrating, we find that the surface area is

$$A = 4\pi R^2 + \sum |c_{lm}|^2 \left[\tfrac{1}{2}l(l+1) - 1\right] \tag{5.502}$$

It should be noted that the $l = 1$ contribution is zero. This is due to the fact that the dipole "distortion" is simply a displacement of the center of the drop.

Taking the surface tension to be σ, the surface energy associated with the distortion is

$$V_{\text{surface}} = \sigma \sum (\alpha_{lm}^2 + \beta_{lm}^2)[\tfrac{1}{2}l(l+1) - 1] \qquad (5.503)$$

Finally, we must calculate the dependence of the Coulomb energy on the distortion. This calculation is simplified by noting that we may consider the charge distribution to consist of a uniform *spherical* distribution of radius R_0, plus a *surface* distribution of surface density $\rho_0(r_0 - R_0) = \rho_0 \Sigma c_{lm} Y_{lm}$, ρ_0 being the volume density. We then calculate separately the self-energies of the volume and surface distributions, and their interaction energy. It is, however, easily seen that this latter is zero, since the potential of the volume distribution is spherically symmetrical, whereas the surface charge distribution has no $l = 0$ component. Their integral is therefore zero.

The volume contribution is

$$\rho_0^2 \int_0^R \frac{1}{r} 4\pi r^2 \left[\int_0^r 4\pi r'^2 \, dr' \right] dr = \frac{16}{15} \pi^2 R_0^5 \rho_0^2$$

Now ρ_0 is given by

$$\frac{Ze^2}{\frac{4}{3}\pi R^3} = \rho_0$$

where Ze is the total charge, so the Coulomb energy of the volume distribution is

$$\frac{3}{5} \frac{Z^2 e^2 R_0^5}{R^6} = \frac{3}{5} \frac{Z^2 e^2}{R} - \frac{3Z^2 e^2}{4\pi R^3} \sum |c_{lm}|^2 \qquad (5.504)$$

on using (5.492). The first term is just the Coulomb energy of the undistorted distribution and is therefore neglected.

As for the surface contribution, it is

$$\frac{1}{2} \int \rho_0^2 \frac{1}{|\mathbf{r} - \mathbf{r}'|} \sum c_{lm} Y_{lm}(\Omega) \sum c_{l'm'} Y_{l'm'}(\Omega') R_0^4 \, d\Omega \, d\Omega'$$

But

$$\frac{1}{|\mathbf{r} - \mathbf{r}'|} = \frac{1}{R_0} \sum P_l(\cos \theta)$$

where θ is the angle between Ω and Ω' and by (3.136)

$$P_l(\cos \theta) = \frac{4\pi}{2l+1} \sum_m Y_{lm}(\Omega) Y_{lm}^*(\Omega')$$

Consequently the surface energy is

$$4\pi\rho_0^2 R_0^3 \sum_{l,m} \frac{1}{2l+1}|c_{lm}|^2$$

which is approximately, to the lowest order in the c's,

$$\frac{9Z^2e^2}{4\pi^2R^3}\sum_{l,m}\frac{1}{2l+1}|c_{lm}|^2$$

Hence the complete Coulomb energy is

$$V_{\text{Coulomb}} = -\frac{3Z^2e^2}{4\pi R^3}\sum(\alpha_{lm}^2+\beta_{lm}^2)\left(1-\frac{3}{2l+1}\right) \qquad (5.505)$$

Once again, we note that there is no $l=1$ contribution.

Collecting the various contributions to the energy, (5.497) and (5.503) and (5.505), we get a Lagrangian

$$L = \frac{3M}{8\pi}\sum\frac{1}{l}(\dot\alpha_{lm}^2+\dot\beta_{lm}^2)+\sigma\sum(\alpha_{lm}^2+\beta_{lm}^2)\left[\frac{l(l+1)}{2}-1\right]$$
$$-\frac{3Z^2e^2}{4\pi R^3}\sum(\alpha_{lm}^2+\beta_{lm}^2)\left(1-\frac{3}{2l+1}\right) \qquad (5.506)$$

Provided that the potential energy terms are positive, we see that the motion is that of periodic surface waves corresponding to $l=2$ and higher, with circular frequencies ω_l given by

$$\omega_l^2 = \frac{8\pi l}{3M}\left\{\sigma\left[\frac{l(l+1)}{2}-1\right]-\frac{3Z^2e^2}{4\pi R^3}\left(1-\frac{3}{2l+1}\right)\right\} \qquad (5.507)$$

If, on the other hand,

$$\frac{3}{5}\frac{Z^2e^2}{R} > 4\pi R^2\sigma$$

there is instability. In the case in which the liquid drop is an atomic nucleus, this instability is related to the phenomenon of nuclear fission; however, the fission problem is in fact greatly complicated by the quantum mechanics of the nuclear particles. Furthermore, of course, one cannot discuss fission in terms of small distortions.

What we have dealt with up to this point is the *motion* of the charges in the drop. These determine the charge and current system which gives rise to the radiation. We now calculate this radiation. For this purpose we note that the current distribution is given by

$$\mathbf{J}(\mathbf{r}) = -\rho_0\nabla\phi S(\mathbf{r}) \qquad (5.508)$$

where S is a function which is 1 inside the drop and zero outside. Now

$$\phi = \sum a_{lm}(t) r^l Y_{lm}(\theta, \varphi)$$

$$= - \sum \frac{1}{l R_0^{l-1}} \dot{c}_{lm} r^l Y_{lm}(\theta, \varphi)$$

$$= i \sum \frac{1}{l R_0^{l-1}} \omega_l c_{lm}(0) r^l Y_{lm}(\theta, \varphi) e^{-i\omega_l t} \qquad (5.509)$$

We have taken advantage of the fact, which is evident from (5.506), that the oscillation frequencies do not depend on the quantity m.

The radiation is now determined by the multipole moments. The electric multipole moment is

$$Q_{lm} = \sqrt{\frac{4\pi}{2l+1}} \frac{i\rho_0}{\omega_l} \int \mathbf{\nabla}' \cdot (S \mathbf{\nabla}' \phi_0) r'^l Y_{lm}(\Omega') d^3 \mathbf{r}' \qquad (5.510)$$

where we have substituted for ρ in terms of $\mathbf{j}$ from the continuity equation and ϕ_0 is ϕ without the time-dependent part. Since $\nabla'^2 \phi_0 = 0$, the surviving term is the one involving $\mathbf{\nabla}'S$. Keeping only the *lowest*-order term in the c's, $\mathbf{\nabla}'S = -\delta(r - R_0)(\mathbf{r}/r)$. Thus

$$Q_{lm} = -\frac{i\rho_0}{\omega_l} \sqrt{\frac{4\pi}{2l+1}} \int \left(\frac{\partial \phi_0}{\partial r}\right)_{R_0} R_0^l Y_{lm}(\Omega') R_0^2 \, d\Omega'$$

$$= \rho_0 \sqrt{\frac{4\pi}{2l+1}} R_0^{l+2} c_{lm}(0) \qquad (5.511)$$

and the intensity of the electric 2^l-pole radiation is

$$\mathscr{P}_{r,lm}^{(l)} = \frac{1}{4\pi\epsilon_0} \frac{(l!)^2 2^{2l}}{(2l+1)!^2} \frac{l+1}{l} 3 Z e^2 \frac{\omega_l^{2l+2}}{c^{2l+1}} R_0^{2l-2} |c_{lm}(0)|^2 \qquad (5.512)$$

Turning to the *magnetic* multipole moment, and using (5.487), we see that we must first calculate

$$\mathbf{\nabla}' \cdot \mathbf{r}' \times (\mathbf{\nabla}' \phi \cdot S) = -\mathbf{r}' \times \mathbf{\nabla}' \cdot (S \mathbf{\nabla}' \phi)$$

We note that there is in this no term of first order in the c's; thus, unless there is a distribution of magnetic moments, there is no magnetic multipole radiation from a charged liquid drop.

This problem was first solved by W. Fierz.[3]

Problem 5-31:   Using the results of Problems 5-18 and 5-19, calculate the contribution to the radiation from the liquid drop if it contains a uniformly distributed magnetic moment, the total moment being $\mathbf{M}$.

---

[3]W. Fierz, *Helv. Phys. Acta* **16**, 365 (1943).

Problem 5-32:   We suggest here a problem involving some similar considerations, but without electromagnetic fields; it is merely a problem in gravitation and hydrodynamics.

Suppose one had a planet of the same size as the earth and made entirely of water, and it vibrated under the influence of its internal gravitational forces. Show that the frequencies of its normal modes of vibration would be

$$\omega_l^2 = \frac{8\pi\rho Gl(l-1)}{3(2l+1)}$$

Describe the oscillation of lowest frequency, and show that its period is a bit more than an hour and a half.

29. Radiation from an Accelerating Point Charge

If a charge e moves along a curve

$$\mathbf{r} = \mathbf{r}_0(t) \tag{5.513}$$

the charge density may be expressed as

$$\rho = e\delta(\mathbf{r} - \mathbf{r}_0(t)) \tag{5.514}$$

and the current as

$$\mathbf{J} = e\dot{\mathbf{r}}_0\delta(\mathbf{r} - \mathbf{r}_0(t)) = e\mathbf{v}_0(t)\delta(\mathbf{r} - \mathbf{r}_0(t)) \tag{5.515}$$

Let us consider this motion in a *dielectric* medium, in which the velocity of light is given by $u^2 = 1/\epsilon\mu$. The particle emits radiation for which the scalar and vector potentials are given by (5.413) and (5.414):

$$\phi = \frac{e}{4\pi\epsilon} \int \frac{\delta[\mathbf{r}' - \mathbf{r}_0(t - (|\mathbf{r} - \mathbf{r}'|)/u)]}{|\mathbf{r} - \mathbf{r}'|} d^3r' \tag{5.516}$$

and

$$\mathbf{A} = \frac{e\mu}{4\pi} \int \mathbf{v}_0\left(t - \frac{|\mathbf{r} - \mathbf{r}'|}{u}\right) \frac{\delta[\mathbf{r}' - \mathbf{r}_0(t - (|\mathbf{r} - \mathbf{r}'|)/u)]}{|\mathbf{r} - \mathbf{r}'|} d^3r' \tag{5.517}$$

This may be put in an instructive form by introducing as alternative coordinates

$$\mathbf{r}'' = \mathbf{r}' - \mathbf{r}_0\left(t - \frac{|\mathbf{r} - \mathbf{r}'|}{u}\right) \tag{5.518}$$

It is straightforward to show that the Jacobian J defined by

$$d^3r'' = J d^3r' \tag{5.519}$$

is given by

$$J = 1 - \frac{1}{uR} \mathbf{R} \cdot \mathbf{v}_0\left(t - \frac{R}{u}\right) \tag{5.520}$$

where

$$R = r - r'$$ (5.521)

It follows that

$$\phi = \frac{e}{4\pi\epsilon} \frac{1}{R - (1/u)\mathbf{R} \cdot \mathbf{v}_0(t - (R/u))}$$ (5.522)

$$\mathbf{A} = \frac{e\mu}{4\pi} \frac{\mathbf{V}_0(t - (R/u))}{R - (1/u)\mathbf{R} \cdot \mathbf{v}_0(t - (R/u))}$$ (5.523)

These are known as the Lienard–Wiechert potentials. The sense in which they are defined as functions of $\mathbf{r}$ and t is the following: $\mathbf{r}'$ is a function of $\mathbf{r}$ and t defined by the equation

$$\mathbf{r}' = \mathbf{r}_0\left(t - \frac{|\mathbf{r} - \mathbf{r}'|}{u}\right)$$ (5.524)

Its position may be seen graphically as follows: Let the curve C in Fig. 5.3 represent the trajectory of the particle, which is traversed in the direction indicated by the arrow. P is the point $\mathbf{r}$. The point $\mathbf{r}'$ is at Q on the trajectory C. Q is the point on the trajectory corresponding to the position of the particle at a time earlier than t by the time required for radiation, with velocity u, to travel from Q to P.

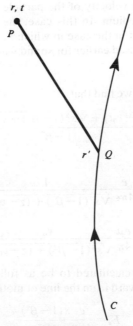

Figure 5.3

The formulas (5.522) and (5.523) for the potentials are very general, but their actual calculation in a given case may be quite involved.

30. Motion in a Straight Line: Čerenkov Radiation

If the particle moves in a straight line with constant velocity, $\mathbf{v}_0$ becomes a constant. The equation for $\mathbf{r}'$ becomes

$$\mathbf{r}' = \mathbf{v}_0 \cdot \left(t - \frac{|\mathbf{r} - \mathbf{r}'|}{u}\right) \tag{5.525}$$

Let us take the Z axis in the direction of motion, and put

$$\beta = \frac{v_0}{u} \tag{5.526}$$

Since

$$\mathbf{R} = \mathbf{r} - \mathbf{v}_0\left(t - \frac{R}{u}\right) \tag{5.527}$$

we can solve for R and $1/u\mathbf{R} \cdot \mathbf{v}_0$ and hence obtain the potential. It is necessary to distinguish two cases (i) $\beta < 1$ and (ii) $\beta > 1$.

In the first case, the velocity of the particle is less than the velocity of propagation in the medium. In this case, the field can be propagated *ahead* of the particle. But in the case in which $v_0 > u$, we meet the sort of "bow wave" effect discussed earlier for sound waves.

(I) $\beta < 1$

Putting $s^2 = r^2 - z^2$, we find that

$$R = \frac{\beta(z - v_0 t) + \sqrt{s^2(1 - \beta^2) + (z - v_0 t)^2}}{1 - \beta^2} \tag{5.528}$$

and the potentials are

$$\phi = \frac{e}{4\pi\epsilon} \frac{1}{\sqrt{s^2(1 - \beta^2) + (z - v_0 t)^2}} \tag{5.529}$$

and

$$\mathbf{A} = \frac{e\mu}{4\pi} \frac{\mathbf{v}_0}{\sqrt{s^2(1 - \beta^2) + (z - v_0 t)^2}} \tag{5.530}$$

The fields may then be calculated to be as follows: the component of electric field radially outward from the line of motion is

$$E_s = \frac{e}{4\pi\epsilon} \frac{s(1 - \beta^2)}{D^3} \tag{5.531}$$

while that in the direction of the line of motion is

$$E_z = \frac{e}{4\pi\epsilon} \frac{(z - v_0 t)(1 - \beta^2)}{D^3} \tag{5.532}$$

The magnetic field is totally transverse (around the direction of motion) and is

$$H_\phi = \frac{ev_0}{4\pi} \frac{s(1 - \beta^2)}{D^3} \tag{5.533}$$

In these formulas

$$D = \sqrt{s^2(1 - \beta^2) + (z - v_0 t)^2} \tag{5.534}$$

There is no radiation from the charge because all components of the Poynting vector drop off as the inverse *fourth* power of distance at large distance; the integral over a distant surface then approaches zero.

(II) $\beta > 1$

We now find that

$$R = \frac{-\sqrt{(z - v_0 t)^2 - s^2(\beta^2 - 1)} - \beta(z - v_0 t)}{\beta^2 - 1} \tag{5.535}$$

There is, of course, only a nonzero solution for the potentials at points $\mathbf{r}'$ satisfying (5.525). This in turn is the case only if the expression for R is positive and real. This is so if $v_0 t - z > 0$ and

$$v_0 t - z \geqslant s\sqrt{\beta^2 - 1} \tag{5.536}$$

Thus, there is radiation only in a cone extending backwards from $v_0 t$ with semivertical angle

$$\alpha = \tan^{-1} \frac{1}{\sqrt{\beta^2 - 1}} = \sin^{-1} \frac{1}{\beta} \tag{5.537}$$

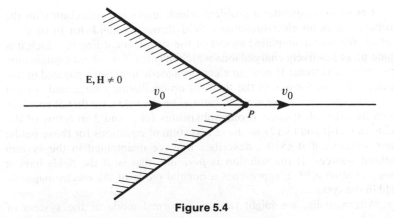

Figure 5.4

The potentials become

$$\phi = \frac{e}{4\pi\epsilon} \frac{1}{\sqrt{(v_0 t - z)^2 - s^2(\beta^2 - 1)}} \qquad (5.538)$$

and

$$\mathbf{A} = \frac{e\mu}{4\pi} v_0 \frac{1}{\sqrt{(v_0 t - z)^2 - s^2(\beta^2 - 1)}} \qquad (5.539)$$

in the cone in question, and zero outside. The potentials, and the fields, become singular on the bounding cone.

The fields are most conveniently expressed in terms of spherical polar coordinates, with origin at the particle; the only nonzero components are the radial component of **E**

$$E_r = -\frac{e}{4\pi\epsilon r^2} \frac{\beta^2 - 1}{(1 - \beta^2 \sin^2\theta)^{3/2}} \qquad (5.540)$$

and the transverse component of **H**

$$H_\varphi = -\frac{e v_0 \sin\theta \,(\beta^2 - 1)}{4\pi r^2 (1 - \beta^2 \sin^2\theta)^{3/2}} \qquad (5.541)$$

These are the potentials of the *Čerenkov radiation*. The situation differs from that in the case $\beta < 1$, in that the fields become large, even at large distances, near the surface of the cone

$$\beta \sin\theta = 1$$

The energy flux is clearly in the θ direction in the wake of the "bow wave."

31. Plasmas in a Magnetic Field

Let us now consider a problem which arises in connection with the propagation of an electromagnetic field through a solid (for instance, a metal). We take a simplified model of the metal, according to which it is made up of positively charged ions which may oscillate about equilibrium, and "free" electrons. If now an electromagnetic field is propagated in this system, the oscillations of the field set up oscillating charge and current distributions. Since these are *determined* by the fields, we do not consider them as external. Rather, if one substitutes for ρ and **J** in terms of the fields in (3.14) and (3.17) we have a system of equations for those fields; their solution, if it exists, describes a wave maintained in the system without sources. If the solution is *periodic*, that is, if the fields have a time variation $e^{-i\omega t}$, it represents a normal mode of the electromagnetic field in the system.

Alternatively, we might talk of a normal mode of the system of charges. If the periodic motion of the charges is first assumed, these produce fields. Those fields in turn determine the motion of the charges; classically, this motion is given by the Lorentz force equation (5.326). If this reproduces the originally assumed *charge* distributions, we will have found normal modes for the *particles* interacting with each other through their electromagnetic fields. If we call this system of charged particles a *plasma*, we shall have found the normal modes of the plasma.

A particularly interesting case is that in which the metal is in a constant external magnetic field. Since this field is constant, it affects the oscillating system only through its effect on the electron motions; that is, it affects the distributions of charge and current.

As far as the ions are concerned, let the dipole moment per unit volume associated with their displacement be $\mathbf{P}$. The potential produced by this polarization is then

$$-\frac{1}{4\pi\epsilon_0}\int \mathbf{P}\cdot\nabla\frac{1}{|\mathbf{r}-\mathbf{r}'|}d^3\mathbf{r}' = \frac{1}{4\pi\epsilon_0}\int \mathbf{P}(\mathbf{r}')\cdot\nabla'\frac{1}{|\mathbf{r}-\mathbf{r}'|}d^3\mathbf{r}'$$

$$= -\frac{1}{4\pi\epsilon_0}\int \nabla'\cdot\mathbf{P}\frac{1}{|\mathbf{r}-\mathbf{r}'|}d^3\mathbf{r}'$$

on integration by parts. But this is the potential of a charge distribution $-\nabla\cdot\mathbf{P}$. We may then write (3.14) as

$$\nabla\cdot(\epsilon_0\mathbf{E}) = \rho - \nabla\cdot\mathbf{P} \tag{5.542}$$

where ρ is now simply the electron density. But if the polarization is proportional to the field producing it, we may put $\mathbf{P} = \chi\mathbf{E}$. Putting $\epsilon_0 + \chi = \epsilon_0\epsilon_L$ to define the specific dielectric constant ϵ_L of the ions (the "lattice") we have

$$\nabla\cdot\epsilon_0\epsilon_L\mathbf{E} = \rho$$

We have shown, then, that the ions of the lattice modify the dielectric constant of the medium through which the electrons interact.

Strictly speaking, the charge distribution ρ should include the static potential of the ions in their undisturbed positions. This, however, clearly makes no contribution to the *oscillating* fields which we are considering.

It is convenient to look for normal modes of a given frequency and wave number; that is, to assume that all quantities depend on space and time coordinates as

$$e^{i(\mathbf{q}\cdot\mathbf{r}-\omega t)}$$

The coefficients will be functions of $\mathbf{q}$ and ω; for example,

$$\mathbf{E} = \mathbf{E_q}(\omega)\, e^{i(\mathbf{q}\cdot\mathbf{r}-\omega t)} \tag{5.543}$$

and similarly for all other quantities appearing in the equations. Maxwell's equations then take the form

$$i\epsilon_0\epsilon_L \mathbf{q} \cdot \mathbf{E_q} = \rho_q \qquad (5.544)$$

$$i\mu_0 \mathbf{q} \cdot \mathbf{H_q} = 0 \qquad (5.545)$$

$$i\mathbf{q} \times \mathbf{E_q} = i\omega\mu_0 \mathbf{H_q} \qquad (5.546)$$

$$i\mathbf{q} \times \mathbf{H_q} = \mathbf{J_q} - i\omega\epsilon_0\epsilon_L \mathbf{E_q} \qquad (5.547)$$

We now assume that the current $\mathbf{J_q}$ depends linearly on the field $\mathbf{E_q}$. However, the relation between them may be, as we see below, a *tensor* one in the presence of an external constant magnetic field. Therefore, we write

$$\mathbf{J_q} = \boldsymbol{\sigma}\mathbf{E_q} \qquad (5.548)$$

in (5.547). This equation then becomes

$$i\mathbf{q} \times \mathbf{H_q} = [-i\omega\epsilon_0\epsilon_L + \boldsymbol{\sigma}]\,\mathbf{E_q} \qquad (5.549)$$

where $\boldsymbol{\sigma}$ is a dyadic.

It is now possible to define the specific dielectric constant of the medium as

$$\epsilon = \epsilon_L + \frac{i\boldsymbol{\sigma}}{\omega\epsilon_0} \qquad (5.550)$$

in terms of which the last of Maxwell's equations (5.547) takes the form

$$i\mathbf{q} \times \mathbf{H_q} = -i\omega\epsilon\epsilon_0 \mathbf{E_q} \qquad (5.551)$$

Similarly, since the equation of continuity

$$\nabla \cdot \mathbf{J} + \frac{\partial\rho}{\partial t} = 0$$

takes the form

$$\mathbf{q} \cdot \mathbf{J_q} = \omega\rho_q \qquad (5.552)$$

the first Maxwell equation (5.544) becomes

$$i\mathbf{q} \cdot \mathbf{D_q} = 0 \qquad (5.553)$$

where

$$\mathbf{D_q} = \epsilon\epsilon_0 \mathbf{E_q} \qquad (5.554)$$

Elimination of $\mathbf{H_q}$ from the last two of Maxwell's equations now gives as the condition for a self-sustaining field in the medium

$$\mathbf{q} \times (\mathbf{q} \times \mathbf{E_q}) + \frac{\omega^2}{c^2}\,\epsilon\,\mathbf{E_q} = 0 \qquad (5.555)$$

The determinant of the coefficient gives the general form of the dispersion relation $\omega = \omega(q)$.

32.   The Dielectric Tensor

Let us suppose that the external magnetic field is in the direction of the unit vector $\mathbf{n}$, and that the dielectric tensor can be expressed in terms of this vector as

$$\epsilon_{ij} = \epsilon_1 \delta_{ij} + (\epsilon_3 - \epsilon_1) n_i n_j + i \epsilon_2 \eta_{ijk} n_k \tag{5.556}$$

η_{ijk} is used to designate the permutation symbol, to distinguish it from the dielectric constants ϵ. Then

$$\mathbf{D} = \epsilon \mathbf{E} = \epsilon_1 \mathbf{E} + (\epsilon_3 - \epsilon_1) \mathbf{n}(\mathbf{n} \cdot \mathbf{E}) + i \epsilon_2 \mathbf{E} \times \mathbf{n} \tag{5.557}$$

Consider now components which are, respectively, right and left transversely polarized and linearly polarized with respect to $\mathbf{n}$, designating them by the subscripts $+, -,$ and 3.
Then

$$D_{\pm} = (\epsilon_1 \mp \epsilon_2) E_{\pm} = \epsilon_{\pm} E_{\pm} \tag{5.558}$$

and

$$D_3 = \epsilon_3 E_3 \tag{5.559}$$

To see how a tensor such as (5.556) may arise, let us calculate the dielectric tensor (or rather, the conductivity tensor, from which it may be derived) for free electrons by means of classical mechanics. Suppose that the position vectors of the electrons are $\mathbf{r}_i(t)$. Then the electron density is

$$\rho = \sum \delta(\mathbf{r} - \mathbf{r}_i(t)) \tag{5.560}$$

and the current is

$$\mathbf{J} = -e \sum_i \mathbf{V}_i(t) \, \delta(\mathbf{r} - \mathbf{r}_i(t)) \tag{5.561}$$

where

$$\mathbf{V}_i = \frac{d\mathbf{r}_i}{dt}$$

Let us expand this current in a Fourier series in a unit volume, the expansion functions being $e^{i\mathbf{q}\cdot\mathbf{r}}$, which satisfy

$$\int e^{-i\mathbf{q}'\cdot\mathbf{r}} e^{i\mathbf{q}\cdot\mathbf{r}} \, d^3\mathbf{r} = \delta_{\mathbf{q}\mathbf{q}'} \tag{5.562}$$

Then we may write, on multiplying through by $e^{-i\mathbf{q}'\cdot\mathbf{r}}$ and integrating

$$\mathbf{J}_{\mathbf{q}'} = \int \mathbf{J}(\mathbf{r}) \, e^{-i\mathbf{q}'\cdot\mathbf{r}} \, d^3\mathbf{r}$$
$$= -e \sum_i \mathbf{V}_i(t) e^{-i\mathbf{q}'\cdot\mathbf{r}_i(t)} \tag{5.563}$$

On differentiating, we obtain

$$\frac{d\mathbf{J}_{\mathbf{q}'}}{dt} = -e \sum_i [\dot{\mathbf{V}}_i - i(\mathbf{q}' \cdot \mathbf{V}_i)\mathbf{V}_i] \, e^{-i\mathbf{q}'\cdot\mathbf{r}_i} \tag{5.564}$$

Making use of the Lorentz equation of motion,

$$\frac{d\mathbf{V}_i}{dt} = -\frac{e}{m}[\mathbf{E}(\mathbf{r}_i) + \mathbf{V}_i \times \mathbf{B}_0] \tag{5.565}$$

Therefore,

$$\frac{d\mathbf{J}_{\mathbf{q}'}}{dt} = \frac{e^2}{m}\sum \mathbf{E}(\mathbf{r}_i)\,e^{-i\mathbf{q}'\cdot\mathbf{r}_i} + \frac{e^2}{m}\sum \mathbf{V}_i \times \mathbf{B}_0\,e^{-i\mathbf{q}'\cdot\mathbf{r}_i} + ie\sum (\mathbf{q}\cdot\mathbf{V}_i)\mathbf{V}_i\,e^{-i\mathbf{q}'\cdot\mathbf{r}_i}$$

Consider the terms on the right one at a time. We calculate the current for a field of the form

$$\mathbf{E} = \mathbf{E}_0\,e^{i(\mathbf{q}\cdot\mathbf{r}-\omega t)} \tag{5.566}$$

Therefore,

$$\sum \mathbf{E}(\mathbf{r}_i)\,e^{-i\mathbf{q}'\cdot\mathbf{r}_i} = \mathbf{E}_0\,e^{-i\omega t}\sum e^{i(\mathbf{q}-\mathbf{q}')\cdot\mathbf{r}_i}$$

It will be shown in Chapter 7 that the value of the sum is much smaller for $\mathbf{q}' \neq \mathbf{q}$ than for $\mathbf{q}' = \mathbf{q}$, the ratio approaching zero as the number of particles approaches infinity. We therefore write the sum approximately as $\delta_{\mathbf{q},\mathbf{q}'}$; this is the classical form of what is known as the "random phase approximation." In this approximation, the only component of current generated by the field (5.566) is $\mathbf{J}_{\mathbf{q}}$.

Writing the second term on the right as $-(eB_0/m)\mathbf{J}_{\mathbf{q}}\times\mathbf{n}$ for the case $\mathbf{q}' = \mathbf{q}$, $\mathbf{n}$ being the unit vector in the direction of $\mathbf{B}_0$, and neglecting the last term, we find that

$$\frac{d\mathbf{J}_{\mathbf{q}}}{dt} = \frac{Ne^2}{m}\mathbf{E}_0\,e^{-i\omega t} - \omega_c\mathbf{J}_{\mathbf{q}}\times\mathbf{n} \tag{5.567}$$

where we have put $\omega_c = eB_0/m$ for the Larmor frequency ("cylotron frequency") of the electrons. If we now put

$$\mathbf{J}_{\mathbf{q}} = \mathbf{J}\,e^{-i\omega t}$$

(5.567) becomes

$$-i\omega\mathbf{J} = \frac{Ne^2}{m}\mathbf{E}_0 - \omega_c\mathbf{J}\times\mathbf{n} \tag{5.568}$$

Taking scalar and vector products with $\mathbf{n}$, and eliminating $\mathbf{J}\cdot\mathbf{n}$ and $\mathbf{J}\times\mathbf{n}$, from the three equations, we may verify that

$$\mathbf{J} = i\frac{Ne^2\omega}{m(\omega^2-\omega_c^2)}\left[\mathbf{E}_0 - \frac{i\omega_c}{\omega}\mathbf{E}_0\times\mathbf{n} - \frac{\omega_c^2}{\omega^2}(\mathbf{E}_0\cdot\mathbf{n})\mathbf{n}\right] \tag{5.569}$$

From this we may calculate the conductivity tensor:

$$\sigma_{ij}(\mathbf{q},\omega) = \frac{iNe^2\omega}{m(\omega^2-\omega_c^2)}\left(\delta_{ij} - i\frac{\omega_c}{\omega}\eta_{ijk}n_k - \frac{\omega_c^2}{\omega^2}n_in_j\right) \tag{5.570}$$

From (5.550) it follows that the dielectric tensor is

$$\epsilon_{ij}(\mathbf{q}, \omega) = \left[\epsilon_L - \frac{Ne^2}{m\epsilon_0} \frac{1}{\omega^2 - \omega_c^2} \right] \delta_{ij} + i \frac{Ne^2}{m\epsilon_0} \frac{\omega_c}{\omega} \frac{1}{\omega^2 - \omega_c^2} \eta_{ijk} n_k$$
$$+ \frac{Ne^2}{m\epsilon_0} \frac{\omega_c^2}{\omega^2} \frac{1}{\omega^2 - \omega_c^2} n_i n_j \tag{5.571}$$

Therefore, in the notation of (5.556),

$$\epsilon_1 = \epsilon_L - \frac{Ne^2}{m\epsilon_0} \frac{1}{\omega^2 - \omega_c^2} \tag{5.572}$$

$$\epsilon_2 = \frac{Ne^2}{m\epsilon_0} \frac{\omega_c}{\omega} \frac{1}{\omega^2 - \omega_c^2} \tag{5.573}$$

and

$$\epsilon_3 = \epsilon_L - \frac{Ne^2}{m\epsilon_0} \frac{1}{\omega^2} \tag{5.574}$$

Before proceeding with the calculation of the normal modes, let us examine the approximation involved in the neglect of the term $ie \Sigma (\mathbf{q} \cdot \mathbf{V}_i)\mathbf{V}_i e^{-i\mathbf{q}' \cdot \mathbf{r}_i}$. In magnitude, the contribution from the ith electron bears to that from the preceding term, $-i\omega_c \mathbf{V}_i \times \mathbf{n} e^{-i\mathbf{q}' \cdot \mathbf{r}_i}$ the ratio $\mathbf{q} \cdot \mathbf{v}_i / \omega_c$. Writing

$$\frac{d\mathbf{J_q}}{dt} = -i\omega \mathbf{J_q} = ie\omega \sum \mathbf{V}_i e^{-i\mathbf{q}' \cdot \mathbf{r}_i}$$

it bears to the ith contribution from *this* term the ratio $\mathbf{q} \cdot \mathbf{V}_i / \omega$. The approximation is then valid provided $\mathbf{q}$ is small enough that $\mathbf{q} \cdot \mathbf{V}_i \ll \omega$ or ω_c. The results which we obtain are then valid only for such values of $\mathbf{q}$.

When there is no magnetic field, it follows from (5.571) that the dielectric tensor is diagonal:

$$\epsilon_{ij} = \left(\epsilon_L - \frac{Ne^2}{m\epsilon_0} \frac{1}{\omega^2} \right) \delta_{ij} \tag{5.575}$$

From (5.555), since in the case of longitudinal modes the first term is zero, the condition that $E_3 \neq 0$ is that $\epsilon_3 = 0$. The frequency is therefore given by

$$\omega^2 = \frac{Ne^2}{m\epsilon_0 \epsilon_L} = \omega_p^2 \tag{5.576}$$

This is the well-known *plasma frequency*. Introducing it into (5.572)–(5.574) they became

$$\epsilon_1 = \epsilon_L \left(1 - \frac{\omega_p^2}{\omega^2 - \omega_c^2} \right) \tag{5.577}$$

$$\epsilon_2 = \epsilon_L \frac{\omega_p^2 \omega_c}{\omega(\omega^2 - \omega_c^2)} \tag{5.578}$$

and

$$\epsilon_3 = \epsilon_L \left(1 - \frac{\omega_p^2}{\omega^2} \right) \tag{5.579}$$

Also,

$$\epsilon_\pm = \epsilon_1 \mp \epsilon_2 = \epsilon_L \left[1 - \frac{\omega_p{}^2}{\omega(\omega \mp \omega_c)} \right] \tag{5.580}$$

Note that, by virtue of (5.544), *longitudinal* waves are associated with a density oscillation, while transverse waves are not.

Let us now look at the normal modes in the case in which the magnetic field $\mathbf{B}_0$ is present. We consider two cases: (*i*) that in which $\mathbf{q}$ is parallel to $\mathbf{B}_0$, (*ii*) that in which $\mathbf{q}$ is *perpendicular* to $\mathbf{B}_0$.

(I) PROPAGATION IN THE DIRECTION OF THE EXTERNAL FIELD $\mathbf{B}_0$.

We start from (5.555). Since $\mathbf{q} = q\mathbf{n}$, the $+$, $-$, and 3-components of that equation are

$$q^2 E_\pm = \frac{\omega^2}{c^2} \epsilon_\pm E_\pm \tag{5.581}$$

and

$$\epsilon_3 E_3 = 0 \tag{5.582}$$

From the latter, we see that the longitudinal mode is exactly as in the case in which the field was absent, that is,

$$\omega = \omega_p.$$

In the transverse case, the dispersion relation is

$$c^2 q^2 = \omega^2 \epsilon_\pm \tag{5.583}$$

for right and left circularly polarized waves, respectively. These are, then, normal modes for which the right-hand side of (5.578) is positive.

Problem 5-33:   Draw a sketch to illustrate the spectrum of normal modes for the case in which $\omega_c > \omega_p$; show that in the case $\omega_c \gg \omega_p$ the normal frequencies are approximately as follows:

(*i*) for right polarization, below ω_c and above $\omega_c(1 + \omega_p{}^2/\omega_c{}^2)$

(*ii*) for left polarization, a band beyond $\omega \approx \omega_p{}^2/\omega_c$.

The situation is illustrated in Fig. 5.5. For the case of right circular polarization, there are normal modes with arbitrarily small frequency. Although the diagram indicates that the frequency spectrum of these modes extends up to ω_c, we must remember that our approximations have limited us to sufficiently small q; as $\omega \to \omega_c$, however, $q \to \infty$. Therefore, these modes cannot be assumed to exist up to the frequency ω_c.

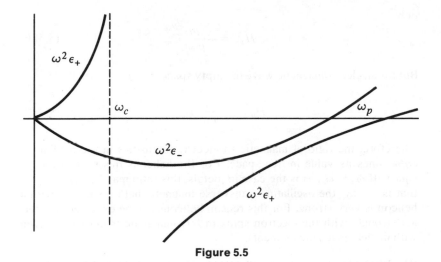

Figure 5.5

For small $\mathbf{q}$, (and consequently small ω)

$$\omega^2\epsilon_+ \approx \epsilon_L \frac{\omega_p^2\omega}{\omega_c} = c^2q^2$$

so that the frequency is given by

$$\omega = \frac{c^2q^2\omega_c}{\epsilon_L\omega_p^2} \qquad (5.584)$$

Substituting the forms of ω_c and ω_p^2, it is found that

$$\omega = c^2q^2\frac{\epsilon_0B_0}{Ne} \qquad (5.585)$$

These transverse, right circularly polarized modes with quadratic dispersion relation are known as *helicons*.

For the case of left circular polarization (l.c.p.), there are no low-frequency modes. In the case of both right and left circular polarization, however, there are higher-frequency modes; for the l.c.p. case their frequency spectrum starts just below ω_p, and for the r.c.p. case, just above.

Let us note the relation between the magnetic and electric fields in these modes. For transverse circularly polarized waves (5.546) gives

$$\pm qE_{\mathbf{q}\pm} = i\omega\mu_0 H_{\mathbf{q}\pm}$$

or

$$H_{q\pm} = \mp \frac{iq}{\mu_0 \omega} E_{q\pm} \tag{5.586}$$

But for an electromagnetic wave in empty space

$$H_{q\pm} = \mp \frac{i}{c\mu_0} E_{q\pm} \tag{5.587}$$

Therefore, the ratio of magnetic to electric intensities in the medium is cq/ω times its value in free space. For the helicons this ratio is $\epsilon_L(\omega_p{}^2/cq\omega_c)$. If $\omega_p \gg \omega_c$, as is the case in metals, this ratio may be very large — that is to say, the oscillating transverse magnetic field associated with a helicon is very strong. For this reason helicons can be expected to interact strongly with the electron spins in ferromagnetic materials, or even with nuclear magnetic moments.

(II) PROPAGATION PERPENDICULAR TO THE EXTERNAL MAGNETIC FIELD B_0.

It is useful in this case to speak of components of fields in the direction of $\mathbf{B}_0$ as *parallel*, and of the other components as *perpendicular*. These are now distinct from *longitudinal* and *transverse* components, contrary to the situation in the previous case. It is convenient, however, to take $\mathbf{B}_0$ still in the 3-direction, and to speak of $+$, $-$, and 3-components. The propagation vector $\mathbf{q}$ may be taken to be in the 1-direction. Then, in (5.555), $\mathbf{q} \times (\mathbf{q} \times \mathbf{E}_q)$ will have components $(0, -q^2 E_2, -q^2 E_3)$ where E_2 and E_3 are the second and third components of $\mathbf{E}_q$. Thus (5.555) gives

$$D_1 = 0 \tag{5.588}$$

$$D_2 = \frac{c^2 q^2}{\omega^2} E_2 \tag{5.589}$$

$$D_3 = \frac{c^2 q^2}{\omega^2} E_3 \tag{5.590}$$

Since $D_3 = \epsilon_3 E_3$, the last equation leads to the relation

$$\epsilon_3 = \frac{c^2 q^2}{\omega^2}$$

Using the form (5.577) for ϵ_3, we may show that

$$\omega^2 = \omega_p{}^2 + c^2 q^2 \tag{5.591}$$

This mode, which is related to the plasma mode in the case of parallel propagation, is now transverse rather than longitudinal, and is no longer associated with a density fluctuation. It is linearly polarized in the direction of the magnetic field.

Equations (5.588) and (5.589) may be written

$$\epsilon_1 E_1 + i\epsilon_2 E_2 = 0,$$

$$\epsilon_1 E_2 - i\epsilon_2 E_1 = \frac{c^2 q^2}{\omega^2} E_2$$

There are nonzero solutions if the determinant of the coefficients is zero; this leads to the equation

$$c^2 q^2 = \frac{\epsilon_+ \epsilon_-}{\epsilon_1} = \frac{\epsilon_+ \epsilon_-}{\frac{1}{2}(\epsilon_+ + \epsilon_-)} \qquad (5.592)$$

There are normal modes at those frequencies for which one or all of ϵ_+, ϵ_-, and ϵ_1, are positive. For small frequencies, only ϵ_- is negative; beyond ω_c (supposed $\ll \omega_p$) all three are. The first to become positive is ϵ_-, so the spectrum begins at the point $\epsilon_- = 0$, that is,

$$\omega = \sqrt{\omega_p{}^2 + \tfrac{1}{2}\omega_c{}^2} - \tfrac{1}{2}\omega_c$$

$$\approx \omega_p - \tfrac{1}{2}\omega_c \qquad (5.593)$$

Since at this point $q = 0$, $D_\pm = 0$. Consequently, only E_- can be $\neq 0$. This band of allowed frequencies terminates when $\epsilon_1 = 0$, that is, when

$$\omega^2 = \omega_p{}^2 + \omega_c{}^2 \qquad (5.594)$$

At this value $E_2 = 0$, and we have a *longitudinal* mode, and corresponding density fluctuations. In fact, throughout the whole band such fluctuations will be present, since there is always a longitudinal component.

There is a further band of normal modes at the frequency for which

$$\epsilon_+ = 0:$$

$$\omega = \sqrt{\omega_p{}^2 + \tfrac{1}{4}\omega_c{}^2} + \tfrac{1}{2}\omega_c$$

$$\approx \omega_p + \tfrac{1}{2}\omega_c \qquad (5.595)$$

At this frequency the field is of the form E_+. Going toward larger frequencies, $\epsilon_+ \to \epsilon_-$ and $\epsilon_2 \to 0$. Thus, the wave becomes primarily transversely polarized in the direction perpendicular to the magnetic field.

33. Excitation of Helicons by a Plane Wave Normally Incident on an Interface

Consider the problem of a plane wave in empty space, impinging normally on a material which may be treated as a free electron gas. If the direction of propagation is the 3-direction, right and left circularly polarized waves, respectively of frequency ω are characterized by the fields

$$\mathbf{E}_\pm = \mathbf{E}'_\pm \, e^{i\omega((z/c)-t)} + \mathbf{E}''_\pm \, e^{-i\omega((z/c)+t)} \qquad (5.596)$$

and, from (5.584)

$$\mathbf{H}_+ = \mp \frac{i}{\mu_0 c} \mathbf{E}'_\pm \, e^{i\omega((z/c)-t)} \pm \frac{i}{\mu_0 c} \mathbf{E}''_\pm \, e^{-i\omega((z/c)+t)} \qquad (5.597)$$

The first terms correspond in each case to the incident wave; the second, formed by changing the sign of q, correspond to the reflected wave.

In the medium, we have

$$\mathbf{E}_\pm = \mathbf{E}'''_\pm \, e^{i(qz-\omega t)} \qquad (5.598)$$

$$\mathbf{H}_\pm = \mp \frac{iq}{\mu_0 \omega} \mathbf{E}'''_\pm \, e^{i(qz-\omega t)} \qquad (5.599)$$

where q is given in terms of ω by (5.581). Letting the interface be at $z = 0$, and applying the boundary conditions of continuity of $\mathbf{E}$ and $\mathbf{H}$,

$$\mathbf{E}'_\pm + \mathbf{E}''_\pm = \mathbf{E}'''_\pm \qquad (5.600)$$

and

$$\mathbf{E}'_\pm - \mathbf{E}''_\pm = \frac{cq}{\omega} \mathbf{E}'''_\pm \qquad (5.601)$$

Solving for the transmitted wave in terms of the incident

$$\mathbf{E}'''_\pm = \frac{2}{1 + (cq/\omega)} \mathbf{E}'_\pm \qquad (5.602)$$

For the corresponding magnetic fields

$$\mathbf{H}'''_\pm = \frac{2cq/\omega}{1 + cq/\omega} \mathbf{H}'_\pm \qquad (5.603)$$

Let us consider a frequency within the allowed band of normal modes. The electric and magnetic fields are then $\pi/2$ out of phase; the Poynting vector is, therefore, simply their product. Thus the ratio

$$\frac{\text{Transmitted intensity}}{\text{Incident intensity}} = \frac{4cq/\omega}{1 + (cq/\omega)^2} = \frac{4\omega cq}{\omega^2 + c^2 q^2} \qquad (5.604)$$

This has its maximum value for $cq = \omega$, i.e.,

$$\omega^2 \mp \omega\omega_c - \frac{\epsilon_L}{\epsilon_L - 1}\, \omega_p{}^2 = 0 \qquad (5.605)$$

For low frequencies, such transmission is only possible for *right* circularly polarized waves.

When we are *not* within the bands of normal modes, q determined by (5.582) is imaginary, and the wave is exponentially attenuated within the material. The electric and magnetic fields within the material are in phase, and there is no energy propagation.

We note that when the magnetic field goes to zero $\epsilon_\pm$ are *negative* from zero frequency up to $\omega = \omega_p$, and no propagation is possible. We see, then, that the propagation of the low-frequency helicons is possible only due to the presence of the magnetic field. The reason for this is physically clear enough: In a magnetic field, the electrons tend to move at right angles to the electric field of the wave. Thus, this field does no work on the electrons and there is no dissipation.

In fact, in all real materials the situation is complicated by the fact that the electrons are scattered and thus acquire a component in the direction of the field, with the result that there is some attenuation. This is not important if the characteristic period of the electrons in the magnetic field is much shorter than the mean time between scattering, that is, if $\omega_c\tau \gg 1$. When $\omega_c\tau \leq 1$, however, there will be substantial attenuation of the helicon waves.

Problem 5-34: Show that the *transmission* of a right circularly polarized wave (helicon) of frequency $\omega \ll \omega_c$ through a slab of material of thickness d is

$$T = \frac{4\omega^2 c^2 q^2}{4\omega^2 c^2 q^2 \cos^2 qd + (q^2 c^2 + \omega^2)^2 \sin^2 qd}$$

Sketch the form of the transmission as a function of the constant external field.

Problem 5-35: Calculate the transmission of a wave similar to that in the previous problem, but with the magnetic field *parallel* to the surface. Show that there is an oscillating surface charge induced on the surface.

Problem 5-36: Recalculate the dielectric constants when there are N_1 carriers having charge $-e$ and mass m_1 and N_2 having charge $+e$ and mass m_2. Show that, if $N_1 = N_2$, the dispersion relation for the waves in the material is linear ($\omega = \text{constant} \times q$) rather than quadratic. (These are known as "Alfven waves.")

Problem 5-37:   Calculate, for the case in which the magnetic field $\mathbf{B}_0$ is parallel to the plane surface of a very thick slab of material, the reflectivity of a wave linearly polarized perpendicular to the direction of $\mathbf{B}_0$, and having a circular frequency between

$$\omega = \sqrt{\omega_p{}^2 + \tfrac{1}{4}\omega_c{}^2} - \tfrac{1}{2}\omega_c$$

and

$$\omega = \sqrt{\omega_p{}^2 + \tfrac{1}{4}\omega_c{}^2} + \tfrac{1}{2}\omega_c$$

Sketch your result.

APPENDIX 5A: Approximate Formulas for Spherical Bessel Functions of Large Order

(i).   Method of Steepest Descents

We shall first derive formulas for the spherical Hankel functions, and deduce ones for the Bessel and Neumann functions.

The method used to derive these formulas will be the "method of steepest descents." We first sketch briefly the main features of this method.

We consider integrals of the form

$$I = \int_c \exp f(z)\, dz \tag{A.1}$$

in the complex plane, "C" designating a contour. We note first that the integrand is largest where $\operatorname{Re} f(z)$ is a maximum.

Because of the exponential dependence, it might be hoped that the greater part of the integral would come from the neighborhood of this maximum. However, this may not be a useful observation for two reasons:

(1) $\operatorname{Re} f(z)$ may not fall off rapidly enough from the point at which it is maximum, or

(2) The imaginary part of $f(z)$ may be rapidly varying, causing rapid oscillations in the integrand which may diminish its value near the point in question.

We may try to improve the situation by replacing the given contour by an alternative one for which these two difficulties do not arise. We may, in fact, choose any alternative contour C^1 such that there are no poles of the integrand between the original contour and the new one (and, of course, the new contour may not cross a branch cut). How, then, should we select this new contour? Clearly, we should choose it, if possible, so that near the maximum of $\operatorname{Re} f(z)$ that quantity decreases as rapidly as possible, while at the same time $\operatorname{Im} f(z)$ varies as little as possible. Fortunately, it is possible to satisfy these conditions simultaneously, by virtue of the following theorems in the theory of functions of a complex variable:

(a) If $f(z) = u(x, y) + iv(x, y)$, the curves $u = $ constant and $v = $ constant are orthogonal at any point, u varying most rapidly along $v = $ constant and vice versa.

(b) If $f(z)$ is analytic on and within a circle of radius R about a point z_0, the average value of the function around the circle is equal to its value at z_0, that is,

$$f(z_0) = \frac{1}{2\pi} \int_0^{2\pi} f(z_0 + Re^{i\theta})\, d\theta \qquad \text{(Ref. 4)} \qquad \text{(A.2)}$$

The second theorem clearly holds for the real and imaginary parts of $f(z)$ separately, and states that neither can have a maximum or a minimum at any point at which $f(z)$ is analytic.

Suppose now we look for points at which the derivative $f'(z)$ is zero. Then, both u and v will have extrema at that point. However, by the second theorem above, this extremum can be neither a maximum nor a minimum relative to all neighboring points. It must, in fact, then, be a "saddle point," such that the function increases for some displacements from the extremum, and decreases for others.

Suppose, now, that we choose our contour through such a point. Let us choose the contour to pass through it along that path on which $u = \operatorname{Re} f(z)$ rises most steeply to a maximum. Along this curve $v = \operatorname{Im} f(z)$ is constant.

If, now, we expand $f(z)$ about the saddle point z_0 along this path,

$$f(z) = f(z_0) + \tfrac{1}{2}(z - z_0)^2 f''(z_0) + \cdots \qquad \text{(A.3)}$$

[where, because $\operatorname{Re} f(z)$ is a *maximum* at $z = z_0$, $\operatorname{Re} f''(z_0) < 0$], we may write the contribution to the integral from the neighborhood of this point approximately as

$$I = e^{f(z_0)} \int e^{-(1/2)(z-z_0)^2[-f''(z_0)]}\, dz \qquad \text{(A.4)}$$

Provided the expansion to the quadratic term in (A.4) provides a good approximation to $f(z)$ over a distance along the curve a few times greater than $1/\sqrt{-f''(z_0)}$, (A.4) will provide a good approximation to the integral I.

---

[4] For a discussion of these theorems, see, for example, Maurice Heins, *Topics in Complex Function Theory*. New York: Holt, Rinehart and Winston, 1962, chapter 4, pp. 55–57.

The above has assumed that the deformed contour may only go through *one* saddle point. If there are several, the contour may perhaps be chosen in alternative ways; one must remember always, however, that the contribution from the part of the contour away from the saddle point must be negligible. A contour passing through more than one saddle point is also possible, the contributions being added.

If the subsequent terms in the expansion, the first of which is $\frac{1}{6}(z-z_0)^3 f'''(z_0)$, are negligible within the range $1/\sqrt{-f''(z_0)}$, we can with negligible error replace the limits on the integral (A.4) by $-\infty$ and ∞ along the line in the complex plane corresponding to the direction of the contour at z_0. Then, the integral may be evaluated to give, approximately,

$$I = e^{f(z_0)} \sqrt{\frac{2\pi}{-f''(z_0)}} \tag{A.5}$$

The condition that this be a good approximation may, for practical purposes, be formulated as

$$\frac{f'''(z_0)}{[-f''(z_0)]^{3/2}} \ll 1 \tag{A.6}$$

[It should be noted, incidentally, that since on the contour $\operatorname{Im} f(z) = 0$, $-f''(z_0) = -\operatorname{Re} f''(z_0) = -u''(z_0)$.]

(ii).   Example: Asymptotic Expansion of the Γ Function

The Γ function is defined by

$$\Gamma(n) = \int_0^\infty x^{n-1} e^{-x} \, dx \tag{A.7}$$

Replacing x by the complex variable z,

$$f(z) = (n-1)\ln z - z \tag{A.8}$$

Then

$$f'(z) = \frac{n-1}{z} - 1 \tag{A.9}$$

which is equal to zero at

$$z = z_0 = n - 1 \tag{A.10}$$

Also

$$f''(z_0) = -\frac{1}{n-1} \tag{A.11}$$

$$f'''(z_0) = \frac{2}{(n-1)^2} \tag{A.12}$$

$$f^{(\text{iv})}(z_0) = -\frac{6}{(n-1)^3} \tag{A.13}$$

Thus, from (A.5)

$$\Gamma(n) \simeq \exp\left[(n-1)\ln(n-1) - (n-1)\right]\sqrt{2\pi(n-1)}$$
$$= \sqrt{2\pi(n-1)}\,(n-1)^{n-1}\,e^{-(n-1)} \tag{A.14}$$

or

$$\ln\Gamma(n+1) = (n+\tfrac{1}{2})\ln n - n + \tfrac{1}{2}\ln 2\pi \tag{A.15}$$

which is the well-known "Stirling's formula." We have, in this case, not had to deform the contour at all, since z_0, being real, already lies on the contour of integration.

(iii). The Spherical Hankel Functions

It is most convenient to work with the Hankel functions, because by taking real and imaginary parts we can extract both the spherical Bessel and Neumann functions. From (4.367) and (4.376) we see that $h_l^{(1)}(x)$ may be expressed as

$$h_l^{(1)}(x) = -i\left(\frac{x}{2}\right)^l \frac{1}{l!}\, e^{ix} \int_0^\infty e^{-\sigma x}\sigma^l(\sigma-2i)^l\, d\sigma \tag{A.16}$$

Let us concern ourselves for the present only with the integral. It may be written

$$I = \int_0^\infty e^{f(\sigma)}\, d\sigma$$

where

$$f(\sigma) = -\sigma x + l\ln\sigma + l\ln(\sigma-2i) \tag{A.17}$$

Thus

$$f'(\sigma) = -x + \frac{l}{\sigma} + \frac{l}{\sigma-2i} \tag{A.18}$$

and

$$f''(\sigma) = -\frac{l}{\sigma^2} - \frac{l}{(\sigma-2i)^2} \tag{A.19}$$

The "saddle points" are given by $f'(\sigma) = 0$ or

$$x\sigma^2 - 2(l+ix)\sigma + 2li = 0 \tag{A.20}$$

The solutions are

$$\sigma_\pm = \frac{l + ix \pm \sqrt{l^2 - x^2}}{x} \tag{A.21}$$

Let us consider first the case $l > x$. It is convenient to write

$$l = x \cosh u \tag{A.22}$$

the "saddle points" are then given by

$$\sigma_\pm = i + e^{\pm u} \tag{A.23}$$

It is easily verified that

$$f(\sigma_\pm) = -ix - xe^{\pm u} + l \ln (e^{\pm 2u} + 1) \tag{A.24}$$

and

$$f''(\sigma_\pm) = \mp le^{\mp u} \frac{\sinh u}{\cosh^2 u} \tag{A.25}$$

How do we choose a deformed contour so as to get the best possible approximation to the function from the integrals near the "saddle points"? Consider first σ_-. At that point f'' is positive and real; the function is given approximately by

$$f(\sigma) = f(\sigma_-) + \tfrac{1}{2}(\sigma - \sigma_-)^2 f''(\sigma_-) \tag{A.26}$$

If σ goes through σ_- following a path parallel to the real axis, it passes through a *minimum* at σ_-. On the other hand, if the path is parallel to the imaginary axis, it goes through a *maximum* at σ_-, and the imaginary part of f is constant along that path. Thus, we want to pass through σ_- along a course parallel to the *imaginary* axis.

Next consider σ_+. Since $f''(\sigma_+)$ is negative, the proper path through this point is parallel to the real axis, Figure A.1 gives a sketch of a possible contour which passes through both saddle points.

The contribution to the integral from σ_- is

$$\frac{i}{2} e^{-ix} e^{-xe^{-u}} e^{-lu} (2 \cosh u)^{l+1} \sqrt{\frac{2\pi}{le^u \sinh u}}$$

The contribution to $h_l^{(1)}$ is then

$$\frac{l^{l+1}}{l!} \frac{1}{x} e^{-xe^{-u}} e^{-lu} \sqrt{\frac{2\pi}{le^u \sinh u}} \tag{A.27}$$

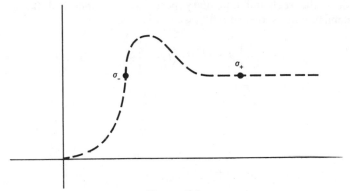

Figure A.1

For small x this should, of course, go like x^l. But if $x \ll l$,

$$\cosh u \approx \sinh u \approx \tfrac{1}{2} e^u \approx \frac{l}{x} \tag{A.28}$$

so that (A.27) becomes

$$\sqrt{\frac{\pi}{l}} \left(\frac{x}{2}\right)^l \frac{1}{l!} \, e^{-x^2/2l} \tag{A.29}$$

Let us turn now to the contribution to the integral from σ_+. It is

$$\tfrac{1}{2} e^{-ix} e^{-xe^u} e^{lu} (2 \cosh u)^{l+1} \sqrt{\frac{2\pi}{le^{-u} \sinh u}}$$

This therefore contributes to $h_l^{(1)}$ the term

$$-\frac{i}{x} \frac{l^{l+1}}{l!} \, e^{-xe^u} e^{lu} \sqrt{\frac{2\pi}{le^{-u} \sinh u}} \tag{A.30}$$

Let us see how this behaves for $x \ll l$. Using (A.28) we have

$$-i \left(\frac{2l}{x}\right)^{l+1} \frac{e^{-2l}}{l!} \sqrt{\frac{\pi}{l}} \tag{A.31}$$

Combining (A.27) and (A.30) or, for $x \ll l$, (A.29) and (A.31), we have approximately, for $x < l$,

$$h_l^{(1)}(x) \doteq \frac{l^{l+1}}{l!} \frac{1}{x} \, e^{-xe^{-u}} e^{-lu} \sqrt{\frac{2\pi}{le^u \sinh u}}$$

$$-\frac{i}{x} \frac{l^{l+1}}{l!} \, e^{-xe^u} e^{lu} \sqrt{\frac{2\pi}{le^{-u} \sinh u}} \tag{A.32}$$

$$\approx \sqrt{\frac{\pi}{l}} \left(\frac{x}{2}\right)^l \frac{1}{l!} \, e^{-x^2/2l} - i \left(\frac{2l}{x}\right)^{l+1} \frac{e^{-2l}}{l!} \sqrt{\frac{\pi}{l}} \tag{A.33}$$

But since the real and imaginary parts are the spherical Bessel and Neumann functions, respectively, we have

$$j_l(x) \doteq \frac{l^{l+1}}{l!} \frac{1}{x} e^{-xe^{-u}} e^{-lu} \sqrt{\frac{2\pi}{le^u \sinh u}}, \qquad (x < l) \tag{A.34}$$

$$\approx \sqrt{\frac{\pi}{l}} \frac{1}{l!} \left(\frac{x}{2}\right)^l e^{-x^2/2l}, \qquad (x \ll l) \tag{A.35}$$

and

$$n_l(x) \doteq -\frac{l^{l+1}}{l!} \frac{1}{x} e^{-xe^u} e^{lu} \sqrt{\frac{2\pi}{le^{-u} \sinh u}}, \qquad (x < l) \tag{A.36}$$

$$\approx -\frac{e^{-2l}}{l!} \left(\frac{2}{x}\right)^{l+1} \sqrt{\frac{\pi}{l}} l^{2l+1}, \qquad (x \ll l) \tag{A.37}$$

We see, then, that the major contribution to the *real* part of h_l comes from the neighborhood of σ_-, and to the *imaginary* part from σ_+. Provided x is not too close to l, these two points remain sufficiently well separated that the chosen contour can be used. As $x \to l$, the two singularities approach each other, and the method runs into rather serious difficulties. At $x = l$, not only is $f'(\sigma)$ zero at $\sigma = 1+i$, but $f''(\sigma)$ is also, and the leading term in the expansion of $f(\sigma)$ is a cubic one. Thus, for x near l, it is essential to take account of the cubic terms. The situation is then more difficult, but formulas for large l may still be obtained. They are, however, not of sufficient interest in the problems we are considering to be included here.

It is perhaps logical to write the formulas for j_l and n_l only in terms of u and l. Using (A.22) to eliminate x, we find that

$$j_l(x) \simeq \sqrt{\frac{2\pi}{le^u \sinh u}} e^{-l} \frac{l^{l+1}}{l!} \frac{\cosh u}{l} e^{-l(u-\tanh u)} \tag{A.38}$$

and

$$n_l(x) \simeq -\sqrt{\frac{2\pi}{le^{-u} \sinh u}} e^{-l} \frac{l^{l+1}}{l!} \frac{\cosh u}{l} e^{l(u-\tanh u)} \tag{A.39}$$

Using (A.15) for the factorials of large integers, these become

$$j_l(x) \simeq \frac{\cosh u}{l\sqrt{e^u \sinh u}} e^{-l(u-\tanh u)} \tag{A.40}$$

and

$$n_l(x) \simeq -\frac{\cosh u}{l\sqrt{e^{-u} \sinh u}} e^{l(u-\tanh u)} \tag{A.41}$$

To see how rapidly j_l decreases as x decreases below $x = l$, put $\tanh u = y$. Then

$$l(u - \tanh u) = l(\tanh^{-1} y - y)$$

$$\doteq \tfrac{1}{3} l y^3$$

for y not too large.

Putting $x = l - x'$, where $x' \ll l$,

$$y = \tanh u$$

$$= \sqrt{1 - \frac{x^2}{l^2}}$$

$$\doteq \left(2\frac{x'}{l}\right)^{1/2}$$

Therefore, approximately

$$(u - \tanh u) = \frac{1}{3} l \left(\frac{2x'}{l}\right)^{3/2} = \frac{2^{3/2}}{3} \sqrt{\frac{x'^3}{l}}$$

Thus, for the exponent to be equal to γ, we must have

$$x' = \left(\frac{9}{8}\gamma^2\right)^{1/3} l^{1/3}$$

What are the conditions that these formulas yield reasonably good approximation? Basically, it is that f'' be large near the saddle points, since then the integrand drops off rapidly on each side of the pass. We note that

$$\cosh u = \frac{l}{x}$$

$$\sinh u = \sqrt{\left(\frac{l}{x}\right)^2 - 1}$$

$$e^u = \left[\frac{l}{x} + \sqrt{\left(\frac{l}{x}\right)^2 - 1}\right]$$

$$e^{-u} = \left[\frac{l}{x} - \sqrt{\left(\frac{l}{x}\right)^2 - 1}\right]$$

Therefore, the approximation to j_l depends on $\sqrt{l^2 - x^2}[1 + \sqrt{1 - (x^2/l^2)}]$ being large, while that to n_l depends on $\sqrt{l^2 - x^2}[1 - \sqrt{1 - (x^2/l^2)}]$ being large. Since these quantities are, in fact, exponents in Gaussian functions in the integrand, the values do not in fact have to be very large. The approximation to n_l appears to have a considerably narrower range of validity than has that to j_l. Although it appears from the above considerations that (A.37) may not be reliable, it does in fact give the correct behavior for small x. In fact, using Stirling's formula for large l, it becomes

$$n_l(x) = -\frac{1}{x^{l+1}} \frac{(2l)!}{2^l l!} \tag{A.42}$$

and this is precisely what is obtained from the series expansion [see (4.377)].

We have pointed out that when x is in the neighborhood of l, the two "saddle points" are close to each other, and the situation becomes complicated. However, once x becomes substantially larger than l, the method of steepest descents again gives reasonable results. We start again from (A.21) but now put

$$l = x \cos v \tag{A.43}$$

In terms of v the saddle points are then

$$\sigma_{\pm} = i + e^{\pm iv} \tag{A.44}$$

It is also easily verified that

$$f(\sigma_{\pm}) = -ix - x\, e^{\pm iv} \pm ilv + l \ln{(2\cos v)} \tag{A.45}$$

and

$$f''(\sigma_{\pm}) = \mp il\, e^{\mp iv} \frac{\sin v}{\cos^2 v} \tag{A.46}$$

How, then, in *this* case, do we define the deformed contours? They should pass through the points (A.44) in the directions along which the imaginary part of $f(\sigma)$ is constant. Now the poles are located as indicated in Fig. A.2.

To see in what directions the contours should pass through these points, we put $z = \sigma - \sigma_{\pm} = |z| e^{i\delta_{\pm}}$. The dependence of f on z near this point is then given by $\frac{1}{2}|z|^2 e^{2i\delta_{\pm}} f''(\sigma_{\pm})$. In order that the imaginary part should not depend on z and for the coefficient of $|z|^2$ to be negative, we must take the total phase to be π or $-\pi$. Therefore let us take

$$\mp \frac{\pi}{2} \mp v + 2\delta_{\pm} = \mp \pi$$

so that $\delta_{\pm} = \pm (v/2) \mp (\pi/4)$. Since we may take (for positive x) $v \lesssim (\pi/2)$.

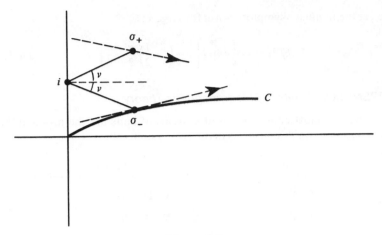

Figure A.2

the directions which must be traversed (or their opposites) are as shown in the figure; the angles which the curves make with the horizontal are $(\pi/4) - (v/2)$.

A simplest way in which to get an asymptotic evaluation is to choose the contour C through σ_-. The value of the integral is, then,

$$I = \sqrt{\frac{2\pi}{l \sin v}}\, 2^l \cos^{l+1} v \exp\left[-ix - x\, e^{-iv} - ilv\right] e^{i((\pi/4)-(v/2))} \quad (A.47)$$

It follows from (A.16) that

$$h_l^{(1)}(x) = -i\left(\frac{x}{2}\right)^l \frac{1}{l!}\, e^{ix} \sqrt{\frac{2\pi}{l \sin v}}\, 2^l \cos^{l+1} v \exp$$

$$\left\{-ix - x\, e^{-iv} - i\left[\left(l+\frac{1}{2}\right)v - \frac{\pi}{4}\right]\right\} \quad (A.48)$$

Using Stirling's formula for $l!$, substituting $x = l \sec v$, and simplifying we obtain

$$h_l^{(1)}(x) \approx \frac{\cos v}{l\sqrt{\sin v}} \exp i\left[l \tan v - \left(l+\frac{1}{2}\right)v - \frac{\pi}{4}\right] \quad (A.49)$$

Real and imaginary parts may now be identified to give $j_l(x)$ and $n_l(x)$. If

$$\tan v \approx \frac{x}{l}, \quad v \approx \frac{\pi}{2}, \quad \sin v \approx 1$$

we get the familiar asymptotic form for large x:

$$h_l^{(1)}(x) = \frac{1}{x} \exp i \left[x - \left(l + \frac{1}{2} \right) \frac{\pi}{2} - \frac{\pi}{4} \right] \tag{A.50}$$

APPENDIX 5B: Units in Electromagnetic Theory

The two most commonly used systems of units are the mks and the Gaussian. Maxwell's equations in the two systems are

mks	Gaussian
$\nabla \cdot \mathbf{D} = \rho$	$\nabla \cdot \mathbf{D} = 4\pi\rho$
$\nabla \cdot \mathbf{B} = 0$	$\nabla \cdot \mathbf{B} = 0$
$\nabla \times \mathbf{E} = -\dfrac{\partial \mathbf{B}}{\partial t}$	$\nabla \times \mathbf{E} = -\dfrac{1}{c}\dfrac{\partial \mathbf{B}}{\partial t}$
$\nabla \times \mathbf{H} = \mathbf{J} + \dfrac{\partial \mathbf{D}}{\partial t}$	$\nabla \times \mathbf{H} = \dfrac{4\pi}{c}\mathbf{J} + \dfrac{1}{c}\dfrac{\partial \mathbf{D}}{\partial t}$
$\mathbf{D} = \epsilon\epsilon_0 \mathbf{E}$	$\mathbf{D} = \epsilon \mathbf{E}$
$\mathbf{B} = \mu\mu_0 \mathbf{H}$	$\mathbf{B} = \mu \mathbf{H}$

Lorentz's equations of motion for a charged particle in an electromagnetic field are

$$m\frac{d\mathbf{V}}{dt} = e(\mathbf{E} + \mathbf{V} \times \mathbf{B}), \qquad m\frac{d\mathbf{V}}{dt} = e\left(\mathbf{E} + \frac{1}{c}\mathbf{V} \times \mathbf{B}\right)$$

Physical quantity	mks unit	Gaussian unit	mks unit ÷ Gaussian unit
length	meter	centimetre	100
mass	kilogram	gram	100
time	second	second	1
force	newton	dyne	10^5
energy	joule	erg	10^7
charge	coulomb	statcoulomb	3×10^9
charge density	coulomb/meter^3	statcoulomb/cm^3	3×10^3
current	ampere	statampere	3×10^9
current density	ampere/meter^2	statampere/cm^2	3×10^5

Physical quantity	mks unit	Gaussian unit	mks unit ÷ Gaussian unit
electric field	volt/meter	statvolt/cm	$\frac{1}{3} \times 10^{-4}$
potential	volt	statvolt	$\frac{1}{300}$
conductivity	mho/meter	$\sec^{-1}$	9×10^{9}
magnetic induction	weber/meter^2	gauss	10^{4}
magnetic field	ampere-turn/ meter	oersted	$4\pi \times 10^{-3}$

PRELUDE TO CHAPTER 6

This chapter is concerned with two different, but often related, types of problem. One concerns the diffusion equation, which is linear in the time derivative and so describes irreversible processes. The heat conduction equation has the same form. The second is the equation which appears in the theory of diffusion of thermal neutrons in reactors, of dissipative media and skin effect in electromagnetic theory, and of superconducting electrodynamics. These latter phenomena are all characterized by exponential decay of the relevant physical quantity with some coordinate.

The chapter opens with a discussion of time-irreversible phenomena, illustrated with the linear differential equation of second order. The diffusion equation is then derived from the continuity equation and Fick's law, which states that the diffusion current is proportional to the density gradient. The heat conduction equation is derived in a similar way, and leads to the same kind of differential equation. Boundary conditions are discussed for each problem.

The Green's function for an infinite medium is derived by transform methods, as was done in the case of the wave equation. The Green's functions for semi-infinite media with various boundary conditions are derived by the method of images. The same method is used for a slab (finite one-dimensional medium).

The solution of the latter problem is also obtained by Fourier series. This solution can be obtained from the image one, or conversely, by the use of Poisson's summation formula (Chapter 4).

The Green's function is calculated for spherical geometry, and the relation to the general formulas relating point, plane, and spherical shell Green's functions, introduced in Chapter 3, is discussed, with particular reference to the role of boundary conditions.

The next problem treated is analogous to one discussed in the preceding chapter for waves — that of the solution determined by given initial and boundary conditions. Again, it is shown that each condition effectively provides a distribution of *sources*, described by integrals over Green's functions; once again, the Green's functions are seen to be the same as those for real sources.

It is then noted that the time-dependent Schrödinger wave equation differs from the heat conduction equation only in that its time derivative term has an imaginary rather than a real coefficient. Thus, the Green's function for this problem may be derived in a manner quite parallel to that developed for the heat conduction equation.

We now turn to the problem of neutron diffusion in chain reactors. A simple standard model is used— high-energy neutrons are assumed to slow down to thermal equilibrium with their surroundings, to diffuse in this state, and then to be captured; they may or may not then give rise to fission which produces new high-energy neutrons. (But energetic and thermal neutrons may of course be lost by escape.) The model is discussed by treating the coupled equations of "fast" and "thermal" neutrons. The criticality conditions are derived in general, and the influence of fission neutrons emitted after a delay on the rate of change of neutron intensity is considered.

Fourier transform theory is next used to introduce the notions of diffusion length of thermal neutrons (root mean square distance travelled during diffusion), slowing-down length (root mean square distance travelled during the moderation of slowing-down process), and "migration length" (rms distance traveled from birth to death). The influence of fission on these concepts is also determined.

We then turn to the propagation of electromagnetic waves in a dissipative medium. The equation here is both wavelike and diffusionlike, containing both first- and second-order derivatives in time. The appropriate Green's function is derived. The problem of propagation into a conducting medium is also discussed; it is shown that there is exponential decay of the fields into the medium, characterized by the so-called "skin depth," which depends on the frequency of the wave. The skin effect phenomenon is discussed in general by means of an integral equation derived by considering the field *outside* the absorbing medium as a source.

The electrodynamics of superconducting media are shown to be described, under suitable assumptions, by an equation which is *wavelike* in time and neutron-diffusion-like in space. A generalization to take account of the "coherence effects" first noted by Pippard is also introduced; it is found that it gives rise to no fundamental difficulties (but practical complications) when the Green's function is determined by Fourier transform methods. If the "coherence length" goes to zero (London limit) the Green's functions are found to take on relatively simple forms involving Bessel functions.

Finally, it is noted that the meson field of Yukawa has the same mathematical form as that of superconducting electrodynamics in the London limit, so that the preceding results can be carried over to classical meson theory.

REFERENCES

Carslaw, H. S. and J. C. Jaeger, *Conduction of Heat in Solids*. New York: Oxford University Press, 1959, 2nd ed.

London, F., *Superfluids*. New York: Dover Publications, 1961, 2nd ed., Vol. 1.

Morse, P. M. and H. Feshbach, *Methods of Theoretical Physics*. New York: McGraw-Hill, 1953.

Sommerfeld, A., *Partial Differential Equations in Physics*. New York: Academic, 1949, Chapter 3.

Wallace, P. R., *Nucleonics* **4**, Nos. 2, 3 (1949).

Wigner, E. P. and A. M. Weinberg, *The Physical Theory of Neutron Chain Reactors*. Chicago: University of Chicago Press, 1958.

6
PROBLEMS OF DIFFUSION AND ATTENUATION

"You can never plan the future by the past"
Edmund Burke

1. Introduction

The problems of wave propagation with which the last chapter was concerned had the property that the basic equations were time-reversible, that is, were unaltered when the time variable t was replaced by $-t$. Nevertheless, the direction of time had to be built into a theory which purported to describe nature. This was accomplished by assumptions extraneous to the field equations themselves; for instance, in problems of radiation a selection was made of *retarded* potentials, which represent a limitation to solutions such that a source at t' produces a field only at time $t > t'$. Mathematically, *advanced* potentials, in which t is replaced by $-t$, are equally possible, as are linear combinations of the two types.

A very simple example will illustrate the point in question. Consider Green's function for the harmonic oscillator equation; it is the solution of

$$\frac{d^2g}{dt^2} + \lambda^2 g = -\delta(t - t') \tag{6.1}$$

If we take the Fourier transform of this equation, which is of course time-reversible, we obtain

$$(\omega^2 - \lambda^2)\bar{g}(\omega) = \frac{1}{2\pi} e^{-i\omega t'} \tag{6.2}$$

where $\bar{g}(\omega)$ is the Fourier transform of $g(t)$. If now we try to invert the transform to obtain $g(t)$,

$$g(t-t') = \frac{1}{2\pi} \int_{-\infty}^{\infty} \frac{1}{\omega^2 - \lambda^2} e^{i\omega(t-t')} d\omega \tag{6.3}$$

The problem arises that the transform has poles on the real axis. A formal way of overcoming the difficulty is to displace the contour slightly either above or below the real axis, treating ω as a complex variable. If the contour is displaced *upward* by ϵ, the poles lie *below* the contour. If $t > t'$, the integral can be evaluated by deforming the contour around the *upper* half-plane, and the value of the integral is found to be zero. For $t < t'$, it must be deformed around the *lower* half-plane, and $g(t-t')$ is found to be

$$g = \frac{1}{\lambda} \sin \lambda(t-t') \tag{6.4}$$

This represents an anticausal situation; the source makes itself felt only at earlier times.

If, on the other hand, the contour in (6.3) is displaced *downward* by ϵ, $g = 0$ for $t < t'$, and becomes, for $t > t'$

$$g = \frac{1}{\lambda} \sin \lambda(t'-t) \tag{6.5}$$

This solution is then causal.

The arbitrariness does not arise in real systems since there must always be some dissipative mechanism present. This can be represented by a frictional resistance term $\kappa(dy/dt)$. Such a term is of course not time-reversible. With $\kappa > 0$, the term $\kappa(dy/dt)$ added to the right-hand side of (6.1) gives rise to *dissipation* of energy from the oscillator with increasing t. A reversal of the sign of this term would correspond to an *increase* in energy with time. Thus, the dissipative process serves to define a direction of time for the problem. But it also removes all ambiguity from the mathematical problem. Taking the Fourier transform of the modified problem, we find that

$$(\omega^2 - ik\omega - \lambda^2)\bar{g} = \frac{1}{2\pi}$$

or

$$\bar{g} = \frac{1}{2\pi} \frac{1}{(\omega - (i\kappa/2))^2 + (\lambda^2 - (\kappa^2/4))} \tag{6.6}$$

We assume that $\lambda^2 - (\kappa^2/4) > 0$. We see that the poles of the integrand are in the *upper* half-plane. But for $t > t'$ the contour must be taken around that half-plane, while for $t < t'$ it must be taken around the lower. It then follows *automatically* that $g = 0$ for $t' > t$ and is different from zero for $t' < t$, that is, it is causal.

It is primarily with this sort of problem, involving terms of the first order in time and therefore not time-reversible, that we deal in the present chapter. We are concerned primarily with problems of thermal conduction; of diffusion, including neutron diffusion; and of electromagnetic wave propagation with losses. However, at the end of the chapter, we also treat problems of spatial attenuation, which are mathematically related in some of their features to the above problems. A problem of particular interest is that of superconducting electrodynamics.

2. Diffusion in a Gas or a Solid

We consider the diffusion of a substance in a uniform background, which may be either gaseous or solid. The basic equations are two in number: (*i*) an equation of continuity, which relates changes in density of the diffusing substance to its flow, and (*ii*) a "constitutive" relation which specifies how the flow is determined by the density distribution. The first relation is trivial and contains no reference to physical mechanisms; it simply asserts the indestructibility of matter. The second is an expression of the physical mechanism of the diffusion process.

Let the density (number of particles per unit volume of the substance) be designated $\rho(r, t)$ and the current $\mathbf{j}(\mathbf{r}, t)$, where $\mathbf{j}(\mathbf{r}, t) \cdot \mathbf{n}\, dS$ is the number of particles flowing per unit time across an element of surface dS to which $\mathbf{n}$ is the unit normal vector. The continuity equation is, then,

$$\frac{\partial \rho}{\partial t} + \nabla \cdot \mathbf{j} = 0 \tag{6.7}$$

So far as the constitutive equation is concerned, we assume "Fick's law":

$$\mathbf{J} = -D\nabla\rho \tag{6.8}$$

We do not enter here into the question of the proof of this equation, though with special assumptions it will be proven in the next chapter. In the case of diffusion in a solid, various particular assumptions have to be made, in particular isotropy (without which the D would become a tensor quantity).

Substituting (6.8) in (6.7) leads to the "diffusion equation"

$$\frac{\partial \rho}{\partial t} = D\nabla^2\rho \tag{6.9}$$

This is, of course, of first order in time, and not time-reversible. The way in which the direction of time is specified here is made clear by imagining the physical consequences of reversing the sign of t. This would imply a flow from regions of low density to regions of high density. The direction of time is then specified by the tendency toward increasing entropy in the process of time evolution. It is the *statistical* character of the phenomenon, rather than a mechanism of dissipation, which is crucial.

A word about boundary conditions would be in order. If diffusion takes place within an enclosure, the appropriate condition is that the normal component of current, $-D\mathbf{n} \cdot \nabla\rho$, is zero on the boundary. If escape of particles is possible over the boundary (for example, for diffusing particles of gas in a solid) the situation becomes more complicated. It may be shown in fact that an appropriate boundary condition is to put the density zero at or near the boundary. The difficulty is that Fick's law can in general be proven only if the concentration gradient is not too large. Since at a surface all particles are flowing outward (i.e., the angular distribution is such that the current is zero over 2π steradians) the concentration gradient becomes large, and Fick's law is no longer valid. The same is true in the neighborhood of a source. Although, strictly speaking, the problem should be treated in these cases in a more accurate way (by means of a transport equation), for many purposes it is sufficient to use the approximation of elementary diffusion theory, especially if one is not specifically interested in the details of the solution near sources or boundaries.

At an *interface* between media, the conditions to be applied to the solution are
  (a) that the normal component of current $-D\mathbf{n} \cdot \nabla\rho$ is continuous
  (b) that the density itself is continuous at all points of the surface.

3. Conduction of Heat in a Solid

Consider a solid containing a heat source distribution $S(\mathbf{r}, t)$, that is, $S(\mathbf{r}, t)$ is the amount of heat energy produced per unit volume, per unit time at the point $\mathbf{r}$ and the time t. Let $U(\mathbf{r}, t)$ be the heat energy per unit volume and $\mathbf{j}(\mathbf{r}, t) \cdot \mathbf{n} \, dS$ the heat energy flowing per unit time over area dS with unit normal vector $\mathbf{n}$. The equation of continuity of energy then has the form

$$\frac{\partial U}{\partial t} + \nabla \cdot \mathbf{j} = S \tag{6.10}$$

an equation identical in form to that for diffusion.

Let us write both terms on the left-hand side of (6.10) in terms of the temperature. If the temperature gradients are not too large, it may be shown by the use of the standard solid-state theory,[1] that the heat flux

---

[1]See, for example, Mott, N. F. and H. Jones, *Properties of Metals and Alloys*. New York: Oxford University Press, 1936, pp. 305–307.

is proportional to the negative gradient of temperature

$$\mathbf{j} = -\kappa \nabla T \tag{6.11}$$

κ being the thermal conductivity and T the temperature.

As for the first term, we know that changes in temperature produce proportionate changes in heat content, the proportionality constant being the specific heat C. Thus

$$\frac{\partial U}{\partial t} = C \frac{\partial T}{\partial t} \tag{6.12}$$

Consequently we obtain an equation for the temperature,

$$\frac{\partial T}{\partial t} - \frac{\kappa}{C} \nabla^2 T = \frac{S}{C} \tag{6.13}$$

which again has the form of a diffusion equation.

Possible boundary conditions may be specified as follows:

(a) At an insulating surface, there will be no heat flow perpendicular to the surface, so that

$$\kappa \frac{\partial T}{\partial n} = 0 \tag{6.14}$$

(b) If a surface is held at a constant temperature T_0, then over the surface $T = T_0$. Since (6.13) contains only derivatives, the zero of temperature may be arbitrarily chosen, and it is convenient to choose it so that $T_0 = 0$.

(c) If there is radiation from the surface into an effective vacuum or a gas, a more complicated condition applies. The rate of radiation from a medium at temperature T is known to be given by Stefan's fourth-power law, that is, it is of the form σT^4. If the surface of the medium is at temperature T, and its surroundings at temperature T_0, the net rate of radiation is

$$-\kappa \frac{\partial T}{\partial n} = \sigma(T^4 - T_0^4) \tag{6.15}$$

We must now measure the temperature T on the Kelvin scale. Equation (6.15) leads to very difficult mathematical problems in general. However, if we are concerned only with a modest range of temperature,

$$T^4 - T_0^4 \approx 4T_0^3(T - T_0) \tag{6.16}$$

This is a good approximation provided

$$T - T_0 \ll T_0$$

Using this approximation, the boundary condition becomes

$$-\kappa \frac{\partial}{\partial n}(T - T_0) = 4\sigma T_0^3(T - T_0) \tag{6.17}$$

or

$$\frac{\partial}{\partial n}(T - T_0) = -\alpha(T - T_0) \tag{6.18}$$

where

$$\alpha = \frac{4\sigma T_0^3}{\kappa} \tag{6.19}$$

It is now possible to return to a scale in which $T_0 = 0$, and the boundary condition is of a familiar linear type.

(d) At an interface, T will be continuous and the normal heat flux $\kappa(\partial T/\partial n)$ will be continuous.

4. Green's Functions for Diffusion and Heat Conduction Problems

Let us first consider a point source at time t_0 in an infinite medium. With no loss of generality, we may take $t_0 = 0$. The solution is then the infinite-medium Green's function

$$\frac{\partial G}{\partial t} - \beta \nabla^2 G = \delta(\mathbf{r})\delta(t) \tag{6.20}$$

If we take a Fourier transform in spatial coordinates and a Laplace transform in time, and call the transformed function $G_2(\mathbf{k}, s)$, this function satisfies

$$(s + \beta k^2) G_2(\mathbf{k}, s) = \left(\frac{1}{2\pi}\right)^3$$

or

$$G_2(\mathbf{k}, s) = \frac{1}{(2\pi)^3} \frac{1}{s + \beta k^2} \tag{6.21}$$

If we now take the inverse Laplace transform to obtain $G_1(\mathbf{k}, t)$, we obtain

$$G_1(\mathbf{k}, t) = \frac{1}{(2\pi)^3} e^{-\beta k^2 t} \tag{6.22}$$

The Green's function $G(\mathbf{r}, t)$ is now the inverse Fourier transform

$$G(\mathbf{r}, t) = \frac{1}{(2\pi)^3} \int e^{-\beta k^2 t} e^{i\mathbf{k}\cdot\mathbf{r}} d^3\mathbf{k} \tag{6.23}$$

Integrating in the familiar way over the *angles* of $\mathbf{r}$ gives

$$G(\mathbf{r}, t) = \frac{1}{2\pi^2} \int_0^\infty e^{-\beta k^2 t} \frac{\sin kr}{kr} k^2 \, dk$$

$$= \frac{1}{4\pi^2 r} \operatorname{Im} \int_{-\infty}^\infty e^{-\beta k^2 t} e^{ikr} k \, dk$$

$$= \frac{1}{4\pi^2 r} \operatorname{Im} \frac{1}{i} \frac{\partial}{\partial r} \int_{-\infty}^\infty e^{-\beta k^2 t} e^{ikr} \, dk \tag{6.24}$$

This may be written

$$\frac{1}{4\pi^2 r} \operatorname{Im} \frac{1}{i} \frac{\partial}{\partial r} \int_{-\infty}^{\infty} \exp\left[-\left(k\sqrt{\beta t} - \frac{ir}{2\sqrt{\beta t}}\right)^2\right] dk \, e^{-r^2/4\beta t}$$

The contour between $-\infty$ and ∞ may be replaced by an alternative one displaced parallel to the real axis by $-r/2\beta t$, since there is no singularity of the integrand between them. The integral is then a familiar Gaussian integral, the value of which is $\sqrt{\pi/\beta t}$. Therefore,

$$G(\mathbf{r}, t) = -\frac{1}{4\pi^2 r} \sqrt{\frac{\pi}{\beta t}} \frac{\partial}{\partial r} e^{-r^2/4\beta t}$$

$$= \left(\frac{1}{4\pi\beta t}\right)^{3/2} e^{-r^2/4\beta t} \tag{6.25}$$

The above derivation is of course valid for $t > 0$; for $t < 0$, $G(\mathbf{r}, t) = 0$.

Problem 6-1: Show that the one-dimensional Green's function, representing the solution due to a source of unit strength per unit area on the plane $z = 0$, is

$$G_{pl}(z, t) = \frac{1}{(4\pi\beta t)^{1/2}} e^{-z^2/4\beta t}$$

Problem 6-2: Show that the solution for a spherical shell source at r' is

$$G_{sh}(\mathbf{r}, \mathbf{r}', t) = \frac{1}{(4\pi\beta t)^{1/2}} \frac{1}{2\pi r r'} e^{-(r^2+r'^2)/4\beta t} \sinh \frac{rr'}{2\beta t}$$

---

5. Green's Function in One Dimension—Semi-Infinite and Finite Media

Starting from the result of Problem 6.1, that the one-dimensional infinite-medium Green's function is

$$G(z, t) = \frac{1}{(4\pi\beta t)^{1/2}} e^{-z^2/4\beta t} \tag{6.26}$$

it is possible to derive Green's functions for semi-infinite and finite media with plane boundaries, for each type of boundary condition discussed above. In the next section, similar problems are solved for spherical media.

A. SEMI-INFINITE MEDIUM, SOURCE AT $z = z_0$

(a) Boundary $z = 0$ at constant temperature $T = 0$.

The method here is to introduce an image source of negative strength at $z = -z_0$ to obtain the solution

$$G = G(z - z_0, t) - G(z + z_0, t) \qquad (6.27)$$

(b) Insulated boundary at $z = 0$.

If one introduces at $z = -z_0$ a source *equal* to that at $z = z_0$,

$$G = G(z - z_0, t) + G(z + z_0, t) \qquad (6.28)$$

the flux over the boundary from the image source is equal and opposite to that of the original source, giving a net flux equal to zero.

(c) Radiative boundary condition $(\partial T/\partial z) - \alpha T = 0$ at $z = 0$, the medium being in the region $z > 0$.

In this case the choice of an image system is not obvious. Let us then take a *distribution* of images $f(z)$ for $z < 0$, and try to choose f so as to satisfy boundary conditions. We wish therefore to apply the boundary condition to

$$G = G(z - z_0) + \int_{-\infty}^{0} G(z - z')f(z')\,dz'$$

Imposing the boundary condition, the equation to be satisfied is

$$G'(-z_0) - \int_{-\infty}^{0} G'(z')f(z')\,dz'$$

$$- \alpha \left[G(-z_0) + \int_{-\infty}^{0} G(z')f(z')\,dz' \right] = 0$$

Let us integrate the second term by parts, taking $f(0) = 0$ and using the fact that $G(-\infty) = 0$. This leads to the equation

$$\int_{-\infty}^{0} G(z')[f'(z') - \alpha f(z')]\,dz' = G'(-z_0) + \alpha G(-z_0)$$

Defining $h(z')$ by

$$f(z') = \delta(z' + z_0) + h(z') \qquad (6.29)$$

we find that $h(z')$ obeys the equation

$$\int_{-\infty}^{0} G(z')[h'(z') - \alpha h(z')]\,dz' = 2\alpha G(-z_0)$$

This may be satisfied provided

$$h'(z') - \alpha h(z') = 2\alpha \delta(z' + z_0) \qquad (6.30)$$

The solution of this equation for the boundary condition $h(0) = 0$ is

$$
\begin{aligned}
h(z') &= 0, & z' &> -z_0 \\
&= -2\alpha e^{\alpha(z'+z_0)}, & z' &< -z_0
\end{aligned}
\tag{6.31}
$$

The Green's function for this problem is, therefore,

$$
G = G(z - z_0) + G(z + z_0) - 2\alpha \int_{-\infty}^{-z_0} G(z - z') \, e^{\alpha(z' + z_0)} \, dz'.
\tag{6.32}
$$

The last integral may be evaluated in terms of error functions: it is

$$
\frac{1}{\sqrt{4\pi\beta t}} \int_{-\infty}^{-z_0} e^{-(z-z')^2/4\beta t} \, e^{\alpha(z'+z_0)} \, dz'
$$

After a little manipulation, this may be put in the form

$$
\tfrac{1}{2} e^{\alpha(z+z_0)} e^{\alpha^2 \beta t} \left[1 - \operatorname{erf} \frac{z + z_0 + 2\alpha\beta t}{2\sqrt{\beta t}} \right]
$$

where

$$
\operatorname{erf} x = \frac{2}{\sqrt{\pi}} \int_0^x e^{-y^2} \, dy
\tag{6.33}
$$

Therefore,

$$
G = G(z - z_0) + G(z + z_0) - \alpha e^{\alpha(z+z_0)} e^{\alpha^2\beta t} \left[1 - \operatorname{erf}\frac{z + z_0 + 2\alpha\beta t}{2\sqrt{\beta t}} \right]
\tag{6.34}
$$

B. FINITE MEDIA BETWEEN $z = 0$ AND $z = a$, SOURCE AT $z = z_0 < a$

For the simple boundary conditions (insulating boundaries, both boundaries held at a constant temperature) the sort of successive image approach used in electrostatics leads directly to finite-medium Green's functions in terms of an infinite series of infinite-medium ones. For example, with $T = 0$ at the boundaries $z = 0$ and $z = a$, we require negative images at $(2na - z)$ and positive ones at $(2na + z)$. The Green's function is, therefore,

$$
G = \sum_{n=-\infty}^{\infty} G_0(z - z_0 - 2na) - \sum_{n=-\infty}^{\infty} G_0(z + z_0 - 2na)
\tag{6.35}
$$

G_0 being the one-dimensional infinite-medium Green's function

$$
G_0 = \frac{1}{(4\pi\beta t)^{1/2}} e^{-z^2/4\beta t}
\tag{6.36}
$$

The function (6.35) may be transformed by Poisson's formula [see Chapter 4, formula (4.82)] into an alternative series. In the present case the alternative form is

$$G = \frac{\pi}{a} \sum_{m=-\infty}^{\infty} g\left(\frac{\pi m}{a}\right)[e^{2\pi im(z-z_0)/2a} - e^{2\pi im(z+z_0)/2a}]$$

$$= \frac{4\pi}{a} \sum_{m=1}^{\infty} g\left(\frac{\pi m}{a}\right) \sin\frac{\pi m z_0}{a} \sin\frac{m\pi z}{a} \tag{6.37}$$

where $g(k)$ is the Fourier transform of G_0,

$$g(k) = \frac{1}{2\pi} e^{-\beta k^2 t} \tag{6.38}$$

Consequently we find that

$$G(z, z_0, t) = \frac{2}{a} \sum_{m=1}^{\infty} e^{-(\pi^2 m^2/a^2)\beta t} \sin\frac{\pi m z_0}{a} \sin\frac{\pi m z}{a} \tag{6.39}$$

Problem 6-3: Obtain (6.39) by solving the problem *directly* by expansion in Fourier series.

Problem 6-4: With the same geometry but with the two boundaries insulated, show that the Green's function may be written as

(i) $$G = \sum_{n=-\infty}^{\infty} G_0(z-z_0-2na) + \sum_{n=-\infty}^{\infty} G(z+z_0-2na)$$

or

(ii) $$G = \frac{1}{a} \sum_{m=-\infty}^{\infty} e^{-(\pi^2 m^2/a^2)\beta t} \cos\frac{\pi m z_0}{a} \cos\frac{\pi m z}{a}$$

Other boundary conditions may be handled by different devices. Let us consider examples

(a) $z = 0$ at $T = 0$; $z = a$ at $T = T_0$.

In this case we note that $T_0(z/a)$ is a solution of the diffusion equation. $(T - T_0(z/a))$ satisfies the boundary condition that it is zero at $x = 0$ and $x = a$ and is, therefore, given by (6.39).

(b) $z = 0$ at $T = 0$; $z = a$ insulated

The solution in this case has the Fourier series expansion

$$T = \sum \gamma_n(t) \sin\frac{(n+\frac{1}{2})\pi z}{a} \tag{6.40}$$

Substituting in the equation, and making use of the fact that

$$\delta(z - z_0) = \sum c_n \sin\frac{(n+\frac{1}{2})\pi z}{a} \tag{6.41}$$

where

$$c_n = \frac{2}{a} \sin \frac{(n+\frac{1}{2})\pi z_0}{a} \qquad (6.42)$$

we find for $\gamma_n(t)$ the equation

$$\frac{d\gamma_n}{dt} + \frac{(n+\frac{1}{2})^2 \pi^2}{a^2} \beta\gamma_n = \delta(t) \frac{2}{a} \sin \frac{(n+\frac{1}{2})\pi z_0}{a}$$

The solution is

$$\gamma_n(t) = \frac{2}{a} \sin \frac{(n+\frac{1}{2})\pi z_0}{a} e^{-[(n+(1/2))^2 \pi^2/a^2]\beta t} \qquad (6.43)$$

Substituting this in (6.40) gives the required solution.

Problem 6-5: Using Poisson's formula, show that an alternative form for the solution is

$$T = \frac{1}{\sqrt{4\pi\beta t}} \sum_{n=-\infty}^{\infty} [e^{-(z-z_0+4na)^2/4\beta t} - e^{-(z+z_0+4na)^2/4\beta t}$$

$$+ e^{-[z+z_0+(4n+2)a]^2/4\beta t} - e^{-[z+z_0+(4n+2)a]^2/4\beta t}]$$

$$= \sum_{n=-\infty}^{\infty} [G_0(z-z_0+4na) + G_0(z+z_0+(4n+2)a)$$

$$- G(z+z_0+4na) - G(z-z_0+(4n+2)a)]$$

Show how this result can be derived directly.

(c) Radiative boundary conditions at $x=0$ and $x=a$

A direct approach to this problem by the method of images is very cumbersome. Instead, let us use the method of separation of variables. Putting

$$T = f(t)g(z) \qquad (6.44)$$

substituting in

$$\frac{\partial T}{\partial t} - \beta \frac{\partial^2 T}{\partial z^2} = 0$$

and dividing by T, we obtain the equation

$$\frac{1}{f}\frac{df}{dt} = \frac{\beta}{g}\frac{d^2 g}{dz^2} \qquad (6.45)$$

Each side of the equation must be a constant. Put

$$\frac{d^2 g}{dz^2} = -\lambda^2 g$$

so that

$$g = A \sin \lambda z + B \cos \lambda z$$

Application of the radiative boundary condition

$$\frac{dg}{dz} \pm \alpha g = 0 \qquad (6.46)$$

at $x = a$ and $x = 0$, respectively, yields a transcendental equation for λ:

$$\tan \lambda a = \frac{2\lambda\alpha}{\lambda^2 - \alpha^2} \qquad (6.47)$$

If we put

$$\tan \frac{\delta}{2} = \frac{\alpha}{\lambda} \qquad (6.48)$$

Equation (6.47) gives

$$\lambda a = \delta \quad \text{or} \quad \lambda a = \delta + \pi \qquad (6.49)$$

The transcendental equation for λ may then be put in the form

$$\tan \frac{\lambda a}{2} = \frac{\alpha}{\lambda} \qquad (6.50)$$

or

$$-\cot \frac{\lambda a}{2} = \frac{\alpha}{\lambda} \qquad (6.51)$$

The nature of the roots is demonstrated in Fig. 6.1. This illustrates the curves corresponding to the right- and left-hand sides of (6.50) and (6.51)

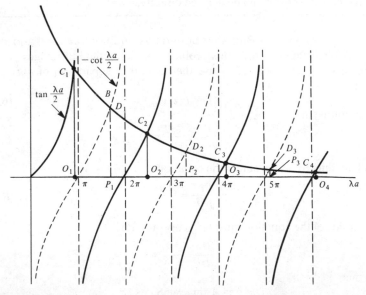

Figure 6.1

intersecting at points $C_1, C_2, C_3, C_4, \ldots; D_1, D_2, D_3, \ldots$ for which λa is given by $O_1, O_2, O_3, O_4, \ldots$ and $P_1, P_2, P_3, \ldots ..$. The location of the roots depends on the magnitude of αa. If the rate of radiation at the surface is rapid, αa is large and the lower roots are close to $\lambda a = n\pi (n = 1, 2, \ldots)$. If it is slow, αa is small and the roots are close to $\lambda a = n\pi (n = 0, 1, 2, \ldots)$. The first case approaches that of constant temperature at the boundaries, the second, that of insulated boundaries.

Since, from the boundary condition at $z = 0$,

$$\frac{B}{A} = \frac{\lambda}{\alpha} = \cot \frac{\delta}{2} \tag{6.52}$$

the case $\lambda a = \delta$ corresponds to the solutions which are *even* with respect to the midpoint of the slab:

$$g = \phi_n = \sqrt{\frac{2}{a}} \cos \lambda_n \left(z - \frac{a}{2} \right)$$

while the case $\lambda a = \delta + \pi$ corresponds to the *odd* solutions

$$g = \phi_n = \sqrt{\frac{2}{a}} \sin \lambda_n \left(z - \frac{a}{2} \right)$$

The solution of the spatial equation can be written

$$g = C \cos \left(\lambda z - \frac{\delta}{2} \right) \tag{6.53}$$

When $\alpha \to \infty, \delta \to \pi$ the solution takes the form

$$g = C \sin \frac{n\pi z}{a}$$

when $\alpha \to 0, \delta \to 0$ it is

$$g = C \cos \frac{n\pi z}{a}$$

both of which clearly satisfy the corresponding boundary conditions.

Labeling the roots of the transcendental equation λ_n in the general case, and calling the corresponding δ's determined by (6.48) δ_n, the functions

$$\phi_n = \sqrt{\frac{2}{a}} \cos (\lambda_n z - \tfrac{1}{2}\delta_n) \tag{6.54}$$

form an orthonormal set.

To find the Green's function for this problem, that is, the solution of

$$\frac{\partial G}{\partial t} - \beta \frac{\partial^2 G}{\partial z^2} = \delta(z - z_0)\delta(t) \tag{6.55}$$

we first expand $\delta(z - z_0)$ in terms of the ϕ_n's,

$$\delta(z - z_0) = \sum \phi_n(z)\phi_n(z_0) \tag{6.56}$$

If we also expand

$$G(z, t) = \sum c_n(t)\phi_n(z)$$

we may substitute in (6.55) and equate separate coefficients of ϕ_n:

$$\dot{c}_n + \beta\lambda_n{}^2 c_n = \phi_n(z_0)\delta(t) \tag{6.57}$$

The solution is

$$c_n(t) = \theta(t)\phi_n(z_0)e^{-\beta\lambda_n{}^2 t}$$

where, as usual, $\theta(t) = 1$ for $t > 0$ and 0 for $t < 0$. The Green's function is, therefore,

$$G(z, t) = \theta(t) \sum_{n=1}^{\infty} \phi_n(z_0)\phi_n(z)e^{-\beta\lambda_n{}^2 t}$$

$$= \theta(t)\frac{2}{a} \sum_{n=1}^{\infty} \cos\left(\lambda_n z_0 - \frac{\delta_n}{2}\right)\cos\left(\lambda_n z - \frac{\delta_n}{2}\right)e^{-\beta\lambda_n{}^2 t} \tag{6.58}$$

Problem 6-6:   A long cylinder of radius a is heated to a temperature T_0 and then cooled by having its surface kept at temperature $T = 0$ from time $t = 0$. Show that the subsequent temperature distribution in the cylinder is

$$T(r, t) = 2T_0 \sum_{n=1}^{\infty} \frac{J_0(\gamma_n(r/a))}{\gamma_n J_1(\gamma_n)} e^{-\gamma_n{}^2\beta t/a^2}$$

where $J_0(\gamma_n) = 0$.

Problem 6-7:   If, in the preceding problem, the cylinder is allowed to cool by radiation, show that the temperature is

$$T(r, t) = 2T_0 \sum_{n=1}^{\infty} \frac{J_1(\lambda_n)J_0(\lambda_n(r/a))e^{-\lambda_n{}^2\beta t/a^2}}{\lambda_n[J_0{}^2(\lambda_n) + J_1{}^2(\lambda_n)]}$$

where the λ_n's are solutions of

$$\lambda J_1(\lambda) = \alpha a J_0(\lambda)$$

Problem 6-8:   If a substance diffuses down a thin tube of length l, there being no diffusion over the sides; and if the rate of influx at the end $x = 0$ is $f(t)$ while the end $x = l$ is held at a fixed concentration C_0, show that the concentration is given by

$$c(x, t) = \left\{ C_0\left[1 - \frac{4}{\pi} \sum_{n=0}^{\infty} \frac{(-1)^n}{2n+1} e^{-(2n+1)^2\pi^2 Dt/4l^2}\right] \right.$$

$$\left. - \frac{1}{l}\int_0^l e^{-(2n+1)^2\pi^2 D(t-t')/4l^2} f(t')\, dt' \right\} \cos\frac{(2n+1)\pi x}{l}$$

6. Green's Functions— Spherical Geometry

We have already noted [Eq. (3.170), Chapter 3] that the solution of a linear equation for a spherical shell source may be written in terms of that for a plane source. This connection between plane and spherical problems may be seen in another way. The Laplacian of a spherically symmetrical function ϕ may be written

$$\nabla^2\phi = \frac{\partial^2\phi}{\partial r^2} + \frac{2}{r}\frac{\partial\phi}{\partial r}$$

$$= \frac{1}{r}\frac{\partial^2}{\partial r^2}(r\phi) \tag{6.59}$$

Therefore, in the case of spherical symmetry, the heat conduction (diffusion) equation is

$$\frac{\partial T}{\partial t} - \beta\frac{1}{r}\frac{\partial^2}{\partial r^2}(rT) = \delta(t)\frac{\delta(r-r')}{4\pi r'^2} \tag{6.60}$$

when there is a unit instantaneous source at time $t = 0$ on the sphere of radius r'. The normalization on the right is such as to make the *total* strength of the shell source unity.

If we introduce

$$\chi = rT \tag{6.61}$$

Equation (6.60) becomes

$$\frac{\partial\chi}{\partial t} - \beta\frac{\partial^2\chi}{\partial r^2} = \delta(t)\frac{1}{4\pi r'}\delta(r-r') \tag{6.62}$$

which is just the one-dimensional problem. However, since T must be finite at $r = 0$, there is the additional "boundary condition" that $\chi = 0$ at $r = 0$. Thus, the problem of a finite sphere is mathematically identical with that of a "one-dimensional" medium confined between two planes, at $r = 0$ and $r = a$, respectively, the former being kept at the fixed temperature $T = 0$. We may, then, immediately write down the solution of a number of problems of spherical geometry in terms of ones already obtained.

It should be noted, however, that it is not simply a matter of the relation between plane and shell source solutions in an infinite medium,

$$\psi_{sh} = \frac{1}{4\pi rr'}[\psi_{pl}(|r-r'|) - \psi_{pl}(r+r')] \tag{6.63}$$

Although there is a correspondence between plane and shell source solutions in finite media, that correspondence must take account of boundary conditions. We must recognize that it is rT in the spherical case which replaces T in the plane one. If we are interested in a sphere of radius a whose surface is held at a fixed temperature, that temperature may be taken to be zero; in this case the problem corresponds to that of a plane source in a finite slab bounded by $z = 0$ and $z = a$, with both surfaces at $T = 0$. Consider, however, the case in which the outer surface of the sphere is insulated. Then

$$\frac{\partial T}{\partial r} = 0 \qquad \text{at} \qquad r = a \tag{6.64}$$

Putting $rT = \chi$ again, this becomes

$$\frac{\partial}{\partial r}\left(\frac{\chi}{r}\right) = 0 \qquad \text{at} \qquad r = a$$

or

$$\chi' - \frac{1}{a}\chi = 0 \qquad \text{at} \qquad r = a \tag{6.65}$$

This has the form of a *radiative* boundary condition for a plane problem, so the analogous problem to be solved is that of a slab with $T = 0$ at $z = 0$ and a radiative boundary condition at $z = a$.

If the boundary condition at $r = a$ in the spherical problem is truly radiative, the same analogy applies, though the constants in the radiative condition are altered. If

$$\frac{\partial T}{\partial r} + \alpha T = 0 \tag{6.66}$$

at $r = a$, in terms of χ this becomes

$$\chi' - \chi\left(\frac{1}{a} - \alpha\right) = 0 \tag{6.67}$$

Since we have not previously solved this problem, let us sketch its solution briefly here.

For $t \neq 0$ and $r \neq r'$, and separating variables as in (6.44) by putting $\chi = g(r)f(t)$, we find that

$$g = \sin \lambda r \tag{6.68}$$

$$f = e^{-\beta\lambda^2 t} \tag{6.69}$$

where, by virtue of the boundary condition (6.67) at $r = a$

$$\lambda a \cot \lambda a = 1 - \alpha a \tag{6.70}$$

Figure 6.2 shows the roots of the transcendental equation (6.70).

The solutions (6.68) are normalized as follows: letting $\tau_n = A_n(1/r)$

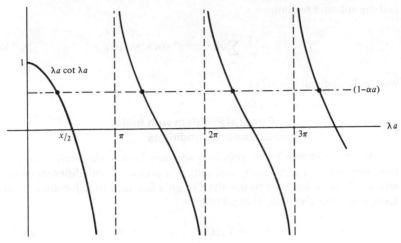

Figure 6.2

$\sin \lambda_n r$, we must have

$$4\pi \int_0^a \tau_n^2 r^2 \, dr = 4\pi A_n^2 \int_0^a \sin^2 \lambda_n r \, dr$$

$$= 2\pi A_n^2 \int_0^a (1 - \cos 2\lambda_n r) \, dr$$

$$= 2\pi A_n^2 \left[a - (1/2\lambda_n) \sin 2\lambda_n a \right]$$

$$= 1$$

Therefore,

$$A_n = \frac{1}{\sqrt{(\pi/\lambda_n)(2\lambda_n a - \sin 2\lambda_n a)}} \tag{6.71}$$

The solution of (6.62) may now be obtained by writing

$$\chi = \sum C_n(t) A_n \sin \lambda_n r \tag{6.72}$$

and making use of the fact that

$$\frac{1}{4\pi} \delta(r - r') = \sum A_n^2 \sin \lambda_n r \sin \lambda_n r' \tag{6.73}$$

Substitution in (6.62) gives the equation for $C_n(t)$,

$$C_n + \beta \lambda_n^2 C_n = \delta(t) A_n \frac{\sin \lambda_n r'}{r'} \tag{6.74}$$

Thus

$$C_n = A_n \frac{\sin \lambda_n r'}{r'} e^{-\beta \lambda_n^2 t}$$

and the solution becomes

$$T = \frac{1}{r}\chi = \frac{1}{rr'} \sum_n A_n{}^2 e^{-\beta\lambda n^2 t} \sin \lambda_n r' \sin \lambda_n r \tag{6.75}$$

where $A_n{}^2$ is given by (6.71).

7. General Problem with Initial and Boundary Conditions

We have shown in the previous sections how to determine Green's functions for heat conduction and diffusion problems with different geometries. We now see how to use the Green's function to solve such problems subject to given initial conditions:

$$T(\mathbf{r}, t) = T_0(\mathbf{r}) \quad \text{at} \quad t = 0 \tag{6.76}$$

and boundary conditions which specify

$$\frac{\partial T}{\partial n} - \gamma T \tag{6.77}$$

over given surfaces S. $\partial/\partial n$ represents the normal derivative over the surface. Particular cases of (6.77) are those in which the temperature or flux of heat are specified over S.

Before proceeding, however, it is necessary to prove the "reciprocity theorem": that the solution at $\mathbf{r}$ due to a unit source at $\mathbf{r}'$ is the same as that at $\mathbf{r}'$ due to a unit source at $\mathbf{r}$, the time interval being the same in each case. In other words, we wish to show that

$$G(\mathbf{r}, \mathbf{r}', t - t') = G(\mathbf{r}', \mathbf{r}, t - t') \tag{6.78}$$

Problem 6-9: Harold Urey, in his book *The Planets*[2] poses the problem of the change in temperature due to surface cooling of a sphere which is heated uniformly by radioactive heating. The equation is

$$\frac{\partial T}{\partial t} = \kappa \left(\frac{\partial^2 T}{\partial r^2} + \frac{2}{r}\frac{\partial T}{\partial r} \right) + \sum_i Q_i e^{-\alpha_i t}$$

Q_i is the heat produced per unit volume by the ith radioactive species, divided by the specific heat C; α_i is the decay constant of that species. Assuming that the temperature in the sphere is uniformly T_0 at $t = 0$, and

[2] New Haven; Yale University Press, 1952, pp. 52–53.

that the surface is at temperature $T = 0$ at all times, show that

$$T = \frac{2a}{r} \left\{ \sum_i \sum_{n=1}^{\infty} \frac{Q_i}{\alpha_i - (\kappa n^2 \pi^2/a^2)} \frac{(-1)^n}{n\pi} \sin\frac{n\pi r}{a} \left[e^{-\alpha_i t} - e^{-(\kappa n^2 \pi^2 t/a^2)} \right] \right.$$

$$\left. - T_0 \sum_{n=1}^{\infty} \frac{(-1)^n}{n\pi} \sin\frac{n\pi r}{a} e^{-(\kappa n^2 \pi^2 t/a^2)} \right\}$$

a being the radius of the sphere.

That the solution is a function only of $t - t'$ is a consequence of time translation invariance. Suppose it is a function of "source time" t' and "observation time" t. It may equally well be written in terms of $t_1 = (t - t')$ and $t_2 = \frac{1}{2}(t + t')$. But if the origin of time is displaced by amount t_0, t_1 is unchanged and t_2 becomes $\frac{1}{2}(t + t') + t_0$. Since, due to time translation invariance, the solution cannot depend on t_0, it must depend only on t_1.

In an *infinite* medium one may similarly show from invariance under spatial translation that G must depend on coordinates only through $\mathbf{r} - \mathbf{r}'$. Since spatial translation invariance breaks down in finite media, the result does not follow in that case; this is verified in the problems considered. In each case, however, the relation (6.78) holds.

The proof of the reciprocity theorem is as follows: From the definitions of the Green's function

$$\frac{\partial}{\partial t'} G(\mathbf{r}, \mathbf{r}', t' - t) - \beta \nabla^2 G(\mathbf{r}, \mathbf{r}', t' - t) = \delta(\mathbf{r} - \mathbf{r}')\delta(t' - t)$$

for $t' > t$, or

$$-\frac{\partial}{\partial t} G(\mathbf{r}, \mathbf{r}', t' - t) - \beta \nabla^2 G(\mathbf{r}, \mathbf{r}', t' - t) = \delta(\mathbf{r} - \mathbf{r}')\delta(t' - t) \quad (6.79)$$

But for $t > t''$

$$\frac{\partial}{\partial t} G(\mathbf{r}, \mathbf{r}'', t - t'') - \beta \nabla^2 G(\mathbf{r}, \mathbf{r}'', t - t'') = \delta(\mathbf{r} - \mathbf{r}'', t - t'') \quad (6.80)$$

We now multiply (6.79) by $G(\mathbf{r}, \mathbf{r}'', t - t'')$ and (6.80) by $G(\mathbf{r}, \mathbf{r}', t' - t)$, subtract the former from the latter, and integrate with respect to $\mathbf{r}$ over the region of definition and over t from below t'' to above t'. The left-hand side becomes

$$\int d^3r \int_{t''_-}^{t'_+} \frac{\partial}{\partial t} \left\{ G(\mathbf{r}, \mathbf{r}'', t - t') G(\mathbf{r}, \mathbf{r}', t' - t) \right\} dt$$

$$+ \beta \int_{-\infty}^{\infty} dt \int dS \left\{ G(\mathbf{r}, \mathbf{r}'', t - t'') \frac{\partial}{\partial n} G(\mathbf{r}, \mathbf{r}', t' - t) \right.$$

$$\left. - G(\mathbf{r}, \mathbf{r}', t' - t) \frac{\partial}{\partial n} G(\mathbf{r}, \mathbf{r}'', t - t'') \right\}$$

The first term is zero because one of the G's is zero at each limit, and the second by virtue of the boundary condition of the form

$$\frac{\partial G}{\partial n} - \alpha G = 0$$

on the boundary of the region of definition of the problem. But the right-hand side is then

$$G(\mathbf{r}'', \mathbf{r}', t' - t'') - G(\mathbf{r}', \mathbf{r}'', t' - t'') = 0$$

so the reciprocity theorem is proven.

Let us return now to the problem of solving the equation, in a specified geometry, subject to the conditions (6.76) and (6.77). For generality, let us also suppose there is also a source distribution $C\sigma(\mathbf{r}, t)$, i.e., that the heat energy produced per unit time in the volume element $d^3\mathbf{r}$ is $C\sigma(\mathbf{r}) \, d^3\mathbf{r}$, where C is the specific heat. The heat conduction equation then becomes

$$\frac{\partial T}{\partial t'} - \beta \nabla'^2 T = \sigma(\mathbf{r}', t') \tag{6.81}$$

On the other hand, the Green's function equation is

$$\frac{\partial}{\partial t} G(\mathbf{r}', \mathbf{r}, t - t') - \beta \nabla'^2 G(\mathbf{r}', \mathbf{r}, t - t') = \delta(\mathbf{r} - \mathbf{r}')\delta(t - t') \tag{6.82}$$

or, in a form more convenient for our purposes

$$-\frac{\partial}{\partial t'} G(\mathbf{r}', \mathbf{r}, t - t') - \beta \nabla'^2 G(\mathbf{r}', \mathbf{r}, t - t') = \delta(\mathbf{r} - \mathbf{r}')\delta(t - t') \tag{6.83}$$

We now multiply (6.81) by $G(\mathbf{r}', \mathbf{r}, t - t')$ and (6.83) by $T(\mathbf{r}', t')$, subtract, integrate over the spatial coordinates $\mathbf{r}'$ and over t' from 0 to $t + \epsilon$ to get

$$T(\mathbf{r}, t) = -\int d^3\mathbf{r}' \int_0^t dt' \left\{ T(\mathbf{r}', t') \frac{\partial}{\partial t'} G(\mathbf{r}, \mathbf{r}', t - t') \right.$$

$$\left. + G(\mathbf{r}, \mathbf{r}', t - t') \frac{\partial}{\partial t'} T(\mathbf{r}', t') \right\}$$

$$+ \beta \int \left\{ G(\mathbf{r}', \mathbf{r}, t - t') \nabla'^2 T - T\nabla'^2 G(\mathbf{r}, \mathbf{r}', t - t') \right\} d^3\mathbf{r}' dt'$$

$$+ \int \sigma(\mathbf{r}', t') G(\mathbf{r}, \mathbf{r}', t - t') \, d^3\mathbf{r}' dt' \tag{6.84}$$

We can do the t' integral in the first term on the right, and convert the second to a surface integral over the boundaries. In this way we obtain

$$T(\mathbf{r}, t) = \int d^3\mathbf{r}' \, T_0(\mathbf{r}') G(\mathbf{r}, \mathbf{r}', t)$$

$$+ \beta \int \left\{ G(\mathbf{r}', \mathbf{r}, t - t') \frac{\partial T}{\partial n'} - T \frac{\partial}{\partial n'} G(\mathbf{r}, \mathbf{r}', t - t') \right\} dS' dt'$$

$$+ \int \sigma(\mathbf{r}', t') G(\mathbf{r}, \mathbf{r}', t - t') \, d^3\mathbf{r}' dt' \tag{6.85}$$

If $(\partial T/\partial n) - \alpha T$ is *given* on the boundaries, the problem is easily solved by using the Green's function satisfying

$$\frac{\partial G}{\partial n} - \alpha G = 0$$

on them. For then the second term on the right is

$$\beta \int G(\mathbf{r}',\mathbf{r},t-t')\left[\frac{\partial T}{\partial n'}-\alpha T\right]dS' = \beta\int G(\mathbf{r},\mathbf{r}',t-t')\left[\frac{\partial T}{\partial n'}-\alpha T\right]dS'$$

the integral being taken over the bounding surfaces. Thus (6.84) finally becomes

$$T(\mathbf{r},t) = \int d^3\mathbf{r}'\, T_0(\mathbf{r}')G(\mathbf{r},\mathbf{r}',t)$$
$$+\beta\int_0^t dt' \int G(\mathbf{r},\mathbf{r}',t-t')\left[\frac{\partial}{\partial n'}T(\mathbf{r}',t)-\alpha T(\mathbf{r}',t')\right]dS'$$
$$+\int_0^t dt' \int \sigma(\mathbf{r}',t')G(\mathbf{r},\mathbf{r}',t-t')\,d^3\mathbf{r}' \qquad (6.86)$$

We see, therefore, that the contributions from initial conditions, from boundaries, and from sources are all expressible in terms of the same Green's function.

Care must be taken, however, for the case in which T is given on the boundary. If we use the Green's function for which $G = 0$ on the boundary, we see from (6.85) that the boundary contribution to the solution is

$$-\beta\int_0^t dt' \int T\frac{\partial}{\partial n'}G(\mathbf{r},\mathbf{r}',t-t')\,dS'$$

where the integral is taken over the boundary, and all $t' < t$.

Problem 6-10: A sphere of specific heat C and radius a is heated by a uniform source of strength S in its interior, and radiates freely into its surroundings which are maintained at a fixed temperature T_0. Show that its equilibrium temperature distribution is

$$T = \frac{1}{r}\sum_n f_n \sin \lambda_n r$$

where the λ_n's are the solutions of the equation $\tan \lambda_n a = \lambda_n a/(1+\gamma a)$ and the f_n's are

$$f_n = -\frac{1}{\kappa\lambda_n^2}\frac{4\gamma a^2 S}{1+\gamma a}\frac{\cos\lambda_n a}{2\lambda_n a - \sin 2\lambda_n a}$$

Problem 6-11: The temperature at the surface of a semi-infinite medium is varied with time, $T = T_0(t)$. Assuming no other sources, determine the temperature inside the medium. Consider the special case of a periodic variation

$$T = T_1 + T_2 \sin \omega t$$

Determine at a depth x the phase lag of the temperature variation compared with that at the surface, and the decrease in the amplitude of the variation. Taking for ordinary soil $\kappa = 2 \times 10^{-3}$ cm²/sec, apply the results obtained to the diurnal and annual variations of the earth's temperature.

8. Green's Functions and the Free-Particle Schrödinger Equation

There is a close formal similarity between the equations of diffusion or heat conduction and the time-dependent Schrödinger equation for a free particle, and this similarity makes it interesting to consider the Green's function for the latter problem. The Schrödinger equation is, as will be shown in Chapter 8,

$$i\hbar \frac{\partial \psi}{\partial t} = -\frac{\hbar^2}{2m} \nabla^2 \psi \tag{6.87}$$

where $\hbar$ is $1/(2\pi)$ times Planck's constant and m is the mass of the particle. Defining the Green's function for the equation by

$$i\hbar \frac{\partial G}{\partial t} + \frac{\hbar^2}{2m} \nabla^2 G = i\hbar \delta(\mathbf{r} - \mathbf{r}') \delta(t - t') \tag{6.88}$$

or

$$\frac{\partial G}{\partial t} - \frac{i\hbar}{2m} \nabla^2 G = \delta(\mathbf{r} - \mathbf{r}') \delta(t - t') \tag{6.89}$$

we note the similarity with (6.20). The real constant β has been replaced by the imaginary one $i\hbar/2m$. From (6.25) we then write down the Green's function for the Schrödinger equation in an infinite medium,

$$G(\mathbf{r} - \mathbf{r}', t - t') = \left(\frac{-2mi}{4\pi\hbar(t - t')} \right)^{3/2} e^{-(m|\mathbf{r} - \mathbf{r}'|^2/2i\hbar(t - t'))} \tag{6.90}$$

The most interesting application of this result is that it enables us to follow the history of the particle, knowing its wave function at an initial time. This is given by a simple adaptation of (6.86):

$$\psi(\mathbf{r}, t) = \int d^3\mathbf{r}' \psi(\mathbf{r}', t') G(\mathbf{r} - \mathbf{r}', t - t') \tag{6.91}$$

A particularly simple example is that in which the particle has initially a Gaussian distribution:

$$\psi(\mathbf{r}', 0) = \left(\frac{\alpha}{\pi} \right)^{3/4} e^{-(1/2)\alpha r'^2} \tag{6.92}$$

where the constant has been chosen so that $\int \psi^2 \, d^3\mathbf{r} = 1$. Then

$$\psi(\mathbf{r}, t) = \left(\frac{\alpha}{\pi}\right)^{3/4} \left(\frac{m}{2\pi i\hbar t}\right)^{3/2} \int e^{-(1/2)\alpha r'^2} e^{-(m|\mathbf{r}-\mathbf{r}'|^2/2i\hbar t)} \, d^3\mathbf{r}' \qquad (6.93)$$

$$= \left(\frac{\alpha}{\pi}\right)^{3/4} \left(\frac{m}{2\pi i\hbar t}\right)^{3/2} e^{-(mr^2/2i\hbar t)} \int e^{-(1/2)\alpha r'^2} e^{-(mr'^2/2i\hbar t)} e^{-(m\mathbf{r}\cdot\mathbf{r}'/i\hbar t)} \, d^3\mathbf{r}'$$

The integral is

$$2\pi \int_{-\infty}^{\infty} e^{-[(1/2)\alpha - (im/2\hbar t)]r'^2} \frac{\sin(mrr'/\hbar t)}{(mrr'/\hbar t)} r'^2 \, dr'$$

$$= \frac{\pi\hbar t}{imr} \int_{-\infty}^{\infty} e^{-(1/2)[\alpha - (im/\hbar t)]r'^2} \left(e^{imrr'/\hbar t} - e^{-imrr'/\hbar t}\right) r' \, dr'$$

$$= \frac{2\pi\hbar t}{imr} \int_{-\infty}^{\infty} e^{-(1/2)[\alpha - (im/\hbar t)]r'^2} e^{imrr'/\hbar t} r' \, dr'$$

$$= 2\pi \left(\frac{\hbar t}{im}\right)^2 \frac{1}{r} \frac{\partial}{\partial r} \int_{-\infty}^{\infty} e^{-(1/2)[\alpha - (im/\hbar t)]r'^2} e^{imrr'/\hbar t} \, dr'$$

The integral may now be converted into a Gaussian integral by completing the square. When this is done, and the result is substituted into the expression for $\psi(\mathbf{r}, t)$, it is found that

$$\psi(\mathbf{r}, t) = \left(\frac{\alpha}{\pi}\right)^{3/4} \left(\frac{1}{1 + (i\alpha\hbar t/m)}\right)^{3/2} \exp \frac{-\frac{1}{2}\alpha r^2}{1 + (\alpha^2\hbar^2 t^2/m^2)} \left(1 - \frac{i\alpha\hbar t}{m}\right) \qquad (6.94)$$

The magnitude squared of this quantity, which gives the probability distribution, is

$$|\psi(\mathbf{r}, t)|^2 = \left[\frac{\alpha}{\pi(1 + (\alpha^2\hbar^2 t^2/m^2))}\right]^{3/2} \exp\left\{\frac{-\alpha r^2}{1 + (\alpha^2\hbar^2 t^2/m^2)}\right\} \qquad (6.95)$$

Thus, the distribution remains Gaussian but broadens with time. The peak also decreases, the area under the curve remaining constant. The time required for the width of the distribution to double is given by

$$\frac{\alpha^2\hbar^2 t^2}{m^2} \simeq 3$$

so that

$$t = \frac{\sqrt{3}\, m}{\alpha\hbar} \qquad (6.96)$$

Problem 6-12: Determine $\psi(\mathbf{r}, t)$ when $\psi(\mathbf{r}, 0) = (\alpha/\pi)^{3/4} e^{-(1/2)\alpha r^2} e^{i\mathbf{k}\cdot\mathbf{r}}$ and show that in this case the peak of the distribution moves with velocity $\hbar\mathbf{k}/m$ but otherwise behaves as in the case treated in the text.

Problem 6-13: By using the Fourier transform method, derive the formulas (6.90) for the Green's function, and Eq. (6.94) for the spread of a Gaussian wave packet.

The problem may also be treated by means of Fourier transforms in both the space and time variables. Letting $g(\mathbf{k}, \omega)$ be the transform of $iG(\mathbf{r}, t)$,

$$g(\mathbf{k}, \omega) = \frac{1}{(\hbar k^2/2m) - \omega}$$

Since the wave function at time t is the spatial convolution of the wave function at $t = 0$ and i times the Green's function, if $\phi(\mathbf{k}, \omega)$ is the transform of $\psi(\mathbf{r}, t)$ and $\phi_0(\mathbf{k})$ is the transform of $\psi(\mathbf{r}, 0) = \psi_0(\mathbf{r})$, by the convolution theorem for transforms

$$\phi(\mathbf{k}, \omega) = \left(\frac{1}{2\pi}\right)^3 \frac{\phi_0(\mathbf{k})}{(\hbar k^2/2m) - \omega}$$

Ambiguities arising from inversion in the time variable may be resolved by imposing the condition that the causality condition must hold [see Chapter 4, pp. 203–204].

9. Diffusion of Neutrons

The elementary theory of the diffusion of neutrons in condensed media has been extensively studied in connection with the development of nuclear chain reactors. The problem is more complicated than the diffusion and heat conduction problems already discussed, due to several factors:

(a) The neutrons are produced at high energies and are slowed down or moderated by collisions with the atoms of the crystal. One must, then, consider not only their total density but their density at each energy; that is, one must treat the moderation process.

(b) Neutrons may be "captured" in the medium, that is, there are "sinks" (negative sources)proportional to the density.

(c) If fissile materials are present, there will also be sources, again proportional to density.

The treatment we give is a simplified and somewhat idealized one. In the first place, we deal only with homogeneous systems. Following usual practice, we divide the neutrons into two categories:

(i) "fast" neutrons which have not yet come into thermal equilibrium with their surroundings, and

(ii) "thermal" neutrons, whose energy should have an equilibrium distribution (a Maxwell distribution) with mean energy[3] $k_B T$ at tempera-

---

[3] k_B is the Boltzmann constant whose value is 1.38×10^{-16} erg/deg.

ture T, but which we in fact consider together, ignoring their energy variation.

The treatment of these "thermal" neutrons is the easier part of the problem. We treat them first, formulating the problem for arbitrary source distributions. When we subsequently turn to the problem of the "fast" neutrons, we will be able to consider those which become fully moderated as constituting the "source" of thermal neutrons.

We also confine our attention primarily to "steady-state" problems, in which the densities do not vary with time. The others are of course mathematically considerably more complicated. The steady-state problems are, however, as we will see, already comparable with the other problems to be discussed in this chapter.

10. Diffusion of "Thermal" Neutrons

We start as usual, with the "equation of continuity": (rate of change of number of neutrons in a given volume) = (rate of production in the volume) − (rate of loss) − (efflux over the boundaries). Letting the number of thermal neutrons per unit volume be ρ, the rate of change of the number in a fixed volume V_0 is $(\partial/\partial t) \int_{(V_0)} \rho \, d^3\mathbf{r}$. Defining the "source density" $S_0(\mathbf{r})$, the rate of production in the volume is $\int_{(V_0)} S_0(\mathbf{r}) \, d^3\mathbf{r}$. If the probability per unit time of "capture" of each neutron is $1/\tau$, the rate of loss is $(1/\tau) \int_{(V_0)} \rho \, d^3\mathbf{r}$. Finally, the rate of efflux is $\int_{\sigma_0} \mathbf{j} \cdot \mathbf{n} \, dS$, a normal surface integral over the surface σ_0 bounding the volume V_0. $\mathbf{j}$ is the thermal neutron current, that is, $\mathbf{j} \cdot \mathbf{n}$ is the number flowing over unit area perpendicular to the unit vector $\mathbf{n}$ per unit time. The surface integral may be converted by Gauss' theorem into a volume integral over V_0, becoming $\int_{(V_0)} \boldsymbol{\nabla} \cdot \mathbf{j} \, d^3\mathbf{r}$. If we then assume Fick's law,

$$\mathbf{j} = -D\boldsymbol{\nabla}\rho \qquad (6.97)$$

the continuity equation takes the form

$$\int_{(V_0)} \left\{ \frac{\partial \rho}{\partial t} - S_0 + \frac{\rho}{\tau} - D\nabla^2\rho \right\} d^3\mathbf{r} = 0 \qquad (6.98)$$

Since this is true for an *arbitrary* volume, the integrand must be zero everywhere, which yields the diffusion equation

$$\frac{\partial \rho}{\partial t} + \frac{\rho}{\tau} - D\nabla^2\rho = S_0 \qquad (6.99)$$

Problem 6-14: Show that the infinite-medium Green's function (point source solution) for Eq. (6.99) is

$$G(\mathbf{r}-\mathbf{r}',t) = \left(\frac{\tau}{4\pi L^2 t}\right)^{3/2} e^{-t/\tau}\, e^{-r^2\tau/4L^2 t}$$

where the "diffusion length" $L^2 = D\tau$.

It may be shown by more detailed and rigorous analysis that Fick's law is valid only so long as the angular distribution of neutron directions at a given point is not too anisotropic. It will thus not be valid near localized sources (where almost all neutrons are flowing *away* from the source), or near boundaries, where almost all are flowing outward. The first restriction is not too important in the problems we consider. As for the second, it may be shown once again that the error introduced by the assumption of Fick's law is appreciable only very close to the surface (within a distance comparable with the "scattering length", the mean distance travelled by neutrons between scatterings); the density further in the interior of the medium may be correctly determined by making a small change in boundary conditions, corresponding to the introduction of an *artificial* boundary slightly outside the actual one. We, therefore, do not concern ourselves further with this point.

The "steady-state" equation is obtained by putting the time-variable term equal to zero. Furthermore, let us divide through by D, and put

$$D\tau = L^2 \tag{6.100}$$

and

$$S = S_0/D \tag{6.101}$$

so that the equation takes the form

$$\nabla^2\rho - \frac{\rho}{L^2} = -S \tag{6.102}$$

This is now a steady-state diffusion equation with the added feature of the *dissipation* term $-\rho/L^2$. The dissipation aspect is one which will characterize all subsequent problems to be discussed in this chapter.

The quantity L defined by (6.100) has the dimensions of a length, and characterizes the scale of variation of the density. It is often referred to as the "diffusion length." The quantity $\nabla^2\rho$ characterizes the "curvature" or "buckling" of the distribution, which is then determined by the source distribution, which tends to *increase* the current, and the dissipation, which tends to decrease it.

It is a trivial matter to verify that the Green's function for the one-dimensional problem in an infinite medium

$$\frac{d^2}{dz^2}G_0(z) - \frac{1}{L^2}G_0(z) = -\delta(z) \quad (6.103)$$

is

$$G_0(z) = \frac{L}{2}e^{-|z|/L} \quad (6.104)$$

By using Eqs. (3.168) and (3.170), we may derive the "point source" and "shell source" Green's functions in an infinite medium:

$$G_{pt}(r) = \frac{1}{4\pi r}e^{-r/L} \quad (6.105)$$

and

$$G_{sh}(r, r') = \frac{1}{8\pi rr'}[e^{-|r-r'|/L} - e^{-|r+r'|/L}] \quad (6.106)$$

For finite media with a linear geometry or a spherical geometry, the method of images may be used to obtain the appropriate Green's functions. In neutron diffusion, the most interesting problems are those in which there is free efflux over the boundary. The boundary condition in this case is that the density goes to zero at the "artificial boundary" which is only a fraction of a scattering length outside the real one. This scattering length is in general only a minute fraction of the dimensions of the media being considered.

In the case of a plane source at $z = z'$ in a medium bounded at $z = 0$ and $z = a$, the Green's function can be written in terms of an infinite series of images

$$G(z, z') = \sum_{n=-\infty}^{\infty}[G_0(z - z' - 2na) - G_0(z + z' - 2na)] \quad (6.107)$$

By Poisson's formula (4.82) this may be put in an alternative form, in terms of the Fourier transform $g_0(\mathbf{k})$ of $G_0(z - z')$:

$$g_0(\mathbf{k}) = \frac{L^2}{2\pi}\frac{1}{1 + k^2L^2}e^{-ikz'} \quad (6.108)$$

The Green's function (6.107) may then be written in the form

$$G(z, z') = \frac{2L^2}{a}\sum_{n=1}^{\infty}\frac{1}{1 + (n^2\pi^2L^2/a^2)}\sin\frac{n\pi z'}{a}\sin\frac{n\pi z}{a} \quad (6.109)$$

Problem 6-15: Find the Green's function for a source at point x_1, y_1, z_1, in the rectangular parallelepiped bounded by $x = 0, a; y = 0, b; z = 0, c;$ (*i*) as a series of images and (*ii*) in a Fourier series form.

Problem 6-16: If, for each neutron captured, k new neutrons are produced, show that the steady-state equation for a reactor is

$$\nabla^2\rho + \frac{k-1}{L^2}\rho = 0$$

Assuming that $\rho = 0$ on the surface of a spherical reactor of radius a, find a such that the density is constant for a given multiplication constant k.

If we consider a reactor like the above surrounded by a nonmultiplying medium with diffusion length L', where

$$\frac{k-1}{L^2} \ll \frac{1}{L'^2}$$

calculate approximately the reduction in a for a given k if the reactor remains steady. [Assume ρ, $(\partial\rho/\partial n)$ continuous at the boundary between reactor and sheath.]

Similarly, the Green's function for a shell source at r' in a spherical medium of radius a is, in terms of images,

$$G_a(r, r') = \frac{L}{4\pi r r'} \sum_{n=-\infty}^{\infty} [e^{-|r-r'+(2na/L)|} - e^{-|r+r'+(2na/L)|}] \qquad (6.110)$$

or in terms of Fourier series,

$$G_a(r, r') = \frac{L^2}{2\pi r r' a} \sum_{n=0}^{\infty} \frac{1}{1+(n^2\pi^2 L^2/a^2)} \sin\frac{n\pi r'}{a} \sin\frac{n\pi r}{a} \qquad (6.111)$$

These solutions complement each other in that the first converges rapidly for $a \gg L$ and the second for $a \ll L$.

A case of some interest in practice is that of a *cylindrical* medium of radius b and height h, that is, extending between $z = 0$ and h. Since we do not know how to do this by the method of images, let us do it directly by the method of expansion in series. We assume a series solution of the form

$$G = \sum_{l=1}^{\infty} \sum_{m=0}^{\infty} \sum_{n=1}^{\infty} A_{lmn} J_m\left(\lambda_{lm}\frac{r}{a}\right) e^{im\phi} \sin\frac{n\pi z}{h} \qquad (6.112)$$

J_m is the Bessel function of integer order m, and the λ_{lm}'s are its zeros. If the unit source has coordinates (r', ϕ', z') it may be represented by a similar expansion.

$$\frac{1}{r'}\delta(r-r')\delta(\phi-\phi')\delta(z-z') = \sum \alpha_{lmn} J_m\left(\lambda_{lm}\frac{r}{a}\right) e^{im\phi} \sin\frac{n\pi z}{h}$$

$$(6.113)$$

Using the orthogonality of the expansion functions, we find that

$$\alpha_{lmn}\left[-\frac{a^2}{2}J_{m-1}(\lambda_{lm})J_{m+1}(\lambda_{lm})\right]\pi h = J_m\left(\lambda_{lm}\frac{r'}{a}\right)e^{-im\phi'}\sin\frac{n\pi z'}{h}$$

or

$$\alpha_{lm} = -2\frac{1}{\pi a^2 h J_{m-1}(\lambda_{lm})J_{m+1}(\lambda_{lm})}J_m\left(\lambda_{lm}\frac{r'}{a}\right)e^{-im\phi'}\sin\frac{n\pi z'}{h} \quad (6.114)$$

Substituting both (6.112) and (6.113) in the Green's function equation

$$\nabla^2 \rho - \frac{\rho}{L^2} = -\delta(\mathbf{r}-\mathbf{r}') = -\frac{1}{r'}\delta(r-r')\delta(\phi-\phi')\delta(z-z') \quad (6.115)$$

we obtain the equation for A_{lmn},

$$A_{lmn} = \frac{\alpha_{lmn}}{(\lambda_{lm}^2/a^2)+(n^2\pi^2/h^2)+(1/L^2)} \quad (6.116)$$

so that the Green's function is given by the expansion

$$G = -\frac{2}{\pi a^2 h}\sum_{l,m,n}\frac{1}{J_{m-1}(\lambda_{lm})J_{m+1}(\lambda_{lm})}$$

$$\times J_m\left(\lambda_{lm}\frac{r'}{a}\right)e^{-im\phi'}\sin\frac{n\pi z'}{h}J_m\left(\lambda_{lm}\frac{r}{a}\right)e^{im\phi}\sin\frac{n\pi z}{h}$$

$$\times\frac{1}{(\lambda_{lm}^2/a^2)+(h^2\pi^2/h^2)+(1/L^2)} \quad (6.117)$$

It is clear that, because of the Bessel functions, this is not in a form which will permit the application of the Poisson formula to obtain a solution in terms of image sources.

11. Moderation of Fast Neutrons

The process of slowing down of neutrons by collision with atoms of the moderating material is a stochastic one, and hence the moderation process should, strictly speaking, be treated by the theory of probability. We should, for instance, to be precise, deal with the *probability* that a neutron loses a certain amount of energy in a given time. We simplify matters, however, by dealing only with *averages*, ignoring the fluctuations for individual neutrons. That this is justified will be seen, from the theory to be developed in the next chapter for stochastic processes, to be assured by the great number of particles being considered.

Elementary dynamics tells us that, in a pure medium, the average *fractional* energy loss per collision is independent of the energy (or velocity) of the particle. In the time interval $(t, t+dt)$ the fractional energy loss $-(dE/E)$ = probability of a collision $P_s(E)dt$, times the fractional energy loss per collision f. [$P_s(E)$ is the probability per unit time of a scattering collision at energy E.] Thus

$$-\frac{dE}{E} = fP_s(E)\, dt \qquad (6.118)$$

This equation establishes a unique relation between energy and time, which is completely determined if the initial energy is known. Thus, instead of specifying the neutrons by their energy, they can be specified by their "age" — the time interval t required for them to reach that energy. We designate this age as ξ.

Let us now specify that $\chi(\mathbf{r}, t, \xi)$ is the density at time t of neutrons in the age interval $(\xi, \xi + d\xi)$. Since the age ξ of each *individual* neutron increases in proportion to time elapsed, χ is also the number of neutrons reaching the age ξ per unit time.

χ must of course be determined by an equation of continuity. We derive it from the statement that the number of neutrons of age $(\xi, \xi + d\xi)$ at time t in a volume V is the number at age $\xi - dt$ at time $t - dt$, minus the number of that age which have flowed out over the boundary, minus the number unproductively captured in the medium, plus the number at that age *produced* in the volume per unit time.

Introducing the notations:

$J(\mathbf{r}, t, \xi)$ = current density of neutrons of age $(\xi, \xi + d\xi)$ at time t
$P_c(\xi)$ = probability per unit time of capture at age ξ
$\sigma_0(\mathbf{r}, t, \xi)$ = source density of neutrons of age ξ at time t

We may express the equation of continuity in the mathematical form

$$\begin{aligned}
\int \chi(\mathbf{r}, t, \xi)\, d^3\mathbf{r}\, d\xi = &\int \chi(\mathbf{r}, t - dt, \xi - dt)\, d^3\mathbf{r}\, d\xi \\
&- dt \int J(\mathbf{r}, t, \xi) \cdot \mathbf{n}\, dS\, d\xi \\
&- dt P_c(\xi) \int \chi(\mathbf{r}, t, \xi)\, d^3\mathbf{r}\, d\xi \\
&+ dt \int \sigma_0(\mathbf{r}, t, \xi)\, d^3\mathbf{r}\, d\xi \qquad (6.119)
\end{aligned}$$

The volume integrals are over the volume V, the surface integral over the surface surrounding it. The fission processes produced by fast neutrons have been neglected.

Introducing again Fick's law for fast neutrons

$$J(\mathbf{r}, t, \xi) = -D_1(\xi)\nabla\chi(\mathbf{r}, t, \xi) \qquad (6.120)$$

and converting the surface integral into a volume integral, we get the differential equation

$$\frac{\partial\chi}{\partial t} + \frac{\partial\chi}{\partial\xi} - D_1\nabla^2\chi + P_c\chi = \sigma_0 \qquad (6.121)$$

A convenient simplification may be made by introducing the "symbolic age"

$$\theta = \int_0^\xi D_1(\xi)\, d\xi \qquad (6.122)$$

which has the dimensions of the square of a length. D_1 may then, in principle, be written as a function of θ. Equation (6.121) becomes

$$\frac{1}{D_1}\frac{\partial \chi}{\partial t} + \frac{\partial \chi}{\partial \theta} - \nabla^2 \chi + \frac{P_c}{D_1}\chi = \sigma \qquad (6.123)$$

where

$$\sigma = \frac{\sigma_0}{D_1} \qquad (6.124)$$

is the source density per unit range of *symbolic* age.

12. Model of a Neutron Chain Reactor

We may now develop a simplified model of a homogeneous chain reactor, based on the following assumptions:

(1) The "fast" neutrons produced in the process of fission are described by (6.123), and reach "thermal" energy at the symbolic age $\theta = \theta_0$. (The assumption of a unique symbolic age for thermal energy implies that all fission neutrons are produced at the same energy — a convenient mathematical idealization. If it is not made, the analysis becomes considerably more cumbersome.)

(2) The neutrons reaching this age are regarded as sources of thermal neutron density. As remarked above, the density χ per unit age interval is the number per unit volume reaching this age per unit time; therefore, $\chi(\mathbf{r}, t, \theta_0)$ is the source σ_0 in the thermal neutron equation. This equation then becomes

$$\frac{1}{D}\frac{\partial \rho}{\partial t} + \frac{\rho}{L^2} - \nabla^2 \rho = \frac{1}{D}\chi(\mathbf{r}, t, \theta_0) \qquad (6.125)$$

(3) Since the "thermal" neutrons which are captured may give rise to fission, and therefore act as sources of fast neutrons (of symbolic age zero), the source term in the fast neutron equation in a reactor isolated from external sources may be expressed in terms of the *thermal* neutron density.

In time-dependent problems, however, it is important to take account of the lapse of time between slow neutron capture and the creation of fission neutrons. The majority of these neutrons are emitted at the instant of splitting of the capturing nucleus. The time interval between capture and fission is very short on the macroscopic time scale; we therefore neglect it and assume that their contribution to the fast neutron source is

$$\sigma_0' = \frac{\rho(\mathbf{r}, t)}{\tau} n_0 \qquad (6.126)$$

n_0 is the average number of "prompt" neutrons arising from capture. (It takes account, of course, of the fact that some of the capture processes are "parasitic" and do not give rise to fission.)

There are, however, some fast neutrons which arise from the radioactive decay of fission products. Since these decays may be relatively long-lived, their contribution to the source does not depend on $\rho(t)$ but on the thermal neutron density for earlier times.

There will in fact be several such "delayed neutron" lifetimes; we, however, idealize the problem by considering only one. (Again, the problem is quite tractable without this assumption, but becomes formally more complicated. Our model is quite adequate to show the physical effect of the delayed neutrons.)

Consider then, fission products with neutron emission lifetime T. Such fission products decay with time according to the law

$$e^{-t/T} \tag{6.127}$$

Suppose that the number of such neutron-active fission products produced per capture is $n - n_0$ (so that n is the *total* number of fission neutrons resulting from a capture). The number produced per unit time at time t' is then

$$\frac{(n - n_0)\rho(\mathbf{r}, t')}{\tau}$$

The *probability* per unit time that any one of them will decay at time $t(> t')$ is $(1/T) e^{-(t-t')/T}$. Consequently, they contribute to the source at time t a term

$$\sigma_0'' = \frac{n - n_0}{\tau} \int_0^t \rho(\mathbf{r}', t') \frac{1}{T} e^{-(t-t')/T} \, dt' \tag{6.128}$$

Here $t' = 0$ is a time in the past at which the system was put into operation. Adding (6.126) and (6.128) and multiplying by $\delta(\xi) = (1/D_1)\delta(\theta)$ we get the total source to be used in the fast neutron equation (6.123). This equation then takes the form

$$\frac{1}{D_1}\frac{\partial \chi}{\partial t} + \frac{\partial \chi}{\partial \theta} - \nabla^2 \chi = \frac{\delta(\theta)}{\tau} \left\{ n_0 \rho(\mathbf{r}, t) + (n - n_0) \int_0^t \rho(\mathbf{r}', t') e^{-(t-t')/T} \frac{dt'}{T} \right\} \tag{6.129}$$

The equations (6.125) and (6.129) therefore constitute the basic equations for our model chain reactor.

In the case of steady-state problems, these equations take the simpler form

$$\nabla^2 \rho - \frac{\rho}{L^2} = -\frac{1}{D} \chi(\mathbf{r}, \theta_0) \tag{6.130}$$

and

$$\frac{\partial \chi}{\partial \theta} - \nabla^2 \chi = \frac{n}{\tau} \rho(\mathbf{r})\delta(\theta) \tag{6.131}$$

13. The Problem of Criticality

We first use the time-dependent equations (6.125) and (6.129) to investigate when the reactor is subcritical and when it is supercritical; that is, when an initial distribution decays and when it "explodes." At the same time, we determine the time scale for these contingencies.

For finite systems, it is not really necessary to solve a particular equation for the spatial variation of the densities. Whatever the geometry, we can solve the eigenvalue problem

$$\nabla^2 \begin{pmatrix} \rho \\ \chi \end{pmatrix} + \beta^2 \begin{pmatrix} \rho \\ \chi \end{pmatrix} = 0 \tag{6.132}$$

subject to the given boundary conditions.

Problem 6-17: (a) Show that, for a system in the form of a rectangular parallelepiped of dimensions $a \times b \times c$, the possible values of β^2 are

$$\beta^2 = \left(\frac{l^2}{a^2} + \frac{m^2}{b^2} + \frac{n^2}{c^2} \right) \pi^2$$

where l, m, and n are arbitrary integers.

(b) Show that, for a cylindrical reactor of height h and radius R

$$\beta^2 = \frac{l^2 \pi^2}{h^2} + \frac{\alpha_{mn}^2}{R^2}$$

where

$$J_m(\alpha_{mn}) = 0$$

l and m being integers.

Each spatial mode determined by (6.132) has its separate evolution in time, determined by diffusion equations. If normalized solutions of (6.132) are designated by f, solutions for the densities may be found in the form

$$\rho = \rho_0(t) f_\beta \tag{6.133}$$

$$\chi = \chi_0(\theta, t) f_\beta \tag{6.134}$$

ρ_0 and χ_0 then satisfy the coupled equations

$$\frac{1}{D} \frac{\partial \rho_0}{\partial t} + \left(\beta^2 + \frac{1}{L^2} \right) \rho_0 = \frac{1}{D} \chi_0 (t, \theta_0) \tag{6.135}$$

and

$$\frac{1}{D_1} \frac{\partial \chi_0}{\partial t} + \frac{\partial \chi_0}{\partial \theta} + \beta^2 \chi_0 = \frac{\delta(\theta)}{T} \left[n_0 \rho_0(t) + (n - n_0) \int_0^t \rho_0(t') e^{-(t-t')/T} dt' \right] \tag{6.136}$$

Let us now take Laplace transforms in time, designating the transforms respectively, as $\bar{\rho}_0(s)$ and $\bar{\chi}_0(s, \theta)$. The *initial* values of ρ_0 and χ_0 are taken to be

$$\rho_0(0) = c \tag{6.137}$$

$$\chi_0(\theta, 0) = \gamma(\theta) \tag{6.138}$$

respectively. The transformed equations are then

$$-\frac{c}{D} + \frac{s}{D}\bar{\rho}_0 + \left(\beta^2 + \frac{1}{L^2}\right)\bar{\rho}_0 = \frac{1}{D}\bar{\chi}_0(s, \theta_0) \tag{6.139}$$

and

$$-\frac{\gamma(\theta)}{D_1} + \frac{s}{D_1}\bar{\chi}_0 + \frac{\partial \bar{\chi}_0}{\partial \theta} + \beta^2\bar{\chi}_0 = \frac{\delta(\theta)}{\tau}\left[\frac{n + n_0 sT}{1 + sT}\right]\bar{\rho}_0 \tag{6.140}$$

Let us now define a Green's function corresponding to (6.140):

$$\frac{\partial g(\theta, \theta')}{\partial \theta} + \left(\beta^2 + \frac{s}{D_1}\right)g(\theta, \theta') = \delta(\theta - \theta') \tag{6.141}$$

It is then easy to show that

$$g(\theta, \theta') = \exp\left[-\beta^2(\theta - \theta') - s\int_{\theta'}^{\theta}\frac{d\theta''}{D_1(\theta'')}\right] \tag{6.142}$$

The solution of (6.140) may be written in terms of this Green's function as

$$\bar{\chi}_0 = \frac{1}{\tau}\bar{\rho}_0\frac{n + n_0 sT}{1 + sT}g(\theta, 0)$$

$$+ \int \frac{\gamma(\theta')}{D_1(\theta')}\exp\left[-\beta^2(\theta - \theta') - s\{\xi(\theta) - \xi(\theta')\}\right]d\theta' \tag{6.143}$$

We have used (6.122) to introduce $\xi(\theta)$.

Substituting from (6.143) in (6.139) we find for $\bar{\rho}_0$ the value

$$\bar{\rho}_0 = \frac{\lambda}{\frac{s}{D} + \beta^2 + \frac{1}{L^2}\left(1 - \frac{n + n_0 sT}{1 + sT}e^{-\beta^2\theta_0 - s\xi(\theta_0)}\right)}$$

$$\times \left\{\frac{c}{D} + \frac{1}{D}\int_0^{\theta_0}\frac{\gamma(\theta')}{D_1(\theta')}\exp\left[-\beta^2(\theta_0 - \theta') - s\{\xi(\theta_0) - \xi(\theta')\}\right]d\theta'\right\}$$

$$\tag{6.144}$$

In general the inversion of this formula to obtain $P_0(t)$ is a formidable task. However, we know from the Laplace inverse theorem (4.182) that the solution is a combination of exponentials whose exponents are the zeros of the denominator of (6.144). Criticality, for a given mode, then depends on having a zero with a positive real part. The physically evident fact that the system cannot be critical if $n < 1$, that is, if less than one neutron is generated for each one absorbed, is reflected in the fact that the magnitude, and consequently the real part of the last term in the denominator is positive for Re $s > 0$; the denominator therefore cannot be zero in this case.

In any case, the last term in the denominator is the only one which can give rise to a negative real part, and it is largest when s is real. Therefore, the root with the largest real part has imaginary part zero. Thus, to look for the root which dominates the behavior at large t, we must look for a *real* zero of the denominator. To have a root with positive real part, it is clearly necessary that n be sufficiently large and/or β^2 sufficiently small. Now there is a root *exactly* at $s = 0$ when

$$1 + \beta^2 L^2 = n e^{-\beta^2 \theta_0} \qquad (6.145)$$

This condition marks the onset of criticality. It may be written alternatively as

$$n = (1 + \beta^2 L^2) e^{\beta^2 \theta_0} \qquad (6.146)$$

A given system and mode becomes critical at this value of n, and is subcritical for smaller n. Alternatively, for fixed n, criticality can ultimately be reached by decreasing β, that is, by *increasing* the size of the system so as to decrease the loss of neutrons over the surface.

Since the right-hand side is a monotonically increasing function of β, criticality is first reached for the *minimum* possible value of β. Thus, it is the lowest mode that most readily becomes critical. Even at the point at which this mode has become critical, the others are highly transient. Thus the density distribution is that of the lowest mode.

It is possible to calculate explicitly the decay or growth rate of the density close to critical conditions. One merely expands the denominator for small s and calculates approximately the root. After some straightforward algebra, we obtain

$$s_0 = \frac{n e^{-\beta^2 \theta_0} - (1 + \beta^2 L^2)}{\tau + e^{-\beta^2 \theta_0} [(n - n_0) T + n \xi(\theta_0)]}$$

$$\simeq \frac{n e^{-\beta^2 \theta_0} - (1 + \beta^2 L^2)}{\tau + (1 + \beta^2 L^2) [(1 - (n_0/n)) T + \xi(\theta_0)]} \qquad (6.147)$$

Since the density evolves in time as $e^{s_0 t}$, we see that the rate is determined primarily by the longest of three times:

(*i*)   the thermal neutron capture lifetime

(*ii*)   the slowing-down time

(*iii*) the mean delay time for emission of fission neutrons after capture [for a number of delay times it may be shown that the term $(1 - (n_0/n)) T$ should be replaced by

$$\frac{1}{n} \sum \delta n_i T_i$$

the T_i's being the delay times and the δn_i's being the contributions to n from each group of neutrons]. Since the numerator is small, the "reaction time" $1/s_0$ is in fact large compared with all the aforementioned times.

A model of a chain reactor proceeding through fission produced by capture of fast neutrons may be generated by taking $\tau = 0$ and replacing $\xi(\theta_0)$ by the mean time for capture of a fast neutron, in Eq. (6.147). We do not pursue this problem except to remark that, since the slowing-down time $\xi(\theta_0)$ is in such cases very short, the reaction time of the system is dominated by the delayed neutrons. Were this not the case, such a system would react too quickly for an automatic control system (e.g., through introduction of a capturing impurity) to work. The delayed neutrons, however, introduce enough "sluggishness" into the system to enable it to be kept under control.

It should be noted that the criticality conditions can be deduced from the steady-state equations. For the steady state is that which marks the division between subcritical and supercritical. Therefore the condition of self-consistency between (6.130) and (6.131) in the case of the "lowest" mode (that is, the mode with smallest β), determines the relation between multiplication constant n and dimensions for the onset of criticality. For larger n or larger dimensions, the system becomes supercritical.

14. Solution of the Steady-State Equations

We have already discussed the calculation of Green's functions for the thermal neutron equation (6.130). Since the fast neutron equation (6.131) is identical with the heat conduction equation, Green's function calculations for (6.131) may be written down directly by analogy with those calculated for that problem.

Consider, as an example, a spherical (and spherically symmetric) system of radius a. Using (6.110), ρ may be written in terms of χ from (6.130):

$$\rho(\mathbf{r}) = \frac{L}{8\pi r D} \int_0^a \chi(\mathbf{r}', \theta_0) \sum_{n=-\infty}^{\infty} \left[e^{-|r-r'+2na|/L} - e^{-|r+r'+2na|/L} \right] 4\pi r' \, dr'$$

$$(6.148)$$

This could then be substituted in (6.131) to yield an integro-differential equation for χ. Similarly, the corresponding Green's function for (6.131) could be calculated, and the result substituted in (6.130) to give an integro-differential equation for ρ. Neither procedure, however, leads to easily manageable equations. But if the Fourier series forms of the Green's functions are used, the problem becomes quite tractable. It becomes equivalent, however, to a calculation in which both ρ and χ are expanded in appropriate Fourier series and the coupled equations (6.130) and (6.131) are used to obtain coupled equations in the coefficients. Since they are homogeneous equations, nonzero solutions are only possible when there is a quite specific relation between n and the dimensional parameters. Although a "steady" solution is possible for densities with the distribution of each Fourier component, those for the higher Fourier components will, as seen in the previous section, correspond to such a combination of high n and large dimensions that all lower components would be extremely unstable.

The problem is in fact then one of obtaining the lowest β such that a solution of

$$\nabla^2 f + \beta^2 f = 0 \tag{6.149}$$

satisfies the boundary conditions of the problem, viz., f finite at $r = 0$ and $f = 0$ at $r = a$. The solution is clearly

$$f = \frac{1}{r} \sin \frac{\pi r}{a} \tag{6.150}$$

with

$$\beta^2 = \left(\frac{\pi}{a}\right)^2 \tag{6.151}$$

Putting

$$\rho = \rho_0 f \tag{6.152}$$

and

$$\chi = \chi_0 f \tag{6.153}$$

we obtain for ρ_0 and χ_0 the equations

$$\left(\beta^2 + \frac{1}{L^2}\right) \rho_0 = \frac{1}{D} \chi_0(\theta_0) \tag{6.154}$$

and

$$\frac{\partial \chi_0}{\partial \theta} + \beta^2 \chi_0 = \frac{n}{\tau} \rho_0 \delta(\theta) \tag{6.155}$$

The solution of the second is

$$\chi = \frac{n}{\tau} \rho_0 e^{-\beta 2\theta} \tag{6.156}$$

Hence from (6.154) the condition that ρ_0 be nonzero (that is, the condition for a stable steady-state solution) is

$$\beta^2 + \frac{1}{L^2} = \frac{n}{D\tau} e^{-\beta 2\theta_0} = \frac{n}{L^2} e^{-\beta 2\theta_0}$$

or

$$n = (1 + \beta^2 L^2) e^{\beta 2\theta_0}$$

$$= \left(1 + \frac{\pi^2 L^2}{a^2}\right) e^{\pi^2 \theta_0 / a^2} \tag{6.157}$$

15.   Diffusion Length, Slowing-Down Length, and Migration Length

Some physical insight into the process of neutron diffusion may be obtained by considering point sources of neutrons and observing the distributions which result. This is, of course, what is given by the Green's functions! We pay particular attention to the extent of the distribution set up by a source, which gives us information about the range of the neutrons in the material.

For this purpose, we consider the nonmultiplicative case $(n = 0)$. The multiplicative term in (6.131), being itself a source term, is suppressed to enable us to see the effect of a single source.

We also consider, in the first instance, infinite media, so that the loss of neutrons over boundaries may be ignored.

We note that, from the Fourier transform of the density, we may derive the spatial *moments* of the distribution, and in particular the second moment, which gives the mean square distance of neutrons from the source. If we take a density $f(\mathbf{r})$ due to a source at the origin, $(2\pi)^3$ times the Fourier transform is

$$F_0(\mathbf{k}) = \int f(\mathbf{r})\, e^{-i\mathbf{k}\cdot\mathbf{r}}\, d^3\mathbf{r} \tag{6.158}$$

Then

$$F_0(0) = \int f(\mathbf{r})\, d^3\mathbf{r} = M_0 \tag{6.159}$$

is the integrated density and

$$[i\nabla_k F_0]_{k=0} = \int \mathbf{r} f(\mathbf{r})\, d^3\mathbf{r} = \mathbf{M}_1 \tag{6.160}$$

$(1/M_0)\mathbf{M}_1$ then gives the mean displacement from the source. Also

$$- [\nabla_k{}^2 F_0]_{k=0} = \int r^2 f(\mathbf{r})\, d^3\mathbf{r} = M_2 \tag{6.161}$$

where (M_2/M_0) is the mean square displacement.

If $f(r)$ is spherically symmetrical, it is easily seen by doing the angular integration in (6.158) with $\mathbf{k}$ as axis that $F_0(\mathbf{k})$ is a function of $k = |\mathbf{k}|$ only. In this case $M_1 = 0$. Also M_2 is six times the coefficient of k^2 in the Taylor's series expansion in k of $(2\pi)^3 F_0 = \int f(\mathbf{r})\, e^{-i\mathbf{k}\cdot\mathbf{r}} d^3\mathbf{r}$. $\mu_2 = M_2/M_0$ is the mean square radius of the density distribution.

Consider now a point source of thermal neutrons at the origin, for which the density ρ is determined by (6.99)

$$\rho - L^2\nabla^2\rho = \tau\delta(\mathbf{r}) \tag{6.162}$$

Taking Fourier transforms, and letting the transform of ρ be $\bar{\rho}(\mathbf{k})$, this yields

$$\bar{\rho}(\mathbf{k}) = \frac{\tau}{1 + k^2 L^2} \tag{6.163}$$

Then

$$M_0 = \tau \qquad (6.164)$$

$$M_2 = 6L^2\tau \qquad (6.165)$$

so that

$$\mu_2^{(t)} = 6L^2 \qquad (6.166)$$

The superscript t on μ_2 designates thermal neutrons.

Consider next fast neutrons, with density given by

$$\frac{\partial \chi}{\partial \theta} - \nabla^2 \chi = \delta(\theta)\delta(\mathbf{r}) \qquad (6.167)$$

The transform $\bar{\chi}$ is given by

$$\frac{\partial \bar{\chi}}{\partial \theta} + k^2 \bar{\chi} = \delta(\theta)$$

which has the solution

$$\bar{\chi} = e^{-k^2\theta} \qquad (6.168)$$

In this case

$$M_0 = 1$$

$$M_2 = 6\theta \qquad (6.169)$$

and

$$\mu_2^{(\theta)} = 6\theta \qquad (6.170)$$

Thus 6θ is the square of the mean distance traveled by fast neutrons in reaching the symbolic age θ. In particular

$$\mu_2^{(\theta_0)} = 6\theta_0 = 6\lambda_0^2 \qquad (6.171)$$

is the mean square distance traveled in becoming thermalized; the length λ_0 defined by (6.171) is designated the "slowing-down" length.

We may finally determine the mean square distance traveled by a neutron from its birth at symbolic age $\theta = 0$ to its capture as a thermal neutron. For this purpose we use the coupled equations (6.167) and (6.130). From the first the transform $\bar{\chi}$ is given by (6.168). The transform of (6.130) then determines $\bar{\rho}$:

$$\left(k^2 + \frac{1}{L^2}\right) \bar{\rho} = \frac{1}{D} \bar{\chi}(\theta_0)$$

$$= \frac{1}{D} e^{-k^2\theta_0}$$

so that

$$\bar{\rho} = \frac{\tau e^{-k^2\theta_0}}{1 + k^2 L^2} \qquad (6.172)$$

Thus, considering the distribution of *thermal* neutrons resulting from a point source of *fast* neutrons at the origin,

$$M_0 = \tau$$
$$M_2 = 6(\theta_0 + L^2)\tau = 6(\lambda_0^2 + L^2)\tau$$

and

$$\mu_2 = 6(\lambda_0^2 + L^2) \tag{6.173}$$

The quantity $\sqrt{6(\lambda_0^2 + L^2)}$ is known as the "migration length." That it is the squares of the lengths and not the lengths themselves which are additive follows naturally from the incoherence of the two processes; on the average, the two displacements are orthogonal.

Pursuing this line of investigation, one other problem is of some interest. It is that of the mean square extent of the distribution from a point source in a *multiplying* medium. This has to do, not with the distance travelled by the original neutron, but by it and its descendants. The relevant equations are (6.130) and (6.131), the latter supplemented by a source term $\delta(r)\delta(\theta)$. Taking transforms of the two equations gives

$$\left(k^2 + \frac{1}{L^2}\right)\bar{\rho} = \frac{1}{D}\bar{\chi}(\theta_0) \tag{6.174}$$

and

$$\frac{\partial\bar{\chi}}{\partial\theta} + k^2\bar{\chi} = \left(\frac{n}{\tau}\bar{\rho} + 1\right)\delta(\theta) \tag{6.175}$$

The solution of the second is

$$\bar{\chi} = \left(1 + \frac{n}{\tau}\bar{\rho}\right)e^{-k^2\theta} \tag{6.176}$$

Substituting this in the first gives

$$\bar{\rho} = \frac{\tau e^{-k^2\theta_0}}{1 - ne^{-k^2\theta_0} + k^2 L^2} \tag{6.177}$$

It then follows that

$$M_0 = \frac{\tau}{1 - n} \quad \text{and} \quad M_2 = \frac{6\tau}{1 - n}\left(\frac{\theta_0}{1 - n} + L^2\right)$$

so that

$$\mu_2 = 6\left(\frac{\theta_0}{1 - n} + L^2\right) \tag{6.178}$$

We see, then, that the total number of neutrons tends to grow indefinitely as $n \to 1$, and that the modified migration length also becomes indefinitely great.

Problem 6-18:   Show that the distributions of fast and thermal neutrons from a monoenergetic source of strength σ_0 at the center of a sphere of radius a are

$$\chi = \frac{\sigma_0}{2a^2r} \sum_{l=1}^{\infty} l \sin\frac{l\pi r}{a} e^{-l^2\pi^2\theta/a^2}$$

and

$$\rho = \frac{\sigma_0\tau}{2a^2r} \sum_{l=1}^{\infty} \frac{l}{1+(l^2\pi^2L^2/a^2)} e^{-l^2\pi^2\theta_0/a^2} \sin\frac{l\pi r}{a}$$

Use Poisson's formula to find series solutions which converge rapidly for $a \gg L$.

Problem 6-19: Show that the critical condition for a spherical reactor of radius a with a hole of radius b at its center is

$$n = \left(1+\frac{x^2L^2}{a^2}\right) e^{x^2\theta_0/a^2}$$

where x satisfies $\tan x(1-b/a) = -xb/a$. Hence show that, for a definite n, it requires a greater amount of material for criticality if there is a hole in the material.

Problem 6-20: For a given multiplication factor, calculate critical conditions for a sphere, a cylinder, and a rectangular parallelepiped. In the latter case, determine the relative dimensions so that criticality can be obtained with the minimum volume of material. Compare the volumes needed for the various shapes. Show that, for sphere, cylinder, and cube, respectively, they are in the ratio $1:(9\sqrt{3}\alpha_0^2/8\pi^2):(9\sqrt{3}/4\pi)$, α_0 being the lowest zero of the zero-order Bessel function.

Problem 6-21: A monoenergetic point source is placed at the center of a nonmultiplying cylindrical medium of radius R and height h. Find the densities of fast and thermal neutrons as functions of position.

---

16. Electromagnetic Waves in a Dissipative Medium

The problem of electromagnetic fields in strongly conducting media leads to equations which are rather similar in mathematical form to those of heat conduction or diffusion. Consider Maxwell's equations in a conducting medium, and with external charge and current densities ρ_0 and $\mathbf{j}_0$:

$$\nabla \cdot \mathbf{E} = \frac{1}{\epsilon} \rho_0 \qquad (3.14)$$

$$\nabla \cdot \mathbf{H} = 0 \qquad (3.15)$$

$$\nabla \times \mathbf{E} = -\mu \frac{\partial H}{\partial t} \tag{3.16}$$

and

$$\nabla \times \mathbf{H} = \sigma \mathbf{E} + \epsilon \frac{\partial \mathbf{E}}{\partial t} + \mathbf{j}_0 \tag{3.17}$$

Using (3.15) and (3.16) to introduce potentials, as in (5.276) and (5.278), we derive the equations

$$\nabla^2 \phi - \nabla \cdot \frac{\partial \mathbf{A}}{\partial t} = \frac{\rho}{\epsilon} \tag{6.179}$$

and

$$\nabla(\nabla \cdot \mathbf{A}) - \nabla^2 \mathbf{A} = -\sigma\mu \left(\frac{\partial \mathbf{A}}{\partial t} + \nabla\phi \right) - \epsilon\mu \frac{\partial}{\partial t} \left(\frac{\partial \mathbf{A}}{\partial t} + \nabla\phi \right) + \mathbf{j}_0 \tag{6.180}$$

If we take the Lorentz condition in the modified form

$$\nabla \cdot \mathbf{A} + \sigma\mu\phi + \epsilon\mu \frac{\partial \phi}{\partial t} = 0 \tag{6.181}$$

we find that the equations for the potentials are

$$-\nabla^2 \phi + \sigma\mu \frac{\partial \phi}{\partial t} + \epsilon\mu \frac{\partial^2 \phi}{\partial t^2} = \frac{\rho_0}{\epsilon} \tag{6.182}$$

and

$$-\nabla^2 \mathbf{A} + \sigma\mu \frac{\partial \mathbf{A}}{\partial t} + \epsilon\mu \frac{\partial^2 \mathbf{A}}{\partial t^2} = \mu \mathbf{j}_0 \tag{6.183}$$

If the conductivity is large enough that the third term is negligible compared with the second, we have equations of the diffusion type. If, then, for example, a charge is introduced at some point in the medium, it dissipates in much the same way as heat does in a thermally conducting medium.

More precisely, however, the problem stands intermediate between one of wave propagation and one of diffusion (or dissipation). To get a better understanding of the situation, let us solve for the Green's function defined by

$$-\nabla^2 G + \sigma\mu \frac{\partial G}{\partial t} + \epsilon\mu \frac{\partial^2 G}{\partial t^2} = \delta(\mathbf{r})\delta(t) \tag{6.184}$$

Let us take Fourier transforms in both space and time variables, designating the double transform $G_{st}(\mathbf{k}, \omega)$ as

$$G_{st}(\mathbf{k}, \omega) = \left(\frac{1}{2\pi} \right)^4 \int G(\mathbf{r}, t)\, e^{-i(\mathbf{k}\cdot\mathbf{r}-\omega t)}\, d^3\mathbf{r}\, dt \tag{6.185}$$

The inverse transform is then

$$G(\mathbf{r}, t) = \int G_{st}(\mathbf{k}, \omega)\, e^{i(\mathbf{k}\cdot\mathbf{r}-\omega t)}\, d^3\mathbf{k}\, d\omega \tag{6.186}$$

The equation for the double transform is

$$[k^2 - i\sigma\mu\omega - \epsilon\mu\omega^2]\, G_{st} = \left(\frac{1}{2\pi}\right)^4 \tag{6.187}$$

Let us first invert only with respect to the space coordinates, to determine

$$G_t(\mathbf{r}, \omega) = \int G_{st}(\mathbf{k}, \omega)\, e^{i\mathbf{k}\cdot\mathbf{r}}\, d^3r$$

$$= \left(\frac{1}{2\pi}\right)^4 \int \frac{1}{k^2 - \sigma\mu i\omega - \epsilon\mu\omega^2}\, e^{i\mathbf{k}\cdot\mathbf{r}}\, d^3k \tag{6.188}$$

$$= \frac{1}{4\pi^3}\int_0^\infty \frac{1}{k^2 - \sigma\mu i\omega - \epsilon\mu\omega^2}\, \frac{\sin kr}{kr}\, k^2\, dk$$

$$= \frac{1}{8\pi^3 r}\int_{-\infty}^\infty \frac{k\sin kr}{k^2 - k_0^2}\, dk \tag{6.189}$$

where

$$k_0^2 = \epsilon\mu\omega^2 + i\sigma\mu\omega \tag{6.190}$$

Writing $\sin kr$ as the sum of complex exponentials of positive and negative exponent, the integral may be evaluated as a contour integral to give

$$G_t(\mathbf{r}, \omega) = \frac{1}{8\pi^2 r}\, e^{ik_0 r} \tag{6.191}$$

Taking the inverse transform in the time variable

$$G(\mathbf{r}, t) = \frac{1}{8\pi^2 r}\int_{-\infty}^\infty e^{ir\sqrt{\epsilon\mu\omega^2 + i\sigma\mu\omega}}\, e^{-i\omega t}\, d\omega \tag{6.192}$$

For the limiting case of zero conductivity, we obtain again the result of (5.93) in the preceding chapter,

$$G(\mathbf{r}, t) = \frac{1}{4\pi r}\, \delta\left(t - \frac{r}{c}\right)$$

In the limiting case $\epsilon = 0$ the equation is seen, from (6.184), to be identical with that of a diffusion problem. While $\epsilon = 0$ is not physically realistic, this limit does give an approximation to a very strongly conducting medium. The corresponding Green's function is

$$G(\mathbf{r}, t) = \frac{1}{8\pi^2 r}\int_{-\infty}^\infty e^{ir\sqrt{i\sigma\mu\omega}}\, e^{-i\omega t}\, d\omega$$

If we make the substitution

$$i\omega = \xi^2 \tag{6.193}$$

this becomes

$$G(\mathbf{r}, t) = \frac{1}{4\pi^2 ri}\int e^{ir\sqrt{\sigma\mu}\xi}\, e^{-\xi^2 t}\, \xi\, d\xi \tag{6.194}$$

The contour is that shown in Fig. 6.3. Since the integrals over arcs such as C_1 and C_2, with radius becoming infinite, are zero, this contour may, however, be deformed to one along the real axis. The integral is then readily evaluated [see (6.24) and (6.25)] to give

$$G(\mathbf{r}, t) = \frac{1}{\sigma\mu} \left(\frac{\sigma\mu}{4\pi t}\right)^{3/2} e^{-r^2\sigma\mu/4t} \tag{6.195}$$

It is also possible to derive explicitly the general form G, starting from (6.192). Putting

$$\Omega = \frac{\sigma}{2\epsilon} \tag{6.196}$$

the Green's function may be written

$$G(\mathbf{r}, t) = \frac{1}{8\pi^2 r} \int_{-\infty}^{\infty} e^{i[(r/c)\sqrt{(\omega + i\Omega)^2 + \Omega^2} - \omega t]} \, d\omega \tag{6.197}$$

Let us first make the substitution

$$\omega + i\Omega = \frac{\Omega}{2}\left(v - \frac{1}{v}\right) \tag{6.198}$$

The path of integration in the v plane follows a curved path above the real axis from $2i - \infty$ to $2i + \infty$ and (6.197) becomes

$$G(\mathbf{r}, t) = \frac{e^{-\Omega t}}{8\pi^2 r} \int_{2i-\infty}^{2i+\infty} \exp i\frac{\Omega}{2}\left[\frac{r}{c}\left(v + \frac{1}{v}\right) - t\left(v - \frac{1}{v}\right)\right]\frac{\Omega}{2}\left(v + \frac{1}{v}\right)\frac{dv}{v}$$

$$= \frac{e^{-\Omega t}}{8\pi^2 r}\frac{c}{i}\frac{\partial}{\partial r}\int_{2i-\infty}^{2i+\infty} \exp i\frac{\Omega}{2}\left[v\left(\frac{r}{c} - t\right) + \frac{1}{v}\left(\frac{r}{c} + t\right)\right]\frac{dv}{v} \tag{6.199}$$

At this point we must distinguish the cases $(r/c) > t$ and $(r/c) < t$. In the former case, the contour may be deformed around the upper half-plane, being closed in the usual way by a semicircle of infinite radius. The integral on this semicircle goes exponentially to zero. Since there are no

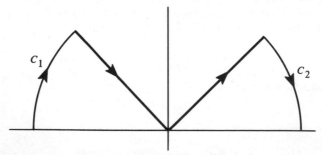

Figure 6.3

poles within the contour, the integral is zero. Thus

$$G(\mathbf{r}, t) = 0, \qquad r > ct \tag{6.200}$$

Consider, then, the second case. It is convenient to make the substitution

$$z = v \sqrt{\frac{ct - r}{ct + r}} \tag{6.201}$$

which leads to the equation

$$G(\mathbf{r}, t) = \frac{e^{-\Omega t}}{8\pi^2 r} \frac{c}{i} \frac{\partial}{\partial r} \int_{i\sqrt{(ct-r)/(ct+r)}-\infty}^{i\sqrt{(ct-r)/(ct+r)}+\infty} \exp\left(-\frac{i\Omega}{2}\right) \sqrt{t^2 - \frac{r^2}{c^2}\left(z - \frac{1}{z}\right)} \frac{dz}{z} S(ct - r) \tag{6.202}$$

where $S = 1$ for positive argument and zero for negative argument. The contour may now be deformed around the *lower* half-plane. From Eq. (4.390) we then obtain for $G(\mathbf{r}, t)$ the explicit form

$$G(\mathbf{r}, t) = \frac{e^{-\Omega t}}{4\pi r} c \frac{\partial}{\partial r} [J_0(i\Omega \sqrt{t^2 - r^2/c^2}) S(ct - r)]$$

$$= \frac{\Omega^2}{4\pi c} e^{-\Omega t} \frac{J_1(i\Omega \sqrt{t^2 - r^2/c^2})}{i\Omega \sqrt{t^2 - r^2/c^2}} + \frac{e^{-\Omega t}}{4\pi r} \delta(t - (r/c)) \tag{6.203}$$

Problem 6-22: Verify that in the limit $\sigma \to 0$, Eq. (6.203) reduces to (5.93) and in the limit $\epsilon \to 0$, to (6.195).

Problem 6-23: Show that the field ϕ generated by a source density $S_0(\mathbf{r}, t)$, whose value and time derivative are given at $t = 0$ and for which $\alpha(\partial\phi/\partial n) + \beta\phi$ is given on specified surfaces, is determined by the equation

$$\phi(\mathbf{r}, t) = \int_0^t dt' \int d^3\mathbf{r}' \, S_0(\mathbf{r}, t')G(\mathbf{r}, \mathbf{r}', t - t')$$

$$+ \sigma\mu \int d^3\mathbf{r}' \, G(\mathbf{r}, \mathbf{r}', t)\phi(\mathbf{r}', 0)$$

$$+ \epsilon\mu \int d^3\mathbf{r}' \left[G(\mathbf{r}, \mathbf{r}', t) \left(\frac{\partial\phi}{\partial t'}\right)_{t'=0} + \frac{\partial}{\partial t} G(\mathbf{r}, \mathbf{r}', t)\phi(\mathbf{r}', 0) \right]$$

$$+ \frac{1}{\alpha} \int_0^t dt' \int dS' \left[\alpha \frac{\partial}{\partial n'} \phi(\mathbf{r}', t') + \beta\phi(\mathbf{r}', t') \right] G(\mathbf{r}, \mathbf{r}', t - t')$$

where the Green's function satisfies the boundary condition

$$\alpha \frac{\partial G}{\partial n} + \beta G = 0$$

over the bounding surfaces.

Problem 6-24:   Show, using the result of the previous problem, that a field $\phi(\mathbf{r}, 0)$ in an infinite source-free medium becomes, after a sufficiently long time t,

$$\phi(\mathbf{r}, t) = \frac{1}{c^2}\left(\frac{\sigma\mu}{4\pi t}\right)^{3/2} \int \phi(\mathbf{r}', 0)\, e^{-(\sigma\mu|\mathbf{r}-\mathbf{r}'|^2/4t)}\, d^3\mathbf{r}'$$

---

17. Propagation into a Conducting Medium

The Green's function for the infinite medium is useful for an understanding of the physical processes involved in a problem; it may also, with the aid of the method of images, serve as a basis for calculating Green's functions for finite media. However, the method of eigenfunction expansions is, for problems of finite media, both more direct and more versatile.

In many wave propagation problems, we are not concerned with the sources, and are interested only in solutions in the source-free region. We may then use the *homogeneous* versions of (6.182) and (6.183) with right-hand sides equal to zero. It is then easily seen that the fields $\mathbf{E}$ and $\mathbf{B}$ themselves satisfy wave equations of the form

$$-\nabla^2\mathbf{E} + \sigma\mu\frac{\partial\mathbf{E}}{\partial t} + \epsilon\mu\frac{\partial^2\mathbf{E}}{\partial t^2} = 0 \tag{6.204}$$

or, using the notation of the previous section,

$$-\nabla^2\mathbf{E} + \frac{\Omega}{c^2}\frac{\partial\mathbf{E}}{\partial t} + \frac{1}{c^2}\frac{\partial^2\mathbf{E}}{\partial t^2} = 0 \tag{6.205}$$

Furthermore, since $\mathbf{j} = \sigma\mathbf{E}$, the same equation is satisfied by the current.

Let us consider first the problem of a plane wave, impinging normally on the plane surface $z = 0$ of a conducting medium lying in the region of positive z. Let the circular frequency of the wave be ω, so that in the medium

$$\mathbf{E} = \mathbf{E}_0(z)\, e^{-i\omega t} \tag{6.206}$$

where $\mathbf{E}_0$ is a vector in the $x - y$ plane. Then, from (6.205)

$$\frac{d^2\mathbf{E}_0}{dz^2} + \frac{\omega^2 + 2i\omega\Omega}{c^2}\, \mathbf{E}_0 = 0 \tag{6.207}$$

The coefficient of $\mathbf{E}_0$ may be written

$$\frac{\omega^2 + 2i\omega\Omega}{c^2} = \frac{1}{\lambda^2}\, e^{2i\gamma} \tag{6.208}$$

where

$$\frac{1}{\lambda^2} = \frac{\omega\sqrt{\omega^2 + 4\Omega^2}}{c^2} \qquad (6.209)$$

and

$$\tan 2\gamma = \frac{2\Omega}{\omega} \qquad (6.210)$$

The solution of (6.207) which remains finite as $z \to \infty$ is

$$\mathbf{E}_0 = \mathbf{E}_1 \, e^{i\cos\gamma(z/\lambda)} \, e^{-\sin\gamma(z/\lambda)} \qquad (6.211)$$

The wave then decays inside the medium with an attenuation length

$$\lambda_0 = \frac{\lambda}{\sin\gamma} \qquad (6.212)$$

This length is also referred to as the "skin depth." By simple algebra it may be shown to be explicitly

$$\lambda_0 = \frac{\sqrt{2}\, c}{\sqrt{\omega\sqrt{\omega^2 + 4\Omega^2} - \omega^2}} \qquad (6.213)$$

For frequency $\omega \ll \Omega$ this is approximately

$$\lambda_0 \approx \frac{c}{\sqrt{\omega\Omega}} \qquad (6.214)$$

while for high frequency ($\omega \gg \Omega$) it becomes

$$\lambda_0 \approx \frac{c}{\Omega} \qquad (6.215)$$

and is then independent of the frequency.

Problem 6-25: An oscillating current of circular frequency ω flows along a cylindrical conducting wire of radius a. Find how the current is distributed over the wire; show, in particular, that if $\omega \ll \Omega$ and a is sufficiently large it decreases approximately exponentially from the surface with a decay length (skin depth)

$$\frac{c}{\sqrt{\Omega\omega}} = \frac{1}{\sqrt{\tfrac{1}{2}\sigma\mu\omega}}$$

---

18. Skin Effect—General Considerations

Consider a conducting medium occupying a region surrounded by a nonconducting one in which there is an arbitrary field. If ψ is a potential (either ϕ or a component of $\mathbf{A}$), or the component of a field, it satisfies the equation

$$-\nabla^2\psi + \sigma\mu\frac{\partial\psi}{\partial t} + \epsilon\mu\frac{\partial^2\psi}{\partial t^2} = \sigma\mu\frac{\partial\psi}{\partial t}\Delta \qquad (6.216)$$

where $\Delta = 1$ in the *external* region, zero in the conducting part. We may now treat the right-hand side as a *source* term, and so write the equation of the problem in a form involving Green's functions:

$$\psi(\mathbf{r},t) = \sigma\mu\int_{-\infty}^{t}dt'\int\frac{\partial\psi(\mathbf{r}',t')}{\partial t'}\Delta(\mathbf{r}')G(\mathbf{r}-\mathbf{r}',t-t')\,d^3\mathbf{r}' \quad (6.217)$$

where G is given by (6.203). The integral on the right is, however, confined to the region *outside* the conductor.

Consider fields with circular frequency ω, that is, of the form

$$\psi(\mathbf{r},t) = e^{-i\omega t}\psi_0(\mathbf{r}) \qquad (6.218)$$

Then

$$\psi_0(\mathbf{r}) = -i\omega\sigma\mu\int_{-\infty}^{t}dt'\int\psi_0(\mathbf{r}')e^{i\omega(t-t')}G(\mathbf{r}-\mathbf{r}',t-t')\Delta(\mathbf{r}')\,d^3\mathbf{r}'$$

But the t' integral is just 2π times $G(\mathbf{r}-\mathbf{r}',\omega)$ as given by (6.191). Thus

$$\psi_0(\mathbf{r}) = -\frac{i\omega\sigma\mu}{4\pi}\int\Delta(\mathbf{r}')\psi_0(\mathbf{r}')\frac{1}{|\mathbf{r}-\mathbf{r}'|}e^{i\sqrt{\epsilon\mu\omega^2+i\sigma\mu\omega}|\mathbf{r}-\mathbf{r}'|}\,d^3\mathbf{r}' \quad (6.219)$$

In the notation of (6.209) and (6.210), this becomes

$$\psi_0(\mathbf{r}) = -\frac{i\omega\sigma\mu}{4\pi}\int\Delta(\mathbf{r}')\psi_0(\mathbf{r}')\frac{1}{|\mathbf{r}-\mathbf{r}'|}\exp\frac{i\cos\gamma}{\lambda}|\mathbf{r}-\mathbf{r}'|$$

$$\times\exp\left(-\frac{\sin\gamma}{\lambda}|\mathbf{r}-\mathbf{r}'|\right)d^3\mathbf{r}' \qquad (6.220)$$

Thus, the field *inside* the conductor can be expressed in terms of that outside, and we see that it decays into the conductor in general with the attenuation length $\lambda_0 = (\lambda/\sin\gamma)$ given by (6.213).

While the formulation of the problem as an integral equation which we have given here is not particularly useful for detailed calculation, it is clearly useful for obtaining an understanding of the physical phenomena involved.

Problem 6-26: Treating the *conduction* term as a source, obtain the alternative integral equation for the problem

$$\psi(\mathbf{r},t) = \chi(\mathbf{r},t) - \frac{\sigma\mu}{4\pi}\int\left[\frac{\partial\psi(\mathbf{r}',t')}{\partial t'}\right]_{t'=t-(|\mathbf{r}-\mathbf{r}'|/c)}$$

$$\times\frac{1}{|\mathbf{r}-\mathbf{r}'|}[1-\Delta(\mathbf{r}')]\,d^3\mathbf{r}'$$

$\chi(r, t)$ being a solution of the wave equation in a completely nonconducting medium. For the case

$$\chi(\mathbf{r}, t) = \chi_0(\mathbf{r})e^{-i\omega t}$$

show that this becomes

$$\psi_0(\mathbf{r}) = \chi_0(\mathbf{r}) + \frac{i\omega\sigma\mu}{4\pi} \int \psi_0(\mathbf{r}') \frac{e^{-i(\omega/c)|\mathbf{r}-\mathbf{r}'|}}{|\mathbf{r}-\mathbf{r}'|} [1 - \Delta(\mathbf{r}')] \, d^3\mathbf{r}'$$

Problem 6-27: Consider a cylindrical wire of conductivity σ and radius a in a nonconducting medium in which a plane wave of circular frequency ω is propagated in the direction of the axis of the wire. If a is considerably smaller than the skin depth, show that $\psi_0(\mathbf{r})$ is given approximately by

$$\psi_0(\mathbf{r}) = \chi_0(\mathbf{r}) + \frac{i\omega\sigma\mu}{4\pi} \int \chi_0(\mathbf{r}') \frac{e^{-i(\omega/c)|\mathbf{r}-\mathbf{r}'|}}{|\mathbf{r}-\mathbf{r}'|} [1 - \Delta(\mathbf{r}')] \, d^3\mathbf{r}'$$

Hence determine approximately the field of the wave perturbed by the conducting wire.

---

19. The Equations of Superconducting Electrodynamics

Other problems, rather similar to those encountered in the theory of diffusion of thermal neutrons, arise in connection with the propagation of electromagnetic waves in superconductors. The similarity extends, however, only to the treatment of stationary problems; the time dependence is quite different in the two cases. Basically, the problem is one of attenuated waves. We consider some of its aspects in the remainder of this chapter.

In the original formulation of London, a superconductor was characterized by a "supercurrent" proportional to the vector potential

$$\mathbf{J}_s = -\frac{1}{\mu\lambda^2}\mathbf{A} \tag{6.221}$$

λ^2 being a constant characteristic of the superconductor. Clearly, however, such a hypothesis is not gauge invariant. The equation obtained by taking the curl of both sides of the equation is, however,

$$\nabla \times \mathbf{J}_s = -\frac{1}{\mu\lambda^2}\mathbf{B} = \frac{1}{\lambda^2}\mathbf{H} \tag{6.222}$$

Gauge invariant results may then be obtained by using the curl of the equation involving the current. The electrodynamic equation

$$\nabla \times \mathbf{H} = \mathbf{J} + \epsilon \frac{\partial \mathbf{E}}{\partial t} \tag{6.223}$$

takes, in the absence of external currents ("sources"), the form

$$\nabla \times \mathbf{H} = \sigma \mathbf{E} - \frac{1}{\lambda^2} \mathbf{A} + \epsilon \frac{\partial \mathbf{E}}{\partial t}$$

Taking the curl and substituting for $\nabla \times \mathbf{E}$ from (3.16) yields

$$-\nabla^2 \mathbf{H} = -\sigma \mu \frac{\partial \mathbf{H}}{\partial t} - \frac{1}{\lambda^2} \mathbf{H} - \epsilon \mu \frac{\partial^2 \mathbf{H}}{\partial t^2} \tag{6.224}$$

For very many problems the conduction current $\sigma \mathbf{E}$ is negligible compared with the supercurrent. For sufficiently high frequencies this is not of course the case. However, we confine ourselves to problems in which its neglect is justified, and deal only with the equation

$$-\nabla^2 \mathbf{H} + \frac{1}{\lambda^2} \mathbf{H} + \epsilon \mu \frac{\partial^2 \mathbf{H}}{\partial t^2} = 0 \tag{6.225}$$

A similar equation may be obtained for $\mathbf{A}$ in the radiation gauge, in which (6.221) is valid, by expressing $\mathbf{H}$ and $\mathbf{E}$ in terms of the vector potential in (3.17). The equation for $\mathbf{E}$ can also be obtained by taking the curl of (3.16):

$$\nabla \times (\nabla \times \mathbf{E}) = -\mu \frac{\partial}{\partial t} \nabla \times \mathbf{H}$$

$$= \frac{1}{\lambda^2} \frac{\partial \mathbf{A}}{\partial t} - \epsilon \mu \frac{\partial^2 \mathbf{E}}{\partial t^2}$$

$$= -\frac{1}{\lambda^2} \mathbf{E} - \frac{1}{c^2} \frac{\partial^2 \mathbf{E}}{\partial t^2}$$

It is, of course, most convenient to work with the vector potential $\mathbf{A}$ in that it yields both the electric and magnetic fields, which are necessarily correctly related to each other. Let us in fact calculate the Green's function for $\mathbf{A}$. If we neglect the conduction current and assume an external (source) current $\mathbf{j}_0$, the equation for $\mathbf{A}$ is

$$-\nabla^2 \mathbf{A} + \frac{1}{\lambda^2} \mathbf{A} + \frac{1}{c^2} \frac{\partial^2 \mathbf{A}}{\partial t^2} = \mu \mathbf{J}_0 \tag{6.226}$$

Problem 6-28: At a large distance from a superconducting sphere of radius a there is a constant magnetic field H_0. Show that the field as a function of position is given outside the superconductor by

$$H = -\nabla \psi$$

where

$$\psi = - H_0 \, r \cos \theta \left[1 + \frac{a^3}{2r^3} \left(1 + \frac{3}{\beta^2 a^2} - 3 \frac{\coth \beta a}{\beta a} \right) \right]$$

while inside it is $\mathbf{H} = \nabla \times \mathbf{A}$, the $\mathbf{A}$ being given by

$$A_\phi = \frac{3}{2} \frac{H_0 a}{\sinh \beta a} \left(\frac{\cosh \beta r}{\beta r} - \frac{\sinh \beta r}{\beta^2 r^2} \right) \sin \theta$$

where $\beta = \dfrac{1}{\lambda}$ and θ the angle between $\mathbf{H}$ and $\mathbf{r}$.

Where is the largest field in the superconductor, and what is its magnitude?

Problem 6-29: Determine the pattern of currents flowing in the superconductor in the previous problem.

Problem 6-30: Show that an electromagnetic wave of sufficiently high frequency impinging normally on the plane surface of a London superconductor can be propagated into it. Calculate the transmission probability through a superconducting slab of thickness $l \gg \lambda$.

Problem 6-31: Verify the self-consistency of a persistent current in a London superconductor as follows:

(a) If there is a magnetic field H_0 in the y direction at the surface $z = 0$ of a semi-infinite superconductor, show that the field inside is given by the vector potential

$$A_x = - \lambda H_0 e^{-z/\lambda}, \qquad A_y = A_z = 0$$

This then gives rise to a surface current distribution in the superconductor of

$$J_x = \frac{H_0}{\lambda \mu} e^{-z/\lambda}$$

(b) Determine the resulting field *outside* the superconductor due to this current, using the relation

$$- \nabla^2 \mathbf{A} = - \mu \mathbf{J}$$

and show that it has the value H_0 at the surface and is in the y direction.

Problem 6-32: Solve the similar problem for a current flowing in the skin depth along a cylindrical superconducting wire of radius $R \gg \lambda$. Verify that the mechanism is as follows: An external field H_φ having value H_0 at the surface penetrates into the wire, setting up a current in the z direction according to London's equation (6.221). This current then *produces* an external field whose surface value is H_0.

We can define the vector Green's function by the equation

$$-\nabla^2\mathbf{G} + \frac{1}{\lambda^2}\mathbf{G} + \frac{1}{c^2}\frac{\partial^2\mathbf{G}}{\partial t^2} = \mathbf{u}\delta(\mathbf{r}-\mathbf{r}')\delta(t-t') \qquad (6.227)$$

$\mathbf{u}$ being an arbitrary unit vector.

Some time after the work of London, it was pointed out by Pippard that the London equation (6.221) should be replaced by a "nonlocal" relation (that is, one in which the current at $\mathbf{r}$ should be determined not only by the field at $\mathbf{r}$, but by that at neighboring points). To describe this "coherence effect" he proposed for the supercurrent the relation

$$\mathbf{J}_s(\mathbf{r}) = C\int\frac{[\mathbf{A}(\mathbf{r}')\cdot(\mathbf{r}-\mathbf{r}')](\mathbf{r}-\mathbf{r}')}{|\mathbf{r}-\mathbf{r}'|^4}e^{-|\mathbf{r}-\mathbf{r}'|/\xi_0}d^3\mathbf{r}' \qquad (6.228)$$

where $\xi_0 = (1/\beta)$ is known as the "coherence length." In this case $(1/\lambda^2)\mathbf{A}$ in (6.226) must be replaced by $\mu\mathbf{J}_s$ as defined. Equation (6.226) and the Green's function equation (6.227) then become integro-differential equations. The constant C is to be determined by the condition that the present form reduces to the previous one when one integrates over the function multiplying $\mathbf{A}$.

The additional difficulties introduced by Pippard's hypothesis seem less formidable if we take Fourier transforms in the Green's function equation

$$G(\mathbf{k},\omega) = \left(\frac{1}{2\pi}\right)^4\int G(\mathbf{r},t)e^{-i(\mathbf{k}\cdot\mathbf{r}-\omega t)}\,d^3\mathbf{r}\,dt \qquad (6.229)$$

Since $\mathbf{J}_s$ is a convolution in the spatial variables, its transform is $(2\pi)^3$ times the product of that of $\mathbf{A}$ and of $(x_ix_j/r^4)e^{-\beta r}$ where x_i ($i = 1, 2, 3$,) are the coordinates. But the transform of the latter is

Problem 6-33: Show that the Fourier transform of $(x_ix_j/r^4)e^{-\beta r}$ is given by Eqs. (6.231) to (6.233).

$$\mathscr{F}\left(\frac{x_ix_j}{r^4}e^{-\beta r}\right) = \frac{\beta}{(2\pi)^2}\left\{\left(\frac{1}{k^2}-\frac{\beta}{k^3}\tan^{-1}\frac{k}{\beta}\right)\left(3\frac{k_ik_j}{k^2}-\delta_{ij}\right)\right.$$

$$\left. +\frac{1}{k\beta}\tan^{-1}\frac{k}{\beta}\left(\delta_{ij}-\frac{k_ik_j}{k^2}\right)\right\} \qquad (6.230)$$

$$= f_1(\mathbf{k})\delta_{ij}+f_2(\mathbf{k})\frac{k_ik_j}{k^2} \qquad (6.231)$$

f_1 and f_2 being given by

$$f_1 = \frac{\beta}{(2\pi)^2}\left[\left(\frac{1}{k\beta}+\frac{\beta}{k^3}\right)\tan^{-1}\frac{k}{\beta}-\frac{1}{k^2}\right] \qquad (6.232)$$

and

$$f_2 = \frac{\beta}{(2\pi)^2}\left[\frac{3}{k^2} - \frac{3\beta}{k^3}\tan^{-1}\frac{k}{\beta} - \frac{1}{k\beta}\tan^{-1}\frac{k}{\beta}\right] \tag{6.233}$$

Since the transform of (6.228) is

$$(2\pi)^3 C \mathbf{A}(\mathbf{k}, \omega)\left[f_1 + \frac{1}{k^2}\mathbf{k}\mathbf{k}f_2\right]$$

and that of the corresponding London term is

$$-\frac{1}{\mu\lambda^2}\mathbf{A}(\mathbf{k}, \omega)$$

we may determine C. For the integral of $(x_i x_j/r^4)e^{-\beta r}$ is $(2\pi)^3$ times its transform for $\mathbf{k} = 0$, viz., $(4\pi/3\beta)\delta_{ij}$. Consequently,

$$C = -\frac{3\beta}{4\pi\lambda^2\mu} \tag{6.234}$$

and the transform of $\mathbf{J}_s(\mathbf{r})$ is

$$\mathscr{F}(\mathbf{J}_s(\mathbf{r})) = -\frac{6\pi^2\beta}{\lambda^2\mu}\left[f_1\mathbf{A}(\mathbf{k}, \omega) + \frac{1}{k^2}f_2\mathbf{k}(\mathbf{k}\cdot\mathbf{A}(\mathbf{k}, \omega))\right] \tag{6.235}$$

$$= -\frac{3}{2\mu\lambda^2\xi_0^2}\left[\bar{f}_1\mathbf{A} + \frac{1}{k^2}\bar{f}_2\mathbf{k}(\mathbf{k}\cdot\mathbf{A})\right] \tag{6.236}$$

where $\bar{f}_1, \bar{f}_2$ are $4\pi^2/\beta$ times f_1 and f_2, respectively.

In the case of the London equation, the transform $\mathbf{G}(\mathbf{k}, \omega)$ of $\mathbf{G}(\mathbf{r}, t)$ is given by

$$\left(k^2 + \frac{1}{\lambda^2} - \frac{\omega^2}{c^2}\right)\mathbf{G}(\mathbf{k}, \omega) = \left(\frac{1}{2\pi}\right)^4\mathbf{u}e^{-i(\mathbf{k}\cdot\mathbf{r}'-\omega t')} \tag{6.237}$$

On the other hand, the Pippard modification leads to

$$\left(k^2 - \frac{\omega^2}{c^2} + \frac{3}{2\lambda^2\xi_0^2}\bar{f}_1\right)\mathbf{G}(\mathbf{k}, \omega) + \frac{3}{2\lambda^2\xi_0^2}\bar{f}_2\frac{\mathbf{k}[\mathbf{k}\cdot\mathbf{G}(\mathbf{k}, \omega)]}{k^2}$$

$$= \left(\frac{1}{2\pi}\right)^4\mathbf{u}e^{-i(\mathbf{k}\cdot\mathbf{r}'-\omega t')} \tag{6.238}$$

Taking the scalar product of the last equation with $\mathbf{k}$, we find

$$\mathbf{k}\cdot\mathbf{G}(\mathbf{k}, \omega) = \left(\frac{1}{2\pi}\right)^4\frac{(\mathbf{u}\cdot\mathbf{k})e^{-i(\mathbf{k}\cdot\mathbf{r}'-\omega t')}}{k^2 - (\omega^2/c^2) + (3/2\lambda^2\xi_0^2)(\bar{f}_1+\bar{f}_2)} \tag{6.239}$$

If we substitute this in (6.238) we obtain the Fourier transform of the Green's function:

$$G(\mathbf{k}, \omega) = \frac{1}{k^2 - (\omega^2/c^2) + (3/2\lambda^2\xi_0^2)\bar{f}_1(\mathbf{k})}$$

$$\times \left\{ \mathbf{u} - \frac{3}{2\lambda^2\xi_0^2} \frac{\bar{f}_2(\mathbf{k})}{k^2} \frac{\mathbf{k}(\mathbf{k} \cdot \mathbf{u})}{k^2 - (\omega^2/c^2) + (3/2\lambda^2\xi_0^2)(\bar{f}_1 + \bar{f}_2)} \right\}$$

$$\times \left(\frac{1}{2\pi}\right)^4 e^{-i(\mathbf{k}\cdot\mathbf{r}' - \omega t')} \tag{6.240}$$

Clearly, the inversion of this formula presents some difficulty in general. For practical purposes, it could be carried out by doing the angular integral explicitly in the Fourier integral, and doing the $\mathbf{k}$ integral numerically.

Though explicit solutions are difficult to obtain in the Pippard model, they are not in the London one. Let us first consider the Green's function for the static case. This is obtained from (6.237) by multiplying by 2π and putting $\omega = 0$. The Green's function may be found by evaluating the Fourier integral by contour integration, to give

$$G(\mathbf{r}, t) = \frac{1}{4\pi r} e^{-r/\lambda} \mathbf{u} \tag{6.241}$$

Problem 6-34:   Derive the integral equation

$$\mathbf{A}(\mathbf{r}') = \frac{1}{4\pi\lambda^2} \int \mathbf{A}(\mathbf{r}') \frac{e^{-|\mathbf{r}-\mathbf{r}'|/\lambda}}{|\mathbf{r}-\mathbf{r}'|} \Delta(\mathbf{r}') \, d^3\mathbf{r}$$

for the vector potential in a medium consisting of a superconductor embedded in a normal material. $\Delta = 1$ in the normal material, 0 in the superconductor. Hence show that fields penetrate into the superconductor only for distances comparable with λ (the "penetration depth").

Let us turn next to the dynamic problem. By inverting (6.237) we obtain

$$G(\mathbf{r}, t) = \left(\frac{1}{2\pi}\right)^4 \mathbf{u} \int \frac{1}{k^2 + (1/\lambda^2) - (\omega^2/c^2)} e^{i(\mathbf{k}\cdot\mathbf{r} - \omega t)} \, d^3\mathbf{k} \, d\omega \tag{6.242}$$

If we try to invert with respect to the time variable we meet the difficulty encountered previously for ordinary electrodynamics, of having poles on the contour of integration. It is in fact convenient to have recourse to the conduction current to circumvent this difficulty. It introduces, on the left-hand side of (6.226), the additional term $\mu\sigma(\partial\mathbf{A}/\partial t)$. A similar term in the G equation adds to the denominator of the integral of (6.242) the quantity $-i\omega\sigma\mu$, σ being the conductivity. The poles in the ω integral then have negative imaginary parts. Consequently, for $t > 0$, when we deform the contour of integration around the lower half-plane, we get a nonzero result. On the other hand, as would be expected, for $t < 0$ the result is zero, since the contour must then circle the upper half-plane, in which there are no singularities. The dissipative term has once again distinguished the direction of time, and ensured the satisfaction of the causality requirement.

After evaluating the integral, it is of course possible to let $\sigma \to 0$. If we proceed in this way, we obtain

$$G(\mathbf{r}, t) = \frac{c}{(2\pi)^3} \mathbf{u} \int \frac{1}{\sqrt{k^2 + (1/\lambda^2)}} \sin c \sqrt{k^2 + \frac{1}{\lambda^2}} \, t \, e^{i\mathbf{k}\cdot\mathbf{r}} \, d^3k \quad (6.243)$$

We can now integrate over the angles of $\mathbf{k}$, leaving an integral in the magnitude k only:

$$G(\mathbf{r}, t) = \frac{c}{2\pi^2 r} \mathbf{u} \int_0^\infty \frac{1}{\sqrt{k^2 + (1/\lambda^2)}} \sin c \sqrt{k^2 + (1/\lambda^2)} \, t \sin kr \, k \, dk$$

$$= -\frac{c}{4\pi^2 r} \mathbf{u} \frac{\partial}{\partial r} \int_{-\infty}^\infty \frac{1}{\sqrt{k^2 + (1/\lambda^2)}} \sin ct \sqrt{k^2 + (1/\lambda^2)} \cos kr \, dk$$

$$(6.244)$$

The advantage of the second form is that the integral is convergent, whereas in the case of the first, the behavior is delta-function-like, as is evident from considering the behavior of the integrand for large k.

Let us make, in (6.244), the substitution

$$k = \frac{1}{2\lambda}\left(\zeta - \frac{1}{\zeta}\right) \quad (6.245)$$

which transforms the integral [which we call $I(\mathbf{r}, t)$] to the form

$$I(\mathbf{r}, t) = \int_0^\infty \sin\frac{ct}{2\lambda}\left(\zeta + \frac{1}{\zeta}\right)\cos\frac{r}{2\lambda}\left(\zeta - \frac{1}{\zeta}\right)\frac{d\zeta}{\zeta} \quad (6.246)$$

$$= \frac{1}{2}\int_0^\infty \left\{\sin\left[\frac{ct}{2\lambda}\left(\zeta + \frac{1}{\zeta}\right) + \frac{r}{2\lambda}\left(\zeta - \frac{1}{\zeta}\right)\right] + \sin\left[\frac{ct}{2\lambda}\left(\zeta + \frac{1}{\zeta}\right)\right.\right.$$

$$\left.\left. - \frac{r}{2\lambda}\left(\zeta - \frac{1}{\zeta}\right)\right]\right\}\frac{d\zeta}{\zeta}$$

$$= \frac{1}{4}\text{Im}\int_{-\infty}^\infty \left\{\exp\frac{i}{2\lambda}\left[\zeta(ct + r) + \frac{1}{\zeta}(ct - r)\right]\right.$$

$$\left. + \exp\frac{i}{2\lambda}\left[\zeta(ct - r) + \frac{1}{\zeta}(ct + r)\right]\right\}\frac{d\zeta}{\zeta} \quad (6.247)$$

We must now consider two cases separately, (*i*) $r > ct$ and (*ii*) $r < ct$. (*i*) Make, in the first term of the integral, the substitution

$$\zeta\sqrt{\frac{r+ct}{r-ct}} = \eta$$

and in the second

$$\zeta\sqrt{\frac{r-ct}{r+ct}} = \eta$$

Then

$$I(\mathbf{r}, t) = \frac{1}{4}\operatorname{Im}\int_{-\infty}^{\infty}\left\{\exp\frac{i}{2\lambda}\sqrt{r^2-c^2t^2}\left(\eta-\frac{1}{\eta}\right)\right.$$

$$\left. + \exp\left(\frac{-i}{2\lambda}\right)\sqrt{r^2-c^2t^2}\left(\eta-\frac{1}{\eta}\right)\right\}\frac{d\eta}{\eta} \qquad (6.248)$$

Since the integrand is real, for $r > ct$

$$I(\mathbf{r}, t) = 0 \qquad (6.249)$$

in keeping with the requirement of the causality condition.

(*ii*) When $r < ct$, we put in the first term

$$\zeta\sqrt{\frac{ct+r}{ct-r}} = \eta$$

and in the second

$$\zeta\sqrt{\frac{ct-r}{ct+r}} = \eta$$

to obtain

$$I(\mathbf{r}, t) = \frac{1}{2}\operatorname{Im}\int_{-\infty}^{\infty}\exp\frac{i}{2\lambda}\sqrt{c^2t^2-r^2}\left(\eta+\frac{1}{\eta}\right)\frac{d\eta}{\eta} \qquad (6.250)$$

We note, with respect to this formula, that when η has a small *positive* imaginary part, the integral becomes very large; a small *negative* imaginary part, however, leaves it perfectly well-behaved. It is therefore expedient to deform the contour so that, in the neighborhood of the origin, it makes a small semicircular detour into the *lower* half-plane as shown in Fig. 6.4. Since the integrand becomes vanishingly small as the radius ϵ of the "detour" $\rightarrow 0$, this does not affect the value of the integral. It may now be deformed around the *upper* half-plane, the integral in a semicircle of infinite radius going to zero exponentially.

The resulting contour integral around the origin can be shown on using (4.390) to be

$$I(r, t) = \frac{1}{2}\operatorname{Im} 2\pi i J_0\left(\frac{1}{\lambda}\sqrt{c^2t^2-r^2}\right)$$

$$= \pi J_0\left(\frac{1}{\lambda}\sqrt{c^2t^2-r^2}\right) \qquad (6.251)$$

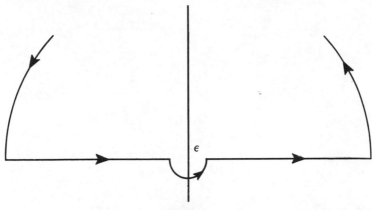

Figure 6.4

We can then write down, for all ranges of the variables, the formula for the Green's function

$$G(\mathbf{r}, t) = -\frac{c}{4\pi r} \mathbf{u} \frac{\partial}{\partial r} S(ct - r) J_0\left(\frac{1}{\lambda} \sqrt{c^2 t^2 - r^2}\right) \qquad (6.252)$$

where S is, as usual, 1 for positive argument, 0 for negative argument. On differentiation, this becomes

$$G(\mathbf{r}, t) = \mathbf{u}\left[\frac{1}{4\pi r} \delta\left(t - \frac{r}{c}\right) - \frac{c}{4\pi \lambda^2} S(ct - r) \frac{J_1((1/\lambda)\sqrt{c^2 t^2 - r^2})}{(1/\lambda)\sqrt{c^2 t^2 - r^2}}\right] \qquad (6.253)$$

20. Meson Equation

The eq. (6.226) has the same form as that for a vector meson field in particle physics. The corresponding scalar (or pseudoscalar) equation would describe corresponding meson fields. We can, therefore, use our results to obtain the Green's function of the classical meson field. For the meson field, the constant

$$\lambda = \frac{\hbar}{mc} \qquad (6.254)$$

$\hbar$ being $1/2\pi$ times Planck's constant, m the meson mass, and c the velocity of light.

It should be noted that, either in the case of the meson field or that of the superconducting electrodynamics, a disturbance is propagated with a continuous spectrum of velocities extending from zero up to the velocity of light.

Fig. 0.8.4

PRELUDE TO CHAPTER 7

The theme of this chapter is *probability and random processes in physics*.

After some preliminary remarks, we introduce first the notion of probability generating functions (p.g.f.) for problems in which the distribution of the random variable is discrete, and obtain formulas for the moments of the distribution in terms of it. As a simple illustration, we consider the problem of coin tossing. It is shown that, as the number of trials becomes very large ("large number limit") the distribution approaches a Gaussian.

A generalization is made to a similar experiment in which the probabilities of the two outcomes of a single trial are not equal.

A related *physical* problem is that of density fluctuations in a gas. The relation of the extent of density fluctuations to the size of the volume sampled is calculated.

The problem of the Poisson distribution, which is the distribution of the number of random events taking place in a given time interval when there is a given probability per unit time per event, is then solved. Again, it is shown that the distribution is Gaussian in the large number limit. The problem is then modified, in the case of radioactive counting, to take account of the "dead time" of the counter following the registration of a count.

The problems considered up to this point have concerned *discrete* random variables. When the result of the probabilistic experiment has a *continuous* spectrum, modified methods must be used. It is shown that the Fourier transform of the probability distribution is then a p.g.f.

We next calculate the distribution of the *sum* of independent random variables. The problem is easily solved using Fourier transforms and the convolution theorem. In the large number limit, a Gaussian distribution is determined, and the root mean square (rms)

413

deviation calculated in terms of those for the individual variables. This is the expression of the so-called "central limit theorem." The corresponding problem is solved for the case in which the independent variables are *vectors*.

In the next section, the result just mentioned is combined with the Poisson distribution to provide a solution to the "random walk" (random flight) problem. The solution obtained is, at large distances from the origin, that of a diffusion equation, as might be expected.

We then consider the problem of fluctuations in systems involving *multiplication* (for example, neutron chain reactors, cosmic ray showers, biological populations). We solve the cosmic ray shower problem according to a simple model. It is shown that the fluctuations are no longer relatively negligible in the large number limit. This depends on the fact that a fluctuation can propagate itself.

The next section is concerned with the Markoff method, which relates probabilities to volumes in a "sample space" using Fourier transform methods. As applications, the two-dimensional random walk problem is solved, and the three-dimensional one re-solved. A further application is that of the random distribution of a fixed amount of energy among an arbitrary number of particles (this problem can arise in nuclear reaction theory, or in the statistical theory of the microcanonical ensemble).

Finally, we include a brief discussion on stochastic processes, using the relation developed in Chapter 4 between the autocorrelation function and the power spectrum of a time series.

REFERENCES

Feller, W., *Probability Theory and Its Applications*. New York: Wiley, 1950.

Miller, K. S., *Engineering Mathematics*. New York: Rinehart, 1956, Chapter 7.

Parzen, E., *Modern Probability Theory and Its Applications*. New York: Wiley, 1960.

Wax, N., *Selected Papers on Noise and Stochastic Processes*. New York: Dover, 1954.

PROBABILITY AND STOCHASTIC PROCESSES

"Man can believe the impossible, but can never believe the improbable"

Oscar Wilde

1. Introduction

We discuss in this chapter some elementary notions of probability and of random (stochastic) processes. All of the problems to be considered are characterized by variables whose values in a specific experiment or trial are uncertain, but are governed by an a priori probability distribution. For purposes of illustration, it is useful to think of the problem of tossing a coin. An experiment has two possible results – the coin may come up heads or tails. The usual assumption is that either outcome is equally likely. (Though the outcome of a particular trial would presumably be completely predictable from complete information about the conduct of the experiment.) What the assumption means is that our knowledge is quite incomplete, but the variation in the conditions of repeated trials are sufficient that both outcomes are equally likely. Presumably, if one flipped coins by machine, and eliminated all variation in external conditions, one would always get the *same* result. The randomness in the usual coin-flipping is introduced by variations in the behavior of the person performing the trials, and the fact that the outcome of individual experiments is sufficiently sensitive to the range of variation which the performer displays. We might be uncertain whether, in fact, a bias might not exist in the actual range of variation of performance of an individual. How could such a bias be detected? Strictly speaking it could *not* be, at least with certainty. Suppose the experiment were repeated a very large number of times, and one result was found to take place significantly more often than the other; for instance, heads might turn up 10% more often than tails in a thousand trials. If the result of an individual experiment were *assumed* to be perfectly random, there would be a small but finite probability of the outcome in question. Therefore, we would not have demonstrated a bias. But the *probability* of the outcome arising from a perfectly random process would be very small (about 1 part in 10^{18} in this case). The probability would be further reduced if the same 10% preference persisted through many *more* repetitions of the experiment, but could obviously never be reduced to zero.

In some sense, every experiment in probability is a test of assumptions about *a priori* probabilities. The results enable us to evaluate the validity of these assumptions. A commonplace example would be the throwing of loaded dice. If the probability for the distribution obtained in a series of trials would be negligibly small for *unloaded* dice, we can assume loading with a high degree of probability.

There are many phenomena in physics concerning which we do not have complete information. These may be analyzed in terms of plausible assumptions about a priori probabilities, and the credibility to be assigned to these assumptions may thus be evaluated. Some concepts of physics are intrinsically probabilistic (e.g., "local" density of a gas in a large container, pressure, temperature, etc.). In fact in macroscopic matter we never have complete knowledge, and our theories must be formulated probabilistically, making what seem tentatively to be reasonable assumptions about a priori probabilities. The success of the resulting theories in describing nature constitutes a test of our assumptions.

2. Probability Generating Functions

Consider an experiment involving the determination of a variable z whose possible values are a set of consecutive integers $s = 0, 1, 2, \ldots, n$. Let the *a priori* probabilities of the various outcomes be designated p_s. Then the function

$$f(x) = p_0 + p_1 x + p_2 x^2 + \cdots + p_n x^n = \sum_{s=1}^{n} p_s x^s \qquad (7.1)$$

is designated the "probability generating function" (p.g.f.) of z. We note certain properties of this function.

(a) $$f(1) = p_0 + p_1 + p_2 + \cdots + p_n = \sum p_s = 1 \qquad (7.2)$$

(b)
$$\left(\frac{df}{dx}\right)_{x=1} = f'(1) = \sum_{s=1}^{n} s p_s \tag{7.3}$$

is the mean value of z, which we call $\langle z \rangle$.

(c) In terms of the p.g.f., the probabilities of the different values of z are

$$p_s = \frac{1}{s!}\left(\frac{d^s f}{dx^s}\right)_{x=0} \tag{7.4}$$

(d) The "rth moment" of the distribution function for z is

$$M_r = \sum s^r p_s = \left[\left(x\frac{d}{dx}\right)\left(x\frac{d}{dx}\right)\cdots(r \text{ times})\cdots\left(x\frac{d}{dx}\right)f\right]_{x=1} \tag{7.5}$$

this moment may also be written

$$M_r = \langle z^r \rangle \tag{7.6}$$

Notice that

$$\langle z - \langle z \rangle \rangle = 0 \tag{7.7}$$

by definition. A measure of the dispersion of values about the mean is given by the "root mean square deviation" (rms deviation)

$$\Delta z = \sqrt{\langle (z - \langle z \rangle)^2 \rangle} \tag{7.8}$$

In terms of the p.g.f., the rms deviation is obtained as follows:

$$f''(1) = \sum s(s-1)p_s = \langle z^2 \rangle - \langle z \rangle$$

But

$$\langle (z - \langle z \rangle)^2 \rangle = \langle z^2 \rangle - 2\langle z \rangle^2 + \langle z \rangle^2 = \langle z^2 \rangle - \langle z \rangle^2 \tag{7.9}$$

Thus

$$\Delta z = \sqrt{f''(1) + f'(1) - [f'(1)]^2} \tag{7.10}$$

Let us illustrate the above with respect to the coin-tossing problem. Let an experiment consist of n tosses of a coin, and define the variable z as the number of heads which come up. The probability generating function is

$$f(x) = \frac{(1+x)^n}{2^n} = \frac{1}{2^n}(1+x)(1+x)\cdots(n \text{ times})\cdots(1+x) \tag{7.11}$$

This is demonstrated by the following argument:

(*i*) If we consider every possible outcome of a series of n tosses we assume that each is equally probable (assumption of equal a priori probability).

(*ii*) The probability that the outcome of a *given* experiment is $z = s$ is equal to the number of experiments (out of all the possible ones) for which $z = s$, divided by the total number of possible outcomes.

Since there are two possibilities for the result of each toss, the total number of results from the experiment is 2^n.

(*iii*) $f(x)$ is evaluated in a power series in x in the following manner: All possible products are formed by choosing from each factor $(1+x)$ either the 1 or the x. The choice of a 1 may be taken to correspond to the throw of a tail; that of an x to the throw of a head. All of the resulting terms containing x^s correspond to throwing s heads. The coefficient of x^s in the expansion of x^s is simply the *number* of terms in x^s. Therefore,

$$p_s = \text{coefficient of } x^s \text{ in } f(x) = \frac{(1+x)^n}{2^n} \qquad (7.12)$$

From this statement, (7.11) follows immediately.

A trivial result is that the mean number of heads is

$$f'(1) = \frac{n}{2} \qquad (7.13)$$

Note, however, that this seems equivalent to saying that, on an individual toss, heads and tails were equally likely. This, of course, was implied in our original assumption that all outcomes of the experiment were equally likely. If the coin had been subtly weighted so that the a priori probability of getting a head was greater than that of getting a tail on a particular toss, the various outcomes of the experiment of n tosses would not be equally probable; for example, the outcome of n heads would be greater than that of n tails. We see below how the p.g.f. would have to be modified in such a case.

Before going into that problem, however, let us calculate the root mean square deviation:

$$(\Delta z)^2 = f''(1) + f'(1) - [f'(1)]^2$$

$$= \frac{n(n-1)}{4} + \frac{n}{2} - \frac{n^2}{4}$$

$$= \frac{n}{4}$$

So that

$$\Delta z = \tfrac{1}{2}\sqrt{n} \qquad (7.14)$$

It should be noted that the root mean square deviation bears to the mean the ratio

$$\frac{\Delta z}{\langle z \rangle} = \frac{1}{\sqrt{n}} \qquad (7.15)$$

CASE OF LARGE n

The probability of a given outcome $z = s$ may be written explicitly in terms of a binomial coefficient:

$$p_s = \frac{n!}{s!(n-s)!}\frac{1}{2^n} \qquad (7.16)$$

For large values of n and s we may use Stirling's formula (4.245) to write

$$
\begin{aligned}
\ln p_s &= \ln n! - \ln s! - \ln (n-s)! - n \ln 2 \\
&\doteq (n+\tfrac{1}{2}) \ln n - n + \tfrac{1}{2} \ln 2\pi - (s+\tfrac{1}{2}) \ln s \\
&\quad + s - \tfrac{1}{2} \ln 2\pi - (n-s+\tfrac{1}{2}) \ln (n-s) \\
&\quad + (n-s) - \tfrac{1}{2} \ln 2\pi - n \ln 2
\end{aligned}
\tag{7.17}
$$

The probability is a maximum where $(\partial/\partial s) \ln p_s = 0$. But

$$
\frac{\partial \ln p_s}{\partial s} = -\left(s+\frac{1}{2}\right)\frac{1}{s} - \ln s + \left(n-s+\frac{1}{2}\right)\frac{1}{n-s} + \ln (n-s)
$$

Neglecting, as in Stirling's formula, terms which approach zero as n, $s \to \infty$ we then have

$$
\ln \frac{n-s}{s} = 0
$$

or

$$
s = \frac{n}{2}
\tag{7.18}
$$

a not very unexpected result. It says that the most likely single outcome of the experiment is also the mean value.

Let us now expand $\ln p_s$ about the mean value

$$
\langle z \rangle = \frac{n}{2}
$$

by putting

$$
s = \frac{n}{2} + \sigma
\tag{7.19}
$$

Then (7.17) becomes

$$
\begin{aligned}
\ln p(\sigma) &= \left(n+\frac{1}{2}\right)\ln n - \left[\frac{n+1}{2}+\sigma\right]\ln\left(\frac{n}{2}+\sigma\right) \\
&\quad -\left[\frac{n+1}{2}-\sigma\right]\ln\left(\frac{n}{2}-\sigma\right) - n\ln 2 - \frac{1}{2}\ln 2\pi
\end{aligned}
$$

If this is expanded in powers of σ, there is no linear term. In fact

$$
\begin{aligned}
\ln p(\sigma) &= \left(n+\frac{1}{2}\right)\ln n - \left[\frac{n+1}{2}+\sigma\right]\left[\ln\frac{n}{2}+\frac{2\sigma}{n}-\frac{1}{2}\left(\frac{2\sigma}{n}\right)^2+\cdots\right] \\
&\quad -\left[\frac{n+1}{2}-\sigma\right]\left(\ln\frac{n}{2}-\frac{2\sigma}{n}-\frac{1}{2}\left(\frac{2\sigma}{n}\right)^2+\cdots\right)-n\ln 2-\frac{1}{2}\ln 2\pi \\
&= \ln 2-\frac{1}{2}\ln 2\pi n-\frac{2}{n}\sigma^2
\end{aligned}
$$

where again we have omitted terms whose ratio to the ones kept approaches zero as $n \to \infty$.

Therefore,

$$p(\sigma) = \sqrt{\frac{2}{\pi n}} \, e^{-(2/n)\sigma^2}$$

or

$$p_s = \sqrt{\frac{2}{\pi n}} \, e^{-(2/n)(s - \langle z \rangle)^2}$$

$$= \frac{1}{\sqrt{2\pi} \, \Delta z} \, e^{-(s - \langle z \rangle)^2/2(\Delta z)^2} \tag{7.20}$$

This is a convenient analytical approximation to the distribution function when n is large, that is, when the experiment consists of a large number of independent random events.

A quantity of interest is that deviation from the mean within which there is a 90% probability that the result of an experiment will lie. Calling this quantity S_0, it is defined by

$$\frac{1}{\sqrt{2\pi} \, \Delta z} \int_{-S_0 + \langle z \rangle}^{S_0 + \langle z \rangle} \exp\left[-\frac{(s - \langle z \rangle)^2}{2(\Delta z)^2}\right] ds = 0.90$$

From tables of the integral, the standard "error integral," it is found that $S_0 = 1.645$.

As a generalization of the coin-tossing problem, consider the problem in which the a priori probabilities of the two outcomes of a single trial are, respectively, p and $q = 1 - p$. The p.g.f. is then

$$f(x) = (q + px)^n = (1 - p + px)^n \tag{7.21}$$

This reduces to the previous case when $p = q = \frac{1}{2}$. That (7.21) is the correct generating function is evident from the fact that the coefficient corresponding to the choice of s x's and $(n - s)$ 1's is weighted with the factor $p^s q^{n-s}$, corresponding to the fact that each event associated with the choice of x had probability p and each one associated with the choice of a 1 had probability q. Let the two events, associated respectively with x and 1, be designated A and B (heads and tails in the previous problem). The mean number of A's in n trials is then

$$f'(1) = np = \langle z \rangle \tag{7.22}$$

and the mean square deviation is

$$\begin{aligned}
\Delta z &= \{f''(1) + f'(1) - [f'(1)]^2\}^{1/2} \\
&= \{n(n-1)p^2 + np - n^2 p^2\}^{1/2} \\
&= \sqrt{np(1-p)} \\
&= \sqrt{npq} \tag{7.23}
\end{aligned}$$

Therefore the rms deviation as a fraction of the mean is

$$\frac{\Delta z}{\langle z \rangle} = \sqrt{\frac{q}{np}} \qquad (7.24)$$

Problem 7-1:   In an election, 55% of the voters favor candidate A and 45% candidate B. A public opinion polling organization samples 1000 randomly chosen voters. What is the rms deviation of the number of voters they report for each candidate? What is it as a percentage of the mean number in each case?

Answer: About 16, or 2.9% for A and 3.5% for B.

Problem 7-2:   Suppose that there are three candidates A, B, and C who are favored by 45, 35, and 20% of the electorate, respectively. Show that the rms deviation Δz_A, Δz_B, and Δz_C of the numbers reported for each candidate in a sample of 1000 are (to the nearest integers) 16, 15, and 13, respectively, so that the 90% "confidence limits" are for A, 42.4 to 47.6%; for B, 32.5 to 37.5%, and for C, 17.9 to 22.1%.

Problem 7-3:   The lifetime of an artificially created radioactive nucleus is determined by a delayed coincidence method. Pulses from creation events are fed into a coincidence circuit after a time delay. Pulses from decays are fed in *without* delay.

Suppose now that the coincidence counting rates c_1 and c_2 are measured for delay times τ_1 and τ_2.

For best results, what proportional times should be spent observing coincidences for each delay?

Show that, if a total time t_0 is available for counting, the root mean square deviation in the derived decay constant in the best circumstances is

$$\frac{1}{\sqrt{t_0}} \frac{c_1^{-1/2} + c_2^{-1/2}}{\tau_2 - \tau_1}$$

---

3.   Density Fluctuations in a Gas

Suppose we have a gas containing N molecules, each of mass m, in a container of volume V. We assume that each molecule is as likely to be at any point in the container as at any other. Consider a fixed subvolume v within the larger container. We may consider the probability of having various numbers n of particles in v, and consequently, the probability distribution for the density of gas in volume v.

Problem 7-4:   Consider two separate subvolumes v in a large volume V ($v \ll V$), the whole containing N molecules. Show that the probability generating function is approximately

$$P(x, y) = p_1(x)p_1(y)$$

where

$$p_1(x) = e^{-(Nv/V)(1-x)}$$

and therefore that the probability distributions in the two subvolumes are effectively independent. Explain why this is so. [The result of this problem illustrates an important principle of statistical mechanics, viz., the *mathematical* independence of subsystems in equilibrium with a much larger system, even though they are interacting and not isolated from each other.]

This is simply an example of the previous problem, where N is the number of "trials," $p = (v/V)$ is the probability of one result (finding a molecule in v) and $q = 1 - (v/V)$ the other (finding a molecule outside v). The root mean square deviation of the number in v is then

$$\Delta = \sqrt{N \frac{v}{V}\left(1 - \frac{v}{V}\right)} \qquad (7.25)$$

while the mean number is

$$\langle n \rangle = N \frac{v}{V} \qquad (7.26)$$

Therefore the mean density in v is

$$\rho = m \frac{\langle n \rangle}{v} = m \frac{N}{V} \qquad (7.27)$$

which is independent of the volume selected for measurement. However, the rms deviation is the fraction

$$\frac{\Delta}{\langle n \rangle} = \sqrt{\frac{1 - (v/V)}{\langle n \rangle}}$$

of the mean. If $v/V \ll 1$, that is, if the volume sampled is only a small fraction of the whole gas, this is

$$\frac{\Delta}{\langle n \rangle} \doteq \frac{1}{\sqrt{\langle n \rangle}} \qquad (7.28)$$

Therefore, so long as the volume v contains many molecules, the fractional density fluctuations are small.

4.   The Poisson Distribution

Consider the problem of a slow radioactive emitter, which emits quanta or particles of radiation at random times, the probability per unit time of an emission being λ. Let us calculate the distribution of *number* of particles emitted in a time interval of length t. This is called a *Poisson distribution*.

We start from the following statement: The probability that n particles have been emitted by time $t + \delta t$ is the probability that n were emitted by time t and *none* in the infinitesimal interval δt thereafter, plus the probability that $(n-1)$ had been emitted by time t, and one was emitted in the subsequent interval δt. [The probability of *two* being emitted in δt is of order $(\delta t)^2$, etc; since our aim is to derive a differential equation for $P_n(t)$ we need only keep terms up to order δt.] The mathematical formulation of the statement is

$$P_n(t + \delta t) = P_n(t)(1 - \lambda \delta t) + P_{n-1}(t)\lambda \delta t + O(\delta t^2) \qquad (7.29)$$

Dividing through by δt and taking the limit as $\delta t \to 0$, we obtain the equation

$$\frac{dP_n(t)}{dt} = -\lambda P_n(t) + \lambda P_{n-1}(t) \qquad (7.30)$$

Giving n all possible values, $n = 0, 1, 2, \ldots$, we have a chain of coupled equations. They may be solved, however, by the simple device of introducing the probability generating function

$$f(x, t) = \sum P_n(t)x^n \qquad (7.31)$$

If we now multiply (7.30) by x^n and add the equations for all possible n's, we get

$$\frac{\partial f(x, t)}{\partial t} = -\lambda(1 - x)f(x, t) \qquad (7.32)$$

The solution of this equation is

$$f(x, t) = f_0(x) \, e^{-\lambda(1-x)t}$$

But at $t = 0$, $P_0 = 1$ and all other P_n's $= 0$, so that $f_0(x) = 1$; therefore,

$$f(x, t) = e^{-\lambda(1-x)t} \qquad (7.33)$$

The P_n's may now be obtained by expanding $f(x, t)$ in powers of x:

$$f(x, t) = \sum P_n(t)x^n = e^{-\lambda t} \sum \frac{(\lambda t)^n}{n!} x^n \qquad (7.34)$$

so that

$$P_n(t) = \frac{(\lambda t)^n}{n!} e^{-\lambda t} \qquad (7.35)$$

which is the Poisson distribution.

The mean value at time t is

$$\left[\frac{\partial f(x, t)}{\partial x}\right]_{x=1} = \lambda t \qquad (7.36)$$

which is in any case intuitively evident.

The mean square deviation is

$$(\Delta n)^2 = \lambda t \tag{7.37}$$

which gives us once again the square root law

$$\frac{\Delta n}{\langle n \rangle} = \frac{1}{\sqrt{\langle n \rangle}} = \frac{1}{\sqrt{\lambda t}} \tag{7.38}$$

The distribution function is sharply peaked and narrow when λt is large. This may be seen by taking the logarithm of (7.35) and using Stirling's approximation

$$\ln P_n(t) = n \ln \lambda t - \lambda t - (n + \tfrac{1}{2}) \ln n + n - \tfrac{1}{2} \ln 2\pi \tag{7.39}$$

Taking $\partial/\partial n$ and omitting terms which approach zero as $n \to \infty$, we find

$$\frac{\partial}{\partial n} \ln P_n(t) = \ln \lambda t - \ln n$$

which is zero at

$$n = \lambda t$$

The distribution function therefore has a maximum at this value. If we put

$$n = \lambda t + \nu \tag{7.40}$$

we can expand $\ln P_n(t)$ in powers of ν:

$$\ln P_n(t) \doteq (\lambda t + \nu) \ln \lambda t - \lambda t + \left(\lambda t + \nu + \frac{1}{2}\right)\left(\ln \lambda t + \frac{\nu}{\lambda t} - \frac{1}{2}\frac{\nu^2}{(\lambda t)^2} + \cdots\right)$$

$$+ \lambda t + \nu - \tfrac{1}{2} \ln 2\pi$$

$$\doteq -\frac{1}{2} \ln \lambda t - \frac{1}{2} \ln 2\pi - \frac{\nu^2}{2\lambda t} \tag{7.41}$$

where we have omitted terms which go to zero as $\lambda t \to \infty$ and higher powers in $\nu/\lambda t$. In this approximation

$$P_n(t) = \frac{1}{\sqrt{2\pi\lambda t}} \exp\left(-\frac{\nu^2}{2\lambda t}\right)$$

$$= \frac{1}{\sqrt{2\pi\lambda t}} \exp\left[-\frac{(n - \lambda t)^2}{2\lambda t}\right] \tag{7.42}$$

which is again a Gaussian distribution whose width is $\simeq \sqrt{\lambda t}$. If we had kept terms $\approx \nu^3/(\lambda t)^2$ Eq. (7.42) would have been multiplied by another negative exponential with an exponent proportional to this quantity. But since the distribution function is already negligible for $n - \lambda t = \nu \geq \sqrt{\lambda t}$, and for smaller values the exponent in the new term is $\simeq (\lambda t)^{3/2}/(\lambda t)^2 = 1/\sqrt{\lambda t}$ which $\to 0$ as λt becomes large, the approximation (7.42) is a good one for large λt.

Problem 7-5:   Suppose that exactly n calls are made through a telephone switchboard in an hour. What is the probability that ν calls will be made in any given minute?

If the number of calls per hour follows a Gaussian distribution with mean n and rms deviation σ, what is the probability for ν calls in a given minute?

Problem 7-6:   Consider a fast radioactive emitter, so that one must take account of the decrease with time of the number of decaying nuclei. If there are N nuclei present initially, show that the probabilities $P_n(t)$ of n decays in time t satisfy

$$\frac{\partial P_n}{\partial t} = -\lambda_0[(N-n)P_n - (N-n+1)P_{n-1}]$$

and that the equation for the generating function is

$$\frac{\partial f(x,t)}{\partial t} = -N\lambda_0(1-x)f(x,t) + \lambda_0 x(1-x)\frac{\partial f(x,t)}{\partial x}$$

where λ_0 is the probability per unit time of the decay of a given nucleus.
    Obtain the solution to the equation in the form

$$f(x,t) = e^{-N\lambda_0 t}[1 + x(e^{\lambda_0 t} - 1)]^N$$

and from it determine $P_n(t)$, $\langle n \rangle$, and Δn.

5.   A Problem in Counting of Radioactive Decays

An elaboration of the problem of fluctuations in rate of radioactive decay is provided by considering the counting of these decays electronically, the counter having a "dead time" τ; that is, after a count is registered, another cannot be detected until a time τ has elapsed. The question is, then, what is the probability of n counts actually being recorded in time t?

Let us introduce a new probability $Q_n(t)$ such that $Q_n(t)\, dt$ is the probability that the nth count takes place in the interval $(t, t + dt)$. Then

$$Q_{n+1}(t) = \int_0^t Q_n(t')p_0(t-t')\, dt' \tag{7.43}$$

where $p_0(t-t')$ is the probability per unit time that the $(n+1)$st count takes place in the time interval $(t, t + dt)$ given that the nth took place at

t'. But we know that

$$p_0(t-t') \, dt = 0, \qquad \text{for} \qquad t-t' < \tau$$

$$= \lambda \, dt \times [\text{probability of no counts in interval } (t-t')]$$

$$= \lambda \, dt \, e^{-\lambda(t-t')} \tag{7.44}$$

If Laplace transform of $Q_n(t)$ is $q_n(s)$, and that of $p_0(t)$ is $\pi_0(s)$, by the convolution theorem the transform of Eq. (7.43) is

$$q_{n+1}(s) = q_n(s)\pi_0(s) \tag{7.45}$$

We may thus determine q_n in terms of q_1:

$$q_n(s) = [\pi_0(s)]^{n-1} q_1(s) \tag{7.46}$$

But the transforms $\pi_0(s)$ and $q_1(s)$ are easily evaluated.

$$\pi_0(s) = \lambda \int_\tau^\infty e^{-\lambda t} e^{-st} \, dt$$

$$= \frac{\lambda e^{-(\lambda+s)\tau}}{\lambda+s} \tag{7.47}$$

and

$$q_1(s) = \int_0^\infty Q_1(t) e^{-st} \, dt$$

$$= \int_0^\infty \lambda e^{-\lambda t} e^{-st} \, dt$$

$$= \frac{\lambda}{\lambda+s} \tag{7.48}$$

Therefore,

$$q_n(s) = \lambda^n \frac{e^{-(n-1)(\lambda+s)\tau}}{(\lambda+s)^n}$$

The inverse transform is easily calculated by the Laplace inverse theorem; it is

$$Q_n(t) = \frac{\lambda^n}{2\pi i} e^{-(n-1)\lambda\tau} \int_{s_0-i\infty}^{s_0+i\infty} \frac{e^{-(n-1)s\tau}}{(\lambda+s)^n} e^{st} \, ds \tag{7.49}$$

s_0 is any quantity $> -\lambda$. The integral may be evaluated by extending the contour into a semicircle of infinite radius on the negative side of s_0. On evaluating the residue at the singularity we find that

$$Q_n(t) = \lambda \frac{\{\lambda[t-(n-1)\tau]\}^{n-1}}{(n-1)!} e^{-\lambda t}, \qquad t > (n-1)\tau$$

$$= 0, \qquad\qquad\qquad\qquad\qquad t < (n-1)\tau \tag{7.50}$$

Putting $\tau = 0$, we obtain the Poisson distribution.

Before continuing, let us note the relation between $Q_n(t)$, which is the probability distribution in time for the nth count, and $P_n(t)$, which is the distribution in number of counts up to a given time. The connection follows from the relation that the probability $Q_n(t) \, dt$ that the nth count takes place in the interval $(t, t+dt)$ is the probability $P_{n-1}(t)$ that up to time $(t-\tau)$ there had been $(n-1)$ counts, multiplied by the probability $\lambda \, dt$ that another count took place in the interval in question. Thus

$$Q_n(t) = \lambda P_{n-1}(t-\tau) \tag{7.51}$$

or

$$P_n(t) = \frac{1}{\lambda} Q_{n+1}(t+\tau)$$

$$= \frac{[\lambda(t-n\tau)]^n}{n!} e^{-\lambda t} \tag{7.52}$$

As $\tau \to 0$ this reduces to the Poisson distribution obtained earlier. The Laplace transform method therefore provides an alternative derivation to that of probability generating functions.

Problem 7-7: For a given t, show that the most likely n is given by the transcendental equation

$$\lambda \tau u = e^{1/u}$$

where $u = (t-n\tau)/n\tau$. For $\lambda \tau \ll 1$, verify that

$$n \approx \frac{\lambda t}{1 + 2\lambda \tau}$$

---

6. Probability Distributions for Continuous Variables

Suppose that we consider a random continuous variable z, such that the probability that its value is in the range $(z, z+dz)$ is $p(z) \, dz$. Then

$$\int_{-\infty}^{\infty} p(z) \, dz = 1 \tag{7.53}$$

[If the range of values of z is limited, we define $p(z) = 0$ outside the limits.] The mean value of z is

$$\langle z \rangle = \int_{-\infty}^{\infty} z p(z) \, dz \tag{7.54}$$

We define the nth moment of the distribution as

$$M_n = \langle z^n \rangle = \int_{-\infty}^{\infty} z^n p(z) \, dz \tag{7.55}$$

Also, the mean value of $f(z)$ is

$$\langle f(z) \rangle = \int_{-\infty}^{\infty} f(z) p(z) \, dz \tag{7.56}$$

Two variables z_1 and z_2 are said to be *independent* if the distribution of one does not depend on the value of the other; the joint distribution $p(z_1, z_2)$ such that $p(z_1, z_2)\, dz_1 dz_2$ is the probability that the first lies in the range $(z_1, z_1 + dz_1)$ and the second in the range $(z_2, z_2 + dz_2)$. If they are independent

$$p(z_1, z_2) = p_1(z_1)p_2(z_2) \qquad (7.57)$$

The mean value of $f_1(z_1) + f_2(z_2)$ is

$$\langle f_1(z_1) + f_2(z_2) \rangle = \langle f_1(z_1) \rangle + \langle f_2(z_2) \rangle \qquad (7.58)$$

and that of $f_1(z_1)f_2(z_2)$ is

$$\langle f_1(z_1)f_2(z_2) \rangle = \langle f_1(z_1) \rangle \langle f_2(z_2) \rangle \qquad (7.59)$$

Problem 7-8:   Suppose that α and β are independent random variables with a Gaussian distribution with rms deviation γ. Define

$$\xi(t) = \alpha \cos \omega t + \beta \sin \omega t$$

and

$$\eta(t) = \frac{d\xi}{dt} = -\alpha\omega \sin \omega t + \beta\omega \cos \omega t$$

Show that $\xi(t)$, $\eta(t)$ are independent random variables. Show also that their distribution functions are

$$p_1(\xi) = \frac{1}{\sqrt{2\pi}\,\gamma}\, e^{-\xi^2/2\gamma^2}$$

$$p_2(\eta) = \frac{1}{\sqrt{2\pi}\,\gamma\omega}\, e^{-\eta^2/2\gamma^2\omega^2}$$

The moments of the distribution of z are easily expressed in terms of the Fourier transform of $p(z)$, which is in fact a sort of generalization of the notion of probability generating functions to continuous variables. Let the Fourier transform of $p(z)$ be

$$\pi(x) = \int p(z)\, e^{izx}\, dz \qquad (7.60)$$

$p(z)$ may be written in terms of its transform as

$$p(z) = \frac{1}{2\pi} \int \pi(x)\, e^{-izx}\, dx \qquad (7.61)$$

We note that $\pi(0) = 1$. Also, from (7.60) it follows that

$$\langle z \rangle = -i\pi'(0) \qquad (7.62)$$

and in general

$$M_n(z) = \langle z^n \rangle = (-i)^n \left[\frac{d^n}{dx^n}\, \pi(x) \right]_{x=0}$$

$$= (-i)^n \pi^{(n)}(0) \qquad (7.63)$$

In particular, the mean square deviation of z from its mean is

$$(\Delta z)^2 = \langle z^2 \rangle - \langle z \rangle^2 = -\pi^{(2)}(0) + [\pi'(0)]^2 \qquad (7.64)$$

We note that

$$\pi(x) = \langle e^{izx} \rangle = 1 + ix\langle z \rangle - \frac{x^2}{2} \langle z^2 \rangle + \cdots$$

$$+ \frac{(ix)^n}{n!} \langle z^n \rangle + \cdots \qquad (7.65)$$

7. Distribution of a Sum of Independent Continuous Random Variables

Consider the distribution of

$$S_N = z_1 + z_2 + \cdots + z_n \qquad (7.66)$$

where z_i is a random variable with a distribution $p_i(z_i)$. We may relate the distribution $P_N(S_N)$ of S_N to that of S_{N-1}:

$$P_N(S_N) = \int P_{N-1}(S_{N-1})p_N(S_N - S_{N-1}) \, dS_{N-1} \qquad (7.67)$$

Letting the transform of P_N be $\Pi_N(x)$, taking the transform of (7.67), and using the convolution theorem, we get the equation

$$\Pi_N(x) = \Pi_{N-1}(x)\pi_N(x) \qquad (7.68)$$

Problem 7-9:   Prove the following:
  (a) If the distribution of z is $p(z)$, that of $Z = f(z)$ is $\bar{p}(Z) = p(z)/f'(z)$ where the right-hand side must be expressed in terms of Z.
  (b) The distribution of $z^2 = Z$ is $p(\sqrt{Z})/2\sqrt{Z}$.
  (c) The distribution of $\sin z = Z$ is $p(\sin^{-1} Z)/\sqrt{1-Z^2}$.

Problem 7-10:   If z_n has a Lorentzian distribution

$$p_n(z_n) = \frac{a_n}{\pi} \frac{1}{a_n{}^2 + z_n{}^2}$$

show that S_N has the Lorentzian distribution

$$P_N = \frac{A}{\pi} \frac{1}{A^2 + S_N{}^2}$$

where

$$A = \sum_{n=1}^{N} a_n$$

Problem 7-11:   If z_n has a Gaussian distribution

$$p_n(z_n) = \frac{1}{\sqrt{2\pi}\,\Delta_n} \exp\left\{ -\left[\frac{(z_n - \langle z_n \rangle)^2}{2\Delta_n{}^2} \right] \right\}$$

show that S_N has the distribution

$$\Pi_N(S_N) = \frac{1}{\sqrt{2\pi}\,\Delta S_N}\exp\left\{-\left[\frac{(S_N - \langle S_N\rangle)^2}{2(\Delta S_N)^2}\right]\right\}$$

where

$$\langle S_N\rangle = \sum_{n=1}^{N}\langle z_n\rangle$$

and

$$(\Delta S_N)^2 = \sum_{n=1}^{N}\Delta_n{}^2$$

It follows that

$$\Pi_N(x) = \pi_0(x)\pi_1(x)\cdots\pi_N(x) \tag{7.69}$$

Taking the inverse transform, we may in principle determine the distribution of the sum. But (7.69) may now be written as

$$\Pi_N(x) = e^{\Sigma\ln\pi_n(x)} \tag{7.70}$$

$$= \exp\sum_n\ln\left[1 + ix\sum\langle z_n\rangle - \frac{x^2}{2}\sum\langle z_n{}^2\rangle - \frac{ix^3}{6}\sum\langle z_n{}^3\rangle + \cdots\right] \tag{7.71}$$

To simplify the calculation slightly, let us choose the random variables z_n so that their mean values are zero (i.e., subtract from each its mean value). Then

$$\Pi_N(x) = \exp\sum_n\ln\left[1 - \frac{x^2}{2}\langle z_n{}^2\rangle - \frac{ix^3}{6}\langle z_n{}^3\rangle + \cdots\right]$$

We note that $\pi_n(x)$ is a maximum at $x = 0$, where it has the value 1. For

$$\left(\frac{\partial\pi_n}{\partial x}\right)_{x=0} = i\int zp(z)\,dz = 0$$

and

$$\left(\frac{\partial^2\pi_n}{\partial x^2}\right)_{x=0} = -\int z^2p(z)\,dz < 0$$

It follows that $\Pi_N(x)$ has a maximum (also unity) at $x = 0$. If we now expand the logarithm in powers of x

$$\Pi_N(x) = \exp\left\{-\frac{x^2}{2}\sum\langle z_n{}^2\rangle - \frac{ix^3}{6}\sum\langle z_n{}^3\rangle + \text{terms of order } x^4, \text{etc.}\right\}$$

$$= \exp\left\{-N\alpha_2\frac{x^2}{2} - iN\alpha_3\frac{x^3}{6} + \cdots\right\} \tag{7.72}$$

where

$$\alpha_2 = \frac{1}{N}\sum\langle z_n{}^2\rangle, \qquad \alpha_3 = \frac{1}{N}\sum\langle z_n{}^3\rangle \tag{7.73}$$

Suppose now that A is a large enough quantity that e^{-A} is negligible; then, $\Pi_N(x)$ has a negligibly small factor if

$$x > \sqrt{\frac{2A}{\alpha_2 N}}$$

But at the largest value of x which we need consider,

$$i\alpha_3 \frac{x^3}{6} = \frac{i}{6}\alpha_3 \left(\frac{2}{\alpha_2}\right)^{3/2} A^{3/2} \frac{1}{N^{3/2}}$$

Continuing to the x^4 term, it is

$$-\alpha_4 x^4/24$$

where

$$\alpha_4 = \frac{1}{N} \sum [3\langle z_n^2\rangle^2 - \langle z_n^4\rangle] \tag{7.74}$$

At $x = \sqrt{2A/\alpha_2 N}$ it has the value $-\frac{1}{6}(\alpha_4/\alpha_2^2)(A^2/N^2)$. We note that, in the limit of large N, the terms subsequent to the x^2 one are all very small in the range in which Π_N is not negligible. Therefore, in this limit

$$\Pi_N(x) = \exp\left[-\frac{x^2(\Delta S_N)^2}{2}\right] \tag{7.75}$$

where

$$(\Delta S_N)^2 = \sum_{n=1}^{N} (\Delta z_n)^2 \tag{7.76}$$

Inverting the Fourier transform, we obtain the distribution of the sum S_N,

$$P_N(S_N) = \frac{1}{\sqrt{2\pi}\Delta S_N} \exp\left[-\frac{S_N^2}{2(\Delta S_N)^2}\right] \tag{7.77}$$

This has been calculated for variables z_n whose mean values $\langle z_n\rangle$ are zero. Otherwise S_N^2 in the above formula must be replaced by $(S_N - \langle S_N\rangle)^2$ where $\langle S_N\rangle = \Sigma \langle z_n\rangle$.

The result (7.77) is generally called the "Central Limit Theorem."

A generalization of the preceding problem is that in which z_n is replaced by a vector $\mathbf{r}_n$. If the distribution function $p_n(\mathbf{r}_n)$ for this function is known, the problem is to find that for $P_N(\mathbf{R}_n)$ where

$$\mathbf{R}_N = \sum \mathbf{r}_n \tag{7.78}$$

If $\pi_n(\mathbf{k})$ is the transform of $p_n(\mathbf{r}_n)$ and $\Pi_N(\mathbf{k})$ that of $P_N(\mathbf{R}_N)$, an equation analogous to (7.69) can be obtained by exactly the same procedure as used in the one-dimensional case:

$$\Pi_N(\mathbf{k}) = \pi_1(\mathbf{k})\pi_2(\mathbf{k})\ldots\pi_N(\mathbf{k}) \tag{7.79}$$

But

$$\pi_n(\mathbf{k}) = \int p_n(\mathbf{r}_n) e^{i\mathbf{k}\cdot\mathbf{r}_n} d^3\mathbf{r}_n \tag{7.80}$$

$$= 1 + i\mathbf{k}\cdot\langle\mathbf{r}_n\rangle - \tfrac{1}{2}\sum_{\alpha,\beta} k_\alpha k_\beta \langle x_{n\alpha}x_{n\beta}\rangle + O(k^3) \tag{7.81}$$

$x_{n\alpha}$ $(\alpha = 1, 2, 3)$ are components of the vector $\mathbf{r}_n$. Therefore,

$$\Pi_N(\mathbf{k}) = \exp \sum_{n=1}^{N} \ln\left\{1 + i\mathbf{k}\cdot\langle\mathbf{r}_n\rangle - \tfrac{1}{2}\sum_{\alpha,\beta} k_\alpha k_\beta \langle x_{n\alpha}x_{n\beta}\rangle + \cdots\right\}$$

If we choose the $\mathbf{r}_n$'s to have mean value zero, to a good approximation for large N

$$\Pi_N(\mathbf{k}) = \exp\left[-\tfrac{1}{2}\sum_{\alpha,\beta} k_\alpha k_\beta M^{(2)}_{N,\alpha\beta}\right] \tag{7.82}$$

where

$$M^{(2)}_{N,\alpha\beta} = \sum_{n=1}^{N} \langle x_{n\alpha}x_{n\beta}\rangle \tag{7.83}$$

To evaluate (7.82) it is necessary to transform axes so as to diagonalize $M^{(2)}_{N,\alpha\beta}$. The inverse transform is then calculated as the product of three one-dimensional Gaussian transforms, one over the components in the direction of each principal axis.

There is some simplification in the case in which $p_n(\mathbf{r}_n)$ depends only on $r_n = |\mathbf{r}_n|$. Then $\langle\mathbf{r}_n\rangle = 0$ automatically and

$$\langle x_{n\alpha}x_{n\beta}\rangle = \int p(r_n) x_{n\alpha}x_{n\beta}\, d^3\mathbf{r}_n$$

$$= \tfrac{1}{3}\langle r_n{}^2\rangle \delta_{\alpha\beta} \tag{7.84}$$

This follows from the fact that, when $\alpha \neq \beta$, the integrand is odd in the variables $x_{n\alpha}$ and $x_{n\beta}$, and also that $\langle x^2_{n1}\rangle = \langle x_{n2}\rangle^2 = \langle x_{n3}\rangle^2 = \tfrac{1}{3}r_n{}^2$ due to spherical symmetry. In this case

$$\Pi_n(\mathbf{k}) = \exp\left[-\tfrac{1}{6}k^2\langle R_N{}^2\rangle\right] \tag{7.85}$$

where

$$\langle R_N{}^2\rangle = \sum_{n=1}^{N} \langle r_n{}^2\rangle \tag{7.86}$$

Taking the inverse transform

$$P_N(\mathbf{R}_N) = \frac{1}{[(2\pi/3)\langle R_N{}^2\rangle]^{3/2}} \exp\left[-\frac{3R_N{}^2}{2\langle R_N{}^2\rangle}\right] \tag{7.87}$$

This is the probability per unit volume at $\mathbf{R}_N$.

The problem just discussed is one version of the so-called "random walk" problem.

8.   Relation between Random Walk Problem and Diffusion

Consider the following problem: A point is moved in random steps, the length of a step being determined by the probability distribution $p(\mathbf{r})$, and the probability per unit time of a step being λ. What is the probability distribution for the resulting total displacement in time t?

If the number of steps taken in time t were known, the solution would follow immediately from (7.87). However, the number of steps is in fact given by the Poisson distribution, the probability of N steps being $[(\lambda t)^N/N!]\, e^{-\lambda t}$. Now the transform of the distribution for N steps is given by (7.79)

$$\Pi_N(\mathbf{k}) = [\pi(\mathbf{k})]^N \qquad (7.88)$$

$\pi(\mathbf{k})$ being the transform of the probability distribution for one step. Thus, the transform of the distribution of displacement in time t is

$$\Pi(\mathbf{k}, t) = \sum_N \frac{(\lambda t)^N}{N!}\, e^{-\lambda t}[\pi(\mathbf{k})]^N$$

$$= e^{-\lambda t[1-\pi(\mathbf{k})]} \qquad (7.89)$$

If λt is large, most of the contribution comes from near $\mathbf{k} = 0$; we can therefore use the expansion of $\pi(\mathbf{k})$ which follows from (7.81):

$$\pi(\mathbf{k}) = 1 - \tfrac{1}{6}k^2\langle r^2\rangle + \cdots \qquad (7.90)$$

so that, approximately,

$$\Pi(\mathbf{k}, t) = \exp\left[-\tfrac{1}{6}\lambda t k^2\langle r^2\rangle\right] \qquad (7.91)$$

This is readily inverted to give the probability distribution $P(\mathbf{R}, t)$ for displacement in time t:

$$P(\mathbf{R}, t) = \frac{1}{[(2\pi/3)\lambda t\langle r^2\rangle]^{3/2}}\, e^{-3R^2/2\langle r^2\rangle\lambda t} \qquad (7.92)$$

This result is clearly the same as that for diffusion from a unit point source, given by (6.25). The identification of the parameters in the two problems

$$\beta = \tfrac{1}{6}\lambda\langle r^2\rangle \qquad (7.93)$$

gives a "first principles" expression for the diffusion constant β.

Problem 7-12:   In *two* dimensions, show that if each step is of length a but in an arbitrary direction,

$$\pi(\mathbf{k}) = J_0(ka)$$

Hence determine the mean square displacement after N steps for *arbitrary* N. Show that this result may also be obtained by elementary means.

Problem 7-13:   By considering the case $N = 1$ show that

$$\left(\frac{1}{2\pi}\right)^2 \int J_0(ka)\, e^{-i\mathbf{k}\cdot\boldsymbol{\rho}}\, d^2\mathbf{k} = \frac{1}{2\pi\rho}\,\delta(\rho - a)$$

Problem 7-14:   By calculating *directly* the probability distribution for $N = 2$ show that

$$\left(\frac{1}{2\pi}\right)^2 \int [J_0(ka)]^2\, e^{-i\mathbf{k}\cdot\boldsymbol{\rho}}\, d^2\mathbf{k} = \frac{1}{4\pi^2\rho\sqrt{1-(\rho^2/4)}}$$

Problem 7-15:   Show that, in three dimensions, if each step is of length "a" but the direction is arbitrary

$$\pi(\mathbf{k}) = \frac{\sin ka}{ka}$$

(a) Show directly that the mean square displacement in N steps is Na^2.

(b) Find explicitly $P_3(\mathbf{R})$.

9. Problems Involving Multiplication

Suppose that we have a system in which, by some method of reproduction, elements give rise to "offspring." If there is a fluctuation in the population, we would expect it somehow to propagate itself. The question poses itself, whether this might not give rise to much larger fluctuations than those discussed up to now, which became proportionately less and less significant in the large number limit.

A simple example may be drawn from physics. Consider a simple model of a cosmic ray shower. An electron or positron, on passing through the atmosphere, has a finite probability per unit length of producing a pair, which we call λ, How many electrons and positrons result from one incident one in a distance l?

Let the probability that there are n at a distance x be $P_n(x)$. The probability that there are n at distance $x + \delta x = $ (probability of n at distance x) $(1 - n\lambda\delta x) + $ (probability of $n - 2$ at distance $x \times [(n-2)\lambda\delta x]$, that is,

$$P_n(x+\delta x) = P_n(x)(1 - n\lambda\delta x) + P_{n-2}(x)(n-2)\lambda\delta x$$

Dividing by δx and taking the limit $\delta x \to 0$ we obtain the differential equation

$$\frac{dP_n}{dx} = -n\lambda P_n + \lambda P_{n-2}(n-2) \tag{7.94}$$

We can introduce the generating function

$$f(\xi) = \sum P_n(x)\xi^n \tag{7.95}$$

and write (7.94) in terms of it by multiplying through by ξ^n:

$$\frac{\partial f(x,\xi)}{\partial x} = -\lambda\xi\frac{\partial f}{\partial\xi} + \lambda\xi^3\frac{\partial f}{\partial\xi} \tag{7.96}$$

The introduction of a new variable

$$u = \int \frac{d\xi}{\xi(1-\xi^2)}$$
$$= \ln\xi - \tfrac{1}{2}\ln(1-\xi^2) \tag{7.97}$$

transforms the equation to

$$\frac{\partial f}{\partial x} + \lambda\frac{\partial f}{\partial u} = 0 \tag{7.98}$$

for which

$$f = F(u - \lambda x) \tag{7.99}$$

is a solution, F being an arbitrary function. To determine the function, we use the initial condition $F(u) = \xi$ when $x = 0$. The problem is then to express ξ in terms of u; but

$$\frac{\xi^2}{1-\xi^2} = e^{2u} \tag{7.100}$$

Therefore,

$$\xi^2 = \frac{e^{2u}}{1+e^{2u}}$$

or

$$\xi = \frac{e^u}{[1+e^{2u}]^{1/2}} \tag{7.101}$$

It follows that the generating function is

$$f(x,\xi) = \frac{\xi}{\sqrt{1-\xi^2}}\frac{e^{-\lambda x}}{[1+\xi^2/(1-\xi^2)\,e^{-2\lambda x}]^{1/2}}$$
$$= \frac{\xi\,e^{-\lambda x}}{[1-\xi^2(1-e^{-2\lambda x})]^{1/2}} \tag{7.102}$$

The probability of n particles at distance l is different from zero only if n is *odd* [that is, $(2s+1)$, say] since only *pairs* are produced. Specifically

$$P_{2s+1} = e^{-\lambda l}\binom{-\tfrac{1}{2}}{s}(1-e^{-2\lambda l})^s \tag{7.103}$$

where

$$\begin{pmatrix} -\frac{1}{2} \\ s \end{pmatrix}$$

is the coefficient of y^s in the expansion of $(1-y)^{-1/2}$. The mean number is

$$\langle n(l) \rangle = e^{2\lambda l} \tag{7.104}$$

A straightforward but slightly tedious calculation also gives

$$\Delta n(l) = \sqrt{2(e^{4\lambda l} - e^{2\lambda l})} \tag{7.105}$$

We see, then, that fluctuations larger than the mean value are quite probable, a peculiar feature of processes involving multiplication.

Problem 7-16: Suppose that a colony of amoeba subdivide in two with a probability λ per unit time, and die with a probability μ per unit time. Starting with 1, determine the probability that n will be alive at time t.

---

10.   The Markoff Method

The Markoff method is one based on Fourier transforms, and on a geometrical representation of the assumption regarding a priori probabilities. To illustrate it, let us return to the "random walk" problems in two and three dimensions.

Consider first the two-dimensional problem. The steps of the walk are characterized by angles θ_n; let us assume that they are all of length "a." Consider the N-dimensional "space" of which all the θ_n's are coordinates; it is a space of volume $(2\pi)^N$. The equations

$$a \sum \cos \theta_n = (X, X+\delta X) \tag{7.106}$$

$$a \sum \sin \theta_n = (Y, Y+\delta Y) \tag{7.107}$$

define a region between infinitesimally separated "surfaces" (of dimensionality $N-2$) in this space. The probability that the coordinates of the displacement resulting from the N random steps lie in the range $(X, X+\delta X; Y, Y+\delta Y)$ is simply the volume of space lying between these surfaces, divided by the total volume of the space. The fact that it is only the volumes which matter is a consequence of the assumption of equal a priori probability for each value of each of the θ_n's. (Otherwise, we would have to calculate volumes *weighted* by the probabilities of the various values of θ_n.) That the probability does depend on the volumes as stated follows from the fact that each point in the θ space represents a possible "walk," all such walks being, in fact, equally probable. The probability required is then that fraction of the total volume corresponding to the walks in question.

The second step in the Markoff method is to express the volumes in terms of Fourier transforms. The volume is in fact the integral over the θ's of a function which is 1 in the regions defined by (7.106) and (7.107) and zero elsewhere. Let us call this function Δ. Its Fourier transform as a function of X and Y is

$$\int \Delta \, e^{i(k_1 X + k_2 Y)} \, dX \, dY = e^{ik_1 \Sigma \, a \cos \theta_n} \, e^{ik_2 \Sigma \, a \sin \theta_n} \delta X \delta Y$$

Therefore by the inverse Fourier theorem,

$$\Delta = \delta X \delta Y \left(\frac{1}{2\pi}\right)^2 \int e^{-ik_1(X - a\Sigma \cos \theta_n)} \, e^{-ik_2(Y - a\Sigma \sin \theta_n)} \, dk_1 \, dk_2 \qquad (7.108)$$

Let us now write

$$k_1 = k \cos \alpha, \qquad k_2 = k \sin \alpha \qquad (7.109)$$

We then have

$$\Delta = \delta X \delta Y \left(\frac{1}{2\pi}\right)^2 \int e^{-i(k_1 X + k_2 Y)} \, e^{ika \Sigma \cos (\theta_n - \alpha)} k \, dk \, d\alpha \qquad (7.110)$$

The integral of Δ over the θ_n's, divided by the volume of θ space, that is, $(2\pi)^N$, is the required probability. It is

$$P_n(X, Y)\delta X \delta Y = \delta X \delta Y \left(\frac{1}{2\pi}\right)^2 \int e^{-i\mathbf{k} \cdot \boldsymbol{\rho}} \, [J_0(ka)]^N k \, dk \, d\alpha \qquad (7.111)$$

where

$$\rho = \sqrt{X^2 + Y^2} \qquad (7.112)$$

Therefore the probability per unit area for displacement $\boldsymbol{\rho}$ is

$$P_N(\boldsymbol{\rho}) = \frac{1}{2\pi} \int J_0(k\rho) \, [J_0(ka)]^N k \, dk \qquad (7.113)$$

This is the same result obtained by previous methods.

Problem 7-17: Find $P_N(\boldsymbol{\rho})$ when the probability of a step making a given angle θ with the x axis proportional to $\cos^2 \theta$, that is, is $2 \cos^2 \theta$. Show that in the large number limit it is

$$P_N(\boldsymbol{\rho}) = \frac{1}{\sqrt{3} N \pi a^2} \exp \left[-\frac{X^2}{3Na^2} - \frac{Y^2}{Na^2} \right]$$

We may also apply the Markoff method to the three-dimensional problem. The generalization of (7.108) to three dimensions is clearly

$$\Delta = \delta X \delta Y \delta Z \left(\frac{1}{2\pi}\right)^3 \int e^{-i\mathbf{k} \cdot (\mathbf{R} - \Sigma \mathbf{r}_n)} \, d^3\mathbf{k} \qquad (7.114)$$

where

$$|\mathbf{r}_n| = a \tag{7.115}$$

To obtain the volume in the sample space corresponding to displacement to a volume $d^3\mathbf{R}$ at $\mathbf{R}$ we integrate over the polar angles of the $\mathbf{r}_n$'s. The total available volume is $(4\pi)^N$. Therefore,

$$P_N(\mathbf{R}) \, d^3\mathbf{R} = \left(\frac{1}{2\pi}\right)^3 \left(\frac{1}{4\pi}\right)^N \int d\Omega_1 \cdots d\Omega_N \int e^{-i\mathbf{k}\cdot\mathbf{R}} \, e^{\Sigma i\mathbf{k}\cdot\mathbf{r}_n} d^3\mathbf{k} \, d^3\mathbf{R} \tag{7.116}$$

where the Ω's are the solid angles of the $\mathbf{r}_n$'s. Evaluation of the angle integrals gives

$$P_N(\mathbf{R}) = \left(\frac{1}{2\pi}\right)^3 \int \left(\frac{\sin ka}{ka}\right)^N e^{-i\mathbf{k}\cdot\mathbf{R}} d^3\mathbf{k} \tag{7.117}$$

as before.

11. Random Distribution of Energy among *N* Particles

Suppose that a system of N particles of mass M has a fixed energy in the range $(E_0, E_0 + \delta E_0)$ but the energy is shared among them randomly. Let us find the probability distribution for the energy of any one particle.

We must, of course, make an assumption about a priori probabilities. We do this in terms of the momenta of the particles. Consider a "phase space" in which the coordinates are the $3N$ momentum components. We shall assume that, *a priori*, every point in this space is equally probable. However, the only available part of the space is that for which

$$\frac{1}{2M} (\mathbf{p}_1{}^2 + \mathbf{p}_2{}^2 + \cdots + \mathbf{p}_N{}^2)$$

lies between E_0 and $E_0 + \delta E_0$. Its volume is that which lies between the hyperspheres

$$\sum_{n=1}^{N} \mathbf{p}_n{}^2 = 2ME_0 \tag{7.118}$$

and

$$\sum_{n=1}^{N} \mathbf{p}_n{}^2 = 2M(E_0 + \delta E_0) \tag{7.119}$$

The probability that the energy of the first particle is between E_1 and $E_1 + dE_1$ is that fraction of the above volume for which

$$E_1 \leq \frac{\mathbf{p}_1{}^2}{2M} \leq E_1 + dE_1 \tag{7.120}$$

Although the volumes in question could be evaluated by the method of Fourier transforms, as in the previous problems, they may in this case be obtained more readily by a recursion method employing Laplace transforms.

The volume within a "sphere" of radius $\sqrt{u}$ in ν dimensions is

$$V_\nu(u) = \int_{(p_1^2 + p_2^2 + \cdots + p_\nu^2 \le u)} dp_1 dp_2 \cdots dp_\nu \qquad (7.121)$$

For a given p_ν the limits of integration of the other $(\nu - 1)$ coordinates are $(0, u - p_\nu^2)$. Therefore,

$$V_\nu(u) = \int V_{\nu-1}(u - p_\nu^2) \, dp_\nu \qquad (7.122)$$

Taking Laplace transforms of both sides of this equation, and defining $U_\nu(s)$ as the transform of $V_\nu(u)$

$$U_\nu(s) = U_{\nu-1}(s) \int e^{-p_\nu^2 s} \, dp_\nu$$

$$= U_{\nu-1}(s) \left(\frac{\pi}{s}\right)^{1/2} \qquad (7.123)$$

Therefore,

$$U_\nu(s) = \left(\frac{\pi}{s}\right)^{1/2(\nu-1)} U_1(\nu) \qquad (7.124)$$

But

$$V_1(u) = \sqrt{u}$$

so that

$$U_1(s) = \int_0^\infty u^{1/2} e^{-su} \, du = \frac{\Gamma(\frac{3}{2})}{s^{3/2}} = \frac{\sqrt{\pi}}{s^{3/2}} \qquad (7.125)$$

Therefore, finally

$$U_\nu(s) = \frac{\pi^{\nu/2}}{s^{\nu/2+1}} \qquad (7.126)$$

The function of which this is the transform is immediately identifiable as

$$V_\nu(u) = \frac{\pi^{\nu/2} u^{\nu/2}}{\Gamma((\nu/2) + 1)} \qquad (7.127)$$

Thus, the volume in the $3N$-dimensional space of the momentum components corresponding to u between $2ME_0$ and $2M(E_0 + \delta E_0)$ is

$$V_T = 2M\delta E_0 \left[\frac{d}{du} V_{3N}(u)\right]_{u=2ME_0}$$

$$= 2M\delta E_0 \pi^{3N/2} \frac{3N}{2} (2ME_0)^{(3N/2)-1} \frac{1}{\Gamma((3N/2) + 1)}$$

$$= (2M\pi)^{3N/2} \frac{1}{\Gamma(3N/2)} E_0^{(3N/2)-1} \delta E_0 \qquad (7.128)$$

That part of the volume corresponding to $p_1{}^2/2M$ in the range $(E_1, E_1 + dE_1)$ is

$$V_p = \frac{d}{dE_0} V_{3(N-1)} \left(2M(E_0 - E_1)\right) \delta E_0 \frac{d^3p_1}{dE_1} dE_1$$

$$= \delta E_0 \frac{\pi^{(3(N-1)/2)}}{\Gamma(3(N-1)/2)} (2M)^{3(N-1)/2} (E_0 - E_1)^{3/2(N-1)-1} 2\pi (2M)^{3/2} E_1 \, dE_1$$

$$(7.129)$$

Therefore the probability that the energy of the first particle lies in the range $(E_1, E_1 + dE_1)$ is

$$p(E_1) \, dE_1 = \frac{V_p}{V_T}$$

$$= \frac{2}{\sqrt{\pi}} \frac{\Gamma(3N/2)}{\Gamma[3(N-1)/2]} \frac{(E_0 - E_1)^{3/2(N-1)-1} E_1{}^{1/2}}{E_0{}^{3N/2-1}} \, dE_1 \quad (7.130)$$

Problem 7-18: Verify directly that $\int P(E_1) \, dE_1 = 1$.

Problem 7-19: Show that $\langle E_1 \rangle = E_0/N$.

Problem 7-20: Show that

$$\Delta E_1 = E_0 \sqrt{\frac{2(N-1)}{N^2(3N+2)}}$$

(Note that the $\sqrt{N}$ law does not apply to this problem.) Prove that the central limit theorem is not valid for this problem. Why?

Problem 7-21: Show that, for large N, the distribution becomes approximately the Maxwell distribution

$$\frac{2}{\sqrt{\pi}} \left(\frac{3N}{2E_0}\right)^{3/2} e^{-3NE_1/2E_0} E_1{}^{1/2}$$

12. Stochastic Processes

By a random or stochastic process $x(t)$ we mean a process in which the variation of x with t does not follow a definite law, but rather contains an element of chance. We designate $x(t)$ as a "signal." The quantity of greatest interest in such a signal is usually its "power spectrum" or intensity distribution, which is $|\phi(\omega)|^2$, ϕ being the Fourier transform of $x(t)$. This power spectrum is related to the autocorrelation function

$$C(\tau) = \frac{\int x(t)x(t+\tau) \, dt}{\int x^2(t) \, dt} = \frac{\tilde{C}(\tau)}{n^2} \quad (7.131)$$

introduced in Chapter 4, by the equation (4.58):

$$|\phi(\omega)|^2 = \frac{n^2}{2\pi} \mathscr{F}(C(\tau)) \tag{7.132}$$

$\mathscr{F}(\ )$ designating the Fourier transform of the quantity in the brackets. Since $n^2 = \int x^2(t)\, dt$ by definition, it in fact cancels out of (7.132).

The preceding considerations apply to *any* signal. The interest in their application to random ones is that it is often relatively easy to determine the autocorrelation function, though not $x(t)$ directly, since $x(t)$ is not analytically defined.

An example of a random signal to which the theory can be applied is the following: $x(t)$ is either equal to x_0 or $-x_0$, the probability of changing from one to the other being λ per unit time. Because $x^2(t) = x_0^2$, which is not integrable, we take the signal to be zero outside the interval $(-T/2, T/2)$; however, T may be taken to be as large as we wish.

Although the value of this signal cannot be specified at any given time, it is rather easy to calculate its autocorrelation function. For in a time τ, the probability of n changes of sign is given by the Poisson distribution $[(\lambda\tau)^n/n!]\, e^{-\lambda\tau}$. Now if n is even the product $x(t)x(t+\tau)$ is a^2; if n is odd, it is $-a^2$. The autocorrelation function is, therefore, given by

$$\tilde{C}(\tau) = a^2 \sum_{n\ even} \frac{(\lambda\tau)^n}{n!} e^{-\lambda\tau} - a^2 \sum_{n\ odd} \frac{(\lambda\tau)^n}{n!} e^{-\lambda\tau} \tag{7.133}$$

The two sums, combined, are simply the expansion of $e^{-2\lambda\tau}$ for $\tau > 0$. Similar considerations hold for $\tau < 0$. Therefore,

$$\tilde{C}(\tau) = a^2 e^{-2\lambda|\tau|} \tag{7.134}$$

The power spectrum is then

$$\phi^2(\omega) = \frac{a^2}{2\pi} \mathscr{F}(e^{-2\lambda|\tau|})$$

$$= \frac{2a^2}{(2\pi)^2} \mathrm{Re} \int_0^\infty e^{-2\lambda\tau} e^{-i\omega\tau}\, d\tau$$

$$= \frac{a^2}{\pi^2} \frac{\lambda}{4\lambda^2 + \omega^2} \tag{7.135}$$

This is known as a "Lorentzian" spectrum.

Problem 7-22:   A random signal is generated in the following way: A time interval $(-T/2, T/2)$ is divided into intervals of length t_0. On each $x(t)$ is equally likely to have the value a or $-a$. Show that the power spectrum is

$$\phi^2(\omega) = \frac{a^2}{\pi^2\omega^2} \sin^2 \frac{\omega t_0}{2}$$

Problem 7-23:   We add the following problem on probabilities as a diversion: A and B agree to play a game of chance with blank cards, the rules of which are as follows: (1) on N cards, A writes *any* arbitrary set of positive numbers, all different; B is not permitted to see what A has written. (2) The cards are shuffled, and A turns them over one at a time from the top of the pack. The aim is for B to try to name the highest-numbered card at the time at which it is turned up.

If B were to bet on his ability to choose correctly the highest card, what odds do you believe that he should demand to have a better than even chance of winning? How do you think these odds should depend on the total number of cards?

B chooses the strategy of letting n_0 cards go by, noting which of them is the largest. He then makes as his choice the first card subsequently turned up which is larger than that one. Show that the optimal value of n_0 is approximately N/e. What are his chances of success in this case?

[Try this out on your friends before they have a chance to work it out. You may find it very profitable.]

PRELUDE TO CHAPTER 8

In this chapter we introduce the mathematical foundations of quantum mechanics. The approach, and the notation, are essentially those of Dirac. The underlying features were introduced in Chapter 2 (Linear Vector Spaces).

We first discuss the basic assumptions of quantum mechanics in terms of linear vector spaces, the representation of dynamical variables by Hermitian operators, of quantum states by eigenvectors of these operators, and of the results of measurement by their eigenvalues. The probabilistic hypothesis is introduced, and quantum mean value defined. The significance of commutators in relation to the simultaneous measurability of two variables is shown, and the generalized uncertainty principle derived.

Unitary operators are defined, and it is shown that the eigenvalue spectrum is unchanged under a unitary transformation.

We next show how to adapt the theory for variables with a continuous spectrum. The most important such variables are coordinates. The coordinate representation is then used to introduce the Schrödinger wave functions (as the amplitudes of coordinate eigenstates). Operators corresponding to classical dynamical variables are derived.

The momentum representation is then introduced; it is shown that this is very similar in structure to the coordinate representation, and is related to it by a Fourier transformation.

We next turn to dynamics, that is, the theory of the time evolution of quantum systems. The time evolution operator is shown to be a unitary operator. The Schrödinger representation (time-dependent state vectors, constant operators) and the Heisenberg representation (constant state vectors, time-dependent operators) are introduced and shown to be equivalent. We then develop the theory of the "interaction representation" which lies between the two. It is used to write in an explicit (but not very useful) way, the state vector at an arbitrary time.

We return to the time evolution operator to study the history of a system with a given (nonsoluble) Hamiltonian. The concept of a "propagator" is introduced through the Schrödinger wave mechanics, and calculated for free particles. A simple diagrammatic description is given for the evolution of a system acted upon by a perturbing potential.

The next sections are concerned with time-dependent perturbation theory, but from a rather general viewpoint. Time-*independent* perturbations are considered first. Starting with the differential equation of the time-evolution operator, we make a Fourier transformation and obtain an algebraic operator equation for the transform. A formal solution can be found as a sort of perturbation series. Several alternative formulations are given.

A practical obstacle to the *use* of the perturbation series for purposes of calculation is that mathematical divergences occur associated with transitions back into the original state in its subsequent history. Using projection operators into and out of this state, we find, after some fairly complicated analysis, that all "closed loop" processes, in which the system returns to its initial state, can be incorporated into a "renormalization" of that state; a perturbation expansion may then be developed ignoring such processes. Finally, by taking inverse Fourier transforms, formal expressions for the matrix elements of the evolution operator between the initial state and *arbitrary* final states may be given.

We then set about to calculate the total transition probability of a system in an arbitrary time interval. Working only to the second order, we find explicitly the shift in energy of the initial state and its "width" (probability per unit time of decay) as a result of the closed loop processes. The decay is in particular shown to be exponential.

We then determine the transition probability to a *definite* final state. For sufficiently large values of t, it is found that the probability peaks for states whose energy is close to that of the initial state. When we then sum, or integrate, over *all* final states, we obtain a formula for the total transition probability in terms of a perturbation expansion.

It is pointed out that this, rather than the second-order calculation carried out earlier, should be used to determine the width of the initial state.

We next carry through the theory for a time-dependent perturbation, specifically, for a periodic one. An explicit formula is obtained for the lowest order of perturbation. Broadening of the initial level is here neglected.

Finally, we turn to the theory of the perturbation of stationary states, when exact states cannot be determined. Complete perturbation series are derived using a method based on projection operators. Two different forms are given, that of Rayleigh–Schrödinger and that of Wigner–Brillouin. The first gives the perturbation energy explicitly; the second gives it implicitly. The Wigner–Brillouin theory is usually (but not always) more accurate to a given order.

Special attention is devoted to the case of the perturbation of degenerate states.

REFERENCES

Davydov, A., *Quantum Mechanics*. New York: Pergamon, 1965.

Gottfried, K., *Quantum Mechanics*. New York: Benjamin, 1965, Vol. 1.

Merzbacher, E., *Quantum Mechanics*. New York: Wiley, 1961.

Messiah, A., *Mécanique Quantique*. Paris: Dunod, 1958.

8
FUNDAMENTAL PRINCIPLES OF QUANTUM MECHANICS

"It is agayns the proces of Nature"
Chaucer, The Frankeleyn's Tale

1. Introduction

The mathematical foundations required for the formulation of quantum mechanics were laid down in Chapter 2 (Linear Vector Spaces). In this and the following chapters we develop the basic structure of quantum mechanics, and apply it to a selection of representative problems. We cannot, of course, in a book of such a broad scope as this one, give a "complete" account of quantum mechanics. Such treatments can, in any case, be found in comprehensive treatises on quantum theory (such as, for example, that of Messiah). Our intention is, rather, to expose the mathematical structure of the theory, and to call attention to certain concepts and techniques of wide use in modern physics. Nor do we concern ourselves with problems of "rigor," such as are dealt with in von Neumann's classic work. Our aim is to provide a practical handbook for the use of those who wish to comprehend and to undertake quantum calculations, rather than a critical examination of fundamentals. Our approach is also more in the spirit of "mathematical physics" than of "theoretical physics" in that it is concerned more with fundamental techniques than with the evolving concepts in terms of which we attempt to understand various aspects of the physical world.

Finally, it must be emphasized that our selection of topics for discussion is somewhat arbitrary. The selection is designed to introduce the reader to as many as possible of the techniques and viewpoints which are likely to be met by physicists in any of the special areas of current interest.

2. Hypotheses of Quantum Mechanics

The underlying structure of the theory is based on the representation of "states" of a system which are to be represented by (i.e., to be in one-to-one correspondence with) vectors in some suitably chosen vector space. The measurement of the attributes of these states ("dynamical variables" or "observables") are described in terms of operations on these vectors. These *operations* are assumed to be linear. The operation of an operator on a vector is intended to describe a *physical* operation (measurement) on the system. The relation between the mathematical operation and the physical "measurement" is assumed to be the following: that *the result of a "measurement" of a dynamical variable must be an eigenvalue of the linear operator representing that dynamical variable. The state in which the dynamical variable has that value is represented by the corresponding eigenvector.* The above statement is, of course, an imprecise one, and raises many questions. For instance, what is the vector space used to describe a given physical system? How are the operators representing given dynamical variables to be chosen? How does one describe the time evolution of a dynamical system? These, and many other questions must be answered if we are to formulate a complete working theory. What we have done, however is to specify the framework within which the theory is to be developed; a framework having its origin in the historical evolution of the subject. The historical process was, however, a tortuous one, marked by contradiction (experimental data in conflict with then existing theoretical concepts), hypothesis (often *ad hoc*, sometimes inspired), synthesis and generalization. The end product is much clearer and more coherent than the process by which it was constructed. We, therefore, forego historical analysis and build our structure inductively but with the wisdom of hindsight, in order to avoid wasting time on transitional steps and misleading sidetracks.

Let us, therefore, set about to construct, within our specified general framework, a detailed theory.

Consider first the question of the vector space by which a physical system can be represented. The major point to be made here is that this space is determined, not by mathematical prescription, but by *physics*. Since states of systems are to be represented by vectors, the vector space must be determined by the totality of attributes of the system. This is, of course, a function of our understanding, and must be modified from time to time in the light of discovery. This point may be readily illustrated by reference to the phenomenon of electron spin. If (to keep matters simple) we consider as our system the hydrogen atom, prior to the discovery of spin the vector space needed to describe the system would have been one based on the independent states of the atom as specified by three "quantum numbers" specifying the energy of the atom, its total angular momentum, and one component thereof. However, once it was recognized that the electron possessed not only these characteristics but also *intrinsic* angular momentum ("spin"), corresponding to each previously known state it was now necessary to recognize *two*, corresponding to the two states of electron spin which were found to exist. Thus, the recognition of a new physical attribute made it necessary to double the "dimensionality" of the vector space used to describe the system. That is, at each point of the original "space" it was now necessary to construct a two-dimensional manifold. Such a situation is of course well known to mathematicians, in whose language one says that the new vector space was the "direct product" of the original one and the new two-dimensional one. From every combination of one vector of the original space and one of the "spin space" we construct a new vector of the product space.

The next question to be answered is, how does one choose the operators to represent the various dynamical variables?

Let us reformulate the problem in a more meaningful way. If we knew the complete spectrum of a dynamical variable, that is, the array of values which could result from a measurement of the variable, we could easily set up a set of vectors and a corresponding operator. Consider, for example, vectors represented by column matrices. If k_n were the possible results of a measurement of the dynamical variable K, the matrix

$$K = \begin{pmatrix} k_1 & 0 & 0 & 0 & \cdots \\ 0 & k_2 & 0 & 0 & \cdots \\ 0 & 0 & k_3 & 0 & \cdots \\ & & & \vdots & \end{pmatrix} \qquad (8.1)$$

could be used to represent the dynamical variable K.

The eigenvalues would then be the results of measurement, as required; the eigenvectors, representing the corresponding states, would be those vectors which have all but one component equal to zero. Conjugate vectors could be represented by row matrices, as shown in Chapter 2. The magnitudes of such vectors are in each case the square of the nonzero element. If the nonzero element is chosen to be *unity*, the magnitude is unity, and the vectors are normalized.

These eigenvectors can be used as a basis for a vector space, in which the general vector is

$$\begin{pmatrix} a_1 \\ a_2 \\ a_3 \\ \vdots \end{pmatrix} = a_1 \begin{pmatrix} 1 \\ 0 \\ 0 \\ \vdots \end{pmatrix} + a_2 \begin{pmatrix} 0 \\ 1 \\ 0 \\ \vdots \end{pmatrix} + a_3 \begin{pmatrix} 0 \\ 0 \\ 1 \\ \vdots \end{pmatrix} + \cdots \tag{8.2}$$

[These vectors in general have an infinite number of components.] If the a's are *complex* numbers, we have a complex vector space. This is required to formulate the quantum theory.

We have interpreted the "unit vectors" as representing states in which the dynamical variable K has specific values. What, then, is the meaning of a general vector; what sort of state does it represent? If we designate the eigenvectors (eigenstates) of K as "pure states" with respect to this dynamical variable, a general state is a *superposition* of pure states. It is necessary to interpret the components of the vector in relation to the measurement of the dynamical variable. For this purpose, we introduce a new and fundamental hypothesis. If we normalize the vector, that is to say, multiply it by such a constant value that the square of the magnitude of the resulting vector is unity, this hypothesis may be stated in the following terms: that the squares of the magnitudes of the components (that is, of the coefficients of the eigenvectors) represent the *probabilities* that a measurement of the dynamical variable K on the system in question will yield the corresponding eigenvalue. Thus, the probability that the result of a measurement yields the value k_n is given by $|a_n|^2$. The normalization of the vector assures the condition

$$\sum |a_n|^2 = 1 \tag{8.3}$$

that the probabilities add up to unity.

In the notation of Dirac, as outlined in Chapter 2, the dynamical variable (operator) K has eigenvectors $|k_n\rangle$ and eigenvalues k_n; that is,

$$K|k_n\rangle = k_n|k_n\rangle \tag{8.4}$$

There is, however, a possibility that there is more than one state in which the operator K has the eigenvalue k_n. In this case, the system is said to have a degenerate spectrum with respect to K. If there were several such states, they would have to be differentiated by some *other* attribute. Thus, there would have to be *another* dynamical variable or observable of the system such that the result of operating on a member of the degenerate set of eigenvectors of K with *its* operator would produce *different* results. Taking account of the totality of independent states of the system, as specified by *all* its attributes, the operator (8.1) may have to be expanded so that some of the eigenvalues k_n are repeated. The expanded set of basis vectors delineates *all* the possible states of the system. They also form a *complete* orthonormal set. An arbitrary state vector for the system, $|\Psi\rangle$, is expansible in terms of these states, which we again designate $|k_n\rangle$:

$$|\Psi\rangle = \sum |k_n\rangle \langle k_n|\Psi\rangle \qquad (8.5)$$

and the *probability* that a measurement of the dynamical variable K in the state $|\Psi\rangle$ is k_n is $|\langle k_n|\Psi\rangle|^2$.

We now demonstrate the following result: that the mean or expectation value of the variable K in the state $|\Psi\rangle$ is

$$\bar{K} = \langle \Psi|K|\Psi\rangle \qquad (8.6)$$

This follows immediately from the expansion (8.5):

$$\langle \Psi|K|\Psi\rangle = \sum_{m,n} \langle \Psi|k_m\rangle \langle k_m|K|k_n\rangle \langle k_n|\Psi\rangle$$

But

$$\langle k_m|K|k_n\rangle = k_n \langle k_m|k_n\rangle$$

$$= k_n \delta_{nm}$$

by virtue of the orthogonality of the states $|k_n\rangle$. Therefore,

$$\bar{K} = \sum_{m,n} \langle \Psi|k_m\rangle k_n \delta_{nm} \langle k_n|\Psi\rangle$$

$$= \sum_n k_n \langle \Psi|k_n\rangle \langle k_n|\Psi\rangle$$

$$= \sum k_n |\langle k_n|\Psi\rangle|^2 \qquad (8.7)$$

by virtue of (2.11). But (8.7) defines the mean value.

By Theorem B of Chapter 2, the fact that the expectation value (8.7) is real permits us to conclude that K is a Hermitian operator. K is in fact *any* operator representing a dynamical variable (and which must therefore have real eigenvalues). *Consequently, all physical dynamical variables must be represented by Hermitian operators.*

All of the foregoing results are valid for any particular dynamical variable. The problem which confronts us in constructing a theory is, however, the following: We must specify how the operators representing *other* dynamical variables operate on the eigenstates of K (or the converse).

It is convenient to formulate this problem in terms of the *commutator*

$$[K_1, K_2] = K_1K_2 - K_2K_1 \qquad (8.8)$$

of pairs of operators K_1 and K_2. We begin by proving the following theorem:

The necessary and sufficient condition that K_1 and K_2 possess *common eigenstates is that their commutator be zero, that is,*

$$K_1 K_2 = K_2 K_1 \qquad (8.9)$$

It is easily seen that if the states $|n\rangle$ are common eigenstates of K_1 and K_2, having the respective sets of eigenvalues k_{1n} and k_{2n}, Eq. (8.9) is satisfied for all operands. For since, by hypothesis, the vectors $|n\rangle$ comprise the totality of physically realizable states for the system, they form a complete set. Hence *any* state vector $|\Psi\rangle$ may be expanded in terms of them,

$$|\Psi\rangle = \sum_n |n\rangle\langle n|\Psi\rangle \qquad (8.10)$$

But then

$$K_1 K_2 \Psi = K_1 K_2 \sum_n |n\rangle\langle n|\Psi\rangle$$

$$= K_1 \sum_n |n\rangle\langle n|\Psi\rangle k_{2n}$$

$$= \sum_n |n\rangle\langle n|\Psi\rangle k_{1n} k_{2n}$$

But $K_2 K_1|\Psi\rangle$ obviously has the same value. Thus we have shown that K_1 and K_2 commute.

Consider now the converse proposition, that *if* K_1 and K_2 commute, they must possess a common set of eigenvectors. The proof is again immediate except in the case of states which belong to a degenerate subset with respect to both K_1 and K_2. Excluding this case, suppose that state $|n\rangle$ is a nondegenerate eigenstate of K_1. Then we know that

$$K_1(K_2|n\rangle) = K_2 K_1|n\rangle$$

$$= k_{1n}(K_2|n\rangle)$$

Thus, $K_2|n\rangle$ is *also* an eigenstate of K_1 with eigenvalue k_{1n}. But since $|n\rangle$ is nondegenerate with respect to K_1, this must be in fact the state $|n\rangle$ itself, or rather, a multiple of it. Therefore,

$$K_2|n\rangle = \text{constant } |n\rangle$$

which says that $|n\rangle$ is also an eigenstate of K_2. Since K_2 is Hermitian, the constant must be real.

If $|n\rangle$ were nondegenerate with respect to K_2, whether or not it was degenerate with respect to K_1, the above argument could be made with the roles of K_1 and K_2 interchanged.

Consider then the only remaining case: Suppose that states $|n_\alpha\rangle$ ($\alpha = 1, 2, \ldots, p$, say) form a degenerate (but orthogonal) set of eigenstates of K_1 with eigenvalue k_n, which are also degenerate with respect to K_2. But then *any* linear combination

$$|\Psi_\beta\rangle = \sum_{\alpha=1} \lambda_{\beta\alpha} |n_\alpha\rangle \qquad (8.11)$$

are eigenstates of both K_1 and K_2. Our theorem is therefore proven.

Let us now interpret physically the above result. Let us call an eigenstate of K_1 a "pure state" with respect to this dynamical variable. If K_1 and K_2 commute, this can also be a "pure state" with respect to K_2. Thus, measurements of K_1 on a system in this state are certain to produce a value k_1 while measurements of K_2 are certain to produce a value k_2. Measurements of these two variables do not "interfere" with each other.

Consider, however, the alternative possibility, in which $|\Psi_n\rangle$ is an eigenstate of K_1 with eigenvalue k_{1n}, but is *not* an eigenstate of K_2 due to the fact that K_1 and K_2 do not commute. Then $|\Psi_n\rangle$ may be expanded in terms of the eigenstates $|\varphi_m\rangle$ of K_2, whose eigenvalues are k_{2m},

$$|\Psi_n\rangle = \sum_m |\varphi_m\rangle\langle\varphi_m|\Psi_n\rangle \qquad (8.12)$$

Then, the result of a measurement of K_2 in an eigenstate of K_1 must be expressed in terms of *probabilities*: In the state in which K_1 is *known* to have the value k_{1n}, the *probability* that K_2 has the value k_{2m} is $|\langle\varphi_m|\Psi_n\rangle|^2$ (we have of course assumed each set of eigenvectors to be normalized). Similarly, if we select a state in which the value of K_2 is known, it is not "pure" with respect to K_1, and various results of a measurement of K_1 are possible, only their respective probabilities being determined. In this case we say that the measurements of these two variables "interfere" with each other.

If a measurement of K_1 produces the value k_{1n}, then conversely, we can surmise that the system is in the state $|\Psi_n\rangle$. Thus, the measurement has, in a sense, *prepared* the system in that state, and the "measurement" of that variable has "selected" the system in the corresponding state. This selection is represented mathematically by the operation of *projection* on the eigenstate. If we start with a state

$$|\Psi\rangle = \sum |\Psi_n\rangle\langle\Psi_n|\Psi\rangle \qquad (8.13)$$

after the measurement which has given the value k_{1n} we have produced the state represented by the term $|\Psi_n\rangle \langle\Psi_n|\Psi\rangle$ in the expansion. The "projection operator"

$$P_n = |\Psi_n\rangle \langle\Psi_n| \qquad (8.14)$$

is then the mathematical operator describing the effect of the measurement on the system.

A simple illustration of the interference of measurements arises in discussion of the Stern–Gerlach experiment. This experiment demonstrates the existence of an intrinsic magnetic moment (and therefore angular momentum) for an electron. One may discuss this experiment in terms of a single electron, taking account of the *probabilities* of different contingencies. Alternatively, one may consider an ensemble, or beam, of electrons with the same momentum, treating them, however, as though they were isolated systems. As is seen later, the electron can exist in two spin, or magnetic moment, states. Its component of spin in an arbitrary direction, which we call the z direction, may be $\frac{1}{2}\hbar$ or $-\frac{1}{2}\hbar$; the magnetic moment e/mc times these values, e being the electron charge, m its mass, and c the velocity of light. $\hbar$ is the Planck constant divided by 2π.

If we take a beam of electrons, with random spin states, and pass them through a magnetic field in the z direction, those with opposite spins interact differently with the field. Those whose magnetic moments in the z direction are positive have their energy lowered in the field, while those whose magnetic moments are negative have theirs raised. If the field is inhomogeneous, the former will be deflected in the direction of *increasing* field, the latter in that of *decreasing* field. This results in a splitting of the beam; it in fact performs a physical separation of the electrons in the two different spin states into two distinct beams.

The electrons in the two beams now are individually "pure" with respect to the z component of spin, or magnetic moment. Let us now try to measure *another* component (say the x component) of the spin, by similarly passing *one* of these beams through an inhomogeneous magnetic field in the x direction. If the x and z components of spin commuted, the beam would be "pure" also with respect to the x component of spin, and no further splitting would occur. In fact, however, it is found that this beam is in turn split into two statistically equal parts. Thus, electrons whose z component of spin is known seem to have *equal* probability of having one or the other x component.

If now we consider one of the beams split by this second field, the electrons in it will be in an eigenstate of the x component of spin. This beam will, then, again split equally when put through a second field in the z direction. Although a selection was originally made of electrons with a given z component of spin, the measurement of the x component so disturbed the beam that after the measurement the two values of z component were equally likely.

The above discussion serves to indicate the importance in quantum mechanics of the *commutator* of two operators. As we have seen, if they commute, they may simultaneously be ascribed exact values. Let us now see in a more precise quantitative way the consequences of having a nonzero commutator.

Let us note first that the commutator as defined by Eq. (8.8) is not a Hermitian operator. To see this, let us first evaluate a general matrix element of the product K_1K_2,

$$\langle n|K_1K_2|n'\rangle = \sum_{n''} \langle n|K_1|n''\rangle \langle n''|K_2|n'\rangle \qquad (8.15)$$

In the case of a Hermitian operator, as was seen in (2.60), any matrix element is equal to its transpose conjugate. But the transpose conjugate of (8.15) is

$$\langle n'|K_1K_2|n\rangle^* = \sum_{n''} \langle n'|K_1|n''\rangle^* \langle n''|K_2|n\rangle^*$$

which is, since K_1 and K_2 are Hermitian,

$$\sum \langle n|K_2|n''\rangle \langle n''|K_1|n'\rangle = \langle n|K_2K_1|n'\rangle$$

Similarly, the transpose conjugate of $\langle n|K_2K_1|n'\rangle$ is $\langle n|K_1K_2|n'\rangle$. Therefore,

$$\langle n'|K_1K_2 - K_2K_1|n\rangle = -\langle n|K_1K_2 - K_2K_1|n'\rangle^* \qquad (8.16)$$

We may, however, very easily create a Hermitian operator by multiplying the commutator by i; that is,

$$C = i(K_1K_2 - K_2K_1) = i[K_1, K_2] \qquad (8.17)$$

is Hermitian.

3. Uncertainty Principle

We now prove the following result, known as the Heisenberg uncertainty principle: The product of the uncertainties of two dynamical variables is greater than or equal to half the magnitude of the expectation value of their Hermitian commutator C, that is,

$$\Delta K_1 \Delta K_2 \geqslant \tfrac{1}{2}|\bar{C}| \qquad (8.18)$$

We must, however, explain what we mean in this context by the "uncertainty" ΔK of a variable K. The meaning is precise—it is the root mean square deviation from the mean. In a state $|\Psi\rangle$, the mean value of K is

$$\bar{K} = \langle \Psi|K|\Psi\rangle$$

The mean value of the operator $(K - \bar{K})^2$ is then the mean square deviation from the mean:

$$(\Delta K)^2 = \langle \Psi|(K - \bar{K})^2|\Psi\rangle \qquad (8.19)$$

and so the uncertainty is

$$\Delta K = \sqrt{\langle \Psi | (K - \bar{K})^2 | \Psi \rangle} \qquad (8.20)$$

If we introduce the operator

$$K' = K - \bar{K} \qquad (8.21)$$

it follows that

$$(\Delta K)^2 = \overline{K'^2}$$

Furthermore,

$$C = i(K_1 K_2 - K_2 K_1)$$
$$= i(K_1' K_2' - K_2' K_1')$$

The uncertainty relation (8.18) may now be proven as follows: If $|\Psi\rangle$ is an arbitrary vector and λ is an arbitrary real constant, the magnitude of the vector

$$(K_1' + i\lambda K_2') | \Psi \rangle$$

is greater than or equal to zero for all λ. Thus

$$\langle \Psi | (K_1' - i\lambda K_2')(K_1' + i\lambda K_2') | \Psi \rangle \geq 0 \qquad (8.22)$$

K_1' and K_2' being Hermitian. But from this it follows that

$$\lambda^2 (\Delta K_2)^2 + \lambda \bar{C} + (\Delta K_1)^2 \geq 0 \qquad (8.23)$$

for all λ. This may be written

$$\left(\lambda \Delta K_2 + \frac{\bar{C}}{2\Delta K_2} \right)^2 + \left[(\Delta K_1)^2 - \frac{\bar{C}^2}{4(\Delta K_2)^2} \right] \geq 0$$

The inequality must therefore hold in particular for the value of λ which makes the first term zero, viz.,

$$\lambda = -\bar{C}/2\Delta K_1 \Delta K_2 \qquad (8.24)$$

It follows that the second term is non-negative, so that

$$\Delta K_1 \Delta K_2 \geq \tfrac{1}{2}|\bar{C}|$$

as was to be shown.

We see then from the preceding considerations, that the result of a measurement of a dynamical variable in a physical system "prepared" by measurement of another variable depends on the commutator of the two variables. The essential substantive hypothesis of quantum mechanics must then concern the specification of such commutators.

It is necessary, in making this hypothesis explicit, to distinguish between "classical" and "nonclassical" dynamical variables. The former are defined in terms of the classical generalized coordinates of the system q_i and their conjugate momenta p_i. Thus, such a variable is a function $K(q_i, p_i)$ of all these coordinates and momenta. Energy, position, momentum, and the angular momentum of a particle about a point are examples of such variables.

The commutation rules for such variables can be expressed in terms of the Poisson bracket of classical mechanics. This we define, for two variables $K_1(q_i, p_i)$ and $K_2(q_i, p_i)$, as

$$(K_1, K_2) = \sum_i \left(\frac{\partial K_1}{\partial q_i} \frac{\partial K_2}{\partial p_i} - \frac{\partial K_2}{\partial q_i} \frac{\partial K_1}{\partial p_i} \right) \tag{8.25}$$

This Poisson bracket is another dynamical variable. The basic hypothesis of quantum mechanics is then this: that the commutator $[K_1, K_2]$ is $i\hbar$ times the dynamical variable (K_1, K_2),

$$[K_1, K_2] = i\hbar(K_1, K_2) \tag{8.26}$$

both sides being operators.[1]

It is important to realise, however, that the operator on the right-hand side of this equation must be chosen to be Hermitian. Consider for example pq, where q is a generalized coordinate and p is its conjugate momentum. If p and q are individually Hermitian, neither pq nor qp is; in fact,

$$(pq)^\dagger = qp \tag{8.27}$$

We therefore represent the classical variable pq by the quantum operator

$$\tfrac{1}{2}(pq + qp)$$

which *is* Hermitian. Equation (8.26) must, therefore, be understood in light of this restriction.

A system is of course specified by a set of independent dynamical variables, equal in number to the number of its "degrees of freedom." We see that the states of the system are determined only by the algebra of the operators as given by (8.26). It may be convenient to use particular representatives of the operators for purposes of calculation, but it is not necessary.

It is shown by a rather tedious calculation that, for classical dynamical variables, (8.26) follows from the relations

$$[q_i, p_j] = i\hbar\delta_{ij} \tag{8.28}$$

---

[1]P. A. M. Dirac, *Principles of Quantum Mechanics*. London: Oxford Clarendon Press, 4th ed., 1958, Chapter 4.

458   Fundamental Principles of Quantum Mechanics

Consequently, these relations may be taken as basic, and determine the commutation relations of all pairs of such classical variables.

The study of quantum systems has, however, revealed that there exist "dynamical variables" which have no classical analogs. A well-known one is the "spin" or intrinsic angular momentum of a particle. Parity, isospin, and strangeness are others which occur in the description of fundamental particles. It is of course necessary to prescribe the commutation rules satisfied by these variables—both those satisfied between these variables themselves and those between each of them and the classical variables.

These nonclassical variables divide into two categories. Consider as an example of one type, the parity operator. This simply changes every classical coordinate and momentum into its negative. It does not, then, represent a new "degree of freedom," but relates to a symmetry property connected with the classical ones. Its commutation rules with these classical variables are then determined by the definition of the operator itself.

The other type of nonclassical operator is one which relates to a new *degree of freedom* having no classical analog (e.g., spin). In this case the appropriate commutation relations must be chosen to represent the physical character of the operators. The new dynamical attribute must be represented by an appropriate vector space at each point of the "classical space," i.e., of the space on which the *classical* dynamical variables operate. Because they operate in different spaces, classical and non-classical variables are represented by independent operators and so commute with each other. The new assumptions which must be made then concern only the commutation of nonclassical operators among themselves.

Consider the case in which the vector space of the system decomposes into two subspaces S_1 and S_2; the system is represented by independent *pairs* of vectors $|n_1\rangle$ and $|n_2\rangle$. We can represent by

$$|n_1 n_2\rangle = |n_1\rangle |n_2\rangle \qquad (8.29)$$

the vectors of the "product space"; $|n_1\rangle$ and $|n_2\rangle$ are vectors within the separate spaces S_1 and S_2. To each new dynamical attribute which we discover there corresponds a new space independent of the others, and a new multiplicity in the "product space" which is the complete Hilbert space of the system.

Problem 8-1:   A particle moves in a potential, and has energy eigenvalues E_n in states $|n\rangle$. Prove that

$$\frac{2m}{\hbar^2} \sum_n |\langle n|x|0\rangle|^2 (E_n - E_0) = 1$$

where $|0\rangle$ is an arbitrary bound state. (Consider the quantity $[x, [x, H]]$.)

Problem 8-2:   Prove that

$$[A,[B,C]]+[B,[C,A]]+[C,[A,B]]=0$$

Problem 8-3:   Prove that

$$[A,BC]=[A,B]C+B[A,C]$$

Problem 8-4:   If $H=p^2/2m+V(x)$ show that

$$\dot{x}=\frac{i}{\hbar}[H,x]=\frac{1}{m}p$$

Hence show that, if $|a\rangle$ and $|b\rangle$ are arbitrary eigenstates,

$$\langle a|x|b\rangle=\frac{i\hbar}{m}\frac{\langle a|p|b\rangle}{E_b-E_a}$$

Problem 8-5:   Using (8-4), prove the "sum rules"

$$\sum_c \langle a|p|c\rangle\langle c|p|b\rangle\left\{\frac{1}{E_c-E_a}+\frac{1}{E_c-E_b}\right\}=m\delta_{ab}$$

and

$$\sum_c \frac{|\langle a|p|c\rangle|^2}{E_c-E_a}=\frac{m}{2}$$

4. Unitary Operators

We define an operator U as unitary when its Hermitian conjugate is its inverse:

$$U^\dagger=U^{-1} \tag{8.30}$$

The following results represent important properties of unitary operators:

(a) Under a unitary operation the magnitude of a vector remains unchanged. For let

$$|\Psi'\rangle=U|\Psi\rangle \tag{8.31}$$

Then

$$\langle\Psi'|=\langle\Psi|U^\dagger=\langle\Psi|U^{-1}$$

Consequently

$$\langle\Psi'|\Psi'\rangle=\langle\Psi|U^{-1}U|\Psi\rangle$$
$$=\langle\Psi|\Psi\rangle \tag{8.32}$$

(b) The inverse of (8.31) is

$$|\Psi\rangle=U^{-1}|\Psi'\rangle=U^\dagger|\Psi'\rangle \tag{8.33}$$

Now the matrix element of an operator K between states $|\Psi_m\rangle$ and $|\Psi_n\rangle$ is

$$\langle\Psi_m|K|\Psi_n\rangle = \langle\Psi'_m|UKU\dagger|\Psi'_n\rangle \tag{8.34}$$

If we define the unitary transform of K to be

$$K' = UKU\dagger = UKU^{-1} \tag{8.35}$$

then the matrix representing a dynamical variable K remains invariant under the unitary transformation.

Problem 8-6:   If λ is a real constant and M a Hermitian operator, the operator

$$U = e^{i\lambda M}$$

is unitary.

Problem 8-7:   The product of two unitary operators U_1 and U_2 is also unitary.

Problem 8-8:   The operator $(1 - iK)/(1 + iK)$, where K is Hermitian, is a unitary operator.

(c) The transformation of an operator K under the unitary transformation generated by the operator $e^{i\lambda M}$ is

$$K' = e^{i\lambda M}Ke^{-i\lambda M}$$

It follows that

$$\frac{\partial K'}{\partial\lambda} = ie^{i\lambda M}[M, K]e^{-i\lambda M} \tag{8.36}$$

which we write as

$$\frac{\partial K'}{\partial\lambda} = i[M, K'] \tag{8.37}$$

(d) A change of basis corresponds to a unitary transformation. Suppose we transform from a basis $|\varphi_n\rangle$ to another basis $|\chi_n\rangle$. The vectors of these two bases are taken to correspond in pairs $|\varphi_n\rangle$ transforming into $|\chi_n\rangle$. The transformation is immediately seen to be represented by the operator

$$W = \sum_n |\chi_n\rangle\langle\varphi_n| \tag{8.38}$$

Its Hermitian conjugate is then

$$W\dagger = \sum_m |\varphi_m\rangle\langle\chi_m| \tag{8.39}$$

Therefore,

$$WW^\dagger = \sum_{n,m} |\chi_n\rangle \langle\varphi_n|\varphi_m\rangle \langle\chi_m| \qquad (8.40)$$
$$= \sum_n |\chi_n\rangle \langle\chi_n|$$
$$= 1$$

so that W is a unitary transformation.

(e) The eigenvalue spectrum of an operator and its expectation value in a given state are invariant under a unitary transformation.

The eigenvalue equation for a variable K is

$$K|\Psi\rangle = k|\Psi\rangle$$

Substituting for $|\Psi\rangle$ in terms of $|\Psi'\rangle$, obtained from it by the unitary transformation U, we obtain

$$KU^{-1}|\Psi'\rangle = kU^{-1}|\Psi'\rangle$$

Multiplying this equation by U gives

$$UKU^{-1}|\Psi'\rangle = K'|\Psi'\rangle = k|\Psi'\rangle$$

by virtue of (8.35). Therefore, a unitary transformation does not change the eigenvalue problem.

As for mean values, the mean value of K is

$$\bar{K} = \langle\Psi|K|\Psi\rangle$$
$$= \langle\Psi'U|U^{-1}K'U|U^{-1}\Psi'\rangle$$
$$= \langle\Psi'|K'|\Psi'\rangle$$

so that the mean value is invariant under a unitary transformation.

5. Representation of Vectors by Eigenvectors of Continuous Coordinates

Consider a coordinate q, whose eigenvalues we assume to be arbitrary within some range (q_1, q_2) (which could be ∞ and $-\infty$, respectively). Designating the eigenvectors by $|q\alpha\rangle$, where α refers to other quantum numbers, we then have

$$q|q'\alpha\rangle = q'|q'\alpha\rangle \qquad (8.41)$$

We henceforth omit the α for convenience.

When vectors are specified by a *continuous* eigenvalue, the problem of normalization becomes a bit more involved than in the discrete case.

The expansion in terms of a *sum* over *discrete* eigenstates becomes one in terms of an *integral* over the *continuous* states:

$$|\Psi\rangle = \int |q\rangle \langle q|\Psi\rangle \, dq \qquad (8.42)$$

The magnitude squared of this vector is

$$\langle\Psi|\Psi\rangle = \int\int \langle\Psi|q'\rangle \langle q'|q\rangle \langle q|\Psi\rangle \, dq \, dq'$$

On the other hand, taking the scalar product of (8.42) with $\langle\Psi|$,

$$\langle\Psi|\Psi\rangle = \int \langle\Psi|q\rangle \langle q|\Psi\rangle \, dq$$

Consequently,

$$\langle\Psi|q\rangle = \int \langle\Psi|q'\rangle \langle q'|q\rangle \, dq' \qquad (8.43)$$

We see, then, that $\langle q'|q\rangle$ is precisely the quantity which is defined as the δ function $\delta(q'-q)$, that is,

$$\langle q'|q\rangle = \delta(q'-q) \qquad (8.44)$$

Thus, the vectors $|q\rangle$ are mutually orthogonal. However, they are not *unit* vectors.

It is an immediate generalization that, if a system is characterized by a number of coordinates $q_i (i = 1, 2, \ldots, N)$, one may write

$$|\Psi\rangle = \int |q_1, \ldots, q_N\rangle \langle q_1 \ldots, q_N|\Psi\rangle \, dq_1 \ldots dq_N \qquad (8.45)$$

where $q_1, \ldots, q_N$ is a simultaneous eigenstate of the coordinates [which commute by virtue of (8.26) and (8.27)]. The normalization of these vectors is given by

$$\langle q_1, q_2, \ldots, q_N|q_1', q_2', \ldots, q_N'\rangle = \delta(q_1 - q_1')\delta(q_2 - q_2')\ldots\delta(q_N - q_N') \qquad (8.46)$$

6. Schrödinger Wave Equation

The wave mechanics of Schrödinger is a particular representation of the general theory of quantum mechanics, in which one chooses as basis vectors the eigenstates of the coordinates of the system considered. We must first specify a representation of the momentum operator p corresponding to the coordinate q.

For this purpose we introduce another operator associated with a "translation," that is to say, a change of q by a constant δ:

$$T_\delta|q\rangle = |q+\delta\rangle \qquad (8.47)$$

If this operates on a general vector $|\Psi\rangle$ we obtain

$$T_\delta|\Psi\rangle = T_\delta \int |q\rangle \langle q|\Psi\rangle \, dq$$

$$= \int |q+\delta\rangle\langle q|\Psi\rangle \, dq$$

$$= \int |q\rangle \langle q-\delta|\Psi\rangle \, dq \qquad (8.48)$$

Let us then introduce the operator

$$\theta = \lim_{\delta\to 0} \frac{T_\delta - 1}{\delta} \qquad (8.49)$$

If this operates on $|\Psi\rangle$, we obtain

$$\theta|\Psi\rangle = \lim_{\delta\to 0} \int |q\rangle \frac{1}{\delta} \{\langle q-\delta|\Psi\rangle - \langle q|\Psi\rangle\} \, dq$$

$$= \int |q\rangle \left(-\frac{\partial}{\partial q}\langle q|\Psi\rangle\right) dq \qquad (8.50)$$

It follows that

$$(\theta q - q\theta)|\Psi\rangle = \int |q\rangle \left\{-\frac{\partial}{\partial q} q\langle q|\Psi\rangle + q\frac{\partial}{\partial q}\langle q|\Psi\rangle\right\} dq$$

$$= \int |q\rangle \langle q|\Psi\rangle \, dq$$

$$= -|\Psi\rangle \qquad (8.51)$$

Therefore, the commutation relation between θ and q is

$$[\theta, q] = -1 \qquad (8.52)$$

Comparing with (8.28), we find that p has the same commutation relation with q as $ih\theta$ has. We may therefore use $ih\theta$ as an operator representative of p. Specifically, then, we may write

$$p|\Psi\rangle = \int |q\rangle \left(-ih\frac{\partial}{\partial q}\right) \langle q|\Psi\rangle \, dq \qquad (8.53)$$

[see (8.50)].

Schrödinger wave mechanics is based on the representation of Hilbert space vectors in terms of the vectors $|q\rangle$. The scalar product $\langle q|\Psi\rangle$ is known as the "Schrödinger wave function." The momentum operator is defined in terms of its operation on this wave function. Since the wave function is the coefficient in the expansion of the vector, the operator on the wave function performs an operation on the vector itself.

From the normalization condition for the state vector $|\Psi\rangle$, namely,

$$\langle\Psi|\Psi\rangle = 1$$

we may, on expanding in terms of coordinate eigenstates, obtain the normalization of the wave function. For

$$\langle \Psi | \Psi \rangle = \int dq \int dq' \, \langle \Psi | q' \rangle \langle q' | q \rangle \langle q | \Psi \rangle$$
$$= \int dq \, \langle \Psi | q \rangle \langle q | \Psi \rangle$$

by virtue of (8.44). Therefore,

$$\langle \Psi | \Psi \rangle = \int dq \, |\langle q | \Psi \rangle|^2$$
$$= 1 \qquad (8.54)$$

If we write an eigenvalue equation in terms of the eigenvectors of $|q\rangle$, this is known as a "Schrödinger wave equation" (though this designation is sometimes confined to the energy equation).

The generalization to a system with N coordinates $q_1, q_2, \ldots, q_N$ is straightforward. Corresponding to each coordinate is a momentum operator on the wave function

$$p_i = - ih \frac{\partial}{\partial q_i} \qquad (8.55)$$

the wave function is

$$\Psi(q_1, \ldots, q_N) = \langle q_1, q_2, \ldots, q_N | \Psi \rangle \qquad (8.56)$$

where $|q_1, \ldots, q_N\rangle$ and $\langle q_1, \ldots, q_N|$ are the ket and bra vectors which are simultaneous eigenvectors of all the coordinates.

The wave equation for the energy may be obtained from the eigenvalue equation

$$H(q_i, p_i) | \Psi \rangle = E | \Psi \rangle \qquad (8.57)$$

if no nonclassical variables are involved. $H(q_i, p_i)$ must, of course, be Hermitian. From general dynamical theory, it must be quadratic in the p_i's. A term

$$f_{ij}(q_1, \ldots, q_N) p_i p_j$$

can be represented by the Hermitian operator

$$\tfrac{1}{2} f_{ij} p_i p_j + p_j p_i f_{ij} \qquad (8.58)$$

When an operator function of $q, f(q)$, operates on $|\Psi\rangle$ we get

$$f(q) | \Psi \rangle = f(q) \int |q'\rangle \langle q' | \Psi \rangle \, dq'$$
$$= \int |q\rangle f(q) \langle q | \Psi \rangle \, dq \qquad (8.59)$$

where the $f(q)$ in the integral is now an ordinary function. (The "q" here may stand for *all* of the coordinates $q_1, \ldots, q_N$.) The operator p_i, operating on (8.59), would give

$$p_i f(q)|\Psi\rangle = \int |q\rangle \left(-i\hbar \frac{\partial}{\partial q_i}\right)(f(q)\langle q|\Psi\rangle)\, dq \qquad (8.60)$$

We may therefore write (8.57) as

$$H(q_i, p_i)\int |q'\rangle \langle q'|\Psi\rangle\, dq' = E\int |q'\rangle \langle q'|\Psi\rangle\, dq'$$

or

$$\int |q'\rangle H\left(q_i', -i\hbar \frac{\partial}{\partial q_i'}\right)\langle q'|\Psi\rangle\, dq' = E\int |q'\rangle \langle q'|\Psi\rangle\, dq' \qquad (8.61)$$

the operator $H(q_i', -i\hbar(\partial/\partial q_i'))$ being so constructed as to be Hermitian; the construction is formally the same as that for the operators on the vectors $|\Psi\rangle$. If we now take scalar products with the bra vector $\langle q|$, and use the property of the δ function that

$$\int F(q')\delta(q' - q)\, dq' = F(q)$$

we obtain

$$H\left(q_i, -i\hbar \frac{\partial}{\partial q_i}\right)\langle q|\Psi\rangle = E\langle q|\Psi\rangle \qquad (8.62)$$

The eigenvalue problem then becomes one of differential equations.

7. Momentum Representation

Let us consider in this section the particular case of Cartesian coordinates x_i.... It is possible to exploit the considerable symmetry of the equations of dynamics between the coordinates and their conjugate momenta, to obtain another representation of the quantum equations in which the eigenstates of the p_i's rather than those of the x_i's are used as basis states. For simplicity in writing, we use the single symbols $\mathbf{x}, \mathbf{p}$ to stand for the $3N$ coordinates and momenta x_i, p_i $(i = 1, 2, \ldots, 3N)$ of the N particles of the system. We also use the short notation

$$\mathbf{x}\cdot\mathbf{p} = \sum_{i=1}^{3N} x_i p_i \qquad (8.63)$$

In the coordinate representation the eigenvalue equation for the momentum is

$$-i\hbar \frac{\partial}{\partial x_i}\langle \mathbf{x}|\mathbf{p}\rangle = p_i\langle \mathbf{x}|\mathbf{p}\rangle \qquad (8.64)$$

This has as a particular solution

$$\langle \mathbf{x} | \mathbf{p} \rangle = A e^{i\mathbf{p} \cdot \mathbf{x}/\hbar} = A e^{i\Sigma p_i \cdot x_i/\hbar} \tag{8.65}$$

where A is a normalization constant. In view of the fact that

$$\langle \mathbf{p} | \mathbf{x} \rangle = A^* e^{-i\mathbf{p} \cdot \mathbf{x}/\hbar} \tag{8.66}$$

we may write

$$\int \langle \mathbf{p}' | \mathbf{x} \rangle \langle \mathbf{x} | \mathbf{p} \rangle \, d^{3N}\mathbf{x} = \int e^{i(\mathbf{p} - \mathbf{p}') \cdot \mathbf{x}/\hbar} \, d^{3N}x \, |A|^2 \tag{8.67}$$

$$= (2\pi\hbar)^{3N} \, \delta(\mathbf{p} - \mathbf{p}') \, |A|^2$$

For normalization, therefore, let us take

$$\langle \mathbf{x} | \mathbf{p} \rangle = \left(\frac{1}{2\pi\hbar} \right)^{3N/2} e^{i\Sigma p_i \cdot x_i/\hbar} \tag{8.68}$$

Let us now write the Schrödinger function $\langle \mathbf{x} | \Psi \rangle$ in terms of momentum eigenfunctions. We know that

$$\langle \mathbf{x}' | \Psi \rangle = \sum_{\mathbf{p}'} \langle \mathbf{x}' | \mathbf{p}' \rangle \langle \mathbf{p}' | \Psi \rangle \tag{8.69}$$

where $\mathbf{p}'$ represents the eigenvalues of momentum of a particle in an infinite volume. Thus

$$\langle \mathbf{x}' | \Psi \rangle = \left(\frac{1}{2\pi\hbar} \right)^{3N/2} \int \langle \mathbf{p}' | \Psi \rangle \exp i \sum \mathbf{p}'_i \cdot \mathbf{x}'_i/\hbar \cdot d^{3N}\mathbf{p}' \tag{8.70}$$

This is a Fourier series expansion, which in the limit has become an integral.

Let us now see how to express dynamical operators in terms of their effect on the *momentum* wave functions, that is, on the amplitudes multiplying the momentum eigenvectors.

We may write

$$x_i | \Psi \rangle = x_i \int d^{3N}\mathbf{x}' \, d^{3N}\mathbf{p}' | \mathbf{x}' \rangle \langle \mathbf{x}' | \mathbf{p}' \rangle \langle \mathbf{p}' | \Psi \rangle$$

$$= \frac{1}{(2\pi\hbar)^{3N/2}} \int d^{3N}\mathbf{x}' \, d^{3N}\mathbf{p}' | \mathbf{x}' \rangle x_i' \, e^{i\Sigma p_j' x_j'/\hbar} \langle \mathbf{p}' | \Psi \rangle$$

$$= \frac{1}{(2\pi\hbar)^{3N/2}} (-i\hbar) \int d^{3N}\mathbf{x}' \, d^{3N}\mathbf{p}' | \mathbf{x}' \rangle \left[\frac{\partial}{\partial p_i'} e^{i\Sigma p_j' x_j'/\hbar} \right] \langle \mathbf{p}' | \Psi \rangle$$

This may be integrated by parts: Assuming that $\langle \mathbf{p}' | \Psi \rangle$ vanishes rapidly enough at infinity that the integral vanishes there, we obtain

$$x_i|\Psi\rangle = \frac{1}{(2\pi\hbar)^{3N/2}} \int d^{3N}\mathbf{x}'\, d^{3N}\mathbf{p}'\,|\mathbf{x}'\rangle\, e^{i\Sigma p_j' x_j'/\hbar} \left(i\hbar\frac{\partial}{\partial p_i'}\right) \langle\mathbf{p}'|\Psi\rangle$$

$$= \int d^{3N}\mathbf{x}'\, d^{3N}\mathbf{p}'\,|\mathbf{x}'\rangle\, \langle\mathbf{x}'|\mathbf{p}'\rangle \left(i\hbar\frac{\partial}{\partial p_i'}\right) \langle\mathbf{p}'|\Psi\rangle$$

$$= \int d^{3N}\mathbf{p}'\,|\mathbf{p}'\rangle \left(i\hbar\frac{\partial}{\partial p_i'}\right) \langle\mathbf{p}'|\Psi\rangle \qquad (8.71)$$

On the other hand, of course,

$$p_i|\Psi\rangle = p_i \int d^{3N}\mathbf{p}'\,|\mathbf{p}'\rangle\, \langle\mathbf{p}'|\Psi\rangle$$

$$= \int d^{3N}\mathbf{p}'\,|\mathbf{p}'\rangle p_i'\langle\mathbf{p}'|\Psi\rangle \qquad (8.72)$$

The situation in the momentum representation is, then, exactly the reverse of that in the coordinate representation. In the latter, position coordinates are represented by *multiplicative* operators on the wave functions, and momentum variables by differential operators, as shown in (8.53). In the momentum representation, momentum operators are multiplicative (8.72) and coordinate operators are represented in terms of derivatives with respect to momentum variables [see (8.71)].

Whether or not the momentum representation is useful for setting up the eigenvalue problem (it is in fact useful only in rather rare cases), the momentum wave function $\langle\mathbf{p}|\Psi\rangle$ may be obtained directly from the coordinate wave function $\langle\mathbf{x}|\Psi\rangle$ and may provide useful information about a system. The relation between the two wave functions is obtained as follows:

$$\langle\mathbf{x}|\Psi\rangle = \int d^{3N}\mathbf{p}'\,\langle\mathbf{x}|\mathbf{p}'\rangle\, \langle\mathbf{p}'|\Psi\rangle$$

$$= \left(\frac{1}{2\pi\hbar}\right)^{3N/2} \int d^{3N}\mathbf{p}'\, e^{i\mathbf{p}'\cdot\mathbf{x}/\hbar}\langle\mathbf{p}'|\Psi\rangle \qquad (8.73)$$

and, conversely,

$$\langle\mathbf{p}|\Psi\rangle = \int d^{3N}\mathbf{x}'\,\langle\mathbf{p}|\mathbf{x}'\rangle\, \langle\mathbf{x}'|\Psi\rangle$$

$$= \left(\frac{1}{2\pi\hbar}\right)^{3N/2} \int d^{3N}\mathbf{x}'\, e^{-i\mathbf{p}\cdot\mathbf{x}'/\hbar}\langle\mathbf{x}'|\Psi\rangle \qquad (8.74)$$

These two wave functions are, then, effectively, Fourier transforms of each other. More precisely, in terms of the definition of transforms given in Chapter 5,

$$\left(\frac{1}{2\pi\hbar}\right)^{3N/2} \langle \mathbf{p} | \Psi \rangle$$

is the Fourier transform of $\langle \mathbf{x} | \Psi \rangle$.

8.   Time Evolution of a System

In classical mechanics, the equations of motion of a dynamical system with Hamiltonian $H(q_j, p_j)$ are

$$\dot{q}_j = \frac{\partial H}{\partial p_j} \tag{8.75}$$

and

$$\dot{p}_j = -\frac{\partial H}{\partial q_j} \tag{8.76}$$

the dots designating derivatives with respect to time. The rate of change of a dynamical variable $K(q_j, p_j, t)$ is

$$\begin{aligned}
\frac{dK}{dt} &= \frac{\partial K}{\partial t} + \sum_j \left(\frac{\partial K}{\partial q_j}\dot{q}_j + \frac{\partial K}{\partial p_j}\dot{p}_j\right) \\
&= \frac{\partial K}{\partial t} + \sum_j \left(\frac{\partial K}{\partial q_j}\frac{\partial H}{\partial p_j} - \frac{\partial K}{\partial p_j}\frac{\partial H}{\partial q_j}\right) \\
&= \frac{\partial K}{\partial t} + (K, H)
\end{aligned} \tag{8.77}$$

where (K, H) is the Poisson bracket of K and H as defined by (8.25). But now, by the hypothesis (8.26) the *commutator* of two variables, K and H in particular, is $i\hbar$ times their Poisson bracket.

Consider for the moment dynamical variables K which do not contain the time explicitly, that is, for which

$$\frac{\partial K}{\partial t} = 0 \tag{8.78}$$

then

$$[K, H] = i\hbar\,(K, H)$$

$$= i\hbar\,\frac{dK}{dt} \tag{8.79}$$

where the right-hand side of this equation represents $i\hbar$ times a new dynamical variable which is the time derivative of K. Thus, if K were a coordinate, dK/dt would be the corresponding velocity, and so forth.

Equation (8.79) may, of course, be inverted to give

$$\frac{dK}{dt} = -\frac{i}{\hbar}[K, H] = \frac{i}{\hbar}[H, K] \tag{8.80}$$

both sides of this equation being operators.

If we consider a system evolving with time, (8.80) may be used to describe its evolution. In fact, it may be verified that

$$K(t) = e^{iHt/\hbar} K(0) e^{-iHt/\hbar} \tag{8.81}$$

defines a time-dependent K operator.

It was on this basis that *Heisenberg* described the time evolution of a system. Using a *constant set of basis vectors*, the time variation of the expectation value of any dynamical variable of the system in any state $|\Psi\rangle$ can be obtained by taking the expectation value with respect to $|\Psi\rangle$ of the *time-dependent operator $K(t)$*:

$$\langle \Psi | K(t) | \Psi \rangle = \langle \Psi | e^{iHt/\hbar} K(0) e^{-iHt/\hbar} | \Psi \rangle \tag{8.82}$$

(The time $t = 0$ is of course arbitrary.)

However, Eq. (8.82) can be interpreted in another way; if we define the *time-dependent state vector*

$$|\Psi(t)\rangle = e^{-iHt/\hbar}|\Psi\rangle = e^{-iHt/\hbar}|\Psi(0)\rangle \tag{8.83}$$

the expectation value of the variable K can be written

$$\langle \Psi(t) | K | \Psi(t) \rangle = \langle \Psi(t) | K(0) | \Psi(t) \tag{8.84}$$

Thus, the evolution of the system with time can be described by using a *fixed operator* and a *time-dependent state vector*. This method of describing time variation is known as the *Schrödinger representation*, as contrasted with that in which the time variation is incorporated in the dynamical variable, viz., the *Heisenberg representation*. It is evident that they provide completely equivalent descriptions of a physical system.

If we use the Schrödinger representation, it is clear that the time-dependent state vector $|\Psi(t)\rangle$ satisfies the equation

$$i\hbar \frac{\partial |\Psi(t)\rangle}{\partial t} = H|\Psi(t)\rangle \tag{8.85}$$

It is clear from (8.83) that if the state of the system, as specified by $|\Psi(0)\rangle$, is known at some initial time, its subsequent state is completely determined. It is therefore appropriate to formulate the problem of the evolution of the system in terms of the first-order differential equation (8.85).

The time-dependent Schrödinger wave equation, that is, the equation satisfied by

$$\Psi(q_i, t) = \langle q_i | \Psi(t) \rangle \tag{8.86}$$

follows from (8.84). For let us expand $|\Psi(t)\rangle$:

$$|\Psi(t)\rangle = \int dq_1 \dots dq_N |q_1, \dots, q_N\rangle \langle q_1, \dots, q_N|\Psi(t)\rangle$$

$$= \int d^N q |q\rangle \langle q|\Psi(t)\rangle \tag{8.87}$$

where q stands for the aggregate of generalized coordinates q_i. As we saw in (8.61)

$$H(q, p)|\Psi(t)\rangle = H(q, p) \int d^N q |q\rangle \langle q|\Psi(t)\rangle$$

$$= \int d^N q |q\rangle\, H\left(q, -i\hbar\frac{\partial}{\partial q}\right) \langle q|\Psi(t)\rangle \tag{8.88}$$

It follows that, on substituting (8.86) into the two sides of (8.84) and equating coefficients of individual $|q\rangle$'s on the two sides, the Schrödinger wave function $\langle q|\Psi(t)\rangle$ satisfies

$$i\hbar\frac{\partial}{\partial t} \langle q|\Psi(t)\rangle = H\left(q, -i\hbar\frac{\partial}{\partial q}\right) \langle q|\Psi(t)\rangle \tag{8.89}$$

The time-dependent equation is not at all necessary, however, since all problems of time development of a system may be treated using either Eq. (8.81) or Eq. (8.83).

Problem 8-9:   Prove the "virial theorem" in quantum mechanics: If V is a homogeneous function of the coordinates of order n, that is, $V(\lambda r) = \lambda^n V(r)$, then the mean values of the kinetic and potential energies (T and V, respectively) of a particle are related by the equation

$$n\langle V \rangle = 2\langle T \rangle$$

[HINT: change the scale of x to write the Schrödinger wave equation in terms of $r' = \lambda r$. Differentiate the equation with respect to λ, multiply through by ψ^*, and integrate.]

---

9. Interaction Representation

For many problems, a representation intermediate between those of Heisenberg and Schrödinger is found useful. It has been developed for problems in which it is possible to find an exact solution for a part of the Hamiltonian. We consider, then, a system described by a Hamiltonian

$$H = H_0 + H_1 \tag{8.90}$$

where we know the eigenstates of H_0:

$$H_0|n\rangle = E_n|n\rangle \tag{8.91}$$

We call H_1 the *interaction* part of the Hamiltonian, since the situation most often encountered is that in which H_0 describes parts or aspects of a system which may be considered separately if certain interactions are ignored, while H_1 describes those interactions.

An equation in which H_1 may be considered as the effective Hamiltonian may be obtained by starting from the equation

$$i\hbar \frac{\partial |\Psi\rangle}{\partial t} = (H_0 + H_1)|\Psi\rangle \tag{8.92}$$

and making the substitution

$$|\psi(t)\rangle = e^{iH_0t/\hbar}|\Psi\rangle \tag{8.93}$$

or rather, its inverse. This leads to

$$e^{-iH_0t/\hbar} \, i\hbar \frac{\partial}{\partial t} |\psi(t)\rangle = H_1 \, e^{-iH_0t/\hbar}|\psi(t)\rangle$$

or

$$i\hbar \frac{\partial |\psi(t)\rangle}{\partial t} = H_1(t)|\psi(t)\rangle \tag{8.94}$$

where

$$H_1(t) = e^{iH_0t/\hbar} H_1 e^{-iH_0t/\hbar} \tag{8.95}$$

Equation (8.94) may be reformulated as an integral equation

$$|\psi(t)\rangle = |\Psi(t_0)\rangle - \frac{i}{\hbar} \int_{t_0}^{t} H_1(t')|\psi(t')\rangle \, dt' \tag{8.96}$$

This equation may be integrated formally in a perturbation series in powers of H_1. A simple way to do this is to imagine H_1 to contain a "strength parameter" λ, such that

$$H_1 = \lambda \bar{H}_1 \tag{8.97}$$

We may then assume $|\Psi(t)\rangle$ to be expanded in a power series in λ, viz.,

$$|\psi(t)\rangle = \sum_{n=0}^{\infty} \lambda^n |\psi_n(t)\rangle \tag{8.98}$$

and equate corresponding powers of λ in (8.96). This leads to the relation

$$|\psi_n(t)\rangle = -\frac{i}{\hbar} \int_{t_0}^{t} dt' \, H_1(t')|\psi_{n-1}(t')\rangle \tag{8.99}$$

But of course

$$|\psi_0(t)\rangle = |\psi(t_0)\rangle \qquad (8.100)$$

We may therefore iterate (8.99) n times to get

$$|\psi_n(t)\rangle = \left(-\frac{i}{\hbar}\right)^n \int_{t_0}^t dt_1 H_1(t_1) \int_{t_0}^{t_1} dt_2 H_1(t_2) \ldots n \text{ times} \ldots$$

$$\int_{t_0}^{t_{n-1}} dt_n H_1(t_n)|\psi(t_0)\rangle \qquad (8.101)$$

Thus the formal perturbation solution

$$|\psi(t)\rangle = \left[1 - \frac{i}{\hbar}\int_{t_0}^t dt' H_1(t') + \left(-\frac{i}{\hbar}\right)^2 \int_{t_0}^t dt_1 H_1(t_1) \int_{t_0}^{t_1} dt_2 H_1(t_2) + \cdots \right.$$

$$+ \left(-\frac{i}{\hbar}\right)^n \int_{t_0}^{t_1} dt' H_1(t') \int_{t_0}^{t_1} dt_2 H_1(t_2) \ldots n \text{ times} \ldots$$

$$\left. \int_{t_0}^{t_{n-1}} dt_n H_1(t_n) + \cdots \right] |\psi(t_0)\rangle \qquad (8.102)$$

can be deduced.

10. Time Evolution Operator

Let us define an operator $U(t, t')$ such that

$$|\Psi(t)\rangle = U(t,t')|\Psi(t')\rangle \qquad (8.103)$$

Substituting in (8.84) we see that $U(t, t')$ must satisfy the equation

$$i\hbar \frac{\partial}{\partial t} U(t, t') = H U(t, t') \qquad (8.104)$$

It is simple to show that if H is Hermitian, $U(t, t')$ is unitary. For in this case the equation conjugate to (8.83) is

$$-i\hbar \frac{\partial}{\partial t} \langle \Psi(t)| = \langle \Psi(t)|H \qquad (8.105)$$

If we now multiply (8.83) by $\langle \Psi(t)|$ and (8.105) by $|\Psi(t)\rangle$ and subtract, we find that

$$i\hbar \left[\langle \Psi(t)| \frac{\partial}{\partial t} \Psi(t)\rangle + \left(\frac{\partial}{\partial t} \langle \Psi(t)|\right)|\Psi(t)\rangle \right] = 0$$

or

$$\frac{\partial}{\partial t} \langle \Psi(t)|\Psi(t)\rangle = 0 \qquad (8.106)$$

Thus, the magnitude of $|\Psi(t)\rangle$ does not vary with time, and

$$\langle \Psi(t)|\Psi(t)\rangle = \langle \Psi(t')|U^\dagger(t,t')U(t,t')|\Psi(t')\rangle$$
$$= \langle \Psi(t')|\Psi(t')\rangle$$

It follows, then, that

$$U^\dagger(t,t')U(t,t') = 1 \tag{8.107}$$

Consider now a special case, the rather trivial one in which we know the eigenstates $|n\rangle$ and the eigenvalues E_n of $H = H_0$.

In this case, it follows immediately from (8.84) that

$$|\Psi(t)\rangle = e^{-iH_0(t-t')/\hbar}|\Psi(t')\rangle \tag{8.108}$$

so that

$$U(t,t') = U_0(t-t') = e^{-iH_0(t-t')/\hbar} \tag{8.109}$$

The unitarity of $U(t,t')$ is then manifest. The matrix elements of U, using the eigenstates of H_0 as a basis, are

$$\langle n|U(t,t')|m\rangle = \delta_{nm}e^{-iE_n(t-t')/\hbar} \tag{8.110}$$

Now by virtue of the completeness condition

$$U - \sum_{n,m} |n\rangle \langle n|U(t,t')|m\rangle \langle m|$$
$$= \sum_{n} |n\rangle \langle n|e^{-iE_n(t-t')/\hbar} \tag{8.111}$$

We may easily determine the time evolution of the wave function. If $|q\rangle$ represent the eigenstates of all the coordinates of the system, multiplying (8.103) by $\langle q|$ and expressing $|\Psi(t')\rangle$ in terms of $|q'\rangle$ yields

$$\langle q|\Psi(t)\rangle = \langle q|U(t,t') \int |q'\rangle\langle q'|\Psi(t')\rangle \, dq'$$
$$= \int \langle q|U(t,t')|q'\rangle \langle q'|\Psi(t')\rangle \, dq' \tag{8.112}$$

or

$$\Psi(q,t) = \int K(q,t;q',t')\Psi(q',t') \, dq' \tag{8.113}$$

The "propagator" $K(q,t;q',t')$ is then

$$K(q,t;q',t') = \langle q|U(t,t')|q'\rangle$$
$$= \sum_{n} \langle q|n\rangle \langle n|q'\rangle \, e^{-iE_n(t-t')/\hbar} \tag{8.114}$$

on substituting (8.111) for U. Thus

$$K(q, t; q', t') = K(q, q'; t - t')$$

$$= \sum_n \phi_n(q) \phi_n^*(q') \, e^{-iEn(t-t')/\hbar} \qquad (8.115)$$

where

$$\phi_n(q) = \langle q|n \rangle \qquad (8.116)$$

Problem 8-10:   Show that, for free particles of mass m, the propagator is

$$K(\mathbf{r}, \mathbf{r}'; t - t') = \left[\frac{m}{\pi i \hbar (t - t')} \right]^{3/2} \exp\left[\frac{-i|\mathbf{r} - \mathbf{r}'|^2 m}{2\hbar(t - t')} \right]$$

Problem 8-11:   Show that the Fourier transform of $K(xt; x'0)$ is

$$K(x, \omega; x'0) = \lim_{\epsilon \to 0} \frac{1}{2\pi i} \sum_n \frac{\phi_n(x) \phi_n^*(x')}{\omega - E_n/\hbar - i\epsilon}$$

so that the energy eigenvalues of the system are given by the poles of K.

Let us now consider again the general case in which

$$H_1 = V(t) \qquad (8.117)$$

and consequently

$$|\psi\rangle = |\psi(t)\rangle \qquad (8.118)$$

Equation (8.102) may be written in terms of the "true wave function" $|\Psi(t)\rangle$ and the potential $V(t)$. If we introduce under the t_j integration, after $H_1(t_j)$, the factor

$$e^{-iH_0 t_j/\hbar} \, e^{iH_0 t_j/\hbar} = 1$$

substitute for $|\psi(t)\rangle$, $|\psi(t_0)\rangle$ in terms of $|\Psi(t)\rangle$ and $|\Psi(t_0)\rangle$, and put

$$H_1(t) = e^{iH_0 t/\hbar} V(t) \, e^{-iH_0 t/\hbar}$$

Eq. (8.101) may be written in terms of U_0, as follows:

$$|\Psi(t)\rangle = \left[U_0(t - t_0) - \frac{i}{\hbar} \int_{t_0}^{t} dt_1 \, U_0(t - t_1) V(t_1) U_0(t_1 - t_0) \right.$$

$$\left. + \left(-\frac{i}{\hbar} \right)^2 \int_{t_0}^{t} dt_1 \, U_0(t - t_1) V(t_1) \int_{t_0}^{t} dt_2 \, U_0(t_1 - t_2) V(t_2) U_0(t_2 - t_0) + \cdots \right.$$

$$+\left(-\frac{i}{\hbar}\right)^n \int_{t_0}^t dt_1 U_0(t-t_1)V(t_1)\int_{t_0}^{t_1} dt_2\ U_0(t_1-t_2)V(t_2)\dots n \text{ times}$$

$$\cdots \int_{t_0}^{t_{n-1}} dt_n\ U_0(t_{n-1}-t_n)V(t_n)U_0(t_n-t_0)+\cdots\Bigg]|\Psi(t_0)\rangle \qquad (8.119)$$

$U_0(t_j-t_{j-1})$ may be referred to as the "propagator" from t_{j-1} to t_j for the Hamiltonian H_0. It is then possible to give a diagrammatic representation for each term of (8.119). The first term corresponds to propagation without interaction. In the second, there is propagation from t_0 to t_1, where the perturbing potential acts to make a transition; the resulting system is then propagated to t. The next corresponds to a second-order transition ("scattering") where the transitions take place at t_2 and t_1, respectively, and so forth. Using a straight line to indicate propagation, and a change of direction to indicate "scattering," these terms may be represented as follows, where a point which is above another corresponds to a later time:

"zero order"   "first order"   "second order".

The square bracket in (8.119) is the operator $U(t, t_0)$. It is immediately verified that it satisfies the integral equation

$$U(t, t_0) = U_0(t-t_0) - \frac{i}{\hbar}\int_{t_0}^t dt_1 U_0(t-t_1)VU(t_1, t_0) \qquad (8.120)$$

Since the time evolution operators are defined only in the "causal" region

$$t > t_1 > t_0$$

we may take them to be zero for $t > t_1 < t_0$. This enables us to extend the range of integration in (8.120) from $-\infty$ to ∞.

11.   Case of a Time-Independent Perturbation

If V is independent of time, a formal solution of (8.120) may be obtained by taking Fourier transforms. The Fourier transform of U_0 may be written

$$u_0(\omega) = \int_{-\infty}^{\infty} U_0(t) \, e^{i\omega t} \, dt \qquad (8.121)$$

the inverse relationship being

$$U_0(t) = \frac{1}{2\pi} \int_{-\infty}^{\infty} u_0(\omega) \, e^{-i\omega t} \, d\omega \qquad (8.122)$$

To ensure convergence at the upper limit, we replace (8.121) by

$$u_0(\omega) = \lim_{\eta \to 0} \int_{-\infty}^{\infty} \theta(t) \, e^{-iH_0 t/\hbar} \, e^{i(\omega + i\eta)t}$$

where $\theta(t) = 1$ for positive argument and zero for negative. Thus

$$u_0(\omega) = i\hbar \lim_{\eta \to 0} \frac{1}{\hbar\omega - H_0 + i\hbar\eta} \qquad (8.123)$$

Since it is common practice nowadays to choose units in which

$$\hbar = c = 1 \qquad (8.124)$$

(c being the velocity of light), we henceforth drop the $\hbar$'s, thus effecting considerable economy in writing.

The factor $i\eta$ in (8.124) assures the "causal" property of $U_0(t)$ (the same device was used in Chapter 4 in a more general context). This is due to the fact that, in evaluating (8.122),

$$U_0(t) = \frac{i}{2\pi} \lim_{\eta \to 0} \int_{-\infty}^{\infty} \frac{1}{\omega - H_0 + i\eta} \, e^{-i\omega t} \, d\omega \qquad (8.125)$$

for $t > 0$ the contour must be deformed around the *lower* half-plane, which encompasses the singularity; whereas, when $t < 0$ it must be deformed about the *upper* half-plane, so that there are no singularities within the contour and the integral is zero.

A word should be said about the interpretation of these remarks, and in fact about the meaning of a quantity like

$$\frac{1}{\omega - H_0 + i\eta}$$

H_0, after all, is an operator, so that this quantity is also an operator. Its meaning is clear if we use a representation in which the basis vectors are the eigenstates of H_0. If these states are designated by $|n\rangle$, and the corresponding energies by E_n, the sequence of calculations proceeding from (8.121) may be interpreted as follows: If $u_0(\omega)$ operates on $|n\rangle$, and the exponential is interpreted in the sense of a power series, then

$$e^{-iH_0 t/\hbar} |n\rangle = e^{-iE_n t/\hbar} |n\rangle$$

At each subsequent step in the calculations, when an operator involving H_0 operates on a basis state $|n\rangle$, it produces the value E_n. Finally, if we know the effect of an operator on the basis vectors, we know its effect on any vector, since the basis vectors form a complete set.

Without loss of generality, we can take $t_0 = 0$. If we then take Fourier transforms of (8.120), and designate the transform of $U(t, 0)$ as $u(\omega)$, we get

$$u(\omega) = u_0(\omega) - iu_0(\omega)Vu(\omega) \qquad (8.126)$$

where we have used the fact that the last term of (8.120) is a convolution. Putting in the form of u_0,

$$\left[1 - \frac{1}{\omega - H_0 + i\eta} V\right] u = \frac{i}{\omega - H_0 + i\eta}$$

Multiplying on the left by $\omega - H_0 + i\eta$, we obtain

$$(\omega - H + i\eta)\, u = i$$

or

$$u = i(\omega - H + i\eta)^{-1} \qquad (8.127)$$

This result, though compact, is completely useless for purposes of calculation, except in so far as it might serve as the basis for an approximation procedure.

Some interesting results can be obtained by making a "perturbation" expansion of (8.127). We write

$$H = H_0 + V$$

and we are then confronted with the problem of expanding the operator

$$\frac{1}{A + B}$$

in a "power series" in B, where in our particular case

$$A = \omega - H_0 + i\eta \qquad (8.128)$$

and

$$B = -V \qquad (8.129)$$

It is easy to show that

$$\frac{1}{A+B} = \frac{1}{A} - \frac{1}{A}B\frac{1}{A} + \frac{1}{A}B\frac{1}{A}B\frac{1}{A} - + \cdots \qquad (8.130)$$

To verify this formula we multiply the operator on the right-hand side by A, then by B, and add. All terms cancel except the initial term (unity). Specifically, then, we have

$$u = \frac{i}{\omega - H_0 + i\eta} + (-i)\frac{i}{\omega - H_0 + i\eta}V\frac{i}{\omega - H_0 + i\eta}$$

$$+ (-i)^2\frac{i}{\omega - H_0 + i\eta}V\frac{i}{\omega - H_0 + i\eta}V\frac{i}{\omega - H_0 + i\eta} + \cdots \qquad (8.131)$$

Let us write this as

$$u(\omega) = \frac{i}{\omega - H_0 + i\eta} - i\frac{i}{\omega - H_0 + i\eta}T\frac{i}{\omega - H_0 + i\eta} \qquad (8.132)$$

$$u(\omega) = u_0(\omega) - iu_0(\omega)Tu_0(\omega) \qquad (8.133)$$

Then T is the operator

$$T = V + V\frac{1}{\omega - H_0 + i\eta}V + V\frac{1}{\omega - H_0 + i\eta}V\frac{1}{\omega - H_0 + i\eta}V + \cdots \qquad (8.134)$$

$$= V - iVu_0T$$

$$= V - iVu(\omega)V$$

$$T = V + V\frac{1}{\omega - H + i\eta}V \qquad (8.135)$$

It should be noted that

$$Tu_0 = Vu \qquad (8.136)$$

The form (8.135) for the T operator is compact, but again of no practical use since eigenstates of H are not known, so that we cannot find a matrix representation of $i/(\omega - H + i\eta) = u(\omega)$. The form (8.134) serves as the basis for a perturbation expansion.

We note that, from (8.103),

$$|\Psi(t)\rangle = U(t,t_0)|\Psi(t_0)\rangle = U(t,0)|\Psi(0)\rangle$$

Now

$$U(t,0) = \frac{i}{2\pi}\lim_{\eta \to 0}\int_{-\infty}^{\infty}\frac{1}{\omega - H + i\eta}e^{-i\omega t}\,d\omega$$

$$= \frac{1}{2\pi}\int_{-\infty}^{\infty}u(\omega)e^{-i\omega t}\,d\omega \qquad (8.137)$$

In what follows, we do not write the $i\eta$ explicitly, but treat the frequency ω as having a small positive imaginary part. We see that the evolution of a system is determined once we know $u(\omega)$. If we consider the evolution of the state $|\Psi(0)\rangle = |0\rangle$ as determined by (8.131), we observe that the transition from $|0\rangle$ to some final state may take place through intermediate steps which involve transitions back *into* the original state. Aside from producing analytic complications, all such processes which start and end in the state $|0\rangle$ describe not so much physical transitions as modifications of the character of the initial state itself. In other words, if after a time t, a system originally in state $|0\rangle$ is found again in that state, we may treat all that has taken place in the intervening time as characteristic only of that state itself.

Let us now see, then, how to separate out all such processes, and incorporate them in a modified ("renormalized") initial state.

For this purpose we introduce P as a projection operator onto the original state, and $Q = 1 - P$ as an operator of projection *out* of that state. Consider now Eq. (8.126), and put

$$u = Pu + Qu \tag{8.138}$$

We note that by orthogonality

$$Qu_0|0\rangle = 0 \tag{8.139}$$

An explicit representation of P and Q is given by

$$P = |0\rangle \langle 0|, \qquad Q = \sum_{r \neq 0} |r\rangle \langle r| \tag{8.140}$$

If, in the equation

$$Pu + Qu = u_0 - iu_0 V(Pu + Qu) \tag{8.141}$$

we separate the components in and perpendicular to $|0\rangle$, we obtain the equations

$$Pu = u_0 - iu_0 PV(Pu) - iu_0 PV(Qu) \tag{8.142}$$

and

$$Qu = -iu_0 QV(Pu) - iU_0 QV(Qu) \tag{8.143}$$

From the first of these equations

$$Pu = (1 + iu_0 PV)^{-1} u_0[1 - iPV(Qu)] \tag{8.144}$$

Thus, the operator

$$u_1 = (1 + iu_0 PV)^{-1} u_0$$

$$= \left(1 - \frac{1}{\omega - H_0} PV\right)^{-1} u_0$$

operates only on $|0\rangle$. But

$$\left(1 - \frac{1}{\omega - H_0} PV\right)^{-1} u_0|0\rangle = \left[1 + \frac{1}{\omega - H_0} PV\right.$$

$$\left. + \frac{1}{\omega - H_0} PV \frac{1}{\omega - H_0} PV + \cdots\right] u_0|0\rangle$$

$$= \sum_{n=0}^{\infty} \left(\frac{\langle 0|V|0\rangle}{\omega - E_0}\right)^n \frac{i}{\omega - E_0} |0\rangle$$

$$= \frac{1}{1 - V_{00}/(\omega - E_0)} \frac{i}{\omega - E_0} |0\rangle$$

$$= \frac{i}{\omega - E_0 - V_{00}} |0\rangle$$

$$= \frac{i}{\omega - H_0 - V_{00}} |0\rangle \qquad (8.145)$$

where

$$V_{00} = \langle 0|V|0\rangle$$

is the mean potential in the initial state. It follows that

$$u_1 = \frac{i}{\omega - H_0 - V_{00}} \qquad (8.146)$$

and that

$$Pu = u_1[1 - iPV(Qu)] \qquad (8.147)$$

or

$$Pu = u_1(1 - iPV\bar{u}) \qquad (8.148)$$

where

$$\bar{u} = Qu \qquad (8.149)$$

In the same notation, Eq. (8.143) may be written

$$(u_0^{-1} + iQV)\bar{u} = -iQV(Pu) \qquad (8.150)$$

We see, that, in principle, (8.148) and (8.150) may be solved for Pu and $\bar{u}$. Equation (8.150) enables us to express $\bar{u}$ in terms of Pu; putting this back in (8.148) then permits a solution for Pu. From Pu we may obtain $\langle 0|u|0\rangle$, which in turn permits us to determine the probability amplitude for a system, starting in state $|0\rangle$ at time $t = 0$, to be still found in it at time t. That is, we may calculate the decay from this state (or the scattering from it, and so forth, depending on the particular problem considered). $\bar{u}$, in turn, permits us to determine the amplitudes for transition to various possible "final" states.

To carry out the calculations explicitly, we first take the diagonal matrix element in the initial state of (8.148):

$$\langle 0|u|0 \rangle = \frac{i}{\omega - E_0 - V_{00}} \left[1 - i \sum_s \langle 0|V|s \rangle \langle s|u|0 \rangle \right] \qquad (8.151)$$

Latin indices such as "s" are taken to comprise all states except the initial state $|0\rangle$.

The matrix elements $\langle s|u|0 \rangle$ may be obtained by writing (8.150) in the form

$$\bar{u} = -i(u_0^{-1} + iQV)^{-1}QV(Pu)$$

and expanding

$$(u_0^{-1} + iQV)^{-1} = u_0 - u_0 iQV u_0 + u_0 iQV u_0 iQV u_0 + \cdots$$

The matrix element in question is then

$$\langle s|u|0 \rangle = \langle s|\bar{u}|0 \rangle$$

$$= \frac{1}{\omega - E_s} \left\{ \langle s|V|0 \rangle + \sum_l \langle s|V|l \rangle \frac{1}{\omega - E_l} \langle l|V|0 \rangle \right.$$

$$\left. + \sum_{l,m} \langle s|V|l \rangle \frac{1}{\omega - E_l} \langle l|V|m \rangle \frac{1}{\omega - E_m} \langle m|V|0 \rangle \cdots \right\}$$

$$\times \langle 0|u|0 \rangle \qquad (8.152)$$

Therefore,

$$\sum_s \langle 0|V|s \rangle \langle s|u|0 \rangle = F(\omega) \langle 0|u|0 \rangle \qquad (8.153)$$

where

$$F(\omega) = \sum_s \frac{\langle 0|V|s \rangle}{\omega - E_s} \left\{ \langle s|V|0 \rangle + \sum_l \langle s|V|l \rangle \frac{1}{\omega - E_l} \langle l|V|0 \rangle \right.$$

$$\left. + \sum_{l,m} \langle s|V|l \rangle \frac{1}{\omega - E_l} \langle l|V|m \rangle \frac{1}{\omega - E_m} \langle m|V|0 \rangle + \cdots \right\} \qquad (8.154)$$

In all the sums the summations are over all states but $|0\rangle$, which is excluded. Substituting (8.154) in (8.151), we obtain

$$\langle 0|u|0 \rangle \left[1 - \frac{F(\omega)}{\omega - E_0 - V_{00}} \right] = \frac{i}{\omega - E_0 - V_{00}}$$

from which it follows that

$$\langle 0|u|0\rangle = \frac{i}{\omega - E_0 - V_{00} - F(\omega)} \qquad (8.155)$$

It is easily seen that

$$\begin{aligned}
V_{00} + F(\omega) &= \left\langle 0 \left| V + V \frac{Q}{\omega - H_0} V + V \frac{Q}{\omega - H_0} V \frac{Q}{\omega - H_0} V + \cdots \right| 0 \right\rangle \\
&= \langle 0|K|0\rangle \\
&= \bar{K}(\omega) \qquad (8.156)
\end{aligned}$$

where

$$K = V + V \frac{Q}{\omega - H_0} V + V \frac{Q}{\omega - H_0} V \frac{Q}{\omega - H_0} V + \cdots \qquad (8.157)$$

K clearly satisfies the equation

$$K = V + V \frac{Q}{\omega - H_0} K \qquad (8.158)$$

It is possible to write $\langle s|u|0\rangle$ in terms of the same operator K:

$$\begin{aligned}
\langle s|u|0\rangle &= \left\langle s \left| \frac{1}{\omega - H_0} K \right| 0 \right\rangle \langle 0|u|0\rangle \\
&= \left\langle s \left| \frac{1}{\omega - H_0} K \right| 0 \right\rangle \frac{i}{\omega - E_0 - \langle 0|K(\omega)|0\rangle} \qquad (8.159)
\end{aligned}$$

The quantity $\langle 0|K(\omega)|0\rangle$ is a sort of "self-energy" of the system at frequency ω.

12. Some Useful Relations

It is easy to make a connection between the matrix elements of T and of u, using (8.133). For if, in this equation, we project into and out of the state $|0\rangle$, putting

$$T = PT + QT = T_1 + T_2 \qquad (8.160)$$

we see immediately that

$$\langle 0|u|0\rangle = \langle 0|u_0|0\rangle \left[1 + \frac{\langle 0|T_1|0\rangle}{\omega - E_0} \right] \qquad (8.161)$$

Comparison with (8.155) and (8.156) reveals that

$$\langle 0|T_1|0\rangle = \frac{\langle 0|K_1|0\rangle}{1 - \dfrac{\langle 0|K_1|0\rangle}{\omega - E_0}} \qquad (8.162)$$

Also, equating Q components of (8.133) gives

$$\langle s|u|0\rangle = \frac{1}{\omega - E_s} \langle s|T_2|0\rangle \langle 0|u_0|0\rangle$$

$$= \frac{1}{\omega - E_s} \langle s|T|0\rangle \langle 0|u_0|0\rangle \qquad (8.163)$$

Comparing this with (8.159) we find that

$$\langle 0|u_0|0\rangle \langle s|T|0\rangle = \langle s|K|0\rangle \langle 0|u|0\rangle \qquad (8.164)$$

From these equations it is straightforward to find $\langle 0|U(t,0)|0\rangle$ and $\langle s|U(t,0)|0\rangle$, and thus the probability that the initial state $|0\rangle$ remains after time t, and the probability that the system has made a transition to state $|s\rangle$. First,

$$\langle 0|u(\omega)|0\rangle = \frac{i}{\omega + i\eta - E_0 - \bar{K}(\omega + i\eta)} \qquad (8.165)$$

whence

$$\langle 0|U(t,0)|0\rangle = \frac{i}{2\pi} \lim_{\eta \to 0} \int_{-\infty}^{\infty} \frac{1}{\omega + i\eta - E_0 - \bar{K}(\omega + i\eta)} e^{-i\omega t}\, d\omega \qquad (8.166)$$

Then

$$\langle s|u(\omega)|0\rangle = \frac{1}{\omega + i\eta - E_s} \langle s|K(\omega + i\eta)|0\rangle \frac{i}{\omega + i\eta - E_0 - \bar{K}(\omega + i\eta)} \qquad (8.167)$$

from which it follows that

$$\langle s|U(t,0)|0\rangle = \frac{i}{2\pi} \lim_{\eta \to 0} \int_{-\infty}^{\infty} \frac{1}{(\omega + i\eta - E_s)} \frac{1}{(\omega + i\eta - E_0 - \bar{K})}$$

$$\times \langle s|K(\omega + i\eta)|0\rangle\, e^{-i\omega t} \qquad (8.168)$$

13. Calculation of Total Transition Probability

It is very difficult to evaluate (8.166) and (8.168) exactly, because of the complicated dependence of $\bar{K}(\omega)$ on ω. In the standard form of perturbation theory given in most text books, $\bar{K}$ is omitted completely in the denominator in (8.168). However, in (8.166), such a neglect does not lead to any decay of the initial system at all, and so is not acceptable. In this case, we may have recourse to an approximation. A simple such approximation, which permits us to complete the calculation, is that in which we evaluate $\bar{K}(\omega)$ only to the second order in the perturbation V:

$$\langle 0|K|0\rangle = \langle 0|V|0\rangle + \left\langle 0\left|V\frac{Q}{\omega + i\eta - H_0}V\right|0\right\rangle$$

$$= \langle 0|V|0\rangle + \sum_s \frac{|\langle 0|V|s\rangle|^2}{\omega + i\eta - E_s} \qquad (8.169)$$

where it is understood that we must, at the end of the calculation, let $\eta \to 0$. But $1/(\omega + i\eta - E_s)$ approaches $1/(\omega - E_s)$ except in a very narrow region near $\omega = E_s$. However, since

$$\frac{1}{\omega - E_s + i\eta} = \frac{(\omega - E_s) - i\eta}{(\omega - E_s)^2 + \eta^2}$$

for $|\omega - E_s| < \xi \ll \eta$ it is approximately equal to its imaginary part

$$\frac{-i\eta}{(\omega - E_s)^2 + \eta^2}$$

This imaginary part $\to 0$ as $\eta \to 0$ except for $\omega \approx E_s$, becomes infinite at $\omega = E_s$, and gives, on integration over E_s,

$$-i\left[\tan^{-1}\frac{E_s - \omega}{\eta}\right]_{-\infty}^{\infty} = -\pi i$$

Thus it is possible to write

$$\frac{1}{\omega - E_s + i\eta} = P\frac{1}{\omega - E_s} - \pi i\delta(E_s - \omega) \qquad (8.170)$$

This signifies that, when we integrate $1/(\omega - E_s + i\eta)$ with a function $A(E_s)$ of E_s, we get a real part which is the principal-part integral

$$P\int \frac{A(E_s)}{\omega - E_s}\,dE_s$$

and imaginary part which is

$$\int A(E_s)\delta(E_s - \omega)\,dE_s = A(\omega)$$

If, now, the states $|s\rangle$ in (8.169) form a continuum, the sum may be replaced by an integral giving

$$\sum_s \frac{|\langle 0|V|s\rangle|^2}{\omega + i\eta - E_s} \to \int \rho(E_s)\frac{|\langle 0|V|s\rangle|^2}{\omega + i\eta - E_s}\,dE_s \qquad (8.171)$$

where $\rho(E_s)$ is the number of states per unit energy interval at energy E_s (density of states at E_s). Then, using (8.170) we obtain the result

$$\int \rho(E_s)\frac{|\langle 0|V|s\rangle|^2}{\omega + i\eta - E_s}\,dE_s = P\int \rho(E_s)\frac{|\langle 0|V|s\rangle|^2}{\omega - E_s}\,dE_s - \pi i\rho(\omega)|\langle 0|V|s\rangle|_\omega^2$$

$$(8.172)$$

$|\langle 0|V|s\rangle|_\omega^2$ represents an average square matrix element over states for which $E_s = \omega$.

In order to evaluate (8.166), we now make another approximation, namely, that $\bar{K}$ is varying slowly enough that $\bar{K}(\omega)$ may be replaced by $\bar{K}(E_0)$. If we put

$$\bar{K}(\omega) = \delta E_0(\omega) - \tfrac{1}{2}i\Gamma_0(\omega) \tag{8.173}$$

this is equivalent to saying that δE_0 and Γ_0 are slowly varying functions of ω at E_0, so that to a good approximation

$$\bar{K}(\omega) \to \delta E_0(E_0) - \tfrac{1}{2}i\Gamma_0(E_0) = \delta E_0 - \tfrac{1}{2}i\Gamma_0 \tag{8.174}$$

δE_0 and Γ_0 being constants given by

$$\delta E_0 = P \int \rho(E_s) \frac{|\langle 0|V|s\rangle|^2}{E_0 - E_s} dE_s \tag{8.175}$$

and

$$\Gamma_0 = 2\pi\rho(E_0)|\langle 0|V|s\rangle|_{E_s = E_0}^2 \tag{8.176}$$

Returning now to (8.166), we can evaluate the integral

$$\langle 0|U(t,0)|0\rangle = \frac{i}{2\pi} \int_{-\infty}^{\infty} \frac{1}{\omega - E_0 - \delta E_0 + i\Gamma_0/2} e^{-i\omega t} \tag{8.177}$$

by deforming the integral around the lower half-plane to obtain

$$\langle 0|U(t,0)|0\rangle = e^{-i(E_0 + \delta E_0)t} e^{-\Gamma_0 t/2} \tag{8.178}$$

It follows that the probability that the system is still in the state $|0\rangle$ at time t is

$$P_{00}(t) = |\langle 0|U(t,0)|0\rangle|^2$$
$$= e^{-\Gamma_0 t} \tag{8.179}$$

so that Γ_0 is the decay constant of the system.

It should be emphasized that the formula (8.176) gives the decay constant only approximately. This formula is generally called the "golden rule" formula, following the terminology first introduced by Fermi.

14. Transition to a Definite Final State

Using the same approximation as in the previous section, we may evaluate $\langle s|U(t,0)|0\rangle$ from (8.168) and so evaluate the probability of transition from state $|0\rangle$ to any arbitrary state in time t. For in this approximation

$$\langle s|U(t,0)|0\rangle = \frac{i}{2\pi} \lim_{\eta\to 0} \int_{-\infty}^{\infty} \frac{1}{\omega+i\eta-E_s} \frac{1}{\omega+i\eta-E_0'+i\Gamma_0/2}$$

$$\times \langle s|K(E_0)|0\rangle \, e^{-i\omega t} \, d\omega \tag{8.180}$$

where

$$E_0' = E_0 + \delta E_0 \tag{8.181}$$

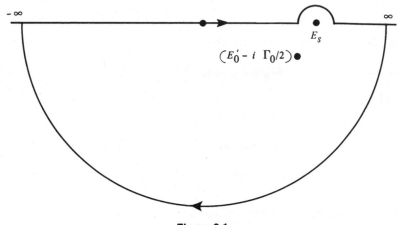

Figure 8.1

The integral may be evaluated by deforming the contour around the lower half-plane (Fig. 8.1), to give

$$\langle s|U(t,0)|0\rangle = \langle s|K(E_0)|0\rangle \, \frac{1}{E_0'-(i\Gamma_0/2)-E_s}$$

$$\times \, [e^{-iE_0't} \, e^{-\Gamma_0 t/2} - e^{-iE_s t}] \tag{8.182}$$

Thus, the probability that the system is in state $|s\rangle$ at time t is

$$P_{0s}(t) = \langle s|K(E_0)|0\rangle \, \frac{1}{(E_0'-E_s)^2+\Gamma_0^2/4}$$

$$\times \, [1+e^{-\Gamma_0 t} - 2e^{-\Gamma_0 t/2} \cos{(E_0'-E_s)t}] \tag{8.183}$$

From this we may find an expression for the total transition probability, provided we may neglect the variation of the matrix element $\langle s|K(E_0)|0\rangle$ and the density of states $\rho(E_s)$ with energy over a range $\simeq \Gamma_0$ about $E_s = E_0$. For if we replace $|\langle s|K(E_0)|0\rangle|^2$ by its average $|\langle s|K(E_0)|0\rangle|^2_{E_0}$ over the region in question, the probability that *some* transition has taken place by time t is

$$\sum P_{0s}(t) = \int \rho(E_s) P_{0s}(t) \, dE_s$$

$$= \rho(E_0) \int P_{0s}(t) \, dE_s$$

$$= \rho(E_0) |\langle s|K(E_0)|0\rangle|^2$$

$$\times \int_{-\infty}^{\infty} \frac{1}{(E_s - E_0')^2 + \Gamma_0^2/4}$$

$$\times [1 + e^{-\Gamma_0 t} - 2e^{-\Gamma_0 t/2} \cos(E_s - E_0')t] \, dE_s \qquad (8.184)$$

But

$$\int_{-\infty}^{\infty} \frac{1}{(E_s - E_0')^2 + \Gamma_0^2/4} \, dE_s = \frac{2\pi}{\Gamma_0} \qquad (8.185)$$

and

$$\int \frac{\cos(E_s - E_0')t}{(E_s - E_0')^2 + \Gamma_0^2/4} \, dE_s = \frac{2\pi}{\Gamma_0} e^{-\Gamma_0 t/2} \qquad (8.186)$$

Therefore,

$$\sum P_{0s}(t) = \frac{2\pi}{\Gamma_0} (1 - e^{-\Gamma_0 t}) \rho(E_0) |\langle s|K(E_0)|0\rangle|^2_{E_0} \qquad (8.187)$$

Finally, we see that

$$\frac{dP_{00}(t)}{dt} = -\frac{d}{dt} \sum P_{0s}(t)$$

$$= -e^{-\Gamma_0 t} 2\pi \rho(E_0) |\langle s|K(E_0)|0\rangle|^2_{E_0} \qquad (8.188)$$

This implies a Γ_0 which is an improvement over that given by (8.176) in that $\langle 0|V|s\rangle$ is replaced by

$$\langle 0|K(E_0)|s\rangle = \langle 0|V|s\rangle + \sum \frac{\langle 0|V|l\rangle \langle l|V|s\rangle}{E_0 - E_l}$$

$$+ \sum_{l,m} \frac{\langle 0|V|l\rangle \langle l|V|m\rangle \langle m|V|s\rangle}{(E_0 - E_l)(E_0 - E_m)} + \cdots \qquad (8.189)$$

This is not unreasonable, since the approximation (8.169) to Γ_0 was based on considering only *direct* transitions from the state $|0\rangle$, whereas (8.189) allows for indirect ones. Better approximations to $\langle 0|K(\omega)|0\rangle$ than that of (8.169) are of course possible. What is important to note is that the calculation of $P_{0s}(t)$ which we have given does not depend on the

488   Fundamental Principles of Quantum Mechanics

approximate form of Γ_0 which we calculated. Thus, while (8.187) is not self-consistent as it stands, it can be made so by calculating Γ_0 more accurately. However, we do not pursue the matter further here.

Problem 8-12:   A quantum-mechanical system in the absence of perturbations can exist in one of two states $|1\rangle$ and $|2\rangle$. In the presence of a perturbation which has no diagonal matrix elements for these states, but only the element $\langle 1|V|2\rangle = \langle 2|V|1\rangle^*$, find the probability that the system, which at time $t = 0$ is known to be in state $|1\rangle$, will be in $|2\rangle$ at time t.

Problem 8-13:   In the preceding problem, suppose that state $|2\rangle$ is at energy ΔE higher than that of state $|1\rangle$. Taking a perturbation

$$V(e^{i\omega t} + e^{-i\omega t})$$

find the probability of transition (*i*) from $|1\rangle$ to $|2\rangle$ and (*ii*) from $|2\rangle$ to $|1\rangle$, in time t.

Problem 8-14:   Solve the above problems if there is a third state $|3\rangle$ higher in energy than both $|1\rangle$ and $|2\rangle$, and such that the only nonzero matrix elements of the perturbation are those between $|1\rangle$ and $|3\rangle$, $|2\rangle$ and $|3\rangle$.

Problem 8-15:   Solve the same problems where the final state in each case is part of a continuum. Neglect transitions between states of the continuum.

Problem 8-16:   Consider an electron in a uniform magnetic field in the z direction. Suppose that its spin is in the x direction at $t = 0$. Find the probability that it is in each of the following states at time t:

(*i*) $S_x = \frac{1}{2}$
(*ii*) $S_x = -\frac{1}{2}$
(*iii*) $S_z = \frac{1}{2}$
(*iv*) $S_y = \frac{1}{2}$

---

15. Case of a Time-Dependent Perturbation

Let the perturbing Hamiltonian be of the form

$$H_1 = V(\mathbf{r}, t) \tag{8.190}$$

Taking the initial time $t_0 = 0$, Eq. (8.120) becomes

$$U(t) = U_0(t) - i \int_0^t dt_1 U_0(t - t_1) V(\mathbf{r}, t_1) U(t_1) \qquad (8.191)$$

where, as earlier, we have put $\hbar = 1$. This equation is difficult to solve, even approximately, in general. A particular case of interest which can be treated is that of a *periodic* potential:

$$V(\mathbf{r}, t) = F(\mathbf{r})\, e^{i\omega t} + F\dagger(\mathbf{r})\, e^{-i\omega t} \qquad (8.192)$$

where F is a stationary operator and $F\dagger$ its Hermitian conjugate.

To solve (8.191), we first take the Fourier transform of the equation. The transform of the integral is

$$\int_{-\infty}^{\infty} dt\, e^{i\omega t} \int_{-\infty}^{\infty} dt_1 U_0(t - t_1) [F e^{i\omega_0 t_1} + F\dagger\, e^{-i\omega_0 t_1}] U(t_1) \qquad (8.193)$$

The limits of the inner integral have been extended from $-\infty$ to ∞, since the evolution operators U_0, U may be taken to be zero for negative argument (this is the expression of the condition of causality).

The variable t may be replaced by a new variable

$$t' = t - t_1 \qquad (8.194)$$

so that (8.193) takes the form

$$\int_{-\infty}^{\infty} dt'\, e^{i\omega t'} U_0(t') \int_{-\infty}^{\infty} e^{i\omega t_1} [F e^{i\omega_0 t_1} + F\dagger\, e^{-i\omega_0 t_1}] U(t_1)\, dt_1$$
$$= u_0(\omega) [Fu(\omega + \omega_0) + F\dagger u(\omega - \omega_0)] \qquad (8.195)$$

Therefore

$$u(\omega) = u_0(\omega)\{1 - i[Fu(\omega + \omega_0) + F\dagger u(\omega - \omega_0)]\} \qquad (8.196)$$

This may be solved by an iterative (perturbation) procedure. Two features of this procedure should be noted:

(*i*) In successive approximations, we get terms involving u_0 with arguments $\omega \pm n\omega_0$, n being an integer. This implies the possibility of successive emission or absorption of amounts $\hbar\omega_0$ of energy. To the lowest order (obtained by replacing u by u_0 on the right), only transitions between the initial energy state E_0 and states with energies $E_0 \pm \hbar\omega_0$ will be accounted for, as we see below.

(*ii*) In higher approximations of *even* order, transitions will be possible back to the initial state (absorption followed by emission and vice versa). Such terms again have the character of a self-energy.

16.   First-Order Perturbation

In the lowest order of perturbation, in which u_0 replaces u on the right-hand side of (8.196), u is given by

$$u(\omega) = u_0(\omega) - iu_0[Fu_0(\omega + \omega_0) + F\dagger u_0(\omega - \omega_0)] \qquad (8.197)$$

In this approximation, the time evolution operator is

$$U(t) = \frac{1}{2\pi} \int_{-\infty}^{\infty} u_0(\omega)\{1 - i[Fu_0(\omega + \omega_0) + F\dagger u_0(\omega - \omega_0)]\}\, e^{-i\omega t}\, d\omega \qquad (8.198)$$

The amplitude for transition from state $|0\rangle$ to state $|s\rangle$ in time t is then

$$\langle s|U(t)|0\rangle = \frac{i}{2\pi} \lim_{\eta \to 0} \int_{-\infty}^{\infty} \frac{1}{\omega - E_s + i\eta}\left[\langle s|F|0\rangle \frac{1}{\omega + \omega_0 - E_0 + i\eta}\right.$$

$$\left. + \langle s|F\dagger|0\rangle \frac{1}{\omega - \omega_0 - E_0 + i\eta}\right] e^{-i\omega t}\, d\omega \qquad (8.199)$$

The integrals may be evaluated by completing the contour with a semi-circle of radius $\to \infty$ around the negative half-plane. In this way we obtain

$$\langle s|U(t)|0\rangle = \langle s|F|0\rangle\, [e^{-iE_s t} - e^{-i(E_0 - \omega_0)t}]\, \frac{1}{E_s - E_0 + \omega_0}$$

$$+ \langle s|F\dagger|0\rangle\, [e^{-iE_s t} - e^{-i(E_0 + \omega_0)t}]\, \frac{1}{E_s - E_0 - \omega_0} \qquad (8.200)$$

The probability that the system is in state $|s\rangle$ at time t is

$$P_{0s}(t) = |\langle s|U(t)|0\rangle|^2 = |\langle s|F|0\rangle|^2\, \frac{\sin^2 \frac{1}{2}(E_s - E_0 - \omega_0)t}{[\frac{1}{2}(E_s - E_0 - \omega_0)]^2}$$

$$+ |\langle s|F\dagger|0\rangle|^2\, \frac{\sin^2 \frac{1}{2}(E_s - E_0 + \omega_0)t}{[\frac{1}{2}(E_s - E_0 + \omega_0)]^2} + R \qquad (8.201)$$

where the remaining terms R, which are cross terms in the magnitude squared of the right-hand side of (8.200), may be written

$$R = \frac{|\langle s|F|0\rangle|^2}{(E_s - E_0)^2 - \omega_0{}^2}\{4\cos 2\delta\,[\sin^2 \frac{1}{2}(E_s - E_0 - \omega_0)t + \sin^2 \frac{1}{2}(E_s - E_0 + \omega_0)t]$$

$$- 2\,[\cos 2\delta - \cos(2\omega_0 t + 2\delta)]$$

$$- 2\sin 2\delta\,[\sin(E_s - E_0 - \omega_0)t - \sin(E_s - E_0 + \omega_0)t] \qquad (8.202)$$

δ being the phase of the matrix element $\langle s|F|0\rangle$. Thus the probability per unit time of *transition* from state $|0\rangle$ to state $|s\rangle$ at time t is

$$\frac{\partial P_{0s}(t)}{\partial t} = 2|\langle s|F|0\rangle|^2\left[\frac{\sin(E_s - E_0 - \omega_0)t}{E_s - E_0 - \omega_0} + \frac{\sin(E_s - E_0 + \omega_0)t}{E_s - E_0 + \omega_0}\right] + \frac{\partial R}{\partial t}$$

$$(8.203)$$

where

$$\frac{\partial R}{\partial t} = 2|\langle s|F|0\rangle|^2 \left\{ \frac{\sin\left[(E_s - E_0 - \omega_0)t - 2\delta\right]}{E_s - E_0 + \omega_0} + \frac{\sin\left[(E_s - E_0 + \omega_0)t + 2\delta\right]}{E_s - E_0 - \omega_0} \right.$$

$$\left. -2\frac{\omega_0}{(E_s - E_0)^2 - \omega_0^2} \sin(2\omega_0 t + 2\delta) \right\} \tag{8.204}$$

Considering first (8.203), we observe that, for large t, the function $(\sin tx)/x$ approaches the δ function $\pi\delta(x)$. Therefore, ignoring $\partial R/\partial t$, the probability of transition per unit time for large t approaches

$$\frac{\partial P_{0s}}{\partial t} = 2\pi|\langle s|F|0\rangle|^2 \left\{ \delta(E_s - E_0 - \omega_0) + \delta(E_s - E_0 + \omega_0) \right\} \tag{8.205}$$

If we make a better approximation in which the level width is taken into account, the δ function is "broadened."

As for the $\partial R/\partial t$ term, it is oscillating and is never large; it decreases rapidly as the energy of the final state becomes substantially different from $E_s = E_0 \pm \hbar\omega_0$.

The relative importance of the two terms is illustrated by considering the case in which there is a continuum of final states $|s\rangle$. If we sum (integrate) over these final states, we obtain from (8.205)

$$\frac{\partial P_{0s}}{\partial t} = 2\pi|\langle s|F|0\rangle|^2_{\text{AV}} \left[\rho(E_0 + \omega_0) + \rho(E_0 - \omega_0) \right] \tag{8.206}$$

where $\rho(E)$ is the density of states. As for the first two terms in (8.204), for large t the sine functions will oscillate rapidly over a range of energy for which the matrix element and density of states generally varies very little; their contribution is therefore very small. The last term oscillates in time and thus does not give rise to "permanent" transitions. For sufficiently large ω_0, and for periods of observation $\gg \pi/\omega_0$ it has no measurable effect. In fact, over periods $\approx \pi/\omega_0$, the energy of the final state can only be determined to within an uncertainty $\approx 2\hbar\omega_0$, and it is for this reason that the term appears.

17. Perturbation Theory for Stationary States

In the majority of quantum physical problems, it is not possible to find the eigenfunctions or eigenvalues exactly. The steady-state problem may, however, often be put in the form

$$(H_0 + \lambda V)|\psi\rangle = E|\psi\rangle \tag{8.207}$$

where the eigenstates of H_0 may be determined exactly; λV is designated the "perturbation." Solutions for the states $|\psi\rangle$ may then be expressed in power series in λ. The solution will, of course, only be valid if the exact solution is an analytic function of λ; where perturbation theory is applied, it will be *assumed* that this is the case. Also, the successive terms in the expansion become more and more complicated, so that the practical use of perturbation theory is confined to cases in which all but the first few terms of the series are negligible.

There are two distinct versions of perturbation theory. The more widely used is associated with the names of Rayleigh and Schrödinger. In this case both $|\psi\rangle$ and E are expressed in series in λ. In the other approach due to Brillouin and Wigner, $|\psi\rangle$ is expressed as a series but the eigenvalues E are determined implicitly.

18. Rayleigh–Schrödinger Perturbation Theory

When the perturbation has zero strength, i.e., when $\lambda = 0$, the states $|n\rangle$ are determined by

$$H_0|n\rangle = E_n|n\rangle \qquad (8.208)$$

As λ is gradually increased, each "unperturbed" state evolves into a "perturbed" one; there is a one-to-one correspondence between the unperturbed and perturbed states. For convenience, we designate the unperturbed state to be considered as $|\psi_0\rangle$, and its energy E_0. Thus, the wave equation for the original state is

$$H_0|\psi_0\rangle = E_0|\psi_0\rangle \qquad (8.209)$$

and that for the corresponding exact state is

$$(H_0 + \lambda V)|\psi\rangle = E|\psi\rangle \qquad (8.210)$$

We designate the energy shift as

$$\Delta E = E - E_0 \qquad (8.211)$$

Then subtracting (8.209) from (8.210), we obtain the equation

$$(E_0 - H_0)(|\psi\rangle - |\psi_0\rangle) = (\lambda V - \Delta E)|\psi\rangle \qquad (8.212)$$

We take $|\psi_0\rangle$ to be normalized, but choose $|\psi\rangle$ so that its component on $|\psi_0\rangle$ has unit amplitude; consequently, $|\psi\rangle$ is *not* normalized to unity. In this case

$$P|\psi\rangle = |\psi_0\rangle \qquad (8.213)$$

where P is the projection operator on the initial state, and (8.212) may be written

$$(E_0 - H_0)Q|\psi\rangle = (\lambda V - \Delta E)|\psi\rangle \qquad (8.214)$$

Q being the operator $(1 - P)$, which consequently projects *out* all components in the direction of $|\psi_0\rangle$.

Since the left-hand side has no component in the direction of $|\psi_0\rangle$, neither has the right; we may therefore place the operator Q in front of it. Then, multiplying across by the operator $(E_0 - H_0)^{-1} = 1/(E_0 - H_0)$ we find that

$$Q|\psi\rangle = \frac{1}{E_0 - H_0} Q(\lambda V - \Delta E)|\psi\rangle$$

or

$$|\psi\rangle = |\psi_0\rangle + \frac{1}{E_0 - H_0} Q(\lambda V - \Delta E)|\psi\rangle \qquad (8.215)$$

The second term on the right represents the perturbation; the equation may be solved by iteration.

ΔE may be obtained in any approximation once $|\psi\rangle$ is known, for on multiplying both sides of (8.214) by the bra vector $\langle\psi_0|$, the left-hand side vanishes and the equation

$$\Delta E = \frac{\langle\psi_0|\lambda V|\psi\rangle}{\langle\psi_0|\psi\rangle} \qquad (8.216)$$

is obtained. With the normalization referred to above, the denominator is unity, so

$$\Delta E = \lambda \langle\psi_0|V|\psi\rangle \qquad (8.217)$$

Thus, if $|\psi\rangle$ is determined up to terms of order λ^{n-1}, ΔE is obtained to order λ^n.

The successive approximations to $|\psi\rangle$ may be expressed explicitly by writing (8.215) in the form

$$\left[1 - \frac{1}{E_0 - H_0} Q(\lambda V - \Delta E)\right]|\psi\rangle = |\psi_0\rangle$$

or

$$|\psi\rangle = \left[1 - \frac{1}{E_0 - H_0} Q(\lambda V - \Delta E)\right]^{-1} |\psi_0\rangle \qquad (8.218)$$

Therefore,

$$|\psi\rangle = |\psi_0\rangle + \frac{Q}{E_0 - H_0} \lambda V|\psi_0\rangle$$

$$+ \frac{Q}{E_0 - H_0} (\lambda V - \Delta E) \frac{Q}{E_0 - H_0} \lambda V|\psi_0\rangle + \cdots \qquad (8.219)$$

We see from (8.217) that ΔE itself is a power series in λ, i.e.,

$$\Delta E = \Sigma \lambda^r \Delta E^{(r)}$$

Equation (8.218) can be expanded as a power series in λ in terms of the eigenstates $|n\rangle$ of H_0, after which λ may be put equal to unity. To the second order in the wave function and the third in the energy

$$|\psi\rangle = |\psi_0\rangle + \sum_{n \neq 0} |\psi_n\rangle \frac{1}{E_0 - E_n} \langle \psi_n|V|\psi_0\rangle$$

$$+ \sum_{n,n'} |\psi_n\rangle \left[\frac{\langle \psi_n|V|\psi_{n'}\rangle}{E_0 - E_n} \frac{\langle \psi_{n'}|V|\psi_0\rangle}{E_0 - E_{n'}} - \langle \psi_0|V|\psi_0\rangle \frac{\langle \psi_n|V|\psi_0\rangle}{(E_0 - E_n)^2} \right] \qquad (8.220)$$

From (8.217) the energy may be expressed in the series:

$$E = E_0 + \langle \psi_0|V|\psi_0\rangle + \sum_n \frac{|\langle \psi_n|V|\psi_0\rangle|^2}{E_0 - E_n}$$

$$+ \sum_{n,n'} \frac{\langle \psi_0|V|\psi_n\rangle\langle \psi_n|V|\psi_{n'}\rangle\langle \psi_{n'}|V|\psi_0\rangle}{(E_0 - E_n)(E_0 - E_{n'})}$$

$$- \langle \psi_0|V|\psi_0\rangle \sum_n \frac{|\langle \psi_0|V|\psi_n\rangle|^2}{(E_0 - E_n)^2} + \cdots \qquad (8.221)$$

At each step in the expansion the nth-order term in the energy is calculated from the $(n - 1)$ term in the wave function.

19. Degenerate Case

The above development runs into difficulty if the state $|\psi_0\rangle$ is degenerate; that is, if any of the states $|n\rangle$ have the energy E_0. The treatment of this case requires a modification of the above derivations. Let us suppose that there are a number of states $|\beta\rangle$ with the same energy E_0, and that we wish to calculate their perturbation. If P is now the projection on the subspace of the states $|\beta\rangle$, following exactly the lines of the previous calculation, we obtain an equation formally identical with (8.214),

$$(E_0 - H_0)(1 - P)|\psi\rangle = (1 - P)(V - \Delta E)|\psi\rangle \qquad (8.222)$$

Before proceeding further, we note that if the states $|\beta\rangle$ are so chosen that there are no matrix elements of V between them, there will not be any zero denominators. Since any linear combination of the $|\beta\rangle$'s is an eigenstate of H_0 with eigenvalue E_0, we first solve within the subspace of the $|\beta\rangle$'s the eigenvalue problem

$$V|\beta\rangle = \epsilon_\beta|\beta\rangle \qquad (8.223)$$

or

$$\sum_{\beta'} |\beta'\rangle \langle \beta'|V|\beta\rangle = \epsilon_\beta|\beta\rangle \qquad (8.224)$$

The ϵ_β's are determined by the secular equation

$$\text{Det } [\langle \beta'|V|\beta \rangle - \epsilon \delta_{\beta\beta'}] \tag{8.225}$$

Since V is a Hermitian operator, the eigenvalues are real and the eigenstates orthogonal within the degenerate subspace.

Reinterpreting $P = P_\beta$ in (8.222) as the operator of the projection on *one* of the states just determined, we have

$$|\psi\rangle = |\psi_\beta\rangle + \frac{1}{E_0 - H_0}(1 - P_\beta)(V - \Delta E)|\psi\rangle$$

By iteration

$$|\psi\rangle = |\psi_\beta\rangle + \frac{1}{E_0 - H_0}(1 - P_\beta)(V - \Delta E)|\psi_\beta\rangle$$

$$+ \frac{1}{E_0 - H_0}(1 - P_\beta)(V - \Delta E)\frac{1}{E_0 - H_0}(1 - P_\beta)(V - \Delta E)|\psi_\beta\rangle + \cdots$$

As before, the last ΔE terms go out because $(1 - P_\beta)|\psi_\beta\rangle = 0$. Also, because $V|\psi_\beta\rangle$ has no components on the other states of the degenerate set, $(1 - P_\beta)$ in the above may be replaced by the operator $(1 - P)$ which projects out of the whole degenerate subspace. Thus, we obtain a formula similar to (8.220):

$$|\psi\rangle = |\beta\rangle + \sum_n{}' |n\rangle \frac{\langle n|V|\beta \rangle}{E_0 - E_n}$$

$$+ \sum_{n,n'} |n\rangle \frac{\langle n|V|n' \rangle \langle n'|V|\beta \rangle}{(E_0 - E_n)(E_0 - E_{n'})} + \cdots \tag{8.226}$$

where the prime on the summation sign Σ signifies that the intermediate states $|n\rangle$, $|n'\rangle$, and so forth have no components in the subspace of the $|\beta\rangle$'s.

Taking the scalar product of (8.214) with $|\beta\rangle$, we find that

$$\Delta E = \frac{\langle \beta|V|\psi \rangle}{\langle \beta|\psi \rangle} \tag{8.227}$$

Using (8.226), we obtain an expression for ΔE; $E = E_0 + \Delta E$ becomes

$$E_\beta = E_0 + \epsilon_\beta + \sum_n{}' \frac{\langle \beta|V|n \rangle \langle n|V|\beta \rangle}{E_0 - E_n}$$

$$+ \sum_{n,n'}{}' \frac{\langle \beta|V|n' \rangle \langle n'|V|n \rangle \langle n|V|\beta \rangle}{(E_0 - E_n)(E_0 - E_{n'})}$$

$$- \sum_n \frac{\langle \beta|V|\beta \rangle \langle \beta|V|n \rangle \langle n|V|\beta \rangle}{(E_0 - E_n)^2} + \cdots \tag{8.228}$$

20. Brillouin–Wigner
Perturbation Theory

We proceed in a manner very similar to that of the Rayleigh–Schrödinger theory, except that, when we subtract (8.209) from (8.210), we write the result in the form

$$(E - H_0)(|\psi\rangle - |\psi_0\rangle) = V|\psi\rangle - \Delta E|\psi_0\rangle \qquad (8.229)$$

Again introducing the operator P which projects on the initial state $|\psi_0\rangle$ and $Q = 1 - P$, and assuming that $|\psi\rangle$ is normalized so that

$$\langle\psi_0|\psi\rangle = 1 \qquad (8.230)$$

(8.229) becomes

$$(E - H_0)Q|\psi\rangle = V|\psi\rangle - \Delta E|\psi_0\rangle \qquad (8.231)$$

Taking the product of each side with the bra vector $\langle\psi_0|$, we find that

$$\Delta E = \langle\psi_0|V|\psi\rangle \qquad (8.232)$$

Since, from (8.231), P operating on the right-hand side gives zero

$$(E - H_0)Q|\psi\rangle = QV|\psi\rangle$$

or

$$(1 - P)|\psi\rangle = \frac{1}{E - H_0} QV|\psi\rangle$$

or, finally,

$$|\psi\rangle = |\psi_0\rangle + \frac{1}{E - H_0} QV|\psi\rangle \qquad (8.233)$$

This leads, exactly as before, to the perturbation series

$$|\psi\rangle = |\psi_0\rangle + \frac{1}{E - H_0} QV|\psi_0\rangle + \frac{1}{E - H_0} QV \frac{1}{E - H_0} QV|\psi_0\rangle + \cdots \qquad (8.234)$$

Substituting this in (8.232) gives

$$\Delta E = \langle\psi_0|V|\psi_0\rangle + \langle\psi_0|V \frac{1}{E - H_0} QV|\psi_0\rangle$$

$$+ \langle\psi_0|V \frac{1}{E - H_0} QV \frac{1}{E - H_0} QV|\psi_0\rangle + \cdots \qquad (8.235)$$

It is interesting to compare the results of second-order perturbation theory in the Rayleigh–Schrödinger and Brillouin–Wigner schemes.

Writing

$$V_{0n} = \langle \psi_0 | V | n \rangle \qquad (8.236)$$

we obtain from (8.221) the Rayleigh–Schrödinger result

$$E = E_0 + V_{00} + \sum_n \frac{|V_{0n}|^2}{E_0 - E_n} \qquad (8.237)$$

while from (8.235) the Brillouin–Wigner method gives

$$E = E_0 + V_{00} + \sum_n \frac{|V_{0n}|^2}{E - E_n} \qquad (8.238)$$

An interesting feature of the Rayleigh–Schrödinger method is the following: The perturbation energy of the ground state to second order is *negative*, even though the perturbation admixes components of higher energy states. This is due to the fact that the modification of the *wave function* by the perturbation is such that the mean value of V in the perturbed state is *less* than in the unperturbed one.

Problem 8-17: (*i*) Calculate to second order in V, the mean value of H_0 in the *perturbed* state, and the amount by which it is greater than the unperturbed ground state energy E_0.

(*ii*) Calculate, also to second order in V, the change in the mean value of V due to the perturbation. Show that it is negative and twice as big as the energy calculated in (*i*). [Caution: Remember that normalization of the perturbed wave function must be taken into account.]

Problem 8-18: Two plane rotators, each with Hamiltonians of the form

$$-\frac{\hbar^2}{2ma^2} \frac{\partial^2}{\partial \theta^2}$$

are coupled by the potential

$$-B \cos (\theta_1 - \theta_2)$$

θ_1 and θ_2 being the angles of the two rotators. Use perturbation theory to estimate the energies of the coupled system.

PRELUDE TO CHAPTER 9

This chapter is devoted to the solution of some of the standard problems of quantum mechanics. The first is the one-dimensional harmonic oscillator, which is analyzed in terms of "ladder operators," one of which increases the quantum number by 1 (creation operator) and one of which decreases it by 1 (annihilation operator). This problem provides the foundation both for the quantum theory of fields and for the "second quantization" formalism which is the most logical basis for many-body problems.

The problem is solved both by operator algebra and by use of the Schrödinger wave equation. The wave functions are expressible in terms of Hermite polynomials, introduced in Chapters 2 and 4.

The second topic of the chapter is the theory of angular momentum, which plays a role in all areas of modern physics. The theory is developed on the basis of operator algebra, using only the commutation rules for the operators. Again, a creation-annihilation formalism can be developed. Integer and half-odd-integer (j) states emerge from the same approach.

Orbital angular momentum, which arises from the separation of the Schrödinger equation in problems of spherically symmetric potentials, is treated in spherical polar coordinates. The eigenfunctions are the spherical harmonics introduced and discussed in Chapter 3.

The states $j = s = \frac{1}{2}$ represent *spin*; the usual Pauli spin operators are derived. It is then shown, however, that the matrix operators of higher order $(2j + 1)$ may be found for all j, including integral ones.

An alternative derivation of angular momentum theory due to Schwinger is then developed. Schwinger's approach makes use of pairs of creation and annihilation operators, combined into so-called "spinor operators." The theory is therefore linked to that of the harmonic oscillator. It is shown that angular momentum states may be expressed in a very simple way in terms of the two creation operators.

We turn next to a very important theory—that of the *addition* of angular momenta. The addition may be that of several particles of a composite system, or of intrinsic and orbital angular momentum of a single particle. The composite angular momentum states are related to those of the components through the *Clebsch–Gordan* coefficients. Our purpose is not to give a general formula for such coefficients, which is very complicated and in any case can be looked up in handbooks, but rather to demonstrate the *principles* underlying the problem, and to give a straightforward method of calculation which can be used in those common problems in which j is not too large.

We then turn to the problems of harmonic oscillators in two and three dimensions. Various approaches are considered. It is possible, using Cartesian coordinates, to decouple these problems into that of two or three one-dimensional ones. The states are then seen to be highly degenerate, and one must seek suitable quantum numbers to define the degenerate states.

In two dimensions the problem is also solved in terms of polar coordinates; the wave function may be expressed as a confluent hypergeometric function.

Finally, an operator method is introduced. Operators are found which independently raise and lower the energy and the angular momentum; all states are then directly derivable from the ground state (quantum numbers zero) by successive use of creation operators. These operators are given a Schrödinger representation, so that the wave functions may be calculated directly by repeated operation on the ground state.

The most important application of the two-dimensional oscillator theory is that of the quantum states of a charged particle in a constant magnetic field.

The three-dimensional oscillator is first treated by a generalization of the method just discussed; and a new third creation and annihilation operator must be introduced. Simultaneous eigenstates characterized by three quantum numbers (number of quanta, two angular momentum quantum numbers) are derived by creation out of the ground state.

The solution is also obtained from the Schrödinger equation, by separation of variables. The angular functions are angular momentum eigenstates; the radial function is a confluent hypergeometric.

The hydrogen atom problem is then studied, using the *factorization method* discussed first in Chapter 3. This method bypasses the usual complications of solution in series. The eigenvalues of energy come out in a very straightforward way, and excited states are again obtained by using appropriate creation operators.

Finally, we consider the problem of scattering of a plane wave by a central potential of short range. The problem is treated in a manner quite parallel to that used for scattering of sound waves in Chapter 5, by using a "partial wave expansion" (expansion in spherical harmonics, i.e., angular momentum eigenstates). It is possible not only to treat pure ("potential") scattering, but also absorption inside a "scattering center." The quantum-mechanical problem is not solved inside this center. The solution *outside* is shown to be completely determined by the logarithmic derivatives of the radial functions of the partial waves at its surface (that is, by the prescription of a sort of "surface impedance").

Once "scattering", "absorption", and "total (= scattering + absorption)" cross sections are determined, it follows that the inequalities developed in Chapter 5 hold once again.

Finally, we consider very low-energy (angular momentum zero) scattering. We see that absorption may be described by giving the potential inside the scattering center an imaginary part. In this case we discuss "resonance scattering," deriving the well-known Breit–Wigner formula of nuclear reaction theory. The conditions which define the location and width of the resonance are determined, and the interference between "resonance" and "potential" scattering demonstrated.

REFERENCES

Gottfried, K., *Quantum Mechanics*. New York: Benjamin, 1965, Vol. 1.

Infeld, L. and T. E. Hull, *Rev. Mod. Phys.* **23**, No. 1 (1951).

Landau, L. D. and E. M. Lifshitz, *Quantum Mechanics*. Reading, Mass.: Addison-Wesley, 1958.

Loeb, A. L. and L. Harris, *Introduction to Wave Mechanics*. New York: McGraw-Hill, 1963.

Mattis, D., *The Theory of Magnetism*. New York: Harper and Row, 1965, Chapter 3.

Messiah, A., *Mécanique Quantique*. Paris: Dunod, 1958.

9

SOME SOLUBLE PROBLEMS OF QUANTUM MECHANICS

"You people speak in terms of circles and ellipses and regular velocities—simple movements that the mind can grasp—very convenient—but suppose almighty God had taken it into his head to make the stars move like that... (He describes an irregular motion with his finger through the air) ... then where would you be?"

Bertold Brecht, Galileo

1. Introduction

In the previous chapter we have discussed some of the general principles of quantum mechanics. In this chapter, we develop the solutions of several of the commonest quantum systems. We begin with the harmonic oscillator, which plays a fundamental role in quantum field theory and provides an introduction to the concept of creation and annihilation operators. The problem is discussed in one, two, and three dimensions. The case of two dimensions is important in connection with the states of electrons in constant magnetic fields. The three-dimensional case arises in the theory of nuclear shell structure.

The discussion of three-dimensional central force problems leads us to simultaneous states of energy and angular momentum. We therefore develop the elements of the theory of angular momentum operators.

We then consider the bound states of the Coulomb potential, and solve the so-called "hydrogenic atom" problem.

Finally, using the methods introduced in Chapter 5 for the problem of the scattering of sound waves, we develop the "partial wave" theory of scattering of particles by a fixed, bounded potential.

2. Harmonic Oscillator in One Dimension

If a particle of mass m is acted upon by a restoring force proportional to its displacement from a fixed point of equilibrium, it is known as a "harmonic oscillator."

Its Hamiltonian is

$$H = \frac{p_x{}^2}{2m} + \tfrac{1}{2}kx^2 \tag{9.1}$$

if the displacement is in one dimension only. In the theory of the vibrating string (Chapter 1) we also met the classical harmonic oscillator problem; the "displacement" in that problem was the amplitude of a normal mode of vibration. It turns up, in fact, in all linear vibration problems including, for example, the vibrations of an electromagnetic field. Such a field, in an enclosure, can also be resolved into normal modes, and the Hamiltonian may be expressed in terms of coordinates which are the amplitudes of these normal modes, and conjugate momenta.

Consequently, the problem of the harmonic oscillator plays an important role in a number of quantum-mechanical problems of physical interest.

The solution of this problem is straightforward, but contains a number of interesting features. We, therefore, explore it thoroughly.

The Schrödinger wave equation of the problem is

$$-\frac{\hbar^2}{2m}\frac{d^2\psi}{dx^2} + \tfrac{1}{2}kx^2\psi = E\psi \tag{9.2}$$

If we introduce a dimensionless coordinate

$$q = \left(\frac{mk}{\hbar^2}\right)^{1/4} x \tag{9.3}$$

and a dimensionless energy

$$\epsilon = \frac{E}{\hbar\sqrt{(k/m)}} = \frac{E}{\hbar\omega} \tag{9.4}$$

this equation takes the simpler form

$$-\frac{1}{2}\frac{d^2\psi}{dq^2} + \frac{1}{2}q^2\psi = \epsilon\psi \tag{9.5}$$

Putting

$$p = -i\frac{\partial}{\partial q} \tag{9.6}$$

it may be written

$$\tfrac{1}{2}(p^2 + q^2)\psi = \epsilon\psi \tag{9.7}$$

The commutation relation of p and q is

$$[p, q] = -i \tag{9.8}$$

It is, therefore, as though we had chosen units for which $\hbar = 1$.

The operator on the left-hand side of (9.7) may be written in the following two ways:

$$\tfrac{1}{2}(p^2 + q^2) = \tfrac{1}{2}(q + ip)(q - ip) - \tfrac{1}{2} \tag{9.9}$$

or

$$\tfrac{1}{2}(p^2 + q^2) = \tfrac{1}{2}(q - ip)(q + ip) + \tfrac{1}{2} \tag{9.10}$$

The eigenvalue equation for ϵ then has the alternative forms

$$\tfrac{1}{2}(q + ip)(q - ip)\psi = (\epsilon + \tfrac{1}{2})\psi \tag{9.11}$$

or

$$\tfrac{1}{2}(q - ip)(q + ip)\psi = (\epsilon - \tfrac{1}{2})\psi \tag{9.12}$$

Let us now operate on (9.11) with $q - ip$, to obtain

$$\tfrac{1}{2}(q - ip)(q + ip)[(q - ip)\psi] = (\epsilon + \tfrac{1}{2})[(q - ip)\psi] \tag{9.13}$$

Comparing this with (9.12), we see that if ψ is an eigenfunction with eigenvalue ϵ, $(q - ip)\psi$ is an eigenfunction with eigenvalue $(\epsilon + 1)$. We write, for brevity

$$\frac{1}{\sqrt{2}}(q - ip) = a\dagger \tag{9.14}$$

Thus, $a\dagger$ is an operator which produces, from one eigenfunction, another for which the eigenvalue is greater by 1. We say, then, that $a\dagger$ has produced one quantum; we may call it a "creation operator."

Let us return now to Eq. (9.12) and operate on it with $(q + ip)$ to obtain

$$(q + ip)(q - ip)[(q + ip)\psi] = (\epsilon - \tfrac{1}{2})[(q + ip)\psi] \tag{9.15}$$

Comparing this with (9.11), we see that if ψ is an eigenfunction corresponding to the eigenvalue ϵ, $(q + ip)\psi$ is an eigenfunction corresponding to the eigenvalue $(\epsilon - 1)$. Thus,

$$a = \frac{1}{\sqrt{2}}(q + ip) \tag{9.16}$$

lowers the quantum number by 1, and is designated an "annihilation operator" or "destruction operator."

Note that the commutator

$$[a, a\dagger] = \tfrac{1}{2}[q + ip, q - ip] = 1 \tag{9.17}$$

The Hamiltonian may be written either

$$H = aa\dagger - \tfrac{1}{2} \qquad (9.18)$$

or

$$H = a\dagger a + \tfrac{1}{2} \qquad (9.19)$$

or, finally

$$H = \tfrac{1}{2}(aa\dagger + a\dagger a) \qquad (9.20)$$

All the above could, of course, have been done in terms of the bra and ket vectors, without using the position representation. If the original coordinate is x and its conjugate momentum p_x, we make the transformation (9.3) and put

$$p = \left(\frac{1}{\hbar^2 mk}\right)^{1/4} p_x \qquad (9.21)$$

and

$$H = \hbar \sqrt{\frac{k}{m}} H_0 \qquad (9.22)$$

Then

$$H_0 = \tfrac{1}{2}(p^2 + q^2) \qquad (9.23)$$

and we may follow through the whole derivation starting from (9.9), using the vector $|\psi\rangle$ in place of the amplitude $\psi(q) = \langle q|\psi\rangle$ in the coordinate representation.

We have seen that by means of the operator a we can in general create from an eigenstate $|\psi\rangle$ with eigenvalue ϵ another eigenstate $a|\psi\rangle$ with eigenvalue $\epsilon - 1$. However, it is easy to show that this reduction of the eigenvalue cannot continue indefinitely, but that there is a state of minimum (positive) eigenvalue. For consider the magnitude squared of $a|\psi\rangle$, namely,

$$\langle\psi|a\dagger a|\psi\rangle$$

where $|\psi\rangle$ is the eigenstate of H_0 with eigenvalue ϵ. Using (9.19)

$$\langle\psi|a\dagger a|\psi\rangle = (\epsilon - \tfrac{1}{2})\langle\psi|\psi\rangle \geqslant 0 \qquad (9.24)$$

Thus,

$$\epsilon \geqslant \tfrac{1}{2} \qquad (9.25)$$

But if $|\psi\rangle$ is the state with the *lowest* eigenvalue, then $a|\psi\rangle$ must not be a nonzero vector. Consequently, the lowest eigenvalue must be

$$\epsilon_0 = \tfrac{1}{2}$$

and the other eigenstates obtained by operating successively with the "creation" operator $a^\dagger$ have eigenvalues

$$\epsilon_n = n + \tfrac{1}{2} \tag{9.26}$$

The corresponding eigenvectors are designated $|\psi_n\rangle$. If we choose these eigenvectors to be normalized, and put

$$|\psi_{n-1}\rangle = A_n a |\psi_n\rangle \tag{9.27}$$

we see from (9.24) that

$$1 = |A_n|^2 n \tag{9.28}$$

We may take A_n to be real; then

$$a|\psi_n\rangle = \sqrt{n}\,|\psi_{n-1}\rangle \tag{9.29}$$

Multiplying by $a^\dagger$

$$a^\dagger a |\psi_n\rangle = n|\psi_n\rangle = \sqrt{n}\, a^\dagger |\psi_{n-1}\rangle$$

so that

$$a^\dagger |\psi_{n-1}\rangle = \sqrt{n}\,|\psi_n\rangle \tag{9.30}$$

From (9.29) and (9.30) we can write the matrix elements of q and p:

$$\langle \psi_m | q | \psi_n \rangle = \frac{1}{\sqrt{2}} \langle \psi_m | a + a^\dagger | \psi_n \rangle$$

$$= \sqrt{\frac{n}{2}}\, \delta_{m,n-1} + \sqrt{\frac{n+1}{2}}\, \delta_{m,n+1} \tag{9.31}$$

and

$$\langle \psi_m | p | \psi_n \rangle = \frac{1}{i\sqrt{2}} \langle \psi_m | a - a^\dagger | \psi_n \rangle$$

$$= -i \sqrt{\frac{n}{2}}\, \delta_{m,n-1} + i \sqrt{\frac{n+1}{2}}\, \delta_{m,n+1} \tag{9.32}$$

Problem 9-1: Use the fact that $q = (1/\sqrt{2})(a + a^\dagger)$ to show that the diagonal matrix elements of even powers of x in the ground state are

$$\langle 0 | x^{2n} | 0 \rangle = \frac{(2n)!}{2^{2n} n!} \left(\frac{\hbar^2}{mk} \right)^{n/2}$$

Show also that

$$\langle s | x^4 | s \rangle = \tfrac{3}{4}(2s^2 + 2s + 1) \frac{\hbar^2}{mk}, \qquad s \geq 1$$

Let us rewrite the energy and the position matrix elements in terms of the original coordinates. The energy levels are

$$E_n = (n+\tfrac{1}{2})\hbar \sqrt{\frac{k}{m}} = (n+\tfrac{1}{2})\hbar\omega \qquad (9.33)$$

Since

$$q = \left(\frac{mk}{\hbar^2}\right)^{1/4} x = \left(\frac{m\omega}{\hbar}\right)^{1/2} x \qquad (9.34)$$

the matrix elements of x are

$$\langle \psi_m | x | \psi_n \rangle = \sqrt{\frac{\hbar}{2m\omega}} \{ \sqrt{n}\,\delta_{m,n-1} + \sqrt{n+1}\,\delta_{m,n+1} \} \qquad (9.35)$$

while those of p_x are

$$\langle \psi_m | p_x | \psi_n \rangle = \frac{i}{\sqrt{2m\hbar\omega}} \{ -\sqrt{n}\,\delta_{m,n-1} + \sqrt{n+1}\,\delta_{m,n+1} \} \qquad (9.36)$$

3. Schrödinger Wave Functions for the Harmonic Oscillator

The problem of the wave functions of the one-dimensional harmonic oscillator may be approached in several ways. If we are interested in generating the functions of low quantum numbers, the simplest method is to solve first the equation for the lowest-energy state

$$a|\psi_0\rangle = 0 \qquad (9.37)$$

in the Schrödinger form

$$\left(q + \frac{\partial}{\partial q}\right)\psi_0 = 0 \qquad (9.38)$$

which has the normalized solution

$$\psi_0 = \left(\frac{1}{\pi}\right)^{1/4} e^{-q^2/2} \qquad (9.39)$$

Then, successive application of the operator

$$a\dagger = \frac{1}{\sqrt{2}}(q - ip) = \frac{1}{\sqrt{2}}\left(q - \frac{\partial}{\partial q}\right) \qquad (9.40)$$

gives the higher eigenfunctions.

Problem 9-2:   Obtain the formulas for the normalized oscillator function up to $n = 4$:

$$\psi_1 = \left(\frac{4}{\pi}\right)^{1/4} q e^{-q^2/2}$$

$$\psi_2 = \left(\frac{1}{4\pi}\right)^{1/4}(2q^2-1)e^{-q^2/2}$$

$$\psi_3 = \left(\frac{1}{9\pi}\right)^{1/4}(2q^3-3q)e^{-q^2/2}$$

$$\psi_4 = \left(\frac{1}{576\pi}\right)^{1/4}(4q^4-12q^2+3)e^{-q^2/2}$$

Problem 9-3: Find the momentum eigenfunctions of the one-dimensional harmonic oscillator. [Rather than trying to do it by Fourier transformation, calculate them from first principles.]

A more direct approach relates the wave functions to Hermite polynomials and confluent hypergeometric functions. Let us substitute

$$\psi = e^{-q^2/2}u \tag{9.41}$$

in the Schrödinger equation (9.5)

$$\frac{1}{2}\frac{d^2\psi}{dq^2} + (\epsilon - \tfrac{1}{2}q^2)\psi = 0 \tag{9.42}$$

We obtain

$$u'' - 2qu' + 2nu = 0 \tag{9.43}$$

where the prime indicates differentiation with respect to q, and we have substituted the known values of ϵ. This is the Hermite equation (4.303) for variable $\sqrt{2}q$, so that

$$\psi_n = C_n e^{-q^2/2}H_n(\sqrt{2}q) \tag{9.44}$$

These may be written in terms of confluent hypergeometric functions by means of Eqs. (4.318) and (4.319).

There is a simple way to find the normalization factors C_n, by using the generating function for Hermite polynomials given by Eq. (4.309)

$$e^{\sqrt{2}qs}e^{-(1/2)s^2} = \sum_{n=0}^{\infty} H_n(\sqrt{2}q)\frac{s^n}{n!} \tag{9.45}$$

Putting

$$\sqrt{2}s = \sigma$$

and multiplying by $e^{-q^2/2}$ this becomes

$$e^{-q^2/2}e^{\sigma q}e^{-\sigma^2/4} = \sum_{n=0}^{\infty}\frac{\psi_n}{C_n}\frac{\sigma^n}{2^{n/2}n!} \tag{9.46}$$

Similarly we may write

$$e^{-q^2/2}e^{\tau q}e^{-\tau^2/4} = \sum_{m=0}^{\infty} \frac{\psi_m}{C_m} \frac{\tau^m}{2^{m/2}m!} \tag{9.47}$$

Let us now multiply together (9.46) and (9.47), and integrate the result with respect to q from $-\infty$ to ∞. On the left-hand side we get

$$e^{-(\sigma^2+\tau^2)/4} \int_{-\infty}^{\infty} e^{-q^2} e^{q(\sigma+\tau)} \, dq = e^{\sigma\tau/2} \int_{-\infty}^{\infty} e^{-[q-(\sigma+\tau)/2]^2} \, dq$$

$$= \sqrt{\pi} e^{\sigma\tau/2}$$

Therefore,

$$\sqrt{\pi}e^{\sigma\tau/2} = \sum_{n,m=0}^{\infty} \frac{\int \psi_n\psi_m \, dq}{C_nC_m} \frac{1}{2^{(m+n)/2}} \frac{\sigma^n\tau^m}{n!m!} \tag{9.48}$$

It follows that all the terms on the right with $m \neq n$ are zero (verifying the known orthogonality of the oscillator functions); on the other hand, equating on the two sides the coefficients of $(\sigma\tau)^n$ yields

$$\frac{\sqrt{\pi}}{2^n n!} = \int_{-\infty}^{\infty} \psi_n^2 \, dq \frac{1}{C_n^2 2^n (n!)^2} \tag{9.49}$$

and therefore

$$C_n = \left[\frac{1}{n!\sqrt{\pi}}\right]^{1/2} \tag{9.50}$$

Thus, the normalized oscillator functions are

$$\psi_n = \left[\frac{1}{n!\sqrt{\pi}}\right]^{1/2} e^{-q^2/2} H_n(\sqrt{2}q) \tag{9.51}$$

4. Elements of the Theory of Angular Momentum

The Poisson brackets of arbitrary components L_i, L_j of the classical angular momentum operator may be calculated to be

$$\{L_i, L_j\} = \epsilon_{ijk}L_k \tag{9.52}$$

where ϵ_{ijk} is the permutation symbol. Using the rule for relating the commutators of quantum operators to Poisson brackets of the corresponding classical variables, we are led to the commutation relations

$$[L_i, L_j] = i\hbar\epsilon_{ijk}L_k \tag{9.53}$$

Written out in full, these relations are

$$[L_x, L_y] = i\hbar L_z \tag{9.54}$$

$$[L_y, L_z] = i\hbar L_x \tag{9.55}$$

and

$$[L_z, L_x] = i\hbar L_y \tag{9.56}$$

As we will see, the eigenstates and eigenvalues of angular momentum may be determined from these commutation relations alone. However, the states determined cannot all be associated with orbital angular momentum about the origin of coordinates, but rather require the introduction into the physics of the problem of an intrinsic angular momentum or "spin." It is convenient to reserve the symbol **L** for orbital angular momentum, and to use **J** to denote the general operator satisfying the commutation relations (9.54) to (9.56). In using these commutation rules alone, we derive a theory of all angular-momentum-like quantities; whether the states deduced exist in physics is a matter of observation rather than of theory.

Let us then, proceed to the determination of angular momentum states.

We note first, of course, that since the components of angular momentum do not commute with each other, simultaneous eigenstates do not in general exist for any two components.

On the other hand, each component commutes with $\mathbf{J}^2 = J_x^2 + J_y^2 + J_z^2$. For

$$[J^2, J_x] = [J_x^2 + J_y^2 + J_z^2, J_x]$$

$$= [J_y^2, J_x] + [J_z^2, J_x]$$

But in general

$$[A^2, B] = A^2 B - B A^2$$

$$= A(AB - BA) + (AB - BA)A$$

$$= A[A, B] + [A, B]A \tag{9.57}$$

Therefore,

$$[J^2, J_x] = J_y(-i\hbar J_z) - i\hbar J_z J_y + J_z(i\hbar J_y) + i\hbar J_y J_z = 0 \tag{9.58}$$

It follows similarly that J_y and J_z commute with $\mathbf{J}^2$.

We arbitrarily choose to work with eigenstates of J^2 and J_z; we designate the eigenvalues J^2 as $j(j+1)\hbar^2$ and those of J as $m\hbar$. No a priori assumptions are made about the numbers j and m. The simultaneous eigenstates are designated $|jm\rangle$.

In a manner suggestive of the calculation of the energy levels of the harmonic oscillator, we may introduce operators which raise or lower the quantum number m for a fixed j. Let us introduce

$$J_{\pm} = J_x \pm iJ_y \tag{9.59}$$

The commutation rules between J_+, J_- and J_z are easily verified to be

$$[J_{\pm}, J_z] = \mp \hbar J_{\pm} \tag{9.60}$$

and

$$[J_+, J_-] = 2\hbar J_z \tag{9.61}$$

J^2 may also be written quite easily in terms of J_+, J_-, and J_z:

$$J^2 = J_+ J_- + J_z^2 - \hbar J_z \tag{9.62}$$

or

$$J^2 = J_- J_+ + J_z^2 + \hbar J_z \tag{9.63}$$

Consider now the equation, which follows from (9.60),

$$[J_+, J_z]|jm\rangle = -\hbar J_+|jm\rangle$$

which, since $|jm\rangle$ is an eigenstate of J_z, becomes

$$m\hbar(J_+|jm\rangle) - J_z(J_+|jm\rangle) = -\hbar(J_+|jm\rangle)$$

or

$$J_z(J_+|jm\rangle) = (m+1)\hbar(J_+|jm\rangle) \tag{9.64}$$

Thus the operator J_+ has *raised* the value of m by 1 without affecting J^2. In fact, from the relation

$$[J^2, J_+] = 0 \tag{9.65}$$

operating on $|jm\rangle$ we see that

$$J^2(J_+|jm\rangle) = j(j+1)\hbar^2(J_+|jm\rangle) \tag{9.66}$$

From (9.64) we may write

$$J_+|jm\rangle = A_{jm}|j, m+1\rangle \tag{9.67}$$

where A_{jm} is a constant which may be taken to be real. Assuming the states $|jm\rangle$ to be normalized, A_{jm} may be calculated by taking the magnitude squared of both sides of (9.67). For this purpose we note that the conjugate of $J_+|\psi\rangle = \langle\psi|J_-$. Therefore,

$$\langle jm|J_- J_+|jm\rangle = A_{jm}^2 \tag{9.68}$$

From (9.63) this may be written

$$\langle jm|J^2 - J_z^2 - \hbar J_z|jm\rangle = A_{jm}^2$$

or

$$A_{jm}^2 = [j(j+1) - m(m+1)]\hbar^2 \tag{9.69}$$

Thus

$$J_+|jm\rangle = \hbar\sqrt{j(j+1) - m(m+1)}\,|j, m+1\rangle \tag{9.70}$$

We may proceed in a precisely similar manner starting from the commutation relation between J_- and J_z, to show that

(a) $\quad J_-|jm\rangle = B_{jm}|j, m-1\rangle \tag{9.71}$

(b) $\quad B_{jm} = \hbar\sqrt{j(j+1) - m(m-1)} \tag{9.72}$

Suppose, then, that we have an arbitrary state $|jm\rangle$. By continued application of the operator J_+, we may obtain states with m increasing by steps of 1, j remaining fixed. However, (9.69) implies that m may never exceed j. The contradiction can only be resolved if at some point $A_{jm} = 0$, or $m = j$; the chain is then broken.

But now let us operate successively on this state of maximum m, viz., $|jj\rangle$, with the operator J_-. m may then be continually decreased by 1, giving values $j-1, j-2, \ldots$. This time, it would seem that m could be *reduced* indefinitely. But that is also not true, since from (9.72) m cannot be less than $-j$. For if it were, B_{jm} would be imaginary, whereas we assume it originally to be real. This time, then, the only way to break the chain at the lower end is to assume that the *minimum* value of m is $-j$, so that no further states may be produced from $|j, -j\rangle$.

The situation, then, is this: that starting from $m = j$, and reducing it by steps of 1, we reach a state $m = -j$. It follows that j must be half an integer, or

$$j = 0, \tfrac{1}{2}, 1, \tfrac{3}{2}, 2, \tfrac{5}{2}, \ldots \tag{9.73}$$

while, for a *given* j, m runs by intervals of unity from $-j$ to j.

Starting from the state $|jj\rangle$ ("stretched state") successive application of the J_- operator produces all the other states $|jm\rangle$.

For j equal to values l it is straightforward to determine Schrödinger wave functions for the states.

The Schrödinger operator for L_z is

$$L_z = -i\hbar\left(x\frac{\partial}{\partial y} - y\frac{\partial}{\partial x}\right) \tag{9.74}$$

Using the relations between Cartesian and polar coordinates

$$x = r\sin\theta\cos\phi, \qquad y = r\sin\theta\sin\phi, \qquad z = r\cos\theta \tag{9.75}$$

we see that

$$\frac{\partial}{\partial \phi} = -r \sin \theta \sin \phi \frac{\partial}{\partial x} + r \sin \theta \cos \phi \frac{\partial}{\partial y}$$

$$= x \frac{\partial}{\partial y} - y \frac{\partial}{\partial x}$$

so that

$$L_z = -i\hbar \frac{\partial}{\partial \phi} \qquad (9.76)$$

But now

$$L_x + iL_y = \hbar \left[z \left(\frac{\partial}{\partial x} + i \frac{\partial}{\partial y} \right) - (x + iy) \frac{\partial}{\partial z} \right] \qquad (9.77)$$

and

$$L_x - iL_y = \hbar \left[(x - iy) \frac{\partial}{\partial z} - z \left(\frac{\partial}{\partial x} - i \frac{\partial}{\partial y} \right) \right] \qquad (9.78)$$

Putting

$$\xi = x + iy = r \sin \theta \, e^{i\phi}, \qquad \xi^* = x - iy = r \sin \theta \, e^{-i\phi} \qquad (9.79)$$

$$\frac{\partial}{\partial x} + i \frac{\partial}{\partial y} = 2 \frac{\partial}{\partial \xi^*}, \qquad \frac{\partial}{\partial x} - i \frac{\partial}{\partial y} = 2 \frac{\partial}{\partial \xi} \qquad (9.80)$$

Therefore,

$$L_x + iL_y = \hbar \left[2z \frac{\partial}{\partial \xi^*} - \xi \frac{\partial}{\partial z} \right] \qquad (9.81)$$

$$L_x - iL_y = \hbar \left[\xi^* \frac{\partial}{\partial z} - 2z \frac{\partial}{\partial \xi} \right] \qquad (9.82)$$

We may now write $\partial/\partial\theta$ and $\partial/\partial\phi$ in terms of ξ, ξ^*, and z:

$$\frac{\partial}{\partial \theta} = r \cos \theta \, e^{i\phi} \frac{\partial}{\partial \xi} + r \cos \theta \, e^{-i\phi} \frac{\partial}{\partial \xi^*} - r \sin \theta \frac{\partial}{\partial z} \qquad (9.83)$$

$$-i \frac{\partial}{\partial \phi} = r \sin \theta \, e^{i\phi} \frac{\partial}{\partial \xi} - r \sin \theta \, e^{-i\phi} \frac{\partial}{\partial \xi^*} \qquad (9.84)$$

If the second of these equations is multiplied by $\cot \theta$ and added to the first, and the result is multiplied by $e^{-i\phi}$, it is found that

$$e^{-i\phi} \left(\frac{\partial}{\partial \theta} - i \cot \theta \frac{\partial}{\partial \phi} \right) = 2z \frac{\partial}{\partial \xi} - \xi^* \frac{\partial}{\partial z} \qquad (9.85)$$

It follows from (9.82) that

$$L_x - iL_y = -\hbar \, e^{-i\phi} \left(\frac{\partial}{\partial \theta} - i \cot \theta \frac{\partial}{\partial \phi} \right) \qquad (9.86)$$

If, on the other hand, after (9.83) is multiplied by $\cot \theta$ (9.84) is *subtracted* from it and the result is multiplied by $e^{i\phi}$, we obtain instead

$$e^{i\phi}\left(\frac{\partial}{\partial \theta} + i \cot \theta \frac{\partial}{\partial \phi}\right) = 2z\frac{\partial}{\partial \xi^*} - \xi\frac{\partial}{\partial z} \tag{9.87}$$

It follows from (9.81) that

$$L_x + iL_y = \hbar\, e^{i\phi}\left(\frac{\partial}{\partial \theta} + i \cot \theta \frac{\partial}{\partial \phi}\right) \tag{9.88}$$

From (9.62) it is possible to calculate the operator for L^2:

$$L^2 = L_+L_- + L_z^2 - \hbar L_z$$

$$= -\hbar^2\, e^{i\phi}\left(\frac{\partial}{\partial \theta} + i \cot \theta \frac{\partial}{\partial \phi}\right)e^{-i\phi}\left(\frac{\partial}{\partial \theta} - i \cot \theta \frac{\partial}{\partial \phi}\right)$$

$$-\hbar^2\frac{\partial^2}{\partial \phi^2} - i\hbar^2\frac{\partial}{\partial \phi}$$

Since there are simultaneous eigenstates of L^2 and L_z, these states may be written

$$\psi_{lm} = S_{lm}(\theta)\, e^{im\phi} \tag{9.89}$$

Substituting this into

$$L^2\psi_{lm} = l(l+1)\hbar^2\psi_{lm}$$

we obtain for S_{lm} the equation

$$-\left[\frac{d}{d\theta} - (m-1)\cot \theta\right]\left[\frac{d}{d\theta} + m\cot \theta\right]S_{lm} = [l(l+1) - m(m-1)]\,S_{lm}$$

Referring back to Eqs. (3.42), (3.43), and (3.44), we see that the wave functions

$$\psi_{lm} = S_{lm}\, e^{im\phi} \tag{9.90}$$

are spherical harmonics; their properties were very thoroughly investigated in Chapter 3. In the notation of that chapter

$$\psi_{lm} = \Theta_{lm}(\theta)\, e^{im\phi} = Y_{lm}(\theta, \phi) \tag{9.91}$$

5. Matrix Representation of Angular Momentum Operators

Let us calculate the matrix elements for the simplest cases: $j = \frac{1}{2}$ and $j = 1$. In the first case we put $\mathbf{J} = \mathbf{s}$, designating it as the spin operator. Describing the state $|\frac{1}{2}, \frac{1}{2}\rangle$ by the column matrix $\binom{1}{0}$ and $|\frac{1}{2}, -\frac{1}{2}\rangle$ by $\binom{0}{1}$, the matrices for s_z, s_+, and s_- become

$$s_z = \begin{pmatrix} 1 & 0 \\ 0 & -1 \end{pmatrix} \frac{\hbar}{2} = \frac{1}{2} \hbar \sigma_z \tag{9.92}$$

$$s_+ = \begin{pmatrix} 0 & 1 \\ 0 & 0 \end{pmatrix} \hbar = \frac{1}{2} \hbar \sigma_+ \tag{9.93}$$

$$s_- = \begin{pmatrix} 0 & 0 \\ 1 & 0 \end{pmatrix} \hbar = \frac{1}{2} \hbar \sigma_- \tag{9.94}$$

From the latter two it follows that

$$s_x = \begin{pmatrix} 0 & 1 \\ 1 & 0 \end{pmatrix} \frac{\hbar}{2} = \frac{1}{2} \hbar \sigma_x \tag{9.95}$$

and

$$s_y = \begin{pmatrix} 0 & -i \\ i & 0 \end{pmatrix} \frac{\hbar}{2} = \frac{1}{2} \hbar \sigma_y \tag{9.96}$$

The matrices for the case $j = 1$ are constructed from matrix elements as follows:

Problem 9-4: The dipole–dipole interaction energy of two particles with magnetic moments $\mu_1 = g_1 \beta \sigma_1$ and $\mu_2 = g_2 \beta \sigma_2$ is

$$V = \frac{1}{r^3} \left\{ \mu_1 \cdot \mu_2 - \frac{3(\mu_1 \cdot \mathbf{r})(\mu_2 \cdot \mathbf{r})}{r^2} \right\}$$

Find the eigenvalues of this operator when the individual spins are $\frac{1}{2}$ and 1, respectively. β is the Bohr magneton $e\hbar/2mc$ and the g's are the "g factors" of the particles.

Problem 9-5: Two "localized" spins σ_1 and σ_2 interact through the "exchange energy" $J\sigma_1 \cdot \sigma_2$, and are placed in an external magnetic field $\mathbf{H}_0$. The g factors of the spins are g_1 and g_2. Determine the eigenstates of the system if each spin has quantum number $\frac{1}{2}$.

Show that, at zero field, the state of lowest energy is the singlet state (antiparallel spins). At what field does the state of aligned spins become the lowest state?

Problem 9-6: Three spins ($s = \frac{1}{2}$) interact with the Hamiltonian

$$H = J(\sigma_1 \cdot \sigma_2 + \sigma_2 \cdot \sigma_3 + \sigma_3 \cdot \sigma_1)$$

Determine the eigenstates of the system and their degeneracies. [Consider the operator $s^2 = (\sigma_1 + \sigma_2 + \sigma_3)^2$.]

$$\langle 1|J_z|1\rangle = \hbar$$

$$\langle 0|J_z|0\rangle = 0$$

$$\langle -1|J_z|-1\rangle = -\hbar$$

where $|m\rangle$ is a shorthand for $|jm\rangle$. For J_+ we have as the only nonzero matrix elements

$$\langle 1|J_+|0\rangle = \sqrt{2}\hbar$$

$$\langle 0|J_+|-1\rangle = \sqrt{2}\hbar$$

Finally for J_- the only nonzero elements are

$$\langle 0|J_-|1\rangle = \sqrt{2}\hbar$$

and

$$\langle -1|J_-|0\rangle = \sqrt{2}\hbar$$

The matrix operators are then easily constructed as follows:

$$J_z = \begin{pmatrix} 1 & 0 & 0 \\ 0 & 0 & 0 \\ 0 & 0 & -1 \end{pmatrix} \hbar$$

$$J_+ = \begin{pmatrix} 0 & 1 & 0 \\ 0 & 0 & 1 \\ 0 & 0 & 0 \end{pmatrix} \hbar\sqrt{2}$$

$$J_- = \begin{pmatrix} 0 & 0 & 0 \\ 1 & 0 & 0 \\ 0 & 1 & 0 \end{pmatrix} \hbar\sqrt{2}$$

from which we obtain

$$J_x = \begin{pmatrix} 0 & 1 & 0 \\ 1 & 0 & 1 \\ 0 & 1 & 0 \end{pmatrix} \frac{\hbar}{\sqrt{2}} \tag{9.97}$$

and

$$J_y = \begin{pmatrix} 0 & -i & 0 \\ i & 0 & -i \\ 0 & i & 0 \end{pmatrix} \frac{\hbar}{\sqrt{2}} \tag{9.98}$$

6. Schwinger Formulation

Schwinger has given another representation of angular momentum operators which is very versatile. It starts from creation and annihilation operators $a\dagger$ and a of the sort introduced in (9.14) and (9.16) for the harmonic oscillator problem; these satisfy the commutation rule

$$[a, a\dagger] = aa\dagger - a\dagger a = 1$$

as already given in (9.17).

If we consider *two* oscillators, we may associate such operators with each; we call them a_1, $a_1\dagger$, a_2 and $a_2\dagger$. The operators a_1, $a_1\dagger$ each commute with a_2 and $a_2\dagger$. Let us form the row matrix

$$(a_1\dagger a_2\dagger) = \mathbf{a}\dagger$$

and the column matrix

$$\begin{pmatrix} a_1 \\ a_2 \end{pmatrix} = \mathbf{a}$$

which we call "spinor operators."

Let us now construct the operators

$$\mathscr{J}_x = \mathbf{a}\dagger S_x \mathbf{a}, \qquad \mathscr{J}_y = \mathbf{a}\dagger S_y \mathbf{a}, \qquad \mathscr{J}_z = \mathbf{a}\dagger S_z \mathbf{a} \qquad (9.99)$$

The operators for $\mathscr{J}_z$, $\mathscr{J}_+$, and $\mathscr{J}_-$ are seen immediately to be

$$\mathscr{J}_z = \tfrac{1}{2}\hbar(a_1\dagger a_1 - a_2\dagger a_2)$$
$$= \tfrac{1}{2}\hbar(n_1 - n_2) \qquad (9.100)$$

where we have put $n = a\dagger a$,

$$\mathscr{J}_+ = \hbar a_1\dagger a_2 \qquad (9.101)$$
$$\mathscr{J}_- = \hbar a_2\dagger a_1 \qquad (9.102)$$

Calculating the commutation rules, we find that

$$\mathscr{J}_z\mathscr{J}_+ - \mathscr{J}_+\mathscr{J}_z = \tfrac{1}{2}\hbar^2[(n_1 - n_2)a_1\dagger a_2 - a_1\dagger a_2(n_1 - n_2)]$$
$$= \hbar^2 a_1\dagger a_2$$
$$= \hbar\mathscr{J}_+ \qquad (9.103)$$

since

$$n_2 a_2 = a_2 n_2 - a_2 \qquad (9.104)$$

and

$$n_1 a_1\dagger = a_1\dagger n_1 + a_1\dagger \qquad (9.105)$$

Similarly,

$$\mathscr{J}_z\mathscr{J}_- - \mathscr{J}_-\mathscr{J}_z = \tfrac{1}{2}\hbar^2[(n_1 - n_2)a_2\dagger a_1 - a_2\dagger a_1(n_1 - n_2)]$$
$$= -\hbar^2 a_2\dagger a_1$$
$$= -\hbar\mathscr{J}_- \qquad (9.106)$$

Finally,

$$\mathscr{J}_+\mathscr{J}_- - \mathscr{J}_-\mathscr{J}_+ = \hbar^2[a_1{}^\dagger a_2 a_2{}^\dagger a_1 - a_2{}^\dagger a_1 a_1{}^\dagger a_2]$$

$$= \hbar^2[(1+n_2)n_1 - n_2(1+n_1)]$$

$$= (n_1 - n_2)\hbar^2$$

$$= 2\hbar\mathscr{J}_z \qquad (9.107)$$

These commutation relations are exactly those of (9.60) and (9.61); the $\mathscr{J}$'s are then representations of angular momentum operators and may be used in the development of angular momentum theory. We henceforth do not distinguish the $\mathscr{J}$'s and the angular momentum operators **J**.

Since the eigenstates are determined exclusively by the commutation relations and are designated by the quantum numbers j and m, the eigenvalues are exactly as already shown; that is, those of $\mathbf{J}^2$ are $j(j+1)\hbar^2$ and those of J_z are $m\hbar$, where $|m| \leq j$.

From (9.107) we see that the quantum number

$$m = \tfrac{1}{2}(n_1 - n_2) \qquad (9.108)$$

Following (9.48)

$$\mathbf{J}^2 = \hbar^2 a_2{}^\dagger a_1 a_1{}^\dagger a_2 + \tfrac{1}{4}\hbar^2(n_1 - n_2)^2 + \tfrac{1}{2}\hbar^2(n_1 - n_2)$$

$$= \hbar^2[n_2(1+n_1) + \tfrac{1}{4}n_1{}^2 - \tfrac{1}{2}n_1 n_2 + \tfrac{1}{4}n_2{}^2 + \tfrac{1}{2}n_1 - \tfrac{1}{2}n_2]$$

$$= \hbar^2[\tfrac{1}{4}(n_1 + n_2)^2 + \tfrac{1}{2}(n_1 + n_2)]$$

which is equal to $j(j+1)\hbar^2$ where

$$j = \tfrac{1}{2}(n_1 + n_2) \qquad (9.109)$$

From (9.93) and (9.94) we see that $a_1{}^\dagger$ increases and a_1 decreases both m and j by $\tfrac{1}{2}$; on the other hand $a_2{}^\dagger$ increases j by $\tfrac{1}{2}$ while decreasing m by $\tfrac{1}{2}$, while a_2 increases m and decreases j by the same amount.

With these prescriptions, we can generate all angular momentum states from the ground state of the two operators, which we designate $|0\rangle$. Recalling that

$$a^\dagger|n\rangle = \sqrt{n+1}\,|n+1\rangle$$

and

$$a|n\rangle = \sqrt{n}\,|n-1\rangle$$

we see that

$$|\tfrac{1}{2}\tfrac{1}{2}\rangle = a_1{}^\dagger|0\rangle \qquad (9.110)$$

and

$$|\tfrac{1}{2}, -\tfrac{1}{2}\rangle = a_2{}^\dagger|0\rangle \qquad (9.111)$$

Similarly,

$$|1,-1\rangle = \tfrac{1}{2}(a_2{}^\dagger)^2|0\rangle \qquad (9.112)$$

$$|1,0\rangle = a_1{}^\dagger a_2{}^\dagger|0\rangle \qquad (9.113)$$

$$|1,+1\rangle = \tfrac{1}{2}(a_2{}^\dagger)^2|0\rangle \qquad (9.114)$$

By the method of induction it is easily seen that

$$|jm\rangle = \frac{(a_1{}^\dagger)^{j+m}(a_2{}^\dagger)^{j-m}}{\sqrt{(j+m)!(j-m)!}}|0\rangle \qquad (9.115)$$

Problem 9-7: Prove (9.115). [HINT: Consider the effect of each of the operators $a_1{}^\dagger$, a_1, $a_2{}^\dagger$, a_2.]

---

7. Addition of Angular Momenta: Clebsch–Gordan Coefficients

In this section we consider the problem of the addition of two independent angular momenta, $\mathbf{J}_1$ and $\mathbf{J}_2$. We designate $|j_1 m_1\rangle$ as eigenstates of $\mathbf{J}_1{}^2$ and J_{1z}, $|j_2 m_2\rangle$ as those of $\mathbf{J}_2{}^2$ and J_{2z}. The hypothesis of independence means that all components of $\mathbf{J}_1$ commute with all components of $\mathbf{J}_2$.

Direct products of the eigenstates of the two angular momenta constitute a complete set of eigenstates of the system, and may be designated $|j_1 j_2 m_1 m_2\rangle$.

However, the components of

$$\mathbf{J} = \mathbf{J}_1 + \mathbf{J}_2 \qquad (9.116)$$

satisfy the general angular momentum commutation rules, so that there exist common eigenstates of $\mathbf{J}^2$ and J_z. We designate these as $|jm\rangle$ where

$$\mathbf{J}^2|jm\rangle = j(j+1)\hbar^2|jm\rangle \qquad (9.117)$$

and

$$J_z|jm\rangle = m\hbar|jm\rangle \qquad (9.118)$$

Strictly speaking, we should indicate that these states are the result of adding angular momenta $\mathbf{J}_1$ and $\mathbf{J}_2$, and writing them $|j_1 j_2 jm\rangle$. We supress the j_1 and j_2 for brevity in writing; we might then just as well do the same for

$$|j_1 j_2 m_1 m_2\rangle$$

and designate them merely as $|m_1 m_2\rangle$.

In summary, $|m_1 m_2\rangle$ are eigenstates of the commuting operators $\mathbf{J}_1^2$, $\mathbf{J}_2^2$, J_{1z}, J_{2z}, while $|jm\rangle$ are eigenstates of $\mathbf{J}_1^2$, $\mathbf{J}_2^2$, $(\mathbf{J}_1 + \mathbf{J}_2)^2$, and $(J_{1z} + J_{2z})$, all pairs of which also commute with each other.

Since each set of states constitutes a complete set or basis for the states of the two angular momenta, each vector of one set may be expressed as a linear combination of those of the other.

Thus we may write

$$|j_1 j_2 jm\rangle = \sum_{m_1 + m_2 = m} |j_1 j_2 m_1 m_2\rangle \langle j_1 j_2 m_1 m_2 | j_1 j_2 jm\rangle \tag{9.119}$$

The coefficients $\langle j_1 j_2 m_1 m_2 | j_1 j_2 jm\rangle$ are known as the "Clebsch–Gordan" coefficients.

The relations may be expressed the other way about, viz,

$$|j_1 j_2 m_1 m_2\rangle = \sum_{j = m_1 + m_2}^{j_1 + j_2} |j_1 j_2 jm\rangle \langle j_1 j_2 jm | j_1 j_2 m_1 m_2\rangle \tag{9.120}$$

where, in all terms on the right, $m = m_1 + m_2$. The coefficients here are the conjugates of those in (9.119); however, since all phases may be chosen $= 0$, they are actually the same quantities.

If we consider the state $|m_1 m_2\rangle = |j_1 j_2\rangle'$, it must be the same as $|jm\rangle = |j_1 + j_2, j_1 + j_2\rangle$. For in general

$$m_1 + m_2 = m \tag{9.121}$$

But j cannot be larger than $j_1 + j_2$, for if it were, m could also be. But we know that its maximum value is $j_1 + j_2$.

Suppose next we consider the states

$$|j_1, j_2 - 1\rangle', \qquad |j_1 - 1, j_2\rangle'$$

They correspond to the same

$$m = j_1 + j_2 - 1$$

We know that there is a state

$$|jm\rangle = |j_1 + j_2, j_1 + j_2 - 1\rangle$$

There must, however, be another state with the same m. It must therefore correspond to

$$j = j_1 + j_2 - 1$$

Proceeding in the same way, we see that there are $(l + 1)$ states $|m_1 m_2\rangle$ with $m_1 + m_2 = j_1 + j_2 - l$. These must be expressible in terms of states

---

[1]Because there is a risk of confusion between the vectors $|jm\rangle$ and $|m_1 m_2\rangle$, we shall distinguish the latter by a prime in what follows.

$|j, m = j_1 + j_2 - l\rangle$ where

$$j = j_1 + j_2, j_1 + j_2 - 1, \ldots, j_1 + j_2 - l \qquad (9.122)$$

This is, however, only true if l is not too large. To see what the largest value of l is, we note that the number of independent states $|m_1 m_2\rangle$ is

$$N(m_1 m_2) = (2j_1 + 1)(2j_2 + 1) \qquad (9.123)$$

The number of states $|jm\rangle$ is

$$N(jm) = \sum_{s=0}^{l} [2(j_1 + j_2 - s) + 1]$$

$$= [2(j_1 + j_2) + 1](l+1) - l(l+1) \qquad (9.124)$$

But we must have $N(jm) = N(m_1 m_2)$, since both sets span the same vector space. Thus

$$4j_1 j_2 + 2(j_1 + j_2) + 1 = 2(j_1 + j_2) + 1 + 2l(j_1 + j_2) - l^2$$

or

$$l^2 - 2l(j_1 + j_2) + 4j_1 j_2 = 0$$

or, finally

$$(l - 2j_1)(l - 2j_2) = 0 \qquad (9.125)$$

However, l cannot be the *larger* of j_1 and j_2, since this would lead to

$$j = j_< - j_>$$

where $j_<$ is the lesser and $j_>$ the larger of the j_1, j_2; however, j *must* be positive. Therefore,

$$l = 2j_<$$

and the lowest value of j appearing in (9.124) is

$$j = j_> + j_< - 2j_< = j_> - j_< = |j_1 - j_2| \qquad (9.126)$$

We have thus established the limits

$$|j_1 - j_2| \leq j \leq j_1 + j_2 \qquad (9.127)$$

this is known as the "triangular inequality."

8.   Techniques of Calculation

The case in which

$$j = j_1 + j_2$$

("stretched case") is easily handled, since we merely start from the relation

$$|j = j_1 + j_2, m = j_1 + j_2\rangle = |m_1 = j_1, m_2 = j_2\rangle'$$

or

$$|j_1 + j_2, j_1 + j_2\rangle = |j_1 j_2\rangle' \qquad (9.128)$$

and operate successively with J_-. Note that on the right, $|j_1 j_2\rangle'$ is really a direct product state; we operate on it with J_{1-}, J_{2-}, the first acting only on the j_1 state and the second on the j_2 state. Using (9.71) and (9.72) on (9.128) yields

$$\sqrt{(j_1 + j_2)(j_1 + j_2 + 1) - (j_1 + j_2)(j_1 + j_2 - 1)} \quad |j_1 + j_2, j_1 + j_2 - 1\rangle$$
$$= \sqrt{j_1(j_1 + 1) - j_1(j_1 - 1)} \quad |j_1 - 1, j_2\rangle' + \sqrt{j_2(j_2 + 1) - j_2(j_2 - 1)}$$
$$\times |j_1, j_2 - 1\rangle'$$

which may be simplified to

$$|j_1 + j_2,\ j_1 + j_2 - 1\rangle = \sqrt{\frac{j_1}{j_1 + j_2}}\,|j_1 - 1, j_2\rangle' + \sqrt{\frac{j_2}{j_1 + j_2}}\,|j_1, j_2 - 1\rangle' \quad (9.129)$$

One may continue to operate with $J_- = J_{1-} + J_{2-}$ until we reach the state $|j_1 + j_2, -(j_1 + j_2)\rangle$.

How, then, do we deal with $j = j_1 + j_2 - l$, where $l > 0$? The simplest procedure is to write for instance

$$|j_1 - l, j_2\rangle' = C_0|j_1 + j_2, j_1 + j_2 - l\rangle + C_1|j_1 + j_2 - 1, j_1 + j_2 - l\rangle$$
$$+ \cdots + C_l|j_1 + j_2 - l, j_1 + j_2 - l\rangle \qquad (9.130)$$

and operate successively on both sides with $J_+ = J_{1+} + J_{2+}$. Operating once we get an equation without C_l; the second time, C_{l-1} also drops out, etc. After l operations, we have only C_0. Working back through the equations, the second to the last permits us to determine C_1, and so on. At the second stage, we are able to write $|j_1 + j_2 - 1, j_1 + j_2 - l\rangle$ in terms of the states $|m_1 m_2\rangle$ because $(j_1 + j_2, j_1 + j_2 - l)$, being a state of the "stretched case," may be so written. Thus, at each stage we calculate the state $|jj\rangle$ in terms of states which have previously been expressed in terms of the product states $|m_1 m_2\rangle$. Then, by successive application of the J_- operator, we obtain all the states $|jm\rangle$.

The preceding ideas are best illustrated by carrying out the addition of angular momenta of low quantum number, viz.,

A	$j_1 = \tfrac{1}{2}$,	$j_2 = \tfrac{1}{2}$
B	$j_1 = 1$,	$j_2 = \tfrac{1}{2}$
C	$j_1 = 1$,	$j_2 = 1$

CASE A

We start from

$$|1\ 1\rangle = |\tfrac{1}{2}\tfrac{1}{2}\rangle' \tag{9.131}$$

where the state on the left is of the type $|jm\rangle$ and that on the right is of the type $|m_1 m_2\rangle'$. Operating with J_- gives

$$\sqrt{2}|1\ 0\rangle = |\tfrac{1}{2}, -\tfrac{1}{2}\rangle' + |-\tfrac{1}{2}, \tfrac{1}{2}\rangle' \tag{9.132}$$

Recognizing the states on the right as product states, and writing $|j_1 = \tfrac{1}{2}, m_1 = \tfrac{1}{2}\rangle$ as α_1 and $|j_1 = \tfrac{1}{2}, m_1 = -\tfrac{1}{2}\rangle$ as β_1, (9.131) becomes

$$|1\ 1\rangle = \alpha_1 \alpha_2 \tag{9.133}$$

and

$$|1\ 0\rangle = \frac{1}{\sqrt{2}}(\alpha_1 \beta_2 + \beta_1 \alpha_2) \tag{9.134}$$

Operating once more with J_- gives

$$|1, -1\rangle = \beta_1 \beta_2 \tag{9.135}$$

The $j = 0$ state may be obtained by putting

$$\alpha_1 \beta_2 = C_0 |1\ 0\rangle + C_1 |0\ 0\rangle$$

and operating with J_+. This gives

$$\alpha_1 \alpha_2 = C_0 \sqrt{2} |1\ 1\rangle$$

so that $C_0 = 1/\sqrt{2}$. But then

$$C_1 |0\ 0\rangle = \alpha_1 \beta_2 - \frac{1}{\sqrt{2}} |1\ 0\rangle$$

$$= \frac{1}{\sqrt{2}}(\alpha_1 \beta_2 - \beta_1 \alpha_2)$$

The normalization clearly requires that $C_1 = 1$, so that

$$|0\ 0\rangle = \frac{1}{\sqrt{2}}(\alpha_1 \beta_2 - \beta_1 \alpha_2)$$

$$= \frac{1}{\sqrt{2}}\{|\tfrac{1}{2}, -\tfrac{1}{2}\rangle' - |-\tfrac{1}{2}, \tfrac{1}{2}\rangle'\} \tag{9.136}$$

CASE B

This time the starting point is the equation

$$|jm\rangle = |\tfrac{3}{2}\tfrac{3}{2}\rangle = |1\ \tfrac{1}{2}\rangle' = |m_1 m_2\rangle'$$

Successive operations with J_- yield

$$|\tfrac{3}{2}\tfrac{1}{2}\rangle = \sqrt{\tfrac{2}{3}}|0\,\tfrac{1}{2}\rangle' + \sqrt{\tfrac{1}{3}}|1,-\tfrac{1}{2}\rangle' \qquad (9.137)$$

$$|\tfrac{3}{2},-\tfrac{1}{2}\rangle = \sqrt{\tfrac{1}{3}}|-1,\tfrac{1}{2}\rangle' + \sqrt{\tfrac{2}{3}}|0,-\tfrac{1}{2}\rangle' \qquad (9.138)$$

and

$$|\tfrac{3}{2},-\tfrac{3}{2}\rangle = |-1,-\tfrac{1}{2}\rangle' \qquad (9.139)$$

The states of $j = \tfrac{1}{2}$ are obtained by starting with

$$|m_1 m_2\rangle' = |0\,\tfrac{1}{2}\rangle' = C_0|\tfrac{3}{2}\tfrac{1}{2}\rangle + C_1|\tfrac{1}{2}\tfrac{1}{2}\rangle \qquad (9.140)$$

Operating with J_+ leads to

$$\sqrt{2}|1\tfrac{1}{2}\rangle' = \sqrt{3}C_0|\tfrac{3}{2}\tfrac{3}{2}\rangle$$

whence

$$C_0 = \sqrt{\tfrac{2}{3}}$$

It follows that

$$C_1|\tfrac{1}{2}\tfrac{1}{2}\rangle = |0\,\tfrac{1}{2}\rangle' - \sqrt{\tfrac{2}{3}}|\tfrac{3}{2}\tfrac{1}{2}\rangle$$

Substituting for the second from (9.137), we find that

$$C_1|\tfrac{1}{2}\tfrac{1}{2}\rangle = \tfrac{1}{3}|0\,\tfrac{1}{2}\rangle' - \frac{\sqrt{2}}{3}|1,-\tfrac{1}{2}\rangle'$$

The normalization condition gives $C_1 = 1/\sqrt{3}$, so that

$$|\tfrac{1}{2}\tfrac{1}{2}\rangle = \frac{1}{\sqrt{3}}|0\,\tfrac{1}{2}\rangle' - \sqrt{\tfrac{2}{3}}|1,-\tfrac{1}{2}\rangle' \qquad (9.141)$$

Operating on this with J_- gives finally

$$|\tfrac{1}{2},-\tfrac{1}{2}\rangle = \frac{1}{\sqrt{3}}\left(|0,-\tfrac{1}{2}\rangle' + \sqrt{\tfrac{2}{3}}|-1,\tfrac{1}{2}\rangle'\right) - \sqrt{\tfrac{2}{3}}\sqrt{2}|0,-\tfrac{1}{2}\rangle'$$

or

$$|\tfrac{1}{2},-\tfrac{1}{2}\rangle = \sqrt{\tfrac{2}{3}}|-1,\tfrac{1}{2}\rangle' - \sqrt{\tfrac{1}{3}}|0,-\tfrac{1}{2}\rangle' \qquad (9.142)$$

CASE C

The $m = j$ state for the stretched case is

$$|jm\rangle = |2\,2\rangle = |1\,1\rangle' = |m_1 m_2\rangle' \qquad (9.143)$$

where, once again, we use a prime to distinguish an $|m_1 m_2\rangle$ state from a $|jm\rangle$ one. Operating once with J_- gives

$$|2\,1\rangle = \frac{1}{\sqrt{2}}|1\,0\rangle' + \frac{1}{\sqrt{2}}|0\,1\rangle' \qquad (9.144)$$

Operating again with J_- yields

$$|2\,0\rangle = \frac{1}{\sqrt{6}}|1,-1\rangle' + \frac{2}{\sqrt{6}}|0\,0\rangle' + \frac{1}{\sqrt{6}}|-1,1\rangle' \qquad (9.145)$$

Continuing, we get the remaining states of the "stretched case."

$$|2,-1\rangle = \frac{1}{\sqrt{2}}|0,-1\rangle' + \frac{1}{\sqrt{2}}|-1,0\rangle' \qquad (9.146)$$

and

$$|2,-2\rangle = |-1,-1\rangle' \qquad (9.147)$$

The starting point for the $j = 1$ states is the equation

$$|1\,0\rangle' = C_0|2\,1\rangle + C_1|1\,1\rangle$$

Operating with J_+ gives

$$\sqrt{2}|1\,1\rangle' = 2C_0|2\,2\rangle$$

so that $C_0 = 1/\sqrt{2}$. Therefore,

$$C_1|1\,1\rangle = |1\,0\rangle' - \frac{1}{\sqrt{2}}\left\{ \frac{1}{\sqrt{2}}|1\,0\rangle' + \frac{1}{\sqrt{2}}|0\,1\rangle' \right\}$$

Since $C_1 = 1$ because of the normalization requirement

$$|1\,1\rangle = \frac{1}{\sqrt{2}}|1\,0\rangle' - \frac{1}{\sqrt{2}}|0\,1\rangle' \qquad (9.148)$$

Operating with J_- we obtain

$$|1\,0\rangle = \frac{1}{\sqrt{2}}|1,-1\rangle' - \frac{1}{\sqrt{2}}|-1,1\rangle' \qquad (9.149)$$

Finally, to obtain the state $|0\,0\rangle$, we put

$$|0\,0\rangle' = C_0|2\,0\rangle + C_1|1\,0\rangle + C_2|0\,0\rangle$$

Operating once with J_+ produces

$$\sqrt{2}(|1\,0\rangle' + |0\,1\rangle') = C_0\sqrt{6}|2\,1\rangle + C_1\sqrt{2}|1\,1\rangle$$

A second operation gives

$$4|1\,1\rangle' = 2C_0\sqrt{6}|2\,2\rangle$$

so that

$$C_0 = \frac{2}{\sqrt{6}}$$

Therefore,

$$C_1|1\,1\rangle = |1\,0\rangle' + |0\,1\rangle' - \sqrt{2}|2\,1\rangle = 0$$

and

$$C_1 = 0$$

Consequently,

$$C_2 |0\,0\rangle = |0\,0\rangle' - \frac{2}{\sqrt{6}} |2\,0\rangle$$

$$= |0\,0\rangle' - \frac{2}{\sqrt{6}} \left\{ \frac{1}{\sqrt{6}} |1,-1\rangle' + \frac{2}{\sqrt{6}} |0\,0\rangle' + \frac{1}{\sqrt{6}} |-1,1\rangle' \right\}$$

The normalization condition gives $C_2 = 1/\sqrt{3}$ and therefore

$$|0\,0\rangle = \frac{1}{\sqrt{3}} |0\,0\rangle' - \frac{1}{\sqrt{3}} |1,-1\rangle' - \frac{1}{\sqrt{3}} |-1,1\rangle' \qquad (9.150)$$

[See Problem 9-8 on page 528.]

9. Two-Dimensional Harmonic Oscillator

If ξ, η are rescaled variables,

$$\xi = \left(\frac{mk}{\hbar^2} \right)^{1/4} x \qquad (9.151)$$

$$\eta = \left(\frac{mk}{\hbar^2} \right)^{1/4} y \qquad (9.152)$$

the oscillator equation

$$-\frac{\hbar^2}{2m} \left(\frac{\partial^2 \psi}{\partial x^2} + \frac{\partial^2 \psi}{\partial y^2} \right) + \frac{1}{2} k(x^2 + y^2)\psi = E\psi \qquad (9.153)$$

may be simplified, as in the one-dimensional case, to

$$-\frac{1}{2} \left(\frac{\partial^2 \psi}{\partial \xi^2} + \frac{\partial^2 \psi}{\partial \eta^2} \right) + \frac{1}{2} (\xi^2 + \eta^2)\psi = \epsilon\psi \qquad (9.154)$$

where again, as in (9.4),

$$\epsilon = \frac{E}{\hbar\omega}$$

Equation (9.154) may be solved in Cartesian coordinates, by separation of variables, that is, by putting

$$\chi = \varphi(\xi)\chi(\eta) \qquad (9.155)$$

which individually solve one-dimensional oscillator equations with

Problem 9-8: Determine the states obtained by adding $j_1 = 2$ and $j_1 = 1$. The results are given in the following table.

	$\lvert 33\rangle$	$\lvert 32\rangle$	$\lvert 31\rangle$	$\lvert 30\rangle$	$\lvert 3,-1\rangle$	$\lvert 3,-2\rangle$	$\lvert 3,-3\rangle$	$\lvert 22\rangle$	$\lvert 21\rangle$	$\lvert 20\rangle$	$\lvert 2,-1\rangle$	$\lvert 2,-2\rangle$	$\lvert 11\rangle$	$\lvert 10\rangle$	$\lvert 1,-1\rangle$
$\lvert 2\,1\rangle'$	1														
$\lvert 2\,0\rangle'$		$\sqrt{\tfrac{1}{3}}$						$\sqrt{\tfrac{2}{3}}$							
$\lvert 2,-1\rangle'$			$\sqrt{\tfrac{1}{15}}$						$\sqrt{\tfrac{1}{3}}$				$\sqrt{\tfrac{3}{5}}$		
$\lvert 1\,1\rangle'$		$\sqrt{\tfrac{2}{3}}$						$-\sqrt{\tfrac{1}{3}}$							
$\lvert 1\,0\rangle'$			$\sqrt{\tfrac{8}{15}}$						$\sqrt{\tfrac{1}{6}}$				$-\sqrt{\tfrac{3}{10}}$		
$\lvert 1,-1\rangle'$				$\sqrt{\tfrac{1}{5}}$						$\sqrt{\tfrac{1}{2}}$				$\sqrt{\tfrac{3}{10}}$	
$\lvert 0,1\rangle'$			$\sqrt{\tfrac{2}{5}}$						$-\sqrt{\tfrac{1}{2}}$				$\sqrt{\tfrac{1}{10}}$		
$\lvert 0\,0\rangle'$				$\sqrt{\tfrac{3}{5}}$										$-\sqrt{\tfrac{2}{5}}$	
$\lvert 0,-1\rangle'$					$\sqrt{\tfrac{2}{5}}$						$\sqrt{\tfrac{1}{2}}$				$\sqrt{\tfrac{1}{10}}$
$\lvert -1,1\rangle'$				$\sqrt{\tfrac{1}{5}}$						$-\sqrt{\tfrac{1}{2}}$				$\sqrt{\tfrac{3}{10}}$	
$\lvert -1,0\rangle'$					$\sqrt{\tfrac{8}{15}}$						$-\sqrt{\tfrac{1}{6}}$				$-\sqrt{\tfrac{3}{10}}$
$\lvert -1,-1\rangle'$						$\sqrt{\tfrac{2}{3}}$						$-\sqrt{\tfrac{1}{3}}$			
$\lvert -2,1\rangle'$					$\sqrt{\tfrac{1}{15}}$						$-\sqrt{\tfrac{1}{3}}$				$\sqrt{\tfrac{3}{5}}$
$\lvert -2,0\rangle'$						$\sqrt{\tfrac{1}{3}}$						$\sqrt{\tfrac{2}{3}}$			
$\lvert -2,-1\rangle'$							1								

energy eigenvalues ϵ_1 and ϵ_2. The total energy is

$$\epsilon = \epsilon_1 + \epsilon_2 \qquad (9.156)$$

where

$$\epsilon_1 = (n_1 + \tfrac{1}{2})\hbar\omega \qquad (9.157)$$

$$\epsilon_2 = (n_2 + \tfrac{1}{2})\hbar\omega \qquad (9.158)$$

so that

$$\epsilon = (n_1 + n_2 + 1)\hbar\omega \qquad (9.159)$$

The energy states are degenerate, all states for which $n_1 + n_2$ has a definite value n having the same energy, although they have different wave functions. For a given n, the degeneracy is $(n+1)$-fold.

If we write the wave equation in cylindrical coordinates,

$$\xi = \rho \cos \varphi, \qquad \eta = \rho \sin \varphi \qquad (9.160)$$

it becomes

$$-\frac{1}{2}\left[\frac{\partial^2\psi}{\partial\rho^2} + \frac{1}{\rho}\frac{\partial\psi}{\partial\rho} + \frac{1}{\rho^2}\frac{\partial^2\psi}{\partial\varphi^2}\right] + \tfrac{1}{2}\rho^2\psi = \epsilon\psi \qquad (9.161)$$

Recalling, from Eq. (9.76), that the operator for the angular momentum about an axis perpendicular to the plane of motion is

$$L_z = -i\hbar\frac{\partial}{\partial\varphi}$$

we see that the equation for a state with angular momentum $m\hbar$ is

$$-\frac{1}{2}\left[\frac{d^2R}{d\rho^2} + \frac{1}{\rho}\frac{dR}{d\rho} - \frac{m^2}{\rho^2}R\right] + \tfrac{1}{2}\rho^2 R = \epsilon R \qquad (9.162)$$

where the wave function is

$$\psi = e^{im\varphi}R(\rho) \qquad (9.163)$$

Putting

$$\frac{\rho^2}{2} = z \qquad (9.164)$$

Eq. (9.162) becomes

$$z^2R'' + zR' + \left(\epsilon z - \frac{m^2}{4} - z^2\right)R = 0 \qquad (9.165)$$

Comparing this with (4.327), we see that the solution may be written in terms of the confluent hypergeometric. The coefficients in that equation, in order that it may reduce to (9.165), are

$$2\alpha - k = 0, \qquad \gamma - 2s = 1$$

$$\alpha(\alpha - k) = -1, \qquad (\gamma - 2s)\alpha + k(s - \beta) = \epsilon$$

$$s(\gamma - s - 1) = \frac{m^2}{4}$$

These have the solutions

$$s = \pm\frac{m}{2}, \qquad \gamma = \pm m + 1, \qquad \alpha = 1$$

$$k = 2, \qquad \beta = \tfrac{1}{2}(1 \pm m - \epsilon)$$

The solution $s = m/2$ is finite at the origin. Thus

$$R = Ce^{-z}z^{m/2}F(\tfrac{1}{2}(1 + m - \epsilon), m + 1, 2z) \tag{9.166}$$

or, expressed in terms of ρ

$$R = Ce^{-\rho^2/2}\rho^m F(\tfrac{1}{2}(1 + m - \epsilon), m + 1, \rho^2) \tag{9.167}$$

The confluent hypergeometric series, if it does not terminate, behaves for large ρ like e^{ρ^2}. It follows that, to behave properly at infinity, the series *must* terminate. That is to say, we must have

$$\tfrac{1}{2}(1 + m - \epsilon) = -n \tag{9.168}$$

where n is an integer. The energy levels are therefore given by

$$\epsilon = 2n + m + 1$$

or

$$E = (2n + m + 1)\hbar\omega \tag{9.169}$$

The radial wave function R_{nm} is then

$$C_{nm}e^{-\rho^2/2}\rho^m F(-n, m + 1, \rho^2) \tag{9.170}$$

Comparing with (4.295) we see that this may be expressed in terms of associated Laguerre polynomials

$$R_{nm} = D_{nm}e^{-\rho^2/2}\rho^m L_n^{(m)}(\rho^2) \tag{9.171}$$

D_{nm} is a normalizing factor.

Finally, the wave function is

$$\psi_{nm} = \frac{D_{nm}}{\sqrt{2\pi}} e^{-\rho^2/2}\rho^m L_n^{(m)}(\rho^2)e^{im\varphi} \tag{9.172}$$

There is also a very convenient solution of the two-dimensional problem in terms of operators. If a_1, $a_1\dagger$ are the annihilation and creation operators of the ξ-oscillator and a_2, $a_2\dagger$ are those of the η-oscillator, we can introduce operators associated with $\xi \pm i\eta$, that is,

$$b_\pm = \frac{1}{\sqrt{2}}(a_1 \mp ia_2) \qquad (9.173)$$

$$b_\pm\dagger = \frac{1}{\sqrt{2}}(a_1\dagger \pm ia_2\dagger) \qquad (9.174)$$

It follows from the commutation relations for the a's that

$$[b_i, b_j] = 0 = [b_i\dagger, b_j\dagger] \qquad (9.175)$$

$$[b_i, b_j\dagger] = \delta_{ij} \qquad (9.176)$$

where i, j may be $+$ or $-$.

We may also calculate

$$N_+ = b_+\dagger b_+ = \frac{1}{2}(a_1\dagger a_1 + a_2\dagger a_2) + \frac{i}{2}(a_1 a_2\dagger - a_2 a_1\dagger) \qquad (9.177)$$

and

$$N_- = b_-\dagger b_- = \frac{1}{2}(a_1\dagger a_1 + a_2\dagger a_2) - \frac{i}{2}(a_1 a_2\dagger - a_2 a_1\dagger) \qquad (9.178)$$

The second operator in the above equations may be expressed in terms of coordinate and momentum operators as

$$\frac{i}{2}(a_1 a_2\dagger - a_2 a_1\dagger) = \frac{i}{4}[(\xi + ip_\xi)(\eta - ip_\eta) - (\eta + ip_\eta)(\xi - ip_\xi)]$$

$$= \frac{1}{2}(\xi p_\eta - \eta p_\xi) \qquad (9.179)$$

$$= \frac{1}{2}L_z$$

L_z being the operator for the component of angular momentum perpendicular to the plane of motion.

In view of the fact that

$$H = a_1\dagger a_1 + a_2\dagger a_2 + 1 \qquad (9.180)$$

we see that

$$H = N_+ + N_- + 1 \qquad (9.181)$$

while

$$L_z = N_+ - N_- \qquad (9.182)$$

Thus, simultaneous eigenstates of N_+ and N_- are eigenstates of the Hamiltonian. Also because the b's and $b\dagger$'s obey the same algebraic relations as the a's, and appear in the same way in the Hamiltonian,

$$b_\pm\dagger|N_\pm\rangle = \sqrt{N_\pm + 1}\,|N_\pm + 1\rangle \qquad (9.183)$$

and

$$b_\pm |N_\pm\rangle = \sqrt{N_\pm}\, |N_\pm - 1\rangle \qquad (9.184)$$

The ground state is

$$|N_+ N_-\rangle = |0\,0\rangle \qquad (9.185)$$

being given by $N_+ = N_- = 0$. The general state is

$$|N_+ N_-\rangle = \frac{1}{\sqrt{N_+! N_-!}} (b_+\dagger)^{N_+} (b_-\dagger)^{N_-} |0\,0\rangle \qquad (9.186)$$

In this state

$$\epsilon = N_+ + N_- + 1$$

or

$$E = (N_+ + N_- + 1)\,\hbar\omega \qquad (9.187)$$

and

$$L_z = (N_+ - N_-)\,\hbar \qquad (9.188)$$

in ordinary units.

By virtue of the easily verified commutation relations

$$[L_z, b_\pm\dagger] = \pm b_\pm\dagger \qquad (9.189)$$

$$[L_z, b_\pm] = \mp b_\pm \qquad (9.190)$$

we see that $b_+\dagger$, b_- increase L_z by one unit while $b_-\dagger$, b_+ decrease it by the same amount.

We may therefore independently raise and lower the energy and the angular momentum; for example, $b_+ b_-$ keeps L_z the same but decreases the energy by two quanta, while $b_+\dagger b_-$ increases L_z by two units while leaving the energy unchanged.

Note, however, that states of odd angular momentum must have odd numbers of quanta, while those with even angular momenta have *even* numbers of quanta.

Schrödinger wave functions for the problem may be obtained explicitly by operator methods. Using the notation of (9.185), the ground state is defined by the equations

$$b_+|0\,0\rangle = b_-|0\,0\rangle = 0 \qquad (9.191)$$

The Schödinger operators may be written in terms of ξ and η as

$$b_\pm = \frac{1}{\sqrt{2}}\,(a_1 \mp i a_2) = \frac{1}{2}\left[\xi + \frac{\partial}{\partial\xi} \mp i\left(\eta + \frac{\partial}{\partial\eta}\right)\right] \qquad (9.192)$$

$$b_\pm\dagger = \frac{1}{\sqrt{2}}\,(a_1\dagger \pm i a_2\dagger) = \frac{1}{2}\left[\xi - \frac{\partial}{\partial\xi} \pm i\left(\eta - \frac{\partial}{\partial\eta}\right)\right] \qquad (9.193)$$

In cylindrical coordinates these become

$$b_\pm = \frac{e^{\mp i\varphi}}{2}\left(\rho + \frac{\partial}{\partial\rho} \mp \frac{i}{\rho}\frac{\partial}{\partial\varphi}\right) \tag{9.194}$$

$$b_\pm\dagger = \frac{e^{\pm i\varphi}}{2}\left(\rho - \frac{\partial}{\partial\rho} \mp \frac{i}{\rho}\frac{\partial}{\partial\varphi}\right) \tag{9.195}$$

Let us designate by χ_{nm} the wave function of the state for which the energy is $n\hbar\omega$ and the angular momentum $m\hbar$. Then

$$b_+\chi_{nm} = \sqrt{\tfrac{1}{2}(n+m)}\,\chi_{n-1,\,m-1} \tag{9.196}$$

$$b_-\chi_{nm} = \sqrt{\tfrac{1}{2}(n-m)}\,\chi_{n-1,\,m+1} \tag{9.197}$$

$$b_+\dagger\chi_{nm} = \sqrt{\tfrac{1}{2}(n+m)+1}\,\chi_{n+1,\,m+1} \tag{9.198}$$

$$b_-\dagger\chi_{nm} = \sqrt{\tfrac{1}{2}(n-m)+1}\,\chi_{n+1,\,m-1} \tag{9.199}$$

if the χ's are normalized. The ground state wave function satisfies the equation

$$\left(\rho + \frac{\partial}{\partial\rho}\right)\chi_{00} = 0 \tag{9.200}$$

for which the normalized solution is

$$\chi_{00} = \frac{1}{\sqrt{\pi}}\,e^{-\rho^2/2} \tag{9.201}$$

The wave functions for the states $n = 1$, $m = \pm 1$ and $n = 2$, $m = 0$, ± 2 are easily calculated to be

$$\chi_{1,\pm1} = \sqrt{\frac{1}{\pi}}\,\rho\,e^{-\rho^2/2}\,e^{\pm i\varphi} \tag{9.202}$$

$$\chi_{2,\pm2} = \frac{1}{\sqrt{2\pi}}\rho^2\,e^{-\rho^2/2}\,e^{\pm 2i\varphi} \tag{9.203}$$

$$\chi_{2,0} = \frac{1}{\sqrt{\pi}}\,(\rho^2 - 1)e^{-\rho^2/2} \tag{9.204}$$

Problem 9-9: Calculate the various states corresponding to $n = 3$, viz., $\chi_{3,\pm3}$ and $\chi_{3,\pm1}$.

---

10. Three-Dimensional Oscillator

Once again, we may make use of the fact that the equation is separable, and find states which are products of three one-dimensional oscillator states. The operators a_1, a_2 for the x and y directions may be supplemented by the operator a_3 for the z direction. Designating the ground state by

$$|n_1 n_2 n_3\rangle = |0\,0\,0\rangle \qquad (9.205)$$

we may write the general state as

$$|n_1 n_2 n_3\rangle = \frac{1}{\sqrt{n_1! n_2! n_3!}} (a_1\dagger)^{n_1}(a_2\dagger)^{n_2}(a_3\dagger)^{n_3}|0\,0\,0\rangle \qquad (9.206)$$

The energy is

$$\epsilon = n_1 + n_2 + n_3 + \tfrac{3}{2} \qquad (9.207)$$

The various states for which $n_1 + n_2 + n_3 = n$ have the same value are degenerate; the number of them is

$$D_n = \frac{(n+2)!}{2!n!} = \frac{(n+1)(n+2)}{2} \qquad (9.208)$$

The individual states (9.206) are not eigenstates of angular momentum, but linear combinations may be found which are. States of odd n have *odd* angular momentum, while those of even n have *even* angular momentum. This follows from the fact that under a parity transformation

$$x, y, z \to -x, -y, -z$$

the wave function of states of odd angular momentum change sign, while those of even angular momentum do not. But the a operators change sign under this transformation; therefore, an even number of them must lead to even angular momentum states and vice versa.

We can easily see that, for a given n, all the angular momentum states $n, n-2, \ldots, 0$ or 1 must appear. For the sum

$$(2n+1) + [2(n-2)+1] + \cdots + 3 \text{ or } 1$$

is

$$\frac{(n+1)(n+2)}{2}$$

Angular momentum eigenstates may be constructed out of a_1, a_2, and a_3 as follows: Introduce

$$b_\pm = \frac{1}{\sqrt{2}}(a_1 \mp ia_2)$$

as in (9.192) and also

$$b_0 = a_3 \qquad (9.209)$$

Commutation rules of the form (9.175) and (9.176) again apply, where now i, j may be $0, \pm 1$.

Just as in the two-dimensional case

$$H = N_+ + N_- + N_0 + \tfrac{3}{2} \qquad (9.210)$$

and

$$L_z = N_+ - N_-$$

In terms of the b operators we may construct simultaneous eigenstates of H and L_z:

$$|N_+ N_- N_0\rangle = \frac{1}{\sqrt{N_+! N_-! N_0!}} (b_+\dagger)^{N_+} (b_-\dagger)^{N_-} (b_0\dagger)^{N_0} |0\,0\,0\rangle \quad (9.211)$$

There remains the problem of finding eigenstates of $\mathbf{L}^2$.

Let us designate by ϕ_{nlm} the Schrödinger wave function for the state with principal quantum number

$$n = N_0 + N_+ + N_- \qquad (9.212)$$

and angular momentum quantum numbers l and m.

The lowest state ϕ_{000} is determined by the conditions

$$b_+\phi_{000} = b_-\phi_{000} = b_0\phi_{000} = 0$$

Using the fact that $b_\pm$ are given by (9.194) and $b_0 = a_3$ by

$$b_0 = z + \frac{\partial}{\partial z}$$

we see immediately that

$$\phi_{000} = \frac{1}{\pi^{3/4}} e^{-(\rho^2 + z^2)/2} = \frac{1}{\pi^{3/4}} e^{-r^2/2} \qquad (9.213)$$

[This also follows from (9.39), since the wave function must be the product of three $n = 0$ one-dimensional ones.]

The wave functions for $n = l = 1$ are

$$\phi_{111} = b_+\dagger\phi_{000}, \qquad \phi_{110} = b_0\dagger\phi_{000}, \qquad \phi_{11,-1} = b_-\dagger\phi_{000} \quad (9.214)$$

Applying the operators to (9.213) it is easily verified that

$$\phi_{111} = \frac{1}{\pi^{3/4}} \sin\theta\, e^{i\varphi} r\, e^{-r^2/2} \qquad (9.215)$$

$$\phi_{110} = \frac{1}{\pi^{3/4}} \cos\theta\, r\, e^{-r^2/2} \qquad (9.216)$$

$$\phi_{11,-1} = \frac{1}{\pi^{3/4}} \sin\theta\, e^{-i\varphi} r\, e^{-r^2/2} \qquad (9.217)$$

It is always possible to get the states for which $n = l$. For

$$\phi_{nnn} = \frac{1}{\sqrt{n!}} (b_+\dagger)^n \phi_{000} \tag{9.218}$$

This follows from the fact that $m = n$ and therefore $l = n$. The other states ϕ_{nnm} may then be obtained by successive application of the L operator.

Since the angular momentum states are already known, (9.218) is in fact only needed to determine the radial function. However, all states can be expressed in terms of the b operators once the angular momentum operators $L_\pm$ are so expressed. Consider, then,

$$L_+ = L_x + iL_y$$

$$= (yp_z - zp_y) + i(zp_x - xp_z)$$

We first express the coordinates and momenta in terms of the a operators:

$$x_j = \frac{1}{\sqrt{2}} (a_j + a_j\dagger) \tag{9.219}$$

$$p_j = \frac{1}{\sqrt{2}i} (a_j - a_j\dagger) \tag{9.220}$$

Thus

$$L_+ = \frac{1}{2i} [(a_2 + a_2\dagger)(a_3 - a_3\dagger) - (a_3 + a_3\dagger)(a_2 - a_2\dagger)]$$

$$+ \frac{1}{2} [(a_3 + a_3\dagger)(a_1 - a_1\dagger) - (a_1 + a_1\dagger)(a_3 - a_3\dagger)]$$

$$= \frac{1}{i} (a_2\dagger a_3 - a_3\dagger a_2) + (a_3\dagger a_1 - a_1\dagger a_3)$$

$$= -a_3(a_1\dagger + ia_2\dagger) + a_3\dagger(a_1 + ia_2)$$

$$= \sqrt{2} (b_0\dagger b_- - b_0 b_+\dagger) \tag{9.221}$$

by virtue of (9.173) and (9.174). Similarly it follows that

$$L_- = \sqrt{2} (b_0\dagger b_+ - b_0 b_-\dagger) \tag{9.222}$$

We may therefore write

$$\phi_{n,n,n-1} = \frac{1}{\sqrt{n(n+1) - n(n-1)}} L_- \phi_{nnn}$$

[by virtue of (9.71) and (9.72)] or

$$\phi_{n,n,n-1} = \frac{1}{\sqrt{2n}} \frac{1}{\sqrt{n!}} \sqrt{2}\,(b_0{}^\dagger b_+ - b_0 b_-{}^\dagger)(b_+{}^\dagger)^n \phi_{000}$$

The b_0 operator on the ground state gives zero, so that the second term goes out. In the first

$$b_+(b_+{}^\dagger)^n \phi_{000} = n(b_+{}^\dagger)^{n-1}\phi_{000}$$

and therefore

$$\phi_{n,n,n-1} = \frac{1}{\sqrt{n-1}}\, b_0{}^\dagger (b_+{}^\dagger)^{n-1}\phi_{000} \qquad (9.223)$$

Let us carry this one step further.

$$\phi_{n,n,n-2} = \frac{1}{\sqrt{n(n+1)-(n-1)(n-2)}}\, L_-\phi_{n,n,n-1}$$

$$= \frac{1}{\sqrt{2n-1}} \frac{1}{\sqrt{(n-1)!}}\,(b_0{}^\dagger b_+ - b_0 b_-{}^\dagger)b_0{}^\dagger(b_+{}^\dagger)^{n-1}\phi_{000}$$

$$= \frac{1}{\sqrt{2n-1}} \frac{1}{\sqrt{(n-2)!}}$$

$$\times [(b_0{}^\dagger)^2 - (n-1)b_-{}^\dagger b_+{}^\dagger](b_+{}^\dagger)^{n-2}\phi_{000} \qquad (9.224)$$

Problem 9-10: Show that the three-dimensional oscillator states with $n=2$ and $l=2$ are

$$\phi_{222} = 2\left(\frac{1}{4\pi}\right)^{3/4} \sin^2\theta\, e^{2i\varphi} r^2 e^{-r^2/2}$$

$$\phi_{221} = 4\left(\frac{1}{4\pi}\right)^{3/4} \sin\theta\cos\theta\, e^{i\varphi} r^2 e^{-r^2/2}$$

$$\phi_{220} = 2\sqrt{\frac{2}{3}}\left(\frac{1}{4\pi}\right)^{3/4} (3\cos^2\theta - 1) r^2 e^{-r^2/2}$$

$$\phi_{22,-1} = \phi_{221}^*$$

$$\phi_{22,-2} = \phi_{222}^*$$

The question then arises, how do we obtain the states for which $l = n-2s$, where s is an integer?

Consider first the case $s=1$. To obtain the state $\phi_{n,n-2,n-2}$, we see that we must combine the following states:

$$N_+ = n-2, \qquad N_- = 0, \qquad N_0 = 2$$

$$N_+ = n-1, \qquad N_- = 1, \qquad N_0 = 0$$

in such a way that L_+ operating on the combined state gives zero. That is to say, we must choose α and β so that

$$(b_0\dagger b_- - b_0 b_+ \dagger)[\alpha(b_+\dagger)^{n-2}(b_0\dagger)^2 + \beta(b_+\dagger)^{n-1}b_-\dagger]\phi_{000} = 0$$

This leads immediately to the condition

$$\beta = 2\alpha$$

so that

$$\phi_{n,n-2,n-2} = \alpha[(b_+\dagger)^{n-2}(b_0\dagger)^2 + 2(b_+\dagger)^{n-1}b_-\dagger]\phi_{000} \qquad (9.225)$$

Since

$$\frac{1}{\sqrt{(n-2)!2!}}(b_+\dagger)^{n-2}(b_0\dagger)^2\phi_{000}$$

and

$$\frac{1}{\sqrt{(n-1)!}}(b_+\dagger)^{n-1}b_-\dagger\phi_{000}$$

are normalized, the normalization condition on (9.225) is

$$\alpha^2[2(n-2)! + 4(n-1)!] = 1$$

so that

$$\alpha = \frac{1}{\sqrt{(n-2)!2(2n-1)}} \qquad (9.226)$$

Therefore, finally,

$$\phi_{n,n-2,n-2} = \frac{1}{\sqrt{(n-2)!2(2n-1)}}$$

$$\times [(b_+\dagger)^{n-2}(b_0\dagger)^2 + 2(b_+\dagger)^{n-1}b_-\dagger]\phi_{000} \qquad (9.227)$$

Problem 9-11:   Show that

$$\phi_{200} = \frac{\sqrt{\frac{2}{3}}}{\pi^{3/4}}(r^2 - 1)e^{-r^2/2}$$

Problem 9-12:   Calculate all of the three-dimensional oscillator functions for $n = 3$.

Answers

$$\phi_{333} = \frac{1}{\sqrt{6}\,\pi^{3/4}} \sin^3\theta\, e^{3i\varphi} r^3\, e^{-r^2/2}$$

$$\phi_{332} = \frac{1}{\pi^{3/4}} \sin^2\theta\, \cos\theta\, e^{2i\varphi} r^3\, e^{-r^2/2}$$

$$\phi_{331} = \frac{1}{\sqrt{10}\,\pi^{3/4}} \sin\theta\, (5\cos^2\theta - 1)\, e^{i\varphi} r^3\, e^{-r^2/2}$$

$$\phi_{330} = \sqrt{\frac{2}{15}}\, \frac{1}{\pi^{3/4}} (5\cos^3\theta - 3\cos\theta) r^3\, e^{-r^2/2}$$

$$\phi_{33,-m} = \phi_{33m}^{*}$$

$$\phi_{311} = \frac{1}{\sqrt{10}}\, \frac{1}{\pi^{3/4}} \sin\theta\, e^{i\varphi}(2r^3 - 5r)\, e^{-r^2/2}$$

$$\phi_{310} = \frac{1}{\sqrt{5}}\, \frac{1}{\pi^{3/4}} \cos\theta\, (2r^3 - 5r)\, e^{-r^2/2}$$

$$\phi_{31,-1} = \phi_{311}^{*}$$

Problem 9-13:   Find the momentum eigenfunctions for the $1s$, $2s$, and $2p$ states of the three-dimensional oscillator.

Since, knowing ϕ_{nll} we may always obtain the ϕ_{nlm} ($m < l$) by using the L_- operator, there remains the problem of finding ϕ_{nll}. We know that $l = n - 2s$, s taking all integer values for which l is $\geqslant 0$.

The state in question must be constructed of a linear combination of n $b_1^\dagger$, $b_-^\dagger$, and $b_0^\dagger$ operators operating on ϕ_{000}. If the numbers of these operators are N_+, N_-, and N_0, respectively, we must have

$$N_+ + N_- + N_0 = n$$

and

$$N_+ - N_- = l = n - 2s$$

for the state in question. These may be satisfied by taking

$$N_0 = 2r$$

$$N_- = s - r$$

$$N_+ = n - s - r$$

where

$$r = 0, 1, 2, \ldots, s.$$

We therefore take

$$\phi_{nll} = \left[\sum_{r=0}^{s} \lambda_r (b_0{\dagger})^{2r} (b_-{\dagger})^{s-r} (b_+{\dagger})^{n-s-r} \right] \phi_{000} \qquad (9.228)$$

The λ_r's are to be determined from the condition that

$$L_+ \phi_{nll} = 0 \qquad (9.229)$$

or

$$(b_0{\dagger}b_- - b_0 b_+{\dagger}) \left[\sum_{r=0}^{s} \lambda_r (b_0{\dagger})^{2r} (b_-{\dagger})^{s-r} (b_+{\dagger})^{n-s-r} \right] \phi_{000} = 0$$

Using the commutation properties of the b operators, this condition takes the form

$$\sum_{r=0}^{s-1} \lambda_r (s-r)(b_0{\dagger})^{2r+1} (b_-{\dagger})^{s-r-1} (b_+{\dagger})^{n-s-r} \phi_{000}$$

$$- \sum_{r=1}^{s} \lambda_r 2r (b_0{\dagger})^{2r-1} (b_-{\dagger})^{s-r} (b_+{\dagger})^{n-s-r+1} \phi_{000} = 0 \qquad (9.230)$$

We obtain in this way a recurrence relation for the λ_r's:

$$\lambda_r = \frac{s-r+1}{2r} \lambda_{r-1} \qquad (9.231)$$

from which we obtain

$$\lambda_r = \frac{s!}{r!\,(s-r)!} \frac{1}{2^r} \lambda_0 \qquad (9.232)$$

It follows that, to within a normalization constant λ_0,

$$\phi_{nll} = \lambda_0 \sum_{r=0}^{s} \frac{s!}{r!\,(s-r)!} \frac{1}{2^{s-r}} (b_0{\dagger})^{2(s-r)} (b_-{\dagger})^{r} (b_+{\dagger})^{n-2s+r} \phi_{000} \qquad (9.233)$$

In all of the preceding discussion of the oscillator problem, lengths are expressed in units of $(\hbar^2/mk)^{1/4}$. It follows that, if normalization is carried out in terms of dimensional coordinates, normalization factors must all be multiplied by

$$\left(\frac{mk}{\hbar^2}\right)^{3/8} = \left(\frac{m\omega}{\hbar}\right)^{3/4} \qquad (9.234)$$

11.   Three-Dimensional Oscillator; Direct Solution

Again in terms of dimensionless coordinates, the Schrödinger wave equation is

$$-\tfrac{1}{2}\nabla^2 \psi + \tfrac{1}{2} r^2 \psi = \epsilon \psi \qquad (9.235)$$

If we put

$$\psi = Y_l^m(\theta, \varphi)\chi(r) \tag{9.236}$$

and proceed as in the treatment of the potential problem in Chapter 3, we obtain for the radial function χ the equation

$$-\frac{1}{r}\frac{d^2}{dr^2}(r\chi) + \frac{l(l+1)}{r^2}\chi + \tfrac{1}{2}r^2\chi = 2\epsilon\chi \tag{9.237}$$

Putting $r\chi = f$, this becomes

$$-f'' + \frac{l(l+1)}{r^2}f + r^2 f = 2\epsilon f \tag{9.238}$$

If we now make the substitution

$$u = r^2 \tag{9.239}$$

we obtain the equation

$$\frac{d^2 f}{du^2} + \frac{1}{2u}\frac{df}{du} + \left(\frac{\epsilon}{2u} - \frac{1}{4} - \frac{l(l+1)}{4u^2}\right)f = 0 \tag{9.240}$$

It follows, on comparing this with Eq. (4.237), that

$$f = e^{-\alpha u}u^s F(\beta, \gamma, u) \tag{9.241}$$

where $\alpha, \beta, \gamma,$ and s are determined by

$$2\alpha - k = 0$$
$$\gamma - 2s = \tfrac{1}{2}$$
$$\alpha(\alpha - k) = -\tfrac{1}{4}$$
$$\gamma\alpha - k\beta - 2s\alpha + ks = \epsilon/2$$
$$s(\gamma - s - 1) = l(l+1)/4$$

and $F(\beta, \gamma, u)$ is the confluent hypergeometric function. The solution of these equations is

$$k = 1, \qquad \alpha = \tfrac{1}{2}$$
$$s = \tfrac{1}{2}(l+1), \qquad \gamma = l + \tfrac{3}{2}$$
$$\beta = \tfrac{1}{2}(l + \tfrac{3}{2}) - (\epsilon/2)$$

Therefore,

$$\chi = e^{-r^2/2}r^l F\left(\frac{1}{2}\left(l + \frac{3}{2}\right) - \frac{\epsilon}{2}, l + \frac{3}{2}, r^2\right) \tag{9.242}$$

In general the confluent hypergeometric behaves for large r^2 like e^{r^2}, so that χ diverges at infinity. The boundary condition (that χ must vanish at $r \to \infty$) may be satisfied provided

$$\tfrac{1}{2}(l+\tfrac{3}{2}) - \frac{\epsilon}{2} = -\nu \qquad (9.243)$$

where ν is an integer, so that

$$\epsilon = 2\nu + l + \tfrac{3}{2} \qquad (9.244)$$

It follows that the (unnormalized) radial wave function is

$$\chi = r^l e^{-r^2/2} F(-\nu, l+\tfrac{3}{2}, r^2)$$

$$= r^l e^{-r^2/2} F\left(-\frac{n-l}{2}, l+\frac{3}{2}, r^2\right) \qquad (9.245)$$

where

$$n = l + 2\nu \qquad (9.246)$$

Comparing this with Eq. (4.295) of Chapter 4, we see that

$$\chi = A r^l e^{-r^2/2} L_{1/2(n-l)}^{l+(1/2)}(r^2) \qquad (9.247)$$

where A is an arbitrary constant and L is the Laguerre polynomial.

Problem 9-14:   Calculate the *normalized* radial wave function for

$n = 2, l = 0$, and $l = 2$; and $n = 3, l = 1$, and $l = 3$.

---

12. Hydrogen Atom

The wave equation for a particle in a central field with potential

$$V = -\frac{Ze^2}{r} \qquad (9.248)$$

is

$$-\frac{\hbar^2}{2m}\nabla^2\psi - \frac{Ze^2}{r}\psi = E\psi \qquad (9.249)$$

If $-e$ is the electron charge, the case in which $Z = 1$ is that of the hydrogen atom. $Z = 2$ corresponds to the helium singly charged ion He^+, $Z = 3$ to the ion Li^{++}, etc. All such single-electron ions are generally designated "hydrogenic."

The problem of these systems is not really that of a one-particle system; the kinetic energy is

$$T = \tfrac{1}{2}m\dot{r}_1^2 + \tfrac{1}{2}M\dot{r}_2^2 \qquad (9.250)$$

where m, M are, respectively, the mass of electron and nucleus. If we introduce the position vector $\mathbf{R}$ of the center of mass and the *relative* position vector $\mathbf{r}$,

$$\mathbf{R} = \frac{1}{m+M}(m\mathbf{r}_1 + M\mathbf{r}_2) \qquad (9.251)$$

and

$$\mathbf{r} = \mathbf{r}_1 - \mathbf{r}_2 \qquad (9.252)$$

we find that

$$T = \frac{1}{2}(m+M)\dot{\mathbf{R}}^2 + \frac{1}{2}\frac{mM}{m+M}\dot{\mathbf{r}}^2 \qquad (9.253)$$

If we take a system in which the mass center is at rest, we may ignore the first term. The problem of relative motion is then formally identical with the problem of a single particle with mass $mM/(m+M)$ (the "reduced mass"). For $M \gg m$, this is very nearly m. In general, the m in (9.249) should be replaced by the reduced mass.

If we make the substitution

$$\psi = \Phi_l(r)Y_l^m(\theta, \varphi) \qquad (9.254)$$

the equation for the radial wave function Φ_l is

$$-\frac{\hbar^2}{2m}\left[\nabla^2\Phi_l - \frac{l(l+1)}{r^2}\Phi_l\right] - \frac{Ze^2}{r}\Phi_l = E\Phi_l \qquad (9.255)$$

in the state in which the angular momentum state is given by the quantum numbers (l, m).

The equation may be simplified by multiplying through by $2/mZ^2e^4$ and introducing the following dimensionless units:

$$\xi = \frac{r}{\hbar^2/mZe^2} \qquad (9.256)$$

and

$$\epsilon = \frac{E}{E_0} \qquad (9.257)$$

where

$$E_0 = \frac{mZ^2e^4}{2\hbar^2} \qquad (9.258)$$

If we now put

$$\xi\Phi_l = f_l$$

the Schrödinger equation takes the form

$$\frac{d^2f_l}{d\xi^2} - \frac{l(l+1)}{\xi^2}f_l + \frac{2}{\xi}f_l = -\epsilon f_l \qquad (9.259)$$

The "factorization method" can be used again in the problem, though in a somewhat different form than in the case of the oscillator problem. Introduce the operators

$$A_l^+ = \frac{d}{d\xi} - \frac{l+1}{\xi} + \frac{1}{l+1} \tag{9.260}$$

and

$$A_l^- = \frac{d}{d\xi} + \frac{l+1}{\xi} - \frac{1}{l+1} \tag{9.261}$$

It is then easily verified that

$$A_l^- A_l^+ f_l = \frac{d^2 f_l}{d\xi^2} - \frac{l(l+1)}{\xi^2} f_l + \frac{2}{\xi} f_l - \frac{1}{(l+1)^2} f_l$$

By virtue of the wave equation, this may be written

$$A_l^- A_l^+ f_l = -\left[\epsilon_l + \frac{1}{(l+1)^2}\right] f_l \tag{9.262}$$

where ϵ_l is the energy, in dimensionless units, for the angular momentum state l.

The equation for the angular momentum state $(l+1)$ can be written

$$A_l^+ A_l^- f_{l+1} = -\left[\epsilon_{l+1} + \left(\frac{1}{l+1}\right)^2\right] f_{l+1} \tag{9.263}$$

Let us now operate on (9.262) with the operator A_l^+, and on (9.263) with A_l^-, to obtain

$$A_l^+ A_l^- [A_l^+ f_l] = -\left[\epsilon_l + \left(\frac{1}{l+1}\right)^2\right][A_l^+ f_l] \tag{9.264}$$

and

$$A_l^- A_l^+ [A_l^- f_{l+1}] = -\left[\epsilon_{l+1} + \left(\frac{1}{l+1}\right)^2\right][A_l^- f_l] \tag{9.265}$$

By comparing (9.265) with (9.262) we see that

$$\epsilon_{l+1} = \epsilon_l \tag{9.266}$$

and

$$A_l^- f_{l+1} = K_l f_l \tag{9.267}$$

where we assume the Φ's to be normalized. Thus, the energy levels do not seem to depend on l; this represents a rather unexpected degeneracy. The quantity K_l may be obtained by normalization. Writing

$$(g_1, g_2) = \int g_1^*(r) g_2(r) \, dr \tag{9.268}$$

we derive from (9.267) that

$$K_l^2 = (A_l^- f_{l+1}, A_l^- f_{l+1}) \qquad (9.269)$$

by virtue of the normalization of f_l, that is,

$$(f_l, f_l) = 1 \qquad (9.270)$$

On integration by parts, and using the condition that the wave function must vanish faster than $1/r$ at infinity, we find that

$$K_l^2 = -(f_{l+1}, A_l^+ A_l^- f_{l+1})$$
$$= \epsilon_{l+1} + \left(\frac{1}{l+1}\right)^2 \qquad (9.271)$$

Similarly a comparison of (9.263) and (9.264) shows that

$$A_l^+ f_l = K_l' f_{l+1} \qquad (9.272)$$

where K_l' is a constant to be determined. The consequence of equating scalar products of the two sides of (9.272) is that

$$K_l'^2 = (A_l^+ f_l, A_l^+ f_l)$$
$$= -(f_l, A_l^- A_l^+ f_l)$$
$$= \epsilon_l + \left(\frac{1}{l+1}\right)^2 \qquad (9.273)$$

so that

$$K_l' = K_l = \sqrt{\epsilon_l + \left(\frac{1}{l+1}\right)^2} \qquad (9.274)$$

We note that, if $\epsilon > 0$, there is no upper limit to l. Since positive energy corresponds to an unbound (scattering) state this is not a particularly surprising result.

If the energy ϵ is negative (that is, if we are concerned with bound states), since K_l^2 and $K_l'^2$ must be positive, l must be bounded. The only way to break the chain of states of increasing l due to successive application of the A_l^+ operator is by having

$$\epsilon = -\frac{1}{n^2} \qquad (9.275)$$

where n is an integer. Then the largest possible value of l is $l = n - 1$.

The equation for the lowest energy state is [from (9.272)]

$$A_0^+ f_0 = 0 \qquad (9.276)$$

More generally, the $l = n - 1$ state is determined by

$$A_{n-1}^+ f_{n,n-1} = 0 \qquad (9.277)$$

where $f_{n,n-1}$ designates the state with energy $\epsilon = -1/n^2$ and $l = n-1$. The differential equation to be solved is

$$\frac{df}{d\xi} - \frac{n}{\xi}f + \frac{1}{n}f = 0 \qquad (9.278)$$

which has the solution

$$f_{n,n-1} = C\xi^n e^{-\xi/n}$$

The normalization factor C is determined by

$$C^2 \int \xi^{2n} e^{-2\xi/n}\, d\xi = 1$$

Since the integral has the value $(2n+2)!(n/2)^{2n+3}$,

$$C = \left(\frac{2}{n}\right)^{n+(1/2)} \frac{1}{\sqrt{2n}}$$

Finally, then, we may write down the wave functions

$$\psi_{nlm} = \psi_{n,n-1,m}$$

$$= \left(\frac{2}{n}\right)^{n+(1/2)} \frac{1}{\sqrt{(2n)!}} \xi^{n-1} e^{-\xi/n} Y_{n-1}^m(\theta,\varphi) \qquad (9.279)$$

The normalized radial functions for lower angular momenta are obtained as follows:

$$f_{nl} = \frac{1}{\sqrt{1/(l+1)^2 - (1/n^2)}} A_l^- f_{n,l+1} \qquad (9.280)$$

so that $\Phi_{n,l-1}$ is expressed in terms of $\Phi_{n,l}$ by the formula

$$\Phi_{n,l-1} = \frac{1}{\sqrt{(1/l^2) - (1/n^2)}} \frac{1}{\xi} A_{l-1}^-(\xi\Phi_{nl}) \qquad (9.281)$$

Problem 9-15:   Show that the following are the radial wave functions for the $n=1$, $n=2$, and $n=3$ states of the hydrogenic atom:

$$\Phi_{10} = 2e^{-\xi}$$

$$\Phi_{2,1} = \frac{1}{\sqrt{4!}} \xi e^{-\xi/2}$$

$$\Phi_{2,0} = \frac{1}{2\sqrt{2}} (\xi - 2) e^{-\xi/2}$$

$$\Phi_{3,2} = \frac{2\sqrt{2}}{81\sqrt{15}}\, \xi^2 e^{-\xi/3}$$

$$\Phi_{3,1} = \frac{2\sqrt{2}}{81\sqrt{3}}\, (6\xi - \xi^2) e^{-\xi/3}$$

$$\Phi_{3,0} = \frac{2}{81\sqrt{3}}\, (27 - 18\xi + 2\xi^2) e^{-\xi/3}$$

Problem 9-16: By comparing (9.259) with (4.327) show that

$$\Phi_l = C_l e^{-\sqrt{-\epsilon}\xi} \xi^l F\left(l+1-\frac{1}{\sqrt{-\epsilon}}, 2(l+1), 2\sqrt{-\epsilon}\xi\right)$$

for $\epsilon < 0$, C_l being a normalization constant and F the confluent hypergeometric function.

Considering the asymptotic form of the confluent hypergeometric function, show that, for Φ_l to vanish as $\xi \to \infty$, we must have

$$\epsilon = -\frac{1}{n^2}$$

where n is an integer $\geq (l+1)$, so that

$$\Phi_l = C_l e^{-\xi/n} \xi^l F\left(l+1-n, 2(l+1), \frac{2\xi}{n}\right)$$

Problem 9-17: Using the expression for Laguerre functions in terms of confluent hypergeometrics given in (4.295), show that

$$\Phi_l = C_l' e^{-\xi/n} \xi^l L^{2l+1}_{n-(l+1)}\left(\frac{2\xi}{n}\right)$$

Use this to verify the results of Problem 9-13.

Problem 9-18: Find the momentum eigenfunctions for the $1s$, $2s$, and $2p$ states of the hydrogen atom.

Problem 9-19: Determine the eigenvalues and eigenfunctions of the positronium atom (electron + positron).

The annihilation rate depends directly on the density at coincidence. Comment on the value of this quantity in the various states.

Problem 9-20: Consider an atom with one $2p$ electron placed in the electric field of a crystal, the field being represented by the potential

$$V = Ax^2 + By^2 - (A+B)z^2$$

Treating the field as a perturbation, determine the levels to first order.

Problem 9-21:   A particle is subject to the central attractive potential $-V_0 e^{-\alpha r}$. Show that the energies of its s states are given by the equation

$$J_{2k/\alpha}\left(\sqrt{\frac{8mV_0}{\hbar^2\alpha^2}}\right) = 0$$

where

$$k = \sqrt{\frac{2m|E|}{\hbar^2}}$$

Thus, show, that, for there to *be* a bound state, it is necessary that

$$V_0 \geqslant \frac{(2.405)^2\hbar^2\alpha^2}{2m}$$

Problem 9-22:   The potential of a diatomic molecule may be represented by

$$V(r) = -D\left(\frac{1}{r} - \frac{1}{2r^2}\right)$$

in appropriate units. Show that the radial wave function has the form

$$R = r^{s-1}e^{-\lambda r}F(s-\gamma^2/\lambda, 2s, 2\lambda r)$$

where F is the confluent hypergeometric function,

$$\lambda = \sqrt{\frac{2\mu|E|}{\hbar^2}}, \qquad \gamma^2 = \frac{\mu D}{\hbar^2}$$

$$s = \tfrac{1}{2} + \sqrt{\gamma^2 + (l+\tfrac{1}{2})^2}$$

μ is the reduced mass and l is the angular momentum. Show that the energy levels are given by

$$E_{nl} = -\frac{\hbar^2}{2\mu}\frac{\gamma^4}{[n+\tfrac{1}{2}+\sqrt{\gamma^2+(l+\tfrac{1}{2})^2}]^2}$$

Problem 9-23:   Another form of potential energy curve for a diatomic molecule is

$$V = D(1-e^{-2\beta R})^2 - D$$

where $R = (r-a)/a$. Show that the s-state wave functions are

$$\psi = \frac{1}{r} e^{-\gamma z/2} z^s F(-n, 2s + 1, z)$$

where

$$z = ae^{-2\beta R}, \qquad s = \sqrt{\frac{\mu a^2 |E|}{2\hbar^2 \beta^2}}$$

$$n = \frac{\mu a^2 D}{\hbar^2 \beta^2 \gamma} - (s+\tfrac{1}{2}), \qquad \gamma = \sqrt{\frac{2\mu D a^2}{\hbar^2 \beta^2}}$$

Show also that the energy levels are

$$E_n = (n+\tfrac{1}{2})\hbar \sqrt{\frac{8\beta^2 D}{\mu a^2}} - (n+\tfrac{1}{2})^2 \frac{2\hbar^2 \beta^2}{\mu a^2}$$

---

13. Positive Energy States and the Problem of Scattering

There are two ways of dealing with problems of scattering. One approach is the time-dependent one, in which the incident particle is described, let us say, by a wave packet, and we follow its evolution in time according to the operator $U(t, t_0)$ introduced in the previous chapter. Normally, the initial time t_0 is taken to be $-\infty$, and the particle is initially free (i.e., localized outside of the potential). The probability of scattering, and the scattering cross section, are then calculated from the value of U at $t = \infty$, when the particle is again far from the scattering center. Thus, the process is described in terms of the "scattering operator" ("scattering matrix," if one deals with matrix elements of the operator)

$$S = U(-\infty, \infty) \tag{9.282}$$

The other approach, which we follow here, is to describe the scattering in terms of a steady-state formalism. (This is of course only possible if the scattering potential is independent of time.) In this approach, we imagine a continuous influx of incident particles, and hence also of scattering ones. Since nothing then varies with time, we can work with stationary state wave functions.

14. Partial-Wave Theory of Scattering by a Potential

We consider then the problem of a particle which is free outside a sphere of radius $r = a$ but which is scattered (and perhaps absorbed) by a potential *inside* the radius. We assume that the *incident* particle has a given momentum **p** and therefore wave number $\mathbf{k} = (1/\hbar)\mathbf{p}$. This may be taken in the z direction. The particle may then be represented by an incident wave function $\psi_{inc} = e^{ikz}$. Following now the formalism of Chapter 5 on scattering of sound waves [formulas (5.202) and following] we expand the incident wave in angular momentum components,

$$e^{ikz} = \sum_{l=0}^{\infty} i^l(2l+1)j_l(kr)P_l(\cos\theta) \tag{9.283}$$

This is a decomposition into angular momentum eigenfunctions, represented by the Legendre polynomials $P_l(\cos\theta)$. The expansion expresses the fact that a plane wave has components of all possible angular momenta about the scattering center.

Using the formula for $j_l(kr)$ in terms of the Hankel functions,

$$j_l(kr) = \tfrac{1}{2}[h_l^{(1)}(kr) + h_l^{(2)}(kr)] \tag{5.204}$$

and recognizing that $h_l^{(1)}$ describes a flow *outward* from the scattering center and $h_l^{(2)}$ one *inward* towards it, we can describe the scattering in terms of a modification of the *outgoing* wave. Thus, we can write the *total* wave function, *including* the scattered wave, as

$$\psi = \tfrac{1}{2}\sum i^l(2l+1)[w_l h_l^{(1)}(kr) + h_l^{(2)}(kr)]\,P_l(\cos\theta) \tag{9.284}$$

the scattered wave being specifically

$$\psi_{sc} = \tfrac{1}{2}\sum i^l(2l+1)(w_l-1)h_l^{(1)}(kr)P_l(\cos\theta) \tag{9.285}$$

which is, of course, entirely outgoing.

The constants w_l are determined by solving the wave equation *inside* the region $r = a$ and fitting ψ and its radial derivative at $r = R$.

Since the wave function must be finite at the origin, $r\psi = \phi$ must be zero there.

If we expand

$$\phi = \sum R_l(r)P_l(\cos\theta) \tag{9.286}$$

in the region of the potential, R_l satisfies

$$-\frac{\hbar^2}{2m}\left[\frac{d^2R_l}{dr^2} - \frac{l(l+1)}{r^2}R_l\right] + VR_l = ER_l \tag{9.287}$$

or

$$\frac{d^2R_l}{dr^2} - \frac{l(l+1)}{r^2}R_l + (k^2 - U)R_l = 0 \tag{9.288}$$

where

$$U = \frac{2m}{\hbar^2}V(r)$$

and

$$k^2 = \frac{2mE}{\hbar^2} \qquad (9.289)$$

Only one of the two solutions of (9.288) — the one which goes as r^l for $r \to 0$ — is physically acceptable. Therefore, there is only one arbitrary multiplication constant in R_l:

$$R_l = A_l g_l(r) \qquad (9.290)$$

Since the radial wave functions for each angular momentum ("partial wave") must be fitted to its outside value at $r = a$, the following equations determine the matching of the wave functions at the boundary:

$$\frac{1}{2} i^l (2l+1)[w_l h_l^{(1)}(ka) + h_l^{(2)}(ka)] = A_l \frac{1}{a} g_l(a) \qquad (9.291)$$

$$\frac{1}{2} i^l (2l+1)[w_l h_l^{(1)'}(ka) + h_l^{(2)'}(ka)]k = A_l \left(\frac{1}{r} g_l(r) \right)'_{r=a}$$

$$= \frac{A_l}{a} \left(g_l'(a) - \frac{1}{a} g_l(a) \right) \qquad (9.292)$$

Dividing the second of these equations by the first, we obtain an equation for w_l,

$$ka \frac{w_l h_l^{(1)'}(ka) + h_l^{(2)'}(ka)}{w_l h_l^{(1)}(ka) + h_l^{(2)}(ka)} = a \frac{g_l'(a)}{g_l(a)} - 1 = \gamma_l(E) \qquad (9.293)$$

We have written $\gamma_l(E)$ in recognition of the fact that it does in fact depend on the energy of the incident particle.

Solving for w_l, we find that

$$w_l = -\frac{ka h_l^{(2)'}(ka) - \gamma_l h_l^{(2)}(ka)}{ka h_l^{(1)'}(ka) - \gamma_l h_l^{(1)}(ka)} \qquad (9.294)$$

Thus, by solving the wave equation inside the region of the potential it is possible to find w_l. The problem of the scattering, absorption, and total cross sections may be solved in *terms* of w_l, which may now be assumed known. However, let us, before proceeding, make one general observation. If γ_l is real, the numerator of (9.294) is the complex conjugate of the denominator and $|w_l| = 1$. In this case, the flux of particles *outward* from the scatterer is the same as the flux *in* toward it; there is, therefore, no absorption. If, on the other hand $|w_l| < 1$, some absorption takes place. It is clearly impossible, on physical grounds, to have $|w_l| > 1$.

Let us now calculate the cross sections in terms of the quantities w_l. The partial cross sections (for a given l) are defined as follows:

Scattering cross section $\sigma_{sc}^{(l)} =$ (flux of scattered wave)/(incident flux per unit area).

Absorption cross section $\sigma_{\text{abs}}^{(l)} = [(\text{flux of convergent wave}) - (\text{flux of divergent wave})]/(\text{incident flux per unit area})$.

Total cross section

$$\sigma_t^{(l)} = \sigma_{\text{sc}}^{(l)} + \sigma_{\text{abs}}^{(l)} \tag{9.295}$$

The incident flux per unit area is the mean value of the velocity operator $(1/m)\mathbf{p}$;

$$\int e^{-ikz} \frac{1}{m} p_z \, e^{ikz} \, dz \, dx \, dy$$

where the x and y integrals are over a unit area. This is

$$f_{\text{inc}} = \frac{\hbar k}{m} \tag{9.296}$$

The total efflux associated with the lth component of the scattered wave is

$$F_{\text{sc}}^{(l)} = \frac{\hbar k}{m} \int |\psi_{\text{sc}}^{(l)}|^2 \, 2\pi r^2 \, d\mu \tag{9.297}$$

where $\mu = \cos \theta$. This should be calculated at large r, where we may use the asymptotic form of the Hankel function. From (9.285)

$$\psi_{\text{sc}}^{(l)} = \frac{1}{2} i^l (2l+1)(w_l - 1)(-i)^{l+1} \frac{e^{ikr}}{kr} P_l(\mu) \tag{9.298}$$

It is an immediate consequence that

$$F_{\text{sc}}^{(l)} = \frac{\hbar k}{m} \frac{1}{4} (2l+1)^2 |(1-w_l)|^2 \frac{1}{k^2 r^2} 2\pi r^2 \frac{2}{2l+1}$$

$$= \frac{\hbar k}{m} \frac{\pi}{k^2} (2l+1)|(1-w_l)|^2 \tag{9.299}$$

Thus the scattering cross section is

$$\sigma_{\text{sc}}^{(l)} = \frac{F_{\text{sc}}^{(l)}}{f_{\text{inc}}} = \frac{\pi}{k^2} (2l+1)|(1-w_l)|^2 \tag{9.300}$$

The terms involved in calculating the absorption cross section may be written down immediately from the above. The converging flux is like the $F_{\text{sc}}^{(l)}$ above except that the factor $(w_l - 1)$ is missing; in the *diverging* flux the $(w_l - 1)$ is replaced by w_l. Therefore,

$$\sigma_{\text{abs}}^{(l)} = \frac{\pi}{k^2} (2l+1)(1 - |w_l|^2) \tag{9.301}$$

For large enough l, $w_l \to 1$ and there is no longer any absorption. If we

write
$$\sigma_{sc} = \sum_l \sigma_{sc}^{(l)} \qquad (9.302)$$

and
$$\sigma_{abs} = \sum_l \sigma_{abs}^{(l)} \qquad (9.303)$$

the series converge. This is because, as l increases, less and less of the wave function is found within the region where the potential is nonzero.

Since the formulas for the partial cross sections are the same as in the problem of sound waves (Chapter 5), the inequalities discussed there [Eqs. (5.233) to (5.247)] again apply. In particular, the total cross section is

$$\sigma_{tot}^{(l)} = \frac{2\pi}{k^2} \operatorname{Re} \sum_l (2l+1)(1-w_l) \qquad (9.304)$$

15.   Low-Energy Scattering

An interesting case which we explore more fully is that in which the energy of the incident particles is low, so that $ka \ll 1$. We discuss the problem of a "hard sphere," as well as that of a constant scattering potential, both real and complex.

First, let us look at the formula (9.294) for $ka \ll 1$. In terms of γ_0, w_0 has the following simple form:

$$w_0 = e^{-2ika} \frac{1+\gamma_0+ika}{1+\gamma_0-ika} \qquad (9.305)$$

If we define an angle β_0 by

$$\tan \beta_0 = \frac{ka}{1+\gamma_0} \qquad (9.306)$$

w_0 can be written

$$w_0 = e^{2i\delta_0} = e^{2i(\beta_0-ka)} \qquad (9.307)$$

so that the $l=0$ phase shift is

$$\delta_0 = \beta_0 - ka = \tan^{-1} \frac{ka}{1+\gamma_0} - ka \qquad (9.308)$$

The scattering cross section is then

$$\sigma_{sc}^{(0)} = \frac{\pi}{k^2} |1-w_0|^2 = \frac{4\pi}{k^2} \sin^2 \delta_0 \qquad (9.309)$$

A specific form is found on substituting from (9.308):

$$\sigma_{sc}^{(0)} = \frac{4\pi}{k^2} \frac{[ka \cos ka - (1+\gamma_0) \sin ka]^2}{(1+\gamma_0)^2 + k^2a^2} \qquad (9.310)$$

For $ka \ll 1$ this becomes approximately

$$\sigma_{sc}^{(0)} = \frac{4\pi a^2 \gamma_0^2}{(1+\gamma_0)^2} \tag{9.311}$$

For the case of a "hard sphere," $\psi = 0$ at $r = a$, so $\gamma_0 \to \infty$. The cross section then is

$$\sigma_{sc}^{(0)} = \frac{4\pi}{k^2} \sin^2 ka \approx 4\pi a^2 \tag{9.312}$$

On the other hand, for a constant potential $V = -V_0$ inside $r = a$,

$$g_0 = \sin Kr \tag{9.313}$$

where

$$K^2 = \frac{2m}{\hbar^2}(E + V_0) \tag{9.314}$$

Then

$$\gamma_0(E) = Ka \cot Ka - 1 \tag{9.315}$$

It appears that there are resonant maxima in $\sigma_{sc}^{(0)}$ when $\gamma_0 \approx -1$, that is, $\cot Ka = 0$. In such a region of energy

$$\sigma_{sc}^{(0)} = \frac{\pi}{k^2} \cos^2 ka \approx \frac{\pi}{k^2} \tag{9.316}$$

Note that the selection of a *complex* value for the scattering potential,

$$V = -V_1 - iV_2 \tag{9.317}$$

leads to an *absorption* in addition to a scattering. K, as defined in (9.314), becomes complex

$$K = K_1 + iK_2 \tag{9.318}$$

and

$$g_0 = \sin K_1 r \cosh K_2 r + i \cos K_1 r \sinh K_2 r \tag{9.319}$$

In this case

$$\gamma_0(E) = \frac{(K_1 a + iK_2 a)(\cos K_1 a \cosh K_2 a - i \sin K_1 a \sinh K_2 a)}{\sin K_1 a \cosh K_2 a + i \cos K_1 a \sinh K_2 a} - 1$$

$$= \gamma_1(E) - i\gamma_2(E) \tag{9.320}$$

It follows from (9.305) that

$$w_0 = e^{-2ika} \frac{1 + \gamma_1 - i(\gamma_2 - ka)}{1 + \gamma_1 - i(\gamma_2 + ka)} \tag{9.321}$$

From (9.310), the absorption cross section is

$$\sigma_{abs}^{(0)} = \frac{\pi}{k^2} (1 - |w_0|^2)$$

$$= \frac{\pi}{k^2} \left[1 - \frac{(1+\gamma_1)^2 + (\gamma_2 - ka)^2}{(1+\gamma_1)^2 + (\gamma_2 + ka)^2} \right]$$

$$= \frac{4\pi a\gamma_2}{k} \frac{1}{(1+\gamma_1)^2 + (\gamma_2 + ka)^2} \tag{9.322}$$

Note that the absorption cross section goes like the inverse of the *velocity* of the incident particles for low incident energy.

Problem 9-24: Show that, for the potential $V = -(V_1 + iV_2)$,

$$\gamma_1 = \frac{K_1 a \sin K_1 a \cos K_1 a + K_2 a \sinh K_2 a \cosh K_2 a}{\sin^2 K_1 a \cosh^2 K_2 a + \cos^2 K_1 a \sinh^2 K_2 a} - 1$$

and

$$\gamma_2 = \frac{K_2 a \sin K_1 a \cos K_1 a - K_1 a \sinh K_2 a \cosh K_2 a}{\sin^2 K_1 a \cosh^2 K_2 a + \cos^2 K_1 a \sinh^2 K_2 a}$$

Problem 9-25: Defining a resonance as an energy E_r such that

$$1 + \gamma_1(E_r) = 0$$

in the neighborhood of the resonance

$$1 + \gamma_1(E) = (E - E_r)\gamma_1'(E_r)$$

Show, then, that

$$\sigma_{abs}^{(0)} = \frac{2\pi a}{k\gamma_1'(E_r)} \frac{\Gamma_r}{(E - E_r)^2 + \frac{1}{4}\Gamma^2}$$

where

$$\Gamma_r = \frac{2\gamma_2(E_r)}{\gamma_1'(E_r)}, \qquad \Gamma = \frac{2(\gamma_2(E_r) + ka)}{\gamma_1'(E_r)}$$

The general formula for the absorption cross section is easily obtained from (9.305) and (9.301),

$$\sigma_{abs}^{(0)} = \frac{\pi}{k^2} \frac{4\gamma_2 ka}{(1+\gamma_1)^2 + (\gamma_2 + ka)^2} \tag{9.323}$$

Expanding about the resonance energy E_r for which

$$1 + \gamma_1(E_r) = 0$$

we obtain the formula

$$\sigma_{\text{abs}}^{(0)} = \frac{\pi}{k^2} \frac{4\gamma_2 ka}{(E - E_r)^2 [\gamma_1'(E_r)]^2 + (\gamma_2 + ka)^2} \tag{9.324}$$

With the definitions

$$\Gamma_2 = \frac{2\gamma_2(E_r)}{\gamma_1'(E_r)} \tag{9.325}$$

and

$$\Gamma_s = \frac{2ka}{\gamma_1'(E_r)} \tag{9.326}$$

this becomes

$$\sigma_{\text{abs}}^{(0)} = \frac{\pi}{k^2} \frac{\Gamma_r \Gamma_s}{(E - E_r)^2 + \frac{1}{4}(\Gamma_r + \Gamma_s)^2} \tag{9.327}$$

The scattering and total cross sections for the $l = 0$ partial wave are obtained from

$$1 - w_0 = e^{-2ika} \left\{ (e^{2ika} - 1) + \left[1 - \frac{1 + \gamma_1 - i(\gamma_2 - ka)}{1 + \gamma_1 - i(\gamma_2 + ka)} \right] \right\}$$

$$= e^{-2ika} \left\{ (e^{2ika} - 1) - \frac{2ika}{1 + \gamma_1 - i(\gamma_2 + ka)} \right\} \tag{9.328}$$

If the second term in the curly bracket is neglected, the cross section is that for a "hard sphere" of radius a, as shown above. Such a term is referred to as the "potential scattering" term.

If, on the other hand, we neglect the *first* term, we get, for $1 + \gamma_1 \approx 0$, a resonant-type scattering. *Near* a resonance, and for not too strong absorption, this term should dominate. (Otherwise, there is interference between the "potential scattered" wave and the "resonant-scattered" one.) When the resonance term *does* dominate, it is clear that

$$\sigma_{\text{sc}}^{(0)} = \frac{\pi}{k^2} \frac{4k^2 a^2}{(1 + \gamma_1)^2 + (\gamma_2 + ka)^2} \tag{9.329}$$

Expanding $1 + \gamma_1 = a(g_0'(a)/g_0(a))$ around the resonance, we then obtain the result

$$\sigma_{\text{sc}}^{(0)} = \frac{\pi}{k^2} \frac{\Gamma_s^2}{(E - E_r)^2 + \frac{1}{4}(\Gamma_r + \Gamma_s)^2} \tag{9.330}$$

We note that this approaches a constant value as the energy $\rightarrow 0$.

Equations (9.327) and (9.330) may be combined to give the total resonance cross section

$$\sigma_{\text{tot}}^{(0)} = \frac{\pi}{k^2} \frac{\Gamma_s \Gamma}{(E - E_r)^2 + \frac{1}{4}\Gamma^2} \tag{9.331}$$

where

$$\Gamma = \Gamma_r + \Gamma_s = 2\left(\frac{\gamma_2(E_r) + ka}{\gamma_1'(E_r)}\right) \tag{9.332}$$

These formulas for the resonance cross sections are the well-known "Breit–Wigner" formulas.

Problem 9-26: Prove that the scattering cross section may in general be written

$$\sigma_{sc}^{(0)} = \frac{4\pi}{k^2}\left\{\sin^2 ka + \frac{1}{4}\frac{\Gamma_s^2}{(E-E_r)^2 + \frac{1}{4}\Gamma^2} - \sin ka \frac{\Gamma_s(E-E_r)}{(E-E_r)^2 + \frac{1}{4}\Gamma^2}\right\}$$

The first term is called "potential scattering" and the second "resonance scattering." The third term then represents an interference between them.

Show that interference enhances the cross section at energies below the resonant energy, but gives rise to a destructive interference above it. For $ka \ll 1$, Γ_s, $\Gamma \ll E_r$ sketch the behavior of the cross section as a function of energy over a region of energy around the resonance extending over a range substantially larger than Γ on each side of it.

Problem 9-27: Find the bound states for a δ function potential in one dimension. $[V = -V_0\delta(x)]$. Calculate also, as a function of energy, the reflection and transmission coefficients for a free particle of energy E and positive momentum, scattered from the potential.

Problem 9-28: Derive a formula for the differential scattering cross section of a particle impinging on a central potential, given that there exists only one very weakly bound state for the system. Sketch the energy dependence of the cross section.

PRELUDE TO CHAPTER 10

This final chapter is concerned with some elementary notions of the quantum mechanics of many-body problems. We stop short of the Green's function methods which dominate the literature of the subject at the present time, but provide the background on the basis of which these methods are built.

Our approach is to consider first the treatment of systems of identical *noninteracting* particles. The method of second quantization is first introduced in this context; the introduction of interactions then represents a separate and distinct step.

Consider the problem first from the viewpoint of Schrödinger wave mechanics. Making the assumption of indistinguishability of identical particles, we see that states must be either symmetric or antisymmetric in exchange of two particles. This leads to the distinction between two *kinds* of particles; when the wave function is symmetric, we call the particles *bosons*; when antisymmetric, *fermions*.

The second quantization theory of bosons is developed as follows: Since, for noninteracting particles, states of many-particle systems may be specified in terms of *individual-particle* states, these many-particle states may be completely specified by designating the *occupation* of the single-particle states. In complete analogy with harmonic oscillator theory, individual-particle creation and annihilation operators may be defined which, operating on a given state, create a system containing either one particle more or less. These operators obey the commutation relations

$$[a_\alpha, a_{\alpha'}{}^\dagger] = \delta_{\alpha\alpha'}$$

a_α being the operator which *annihilates* a particle in the single-particle state $|\alpha\rangle$, $a_{\alpha'}{}^\dagger$, that which *creates* a particle in state $|\alpha'\rangle$. Operators

$$|\psi\rangle = \sum a_\alpha|\alpha\rangle, \qquad \langle\psi| = \sum a_{\alpha'}{}^\dagger\langle\alpha'|$$

can be defined, as well as Schrödinger-like operators $\psi(\mathbf{r})$, $\psi^\dagger(\mathbf{r})$ which in effect annihilate and create a particle *at a given point*, and their commutation relations follow. It is then shown that many-particle operators may be introduced. If the "free" particles are subject to scattering by a potential, an operator may be defined which describes the scattering of a particle in *any* state into *any* other, with an amplitude determined by the operator. Scattering is described in terms of a product of two operators, one of which *annihilates* a particle in one state, the second of which *creates* a particle in another.

At this point the generalization to fermions is given. It is shown that the antisymmetry condition is satisfied if the creation and annihilation operators satisfy anticommutation, rather than commutation rules. The rest of the chapter applies to either bosons or fermions.

Two-particle (interaction) operators are then introduced. These describe the annihilation of two particles in *initial* states, and their creation in *final* states, i.e., they describe the mutual scattering of particles under interaction. It is in fact shown that valid many-particle operators are formed by replacing the expression for *mean* values in Schrödinger *wave functions* by ψ operators. The order of these operators must, of course, be such as to make the resultant operator Hermitian.

The scattering of free particles in momentum eigenstates is then considered as an example, and it is shown that the formalism ensures conservation of momentum provided the interaction is translation invariant.

We next develop, using second quantization, the well-known Hartree–Fock method. It is shown that this approximation neglects all scatterings except those in which two fermi particles are scattered back into their original states, or are exchanged, i.e., each is scattered into the state originally occupied by the other. The basic features of Hartree–Fock theory are developed in some detail.

As a particular result of interest, we show that plane waves (momentum eigenstates) are solutions of the Hartree–Fock equations for translationally invariant potentials (e.g., Coulomb or Yukawa interactions). The various contributions to the energy are calculated.

For such a "fermi gas" the dependence of kinetic and exchange energies on density is calculated (the direct interaction or "Coulomb" energy does not depend on density).

Since the kinetic energy per particle is positive and *increases* with increasing density in such a way that it dominates at high density while the exchange energy which operates between particles of the same spin, is *negative* and *decreases* with increasing density in such a way that it dominates at low density, we ask whether at sufficiently low densities the spins of all particles would not tend all to align, creating spontaneously a ferromagnetic state. It is found that this is true for sufficiently small densities – at least slightly smaller, however, than those existing in any good metal.

As an introduction to the development of density matrix theory we then calculate the number of excited states of a many-body system per unit energy interval.

The theory of statistical operators ("density matrices") is developed, permitting us to calculate *ensemble* averages of dynamical variables which are macroscopically indistinguishable though microscopically different. The statistical operator then gives complete information about macroscopic systems. Its dynamics (time variation) is determined.

The statistical operator is determined for the *canonical ensemble* (for which the number of particles is fixed) in thermal equilibrium with its surroundings, and then for the *grand canonical ensemble*, for which only the *average* number of particles is fixed.

As an illustration of the preceding theory, the ensemble average of the number of particles in an energy state is calculated for both bosons and fermions. These are the familiar *Bose–Einstein* and *Fermi–Dirac* distribution functions.

The linear response theory (originally due to Kubo) is then developed. This is a theory of response, defined by the perturbation of the density matrix in the linear (first-order perturbation) approximation, to an arbitrary external stimulus. This result has a wide range of applicability; of particular interest is the response of an electron system to an electromagnetic perturbation. In particular, we have calculated the transverse conductivity of an electron gas.

Finally, we apply the above result to the problem of propagation of an electromagnetic wave in an electron gas — a problem already discussed in simplified form in Chapter 5. The dispersion relation for the electromagnetic wave is determined (by what is essentially a time-dependent Hartree–Fock approximation), and it is shown that such waves may be propagated in "free-electron" metals at frequencies above the plasma frequency.

REFERENCES

Fano, U., *Rev. Mod. Phys.* **29**, No. 1 (1957).

Kittel, C., *Quantum Theory of Solids*. New York: Wiley, 1953.

Mandl, F., *Introduction to Quantum Field Theory*. New York: Interscience, 1959.

Mattis, D., *Theory of Magnetism*. New York: Harper and Row, 1965, Chapter 8.

Messiah, A., *Mécanique Quantique*. Paris: Dunod, 1958.

Slater, J. C., *Quantum Theory of Matter*. New York: McGraw-Hill, 1951.

Ziman, J., *Elements of Advanced Quantum Theory*. New York: Cambridge Press, 1969.

10

QUANTUM MECHANICS OF MANY-BODY PROBLEMS

"Don't argue about the difficulties. The difficulties will argue for themselves."

Winston Churchill

1. Introduction

In this chapter we develop the bases of the theory of many-body systems, using the so-called "method of second quantization."

To develop an understanding of the technique, it is best to consider first a system of noninteracting particles. In such a case we know that the Hamiltonian for an N-body system has the form

$$H = \sum_{n=1}^{N} H_n \qquad (10.1)$$

where H_n is the Hamiltonian of the nth particle. If we consider identical particles, the Hamiltonians all have the same form, but are functions of the coordinates $\mathbf{r}_n$ and momenta $\mathbf{p}_n$ of the individual particles. The wave function is necessarily constructed from products of single-particle states. If

$$H_n = H^{(0)}(\mathbf{r}_n, \mathbf{p}_n) \qquad (10.2)$$

in the Schrödinger scheme there will be solutions of the form

$$\psi = \psi_{\alpha_1}(\mathbf{r}_1)\psi_{\alpha_2}(\mathbf{r}_2) \ldots \psi_{\alpha_N}(\mathbf{r}_N) \qquad (10.3)$$

563

For if we substitute this into

$$H\psi = E\psi \tag{10.4}$$

and divide through by ψ, we get

$$\sum_n \frac{1}{\psi_{\alpha_n}} H^0(\mathbf{r}_n, \mathbf{p}_n)\psi_{\alpha_n} = E \tag{10.5}$$

Each term on the left-hand side must be a constant, so the ψ_{α_n}'s satisfy

$$H_n\psi_{\alpha_n} = E_{\alpha_n}\psi_{\alpha_n} \tag{10.6}$$

and the total energy is

$$E = \sum E_{\alpha_n} \tag{10.7}$$

Let the eigenstates of the one-particle Hamiltonian be designated by $|\alpha\rangle$ (Schrödinger wave function $\langle\mathbf{r}|\alpha\rangle$) and the energies by E_α. The single-particle states and energies above are then chosen from this set.

It should be noted that in (10.3) equally valid solutions of the same energy are obtained if the α_n's are permuted among themselves in any way. Alternatively, the α_n's may be left as they are and the $\mathbf{r}_n$'s permuted to get the same solution. The most general solution at the energy is an arbitrary linear combination of these various permutations.

However, the choice may be limited if we make the *hypothesis of indistinguishability* of identical particles. This hypothesis implies that, if any two particles are interchanged in the wave function, the resulting distribution function $|\psi|^2$ will be unaltered. Thus, in general it would be necessary that

$$\psi(\ldots \mathbf{r}_i \ldots \mathbf{r}_j \ldots) = e^{i\gamma}\psi(\ldots \mathbf{r}_j \ldots \mathbf{r}_i \ldots) \tag{10.8}$$

If the phase γ applies to *all* interchanges, to permute the ith and jth particles a second time leads to the original wave function; therefore

$$e^{2i\gamma} = 1 \tag{10.9}$$

so that $\gamma = 0$ or π. Thus, interchanging the states occupied by two particles either leaves the wave function unaltered or changes its sign. The second implies that two particles cannot be in the *same* state, since then the wave function would equal its negative.

There appear to be two different kinds of particles in nature: those for which $\gamma = 0$ being known as bosons and those for which $\gamma = \pi$ as fermions.

We now describe our many-particle system in a different way, by simply specifying the occupancy of the single-particle levels $|\alpha\rangle$, that is, by specifying the choice of $\alpha_1, \alpha_2, \ldots, \alpha_N$ from among these levels. A many-particle state of noninteracting particles may then be specified by the state vector

$$|n_1 n_2 \ldots n_\alpha \ldots\rangle$$

For an N-particle system,

$$\sum n_\alpha = N \tag{10.10}$$

A formalism may now be developed in which the filling of single-particle levels is treated analogously to the adding of quanta to an oscillator, as discussed in Chapter 9. (Note then that each single-particle state has its oscillator analog.) We thus introduce for each state $|\alpha\rangle$ in analogy with (9.30) creation operators $a_\alpha^\dagger$ with the property

$$a_\alpha^\dagger| \ldots n_\alpha \ldots\rangle = \sqrt{n_\alpha + 1}| \ldots (n_\alpha + 1) \ldots\rangle \tag{10.11}$$

and annihilation operators a_α such that

$$a_\alpha| \ldots n_\alpha \ldots\rangle = \sqrt{n_\alpha}| \ldots (n_\alpha - 1) \ldots\rangle \tag{10.12}$$

Two results are worth noting:

(a) $\quad a_\alpha a_\alpha^\dagger - a_\alpha^\dagger a_\alpha = 1 \tag{10.13}$

For

$$a_\alpha a_\alpha^\dagger|n_\alpha\rangle = a_\alpha \sqrt{n_\alpha + 1}|n_\alpha + 1\rangle$$
$$= (n_\alpha + 1)|n_\alpha\rangle \tag{10.14}$$

and

$$a_\alpha^\dagger a_\alpha|n_\alpha\rangle = a_\alpha^\dagger \sqrt{n_\alpha}|n_\alpha - 1\rangle$$
$$= n_\alpha|n_\alpha\rangle \tag{10.15}$$

Subtracting, we obtain (10.13). (In the derivation, we have ignored the occupation of the other states, which is irrelevant.) Writing

$$[a_\alpha, a_\alpha^\dagger] = a_\alpha a_\alpha^\dagger - a_\alpha^\dagger a_\alpha = 1$$

and noting that

$$[a_\alpha, a_{\alpha'}^\dagger] = 0, \qquad \alpha' \neq \alpha$$

we obtain commutation relations

$$[a_\alpha, a_{\alpha'}^\dagger] = \delta_{\alpha\alpha'} \tag{10.16}$$

It is also obvious that

$$[a_\alpha^\dagger, a_{\alpha'}^\dagger] = 0 \tag{10.17}$$

and

$$[a_\alpha, a_{\alpha'}] = 0 \tag{10.18}$$

(b) $\quad a_\alpha^\dagger a_\alpha|n_\alpha\rangle = n_\alpha|a_\alpha\rangle$, as obtained in (10.15) above. Thus the states $|n_\alpha\rangle$ are eigenstates of $a_\alpha^\dagger a_\alpha$ with eigenvalue n_α. We therefore call

$a_\alpha^\dagger a_\alpha$ the "number operator"; its eigenvalue is the number of particles in the state α.

Let us now define *operators*

$$|\psi\rangle = \sum a_\alpha |\alpha\rangle \tag{10.19}$$

and

$$\langle\psi| = \sum \langle\alpha|a_\alpha^\dagger \tag{10.20}$$

In terms of Schrödinger wave functions

$$\psi(\mathbf{r}) = \sum a_\alpha \varphi_\alpha(\mathbf{r}) \tag{10.21}$$

and

$$\psi^\dagger(\mathbf{r}) = \sum a_\alpha^\dagger \varphi_\alpha^*(\mathbf{r}) \tag{10.22}$$

where

$$\psi(\mathbf{r}) = \langle\mathbf{r}|\psi\rangle \tag{10.23}$$

and

$$\varphi_\alpha(\mathbf{r}) = \langle\mathbf{r}|\alpha\rangle \tag{10.24}$$

Consider the operator

$$\langle\psi|\psi\rangle = \int \psi^\dagger(\mathbf{r})\psi(\mathbf{r}) \, d^3\mathbf{r} \tag{10.25}$$

This may be expanded to give

$$\sum_{\alpha,\alpha'} \langle\alpha'|\alpha\rangle a_{\alpha'}^\dagger a_\alpha = \sum_\alpha a_\alpha^\dagger a_\alpha$$

$$= \sum_\alpha n_\alpha \tag{10.26}$$

so it is the operator for *total number* of particles. The operator for the *total energy* of the system is

$$\langle\psi|H^{(0)}|\psi\rangle = \int \psi^\dagger(\mathbf{r})H^{(0)}\psi(\mathbf{r}) \, d^3\mathbf{r} \tag{10.27}$$

where $H^{(0)}$ is the one-particle operator on the single-particle states. For, expanding the $|\psi\rangle$ and $\langle\psi|$,

$$\langle\psi|H^{(0)}|\psi\rangle = \sum_{\alpha,\alpha'} a_{\alpha'}^\dagger a_\alpha \langle\alpha'|H^{(0)}|\alpha\rangle$$

$$= \sum_{\alpha,\alpha'} a_{\alpha'}^\dagger a_\alpha E_\alpha \langle\alpha'|\alpha\rangle$$

$$= \sum_\alpha a_\alpha^\dagger a_\alpha E_\alpha$$

$$= \sum_\alpha n_\alpha E_\alpha \tag{10.28}$$

from which the stated result follows.

If the particles are subject to a perturbing potential $V(\mathbf{r})$ the second-quantized operator is, similarly,

$$\langle\psi|V|\psi\rangle = \sum_{\alpha,\alpha'} a_{\alpha'}{}^\dagger a_\alpha \int \varphi_{\alpha'}^*(\mathbf{r}) V(\mathbf{r}) \varphi_\alpha(\mathbf{r}) \, d^3\mathbf{r}$$

$$= \sum_{\alpha,\alpha'} a_{\alpha'}{}^\dagger a_\alpha \langle\alpha'|V|\alpha\rangle \tag{10.29}$$

This is clearly correct, since the potential causes transitions (scatterings) from state $|\alpha\rangle$ to states $|\alpha'\rangle$, with amplitude $\langle\alpha'|V|\alpha\rangle$.

The results obtained heretofore obviously apply to particles any number of which may occupy a given state, that is, bosons. The amendment necessary to make the theory applicable to fermions was first shown by Jordan and Wigner to be the replacement of the *commutators* in (10.16)–(10.18) by *anticommutators*

$$[\xi,\eta]_+ = \xi\eta + \eta\xi \tag{10.30}$$

that is, to assume the relations

$$[c_\alpha, c_{\alpha'}{}^\dagger]_+ = \delta_{\alpha\alpha'} \tag{10.31}$$

$$[c_\alpha, c_{\alpha'}]_+ - [c_\alpha{}^\dagger, c_{\alpha'}{}^\dagger]_+ = 0 \tag{10.32}$$

Let us verify that the operators satisfying these conditions produce states with the desired properties.

(*i*) We note first that, since $c_\alpha{}^\dagger c_\alpha{}^\dagger = 0$, it is not possible to put two particles into any state; nor, since $c_\alpha c_\alpha = 0$, is it possible to annihilate two from a state.

(*ii*) The eigenvalues of $n_\alpha = c_\alpha{}^\dagger c_\alpha$ may be derived as follows:

$$n_\alpha{}^2 = c_\alpha{}^\dagger c_\alpha c_\alpha{}^\dagger c_\alpha$$

$$= c_\alpha{}^\dagger (1 - c_\alpha{}^\dagger c_\alpha) c_\alpha$$

$$= c_\alpha{}^\dagger c_\alpha$$

$$= n_\alpha \tag{10.33}$$

Therefore the eigenvalues of the number operator are $n_\alpha = 0$ or $n_\alpha = 1$. This is the statement, for noninteracting particles, of the Pauli exclusion principle, which says that at most one fermion can be put in a given state.

(*iii*) The effect of $c_\alpha c_\alpha{}^\dagger$ on a state may be calculated as follows: Let

$$c_\alpha{}^\dagger|0\rangle = A_\alpha|1\rangle \tag{10.34}$$

where A_α is a constant which may be taken to be real. Taking the magnitude squared of (10.34)

$$\langle 0|c_\alpha c_\alpha{}^\dagger|0\rangle = A_\alpha{}^2$$

But $c_\alpha c_\alpha\dagger = 1 - n_\alpha$; therefore,

$$A_\alpha^2 = 1, \qquad A_\alpha = \pm 1$$

By a similar argument

$$c_\alpha|1\rangle = B_\alpha|0$$

where

$$B_\alpha^2 = 1$$

It is essential to note that the sign to be chosen depends not only on the state in question, but on other states. For

$$c_{\alpha'}\dagger c_\alpha\dagger|n_{\alpha'} = 0, n_\alpha = 0\rangle = c_{\alpha'}\dagger c_\alpha\dagger|0_{\alpha'}0_\alpha\rangle = -c_\alpha\dagger c_{\alpha'}\dagger|0_{\alpha'}0_\alpha\rangle$$

and

$$c_{\alpha'}c_\alpha|1_{\alpha'}1_\alpha\rangle = -c_\alpha c_{\alpha'}|1_{\alpha'}1_\alpha\rangle$$

These conditions can be met in the following way: Arrange the states $|\alpha\rangle$ in order (say, the order of increasing energy). Then if the operators are applied in this order, take the constant to be 1. If they are applied in any *other* order, the constant will then be $(-1)^p$, where p is the number of permutations required to get from the original order to the one in question.

2. Commutation Rules for the ψ Operators

Once the commutation rules for the operators c_α, $c_\alpha\dagger$ are known, those for $\psi(\mathbf{r})$, $\psi\dagger(\mathbf{r})$ as defined by (10.21) and (10.22) follow. Since ψ, $\psi\dagger$ can be defined for either bosons or fermions, commutation rules may be calculated for either case. Consider, for instance,

$$[\psi(\mathbf{r}), \psi\dagger(\mathbf{r}')]_\eta$$

Where $\eta = \pm$ according as we are dealing with bosons (−) or fermions (+). Expanding the ψ's,[1]

$$[\psi(\mathbf{r}), \psi\dagger(\mathbf{r}')]_\eta = \sum_{\alpha,\alpha'} \varphi_\alpha(\mathbf{r})\varphi_{\alpha'}\dagger(\mathbf{r}')[a_\alpha, a_{\alpha'}\dagger]_\eta$$

$$= \sum_\alpha \varphi_\alpha(\mathbf{r})\varphi_\alpha^*(\mathbf{r}')$$

$$= \sum_\alpha \langle\mathbf{r}|\alpha\rangle\langle\alpha|\mathbf{r}'\rangle$$

$$= \langle\mathbf{r}|\mathbf{r}'\rangle$$

$$= \delta(\mathbf{r} - \mathbf{r}') \tag{10.35}$$

[1]We use a, $a\dagger$ to represent either boson or fermion operators, and c, $c\dagger$ specifically for fermions.

It is also easily verified that

$$[\psi(\mathbf{r}), \psi(\mathbf{r}')]_\eta = 0 \qquad (10.36)$$

and

$$[\psi^\dagger(\mathbf{r}), \psi^\dagger(\mathbf{r}')]_\eta = 0 \qquad (10.37)$$

The symmetry or antisymmetry of wave functions (according as they are bosons or fermions) is easily seen by putting two particles, at $\mathbf{r}$ and $\mathbf{r}'$, into an initial state with no particles present:

$$\psi^\dagger(\mathbf{r}')\psi^\dagger(\mathbf{r})|0\,0\ldots0\rangle = \sum_{\alpha',\alpha} \varphi_{\alpha'}^*(\mathbf{r}')\varphi_\alpha^*(\mathbf{r}) a_{\alpha'}^\dagger a_\alpha^\dagger |0_{\alpha'}0_\alpha\rangle$$

Since α, α' are dummy indices, we may also interchange them to get

$$\psi^\dagger(\mathbf{r}')\psi^\dagger(\mathbf{r})|0\,0\ldots0\rangle = \sum_{\alpha,\alpha'} \varphi_\alpha^*(\mathbf{r}')\varphi_{\alpha'}^*(\mathbf{r}) a_\alpha^\dagger a_{\alpha'}^\dagger |0_{\alpha'}0_\alpha\rangle$$

$$= -\eta \sum_{\alpha,\alpha'} \varphi_\alpha^*(\mathbf{r}')\varphi_{\alpha'}^*(\mathbf{r}) a_{\alpha'}^\dagger a_\alpha^\dagger |0_{\alpha'}0_\alpha\rangle$$

Therefore,

$$\psi^\dagger(\mathbf{r}')\psi^\dagger(\mathbf{r})|0\,0\ldots0\rangle = \tfrac{1}{2} \sum_{\alpha,\alpha'} \{\varphi_\alpha^*(\mathbf{r}')\varphi_\alpha^*(\mathbf{r}) - \eta\varphi_\alpha^*(\mathbf{r}')\varphi_{\alpha'}^*(\mathbf{r})\}$$

$$\times a_{\alpha'}^\dagger a_\alpha^\dagger |0_{\alpha'}0_{\alpha'}\rangle \qquad (10.38)$$

from which it follows that the wave function of the two particles is anti-symmetric for fermions and symmetric for bosons.

3. Representation of Two-Particle Operators

The preceding discussion has been confined to systems of noninter-acting particles. The question, now arises, how do we handle the compli-cations introduced by interactions of the particles? We confine our attention to two-particle interactions, which seem to be most common in familiar physical systems.

We know that, for any but the simplest potentials, it is not possible to find exactly even the state of a two-particle system, while the states of three-particle systems are almost always difficult to determine. We cannot, therefore, expect simple analytic solutions for many-body problems, but will have to develop approximate methods.

Let us consider a system of identical particles in which all pairs interact with a potential $v(\mathbf{r}, \mathbf{r}')$ when the particles of the pair are at positions $\mathbf{r}$ and $\mathbf{r}'$. Our viewpoint is that we can still use single-particle states as a basis, and consider the interactions as producing transitions or scatterings between these states. If the particles are initially in states

$|\alpha\rangle$ and $|\alpha'\rangle$ [with wave functions $\varphi_\alpha(\mathbf{r})$, $\varphi_{\alpha'}(\mathbf{r}')$],

$$v(\mathbf{r}, \mathbf{r}')\varphi_\alpha(\mathbf{r})\varphi_{\alpha'}(\mathbf{r}') = \sum_{\beta,\beta'} \varphi_\beta(\mathbf{r})\varphi_{\beta'}(\mathbf{r}')$$

$$\times \int \varphi_\beta^*(\mathbf{r})\varphi_{\beta'}^*(\mathbf{r}')v(\mathbf{r}, \mathbf{r}')\varphi_\alpha(\mathbf{r})\varphi_{\alpha'}(\mathbf{r}')d^3r\,d^3r' \quad (10.39)$$

Thus, the potential operating on states $|\alpha\rangle$, $|\alpha'\rangle$ produces pairs of states $|\beta\rangle$, $|\beta'\rangle$ with amplitudes

$$\langle\beta'\beta|v|\alpha\alpha'\rangle = \int \varphi_{\beta'}^*(\mathbf{r}')\varphi_\beta^*(\mathbf{r})v(\mathbf{r}, \mathbf{r}')\varphi_\alpha(\mathbf{r})\varphi_{\alpha'}(\mathbf{r}')d^3r\,d^3r' \quad (10.40)$$

We now show that the appropriate "second-quantized" operator for the potential in question is

$$\tfrac{1}{2} \int \psi^\dagger(\mathbf{r}')\psi^\dagger(\mathbf{r})v(\mathbf{r}, \mathbf{r}')\psi(\mathbf{r})\psi(\mathbf{r}')d^3r\,d^3r' \quad (10.41)$$

Because of the commutation (or anticommutation) rules for the ψ and $\psi^\dagger$ operators, the order of the factors in the integrand of (10.41) is important. It is easily seen that the operator (10.41) is Hermitian; if $\psi(\mathbf{r})$ and $\psi(\mathbf{r}')$ were interchanged it would not be. The structure of the operator follows the same pattern as for single-particle operators. We write down the formal expression for the mean value of v, and then interpret the "wave functions" ψ, $\psi^\dagger$ as operators.

If we expand the ψ's and $\psi^\dagger$'s in (10.41) in terms of the single-particle wave functions φ_α and their conjugates, it becomes

$$\tfrac{1}{2} \sum a_{\beta'}{}^\dagger a_\beta{}^\dagger a_\alpha a_{\alpha'} \int \varphi_{\beta'}^*(\mathbf{r}')\varphi_\beta^*(\mathbf{r})v(\mathbf{r}, \mathbf{r}')\varphi_\alpha(\mathbf{r})\varphi_{\alpha'}(\mathbf{r}')\,d^3r\,d^3r'$$

$$= \tfrac{1}{2} \sum \langle\beta'\beta|v|\alpha\alpha'\rangle\, a_{\beta'}{}^\dagger a_\beta{}^\dagger a_\alpha a_{\alpha'} \quad (10.42)$$

This describes the annihilation of states $|\alpha\rangle$ and $|\alpha'\rangle$ and the creation of states $|\beta\rangle$ and $|\beta'\rangle$, with the amplitude $\langle\beta'\beta|v|\alpha\alpha'\rangle$, precisely as in (10.40). The factor of $\tfrac{1}{2}$ allows for the fact that every scattering appears twice in the sum; a given term and that in which both β,β' and α, α' are interchanged are identical. The operator (10.41) then seems to be the correct one. Note, however, that it takes care automatically of the proper symmetry of the states under exchange of coordinates of arbitrary pairs of particles.

A particular example of the preceding discussion is the case of free particles interacting through a potential $v(\mathbf{r} - \mathbf{r}')$ which depends only on the relative position of the particles. Since this potential is invariant under a translation of the coordinate system, we expect the interaction to conserve momentum, and we see that this does indeed follow from the formalism. If the particles are free they are individually in eigenstates of momentum $\mathbf{p} = \hbar\mathbf{k}$, so we designate the states $|\alpha\rangle$ in terms of the wave-number eigenvalue $\mathbf{k}$. The second-quantized Hamiltonian operator consists of the free-particle Hamiltonian and the interaction term. The first is

$$\int \psi\dagger(\mathbf{r})H^{(0)}\psi(\mathbf{r})\, d^3\mathbf{r} = \sum a_{k'}\dagger a_k \frac{1}{\Omega} \int e^{-i\mathbf{k}'\cdot\mathbf{r}} \left(-\frac{\hbar^2}{2m}\nabla^2\right) e^{i\mathbf{k}\cdot\mathbf{r}}$$

$$= \sum \epsilon_k a_k\dagger a_k \tag{10 43}$$

where

$$\epsilon_k = \frac{\hbar^2 k^2}{2m} \tag{10.44}$$

The free-particle wave functions were assumed to be normalized in volume Ω. The interaction term is

$$\frac{1}{2} \int \psi\dagger(\mathbf{r}')\psi\dagger(\mathbf{r})v(\mathbf{r}-\mathbf{r}')\psi(\mathbf{r})\psi(\mathbf{r}')\, d^3\mathbf{r}\, d^3\mathbf{r}'$$

$$= \frac{1}{2\Omega^2} \sum a_{k_4}\dagger a_{k_3}\dagger a_{k_2} a_{k_1} \int e^{-i\mathbf{k}_4\cdot\mathbf{r}'}\, e^{-i\mathbf{k}_3\cdot\mathbf{r}}\, v(\mathbf{r}-\mathbf{r}')\, e^{i\mathbf{k}_2\cdot\mathbf{r}}\, e^{i\mathbf{k}_1\cdot\mathbf{r}'}\, d^3\mathbf{r}\, d^3\mathbf{r}'$$

$$\tag{10.45}$$

This may be simplified by introducing relative and center of mass coordinates such that

$$\boldsymbol{\rho} = \mathbf{r}'-\mathbf{r}, \qquad \mathbf{R} = \tfrac{1}{2}(\mathbf{r}+\mathbf{r}') \tag{10.46}$$

Then

$$\mathbf{r} = \mathbf{R}-\tfrac{1}{2}\boldsymbol{\rho}, \qquad \mathbf{r}' = \mathbf{R}+\tfrac{1}{2}\boldsymbol{\rho} \tag{10.47}$$

Substituting these in (10.45) the interaction operator becomes

$$\frac{1}{2\Omega^2} \sum a_{k_4}\dagger a_{k_3}\dagger a_{k_2} a_{k_1}$$

$$\times \int e^{-i\mathbf{k}_4\cdot(\mathbf{R}+\frac{1}{2}\boldsymbol{\rho})} e^{-i\mathbf{k}_3\cdot(\mathbf{R}-\frac{1}{2}\boldsymbol{\rho})} v(\boldsymbol{\rho}) e^{i\mathbf{k}_2\cdot(\mathbf{R}-\frac{1}{2}\boldsymbol{\rho})} e^{i\mathbf{k}_1\cdot(\mathbf{R}+\frac{1}{2}\boldsymbol{\rho})}\, d^3\boldsymbol{\rho}\, d^3\mathbf{R} \tag{10.48}$$

If we do the $\mathbf{R}$ integral we get $\Omega\delta_{k_1+k_2,\,k_3+k_4}$, which ensures conservation of momentum. But let us now introduce

$$\mathbf{k}_3 = \mathbf{k}_2+\mathbf{q} \tag{10.49}$$

and sum over $\mathbf{q}$ rather than $\mathbf{k}_3$. Also, since

$$\mathbf{k}_3+\mathbf{k}_4 = \mathbf{k}_1+\mathbf{k}_2$$

it follows that

$$\mathbf{k}_4 = \mathbf{k}_1-\mathbf{q} \tag{10.50}$$

The coefficient of ρ in the exponent of the integrand is then

$$\tfrac{1}{2}[-\mathbf{k}_1 + \mathbf{q} + \mathbf{k}_2 + \mathbf{q} - \mathbf{k}_2 + \mathbf{k}_1] = \mathbf{q}$$

Therefore, we obtain for the interaction operator

$$\tfrac{1}{2} \int \psi\dagger(\mathbf{r}')\psi\dagger(\mathbf{r})v(\mathbf{r}-\mathbf{r}')\psi(\mathbf{r})\psi(\mathbf{r}') \; d^3\mathbf{r} \, d^3\mathbf{r}'$$

$$= \tfrac{1}{2} \sum_{\mathbf{k}_1,\mathbf{k}_2,\mathbf{q}} a^\dagger_{\mathbf{k}_1-\mathbf{q}} a^\dagger_{\mathbf{k}_2+\mathbf{q}} a_{\mathbf{k}_2} a_{\mathbf{k}_1} v_{\mathbf{q}} \qquad (10.51)$$

where

$$v_{\mathbf{q}} = \frac{1}{\Omega} \int e^{i\mathbf{q}\cdot\mathbf{r}} v(\boldsymbol{\rho}) \; d^3\boldsymbol{\rho} \qquad (10.52)$$

Equation (10.51) describes a transfer between particles of an amount of momentum $\hbar\mathbf{q}$, with amplitude $v_{\mathbf{q}}$, which is the qth Fourier component of the interaction considered as a function of the separation of the particles.

It is important to note that a term like (10.51) describes multiple scatterings of all orders between particles. This is, of course, the reason why interaction problems are in general difficult. If the interaction energy is sufficiently small, the higher-order processes may be negligible. If it is large, no simple approximation method is available. If it is attractive and large, bound states may occur, necessitating an extension of the theory beyond the scattering formalism.

Problem 10-1: It is sometimes said that a pair of fermions acts like a boson.

To see the extent to which this is true, define pair creation and annihilation operators

$$\alpha\dagger_{kk'} = c_k\dagger c_{k'}\dagger, \qquad \alpha_{kk'} = c_{k'}c_k$$

where the c's are fermion operators and $k \neq k'$.

Verify the commutation relations

$$[\alpha_{kk'}, \alpha\dagger_{kk'}] = 1 - n_k - n_{k'}$$

$$[\alpha_{kk'}, \alpha\dagger_{k''k'''}] = 0 \text{ when } (k, k') \neq (k''.k''')$$

Thus, our operators behave like boson operators only when they operate on the empty state.

Otherwise, show that

$$[\alpha_{kk'}, (1 - n_k - n_{k'})] = -2\alpha_{kk'}$$

$$[\alpha\dagger_{kk'}, (1 - n_k - n_{k'})] = 2\alpha_{kk'}$$

and hence show that $\alpha_{kk'}$, $\alpha\dagger_{kk'}$, $(1 - n_k - n_{k'})$ obey the same commutation rules as the spin operators

$$\sigma_{\pm} = \tfrac{1}{2}(\sigma_x \pm i\sigma_y) \qquad \text{and} \qquad \sigma_z$$

(This result has been used by P. W. Anderson in the theory of super-conductivity.)

---

4. Hartree–Fock Method

Consider a system of identical fermions of spin $\frac{1}{2}$, all in a single-particle potential $V(r)$ and subject to two-body interactions represented by the interaction potential $v(r-r')$. Taking explicit account of spin, and using arbitrary basis functions $|\alpha\sigma\rangle$, where σ is the spin quantum number and α stands for all other quantum numbers, we may write

$$\psi(\mathbf{r},\sigma) = \sum c_{\alpha\sigma}\varphi_{\alpha\sigma}(\mathbf{r}) \qquad (10.53)$$

where

$$\varphi_{\alpha\sigma}(\mathbf{r}) = \langle\mathbf{r}|\alpha\sigma\rangle \qquad (10.54)$$

and $c_{\alpha\sigma}$, $c^\dagger_{\alpha\sigma}$ are the annihilation and creation operators for fermions in states $|\alpha\sigma\rangle$. The corresponding creation operator is

$$\psi^\dagger(\mathbf{r},\sigma) = \sum c^\dagger_{\alpha\sigma}\varphi^*_{\alpha\sigma}(\mathbf{r}) \qquad (10.55)$$

The "wave functions" $\varphi_{\alpha\sigma}$ must satisfy

$$\int \varphi^*_{\alpha\sigma}(\mathbf{r})\varphi_{\alpha'\sigma'}(\mathbf{r})\, d^3\mathbf{r} = \delta_{\alpha\alpha'}\delta_{\sigma\sigma'} \qquad (10.56)$$

The many-body Hamiltonian of the system is then

$$
\begin{aligned}
H &= \sum_{\sigma'} \int \psi^\dagger(\mathbf{r}',\sigma')\left[-\frac{\hbar^2}{2m}\nabla'^2 + V(\mathbf{r}')\right]\psi(\mathbf{r}',\sigma')\, d^3\mathbf{r}' \\
&\quad + \frac{1}{2}\sum_{\sigma,\sigma'} \int \psi^\dagger(\mathbf{r}'',\sigma'')\psi^\dagger(\mathbf{r}',\sigma')v(\mathbf{r}'-\mathbf{r}'')\psi(\mathbf{r}',\sigma')\psi(\mathbf{r}'',\sigma'')\, d^3\mathbf{r}'\, d^3\mathbf{r}'' \\
&= \sum_{\alpha\alpha'\sigma'} c^\dagger_{\alpha'\sigma'}c_{\alpha\sigma'} \int \varphi^*_{\alpha'\sigma'}(\mathbf{r}')\left[-\frac{\hbar^2}{2m}\nabla'^2 + V(\mathbf{r}')\right]\varphi_{\alpha\sigma'}\, d^3\mathbf{r}' \\
&\quad + \frac{1}{2}\sum_{\alpha\alpha'\beta\beta'\sigma'\sigma''} c^\dagger_{\beta'\sigma''}c^\dagger_{\beta\sigma'}c_{\alpha\sigma'}c_{\alpha'\sigma''} \langle\beta'\beta|v|\alpha\alpha'\rangle
\end{aligned}
\qquad (10.57)
$$

using, in the last term, the notation of (10.40).

Problem 10-2: Given a single-particle potential operator containing spin, for example, $\boldsymbol{\sigma}\cdot\mathbf{p}$, show that the second-quantized operator may be taken to be

$$\sum_{\sigma',\sigma''} \int \psi^\dagger(\mathbf{r},\sigma'')[\boldsymbol{\sigma}\cdot\mathbf{p}]\psi(\mathbf{r},\sigma')\, d^3\mathbf{r}$$

and write it in terms of creation and annihilation operators for eigenstates of momentum and spin.

Problem 10-3: Consider an interaction operator containing spin, e.g., the "tensor force" operator

$$u = (\boldsymbol{\sigma}_1 \cdot \mathbf{R})(\boldsymbol{\sigma}_2 \cdot \mathbf{R})v(\mathbf{R})$$

where $\mathbf{R} = \mathbf{r} - \mathbf{r}'$. Show that the interaction operator may be taken to be

$$\frac{1}{2} \sum_{\sigma_1, \sigma'_1} \sum_{\sigma_2, \sigma'_2} \int \psi\dagger(r'', \sigma'_2)\psi\dagger(r', \sigma'_1)u\psi(\mathbf{r}', \sigma_1)(\mathbf{r}'', \sigma_2)\, d^3r'\, d^3r''$$

and write it in terms of creation and annihilation operators for eigenstates of momentum and spin.

To try to find exact solutions for this Hamiltonian is an impossible task. We are therefore led to seek approximate methods which produce adequate solutions for particular purposes. The Hartree–Fock method, which we now develop, provides such an approximation. The concept behind the method is this: Since the major difficulties in dealing with (10.57) arise from the presence of the interaction terms, which contain products of four operators, we replace pairs consisting of a creation and an annihilation operator by their mean values. Thus, we are left only with terms of the sort that occur in problems in which there are only single-particle potentials. The problem is then to solve the resulting single-particle equations. The single-particle states will of course be coupled, since the mean values, which occur in the expansion of the resulting one-particle potentials, will depend on the states of all particles.

Our development of the method depends on the following observation: If we could find single-particle states $|\alpha\sigma\rangle$ for which

$$[H, c_{\alpha\sigma}] = -E_\alpha c_{\alpha\sigma} \qquad (10.58)$$

these states would be such that the annihilation of a particle from one of them in an N-body system of energy E_N would produce an $(N-1)$-body system of energy

$$E_{N-1} = E_N - E_\alpha \qquad (10.59)$$

E_α would then be, in the usual terminology, the "ionization energy" or "work function" of the removed particle. The many-body system could then be considered to be made up of particles in such single-particle states.

The proof of the proposition made in the previous paragraph follows if one lets both sides of (10.58) operate on the N-body state in which there are N_α particles in the single-particle state $|\alpha\sigma\rangle$. (The proof holds for both Bose–Einstein and Fermi–Dirac particles.) Designating the N-body state $N(N_{\alpha\sigma})\rangle$,

$$(Hc_{\alpha\sigma} - c_{\alpha\sigma}H)|N(N_{\alpha\sigma})\rangle = -E_\alpha c_{\alpha\sigma}|N(N_{\alpha\sigma})\rangle$$

so that

$$H\{c_{\alpha\sigma}|N(N_{\alpha\sigma})\rangle\} = (E_0 - E_\alpha)\{c_{\alpha\sigma}|N(N_{\alpha\sigma})\rangle\}$$

or

$$H|(N-1)(N_{\alpha\sigma}-1)\rangle = (E_0 - E_\alpha)|(N-1)(N_{\alpha\sigma}-1)\rangle \quad (10.60)$$

Unfortunately, *exact* states $|\alpha\sigma\rangle$ satisfying (10.58) do not exist, since clearly the commutators of interaction terms contain three-operator products; however, when creation and annihilation pairs are replaced by their mean values, the solution of (10.58) becomes possible.

Rather than working directly with (10.58), however, it is simpler to calculate first the commutator $[H, \psi(\mathbf{r})]$. $[H, c_{\alpha\sigma}]$ can then be made to satisfy (10.58) by properly choosing the basis functions $\varphi_{\alpha\sigma}^*$.

Let us calculate the commutator of each term of H [as defined by the first form given in (10.57)] with $\psi(\mathbf{r}, \sigma)$. The commutator with the first term is

$$K_1 = \left[\sum_{\sigma'} \int \psi^\dagger(\mathbf{r}', \sigma') H_0' \psi(\mathbf{r}', \sigma') d^3r', \psi(\mathbf{r}, \sigma) \right]$$

where

$$H_0' = -\frac{\hbar^2}{2m}\nabla'^2 + V(\mathbf{r}') \quad (10.61)$$

Now

$$[AB, C]_- = ABC - CAB$$
$$= A[B, C]_+ - [A, C]_+ B \quad (10.62)$$

Putting

$$A = \psi^\dagger(\mathbf{r}', \sigma'), \qquad B = H_0'\psi(\mathbf{r}', \sigma'), \qquad C = \psi(\mathbf{r}, \sigma)$$

the quantity K_1 becomes

$$K_1 = -\sum_{\sigma'} \int d^3r' [\psi^\dagger(\mathbf{r}', \sigma'), \psi(\mathbf{r}, \sigma)]_+ H_0'\psi(\mathbf{r}', \sigma')$$
$$= -H_0\psi(\mathbf{r}, \sigma) \quad (10.63)$$

To calculate the second commutator

$$K_2 = \left[\frac{1}{2} \sum_{\sigma', \sigma''} \int \psi^\dagger(\mathbf{r}'', \sigma'')\psi^\dagger(\mathbf{r}', \sigma') v(\mathbf{r}'' - \mathbf{r}') \right.$$
$$\left. \times \psi(\mathbf{r}', \sigma')\psi(\mathbf{r}'', \sigma'') d^3r' \, d^3r'', \psi(\mathbf{r}, \sigma) \right]_-$$

we must calculate

$$[\psi\dagger(x'')\psi\dagger(x')\psi(x')\psi(x''), \ \psi(x)]_-$$

where x stands for $(\mathbf{r}, \sigma)$. We expand this out as follows, using the anti-commutation relations for the ψ's:

$$
\begin{aligned}
[P,\psi(x)]_- &= [\psi\dagger(x'')\psi\dagger(x')\psi(x')\psi(x''), \psi(x)]_- \\
&= \psi\dagger(x'')\psi\dagger(x')\psi(x')\psi(x'')\psi(x) - \psi(x)P \\
&= \psi\dagger(x'')\psi\dagger(x')\psi(x)\psi(x')\psi(x'') - \psi(x)P \\
&= \psi\dagger(x'')[\delta(x'-x) - \psi(x)\psi\dagger(x')]\psi(x')\psi(x'') - \psi(x)P \\
&= -\delta(x'-x)\psi\dagger(x'')\psi(x'')\psi(x') \\
&\quad + [-\delta(x''-x) + \psi(x)\psi\dagger(x'')]\psi\dagger(x')\psi(x')\psi(x'') - \psi(x)P \\
&= -\delta(x'-x)\psi\dagger(x'')\psi(x'')\psi(x') - \delta(x''-x)\psi\dagger(x')\psi(x')\psi(x'')
\end{aligned}
$$

Therefore,

$$K_2 = -\sum_{\sigma'} \int \psi\dagger(\mathbf{r}',\sigma')\psi(\mathbf{r}',\sigma')v(\mathbf{r}'-\mathbf{r}) \ d^3r'\psi(\mathbf{r},\sigma) \quad (10.64)$$

Putting together (10.63) and (10.64), we find that

$$[H,\psi(\mathbf{r},\sigma)] = -\left[H_0' + \sum_{\sigma'} \int \psi\dagger(\mathbf{r}',\sigma')\psi(\mathbf{r}',\sigma')v(\mathbf{r}'-\mathbf{r}) \ d^3r'\right]\psi(\mathbf{r},\sigma)$$

$$(10.65)$$

Making now the Hartree–Fock approximation, the right-hand side becomes

$$-[H_0'\psi(\mathbf{r},\sigma) + \sum_{\sigma'} \int \langle\psi\dagger(\mathbf{r}',\sigma')\psi(\mathbf{r}',\sigma')\rangle v(\mathbf{r}'-\mathbf{r}) \ d^3r'\psi(\mathbf{r},\sigma)$$

$$-\sum_{\sigma'} \int \langle\psi\dagger(\mathbf{r}',\sigma')\psi(\mathbf{r},\sigma)\rangle v(\mathbf{r}'-\mathbf{r})\psi(\mathbf{r}',\sigma') \ d^3r'] \quad (10.66)$$

Problem 10-4: Show that, if H as given by (10.57) is approximated by the operator H^1 obtained by "contracting" all possible pairs consisting of one creation and one annihilation operator (that is, by replacing such pairs by the mean value of their product) in the interaction term, $[H^1, \psi(\mathbf{r},\sigma)]$ is given by (10.66).

Let us now substitute

$$\psi(\mathbf{r},\sigma) = \sum c_{\alpha\sigma}\varphi_{\alpha\sigma}(\mathbf{r}) \quad (10.67)$$

into the Hartree–Fock approximation for $[H,\psi(\mathbf{r},\sigma)]$, the $\varphi_{\alpha\sigma}$'s being the wave functions of an as yet unspecified set of basis states. The second term, which we call the "direct interaction" term becomes

$$\sum_{\sigma'\beta\beta'\alpha} \langle c^{\dagger}{}_{\beta'\sigma'}c_{\beta\sigma'}\rangle c_{\alpha\sigma} \int \varphi_{\beta'\sigma'}^{*}(\mathbf{r}')\varphi_{\beta\sigma'}(\mathbf{r}')v(\mathbf{r}-\mathbf{r}')\,d^3\mathbf{r}'\psi_{\alpha\sigma}(\mathbf{r})$$

$$= \sum_{\sigma'\beta\alpha} n_{\beta\sigma'} \int |\varphi_{\beta\sigma'}(\mathbf{r}')|^2 v(\mathbf{r}-\mathbf{r}')\,d^3\mathbf{r}'\varphi_{\alpha\sigma}(\mathbf{r})c_{\alpha\sigma} \qquad (10.68)$$

The third, or "exchange" term is

$$-\sum_{\beta\beta'\alpha\sigma'} \langle c^{\dagger}{}_{\beta'\sigma'}c_{\beta\sigma}\rangle c_{\alpha\sigma'} \int \varphi_{\beta'\sigma'}^{*}(\mathbf{r}')\varphi_{\beta\sigma}(\mathbf{r})\varphi_{\alpha\sigma'}(\mathbf{r}')v(\mathbf{r}-\mathbf{r}')\,d^3\mathbf{r}'$$

$$= -\sum_{\beta\alpha} n_{\beta\sigma} \int \varphi_{\beta\sigma}^{*}(\mathbf{r}')\varphi_{\alpha\sigma}(\mathbf{r}')v(\mathbf{r}-\mathbf{r}')\,d^3\mathbf{r}'\varphi_{\beta\sigma}(\mathbf{r})c_{\alpha\sigma} \qquad (10.69)$$

If (10.68) and (10.69) are now substituted into (10.65) it follows that

$$[H,\psi(\mathbf{r},\sigma)]_{\text{Hartree–Fock}} = -\sum_{\alpha} c_{\alpha\sigma} \Big\{ H_0\varphi_{\alpha\sigma}(\mathbf{r}) + \sum_{\beta\sigma'} n_{\beta\sigma'}$$

$$\times \int |\varphi_{\beta\sigma'}(\mathbf{r}')|^2 v(\mathbf{r}-\mathbf{r}')\,d^3\mathbf{r}'\varphi_{\alpha\sigma}(\mathbf{r})$$

$$- \sum_{\beta} n_{\beta\sigma} \int \varphi_{\beta\sigma}^{*}(\mathbf{r}')\varphi_{\alpha\sigma}(\mathbf{r}')v(\mathbf{r}-\mathbf{r}')\,d^3\mathbf{r}'\varphi_{\beta\sigma}(\mathbf{r})\Big\}$$

$$(10.70)$$

Substituting on the left-hand side

$$[H,\psi(\mathbf{r},\sigma)] = \sum \varphi_{\alpha\sigma}(\mathbf{r})[H,c_{\alpha\sigma}] \qquad (10.71)$$

We then see, on comparing (10.70) and (10.71), that in the Hartree–Fock approximation it is possible to choose states such that $[H, c_{\alpha\sigma}] = -E_\alpha c_{\alpha\sigma}$ provided that their wave functions satisfy

$$H_0\varphi_{\alpha\sigma}(\mathbf{r}) + \sum_{\beta\sigma'} n_{\beta\sigma'} \int |\varphi_{\beta\sigma'}(\mathbf{r}')|^2 v(\mathbf{r}-\mathbf{r}')\,d^3\mathbf{r}'\varphi_{\alpha\sigma}(\mathbf{r})$$

$$- \sum_{\beta} n_{\beta\sigma} \int \varphi_{\beta\sigma}^{*}(\mathbf{r}')\varphi_{\alpha\sigma}(\mathbf{r}')v(\mathbf{r}-\mathbf{r}')\,d^3\mathbf{r}'\varphi_{\beta\sigma}(\mathbf{r}) = E_\alpha\varphi_{\alpha\sigma}(\mathbf{r}) \qquad (10.72)$$

These are the so-called "Hartree–Fock equations." Let us note several facts about them.

Problem 10-5: Prove that, if a particle in state $|\alpha\sigma\rangle$ is *added* to an $(N-1)$-body system to *create* the N-body system described by (10.72), the energy is *increased* by E_α in the Hartree–Fock approximation.

(a) They are *coupled* equations; that is, they are equations for the set of wave functions of the basis states which has the property that, in the approximation in question, the many-body system can be described in terms of occupancy of these states.

(b) The states of the basis are mutually orthogonal. To show this, multiply Eq. (10.72) by φ_β^*, multiply the equation for φ_β^*, viz.,

$$\left[-\frac{\hbar^2}{2m}\nabla'^2 + V\right]\varphi_\beta^*(\mathbf{r}) + \sum_{\alpha'} n_{\alpha'}\int |\varphi_{\alpha'}(\mathbf{r}')|^2 v(\mathbf{r}-\mathbf{r}')\,d^3r'\,\varphi_\beta^*(\mathbf{r})$$

$$-\sum_{\alpha'} n_{\alpha'}\int \varphi_{\alpha'}(\mathbf{r}')\varphi_\beta^*(\mathbf{r}')v(\mathbf{r}-\mathbf{r}')\,d^3r'\,\varphi_\alpha^*(\mathbf{r}) = E_\beta\varphi_\beta^*(\mathbf{r}) \qquad (10.73)$$

by $\varphi_\alpha(\mathbf{r})$, subtract one equation from the other, and integrate over $\mathbf{r}$. By virtue of the Hermitian property of the operator $(-\hbar^2/2m)\nabla^2 + V$, the left-hand sides cancel, leaving

$$(E_\alpha - E_\beta)\int \varphi_\alpha(\mathbf{r})\varphi_\beta^*(\mathbf{r})\,d^3r = 0$$

so that states of different energy are orthogonal.

(c) While the β sums in (10.71) range over *all* occupied states, including $|\alpha\rangle$, the $\beta = \alpha$ terms cancel.

(d) The sum in the first summation, the "direct-interaction" term, is over states of both spins. In the second, the "exchange" term, the sum is over all states whose spin is the same as that of the state whose wave function is being determined.

(e) The energy of the system is *not* the sum of the energies of the individual particles. This is physically evident, from the interpretation of the E_α's as ionization energies of the individual particles from the N-particle system. The total energy can be calculated as the mean value of the Hamiltonian (10.57). In the interaction term, the only combinations of operators which have nonzero mean value are those for which either

(i) $\beta' = \alpha'$, $\quad \alpha = \beta \quad$ or $\quad$ (ii) $\beta' = \alpha$, $\quad \beta = \alpha'$, $\quad \sigma' = \sigma''$.

Thus the mean value of the interaction term is

$$\frac{1}{2}\sum \langle c^\dagger_{\alpha'\sigma''}c^\dagger_{\alpha\sigma'}c_{\alpha\sigma'}c_{\alpha'\sigma''}\rangle\,\langle\alpha'\alpha|v|\alpha\alpha'\rangle$$

$$+\frac{1}{2}\sum \langle c^\dagger_{\alpha\sigma''}c^\dagger_{\alpha'\sigma'}c_{\alpha\sigma'}c_{\alpha'\sigma''}\rangle\,\langle\alpha\alpha'|v|\alpha\alpha'\rangle \qquad (10.74)$$

Both the mean values can be reduced to number operators:

$$c^\dagger_{\alpha'\sigma''}c^\dagger_{\alpha\sigma'}c_{\alpha\sigma'}c_{\alpha'\sigma''} = -c^\dagger_{\alpha'\sigma''}c^\dagger_{\alpha\sigma'}c_{\alpha'\sigma''}c_{\alpha\sigma'}$$

$$= c^\dagger_{\alpha'\sigma''}c_{\alpha'\sigma''}c^\dagger_{\alpha\sigma'}c_{\alpha\sigma'} - \delta_{\alpha\alpha'}\delta_{\sigma\sigma'}c^\dagger_{\alpha'\sigma''}c_{\alpha\sigma'}$$

which has mean value

$$n_{\alpha'\sigma''}n_{\alpha\sigma'} - \delta_{\alpha\alpha'}\delta_{\sigma'\sigma''}n_{\alpha\sigma'}$$

and

$$c^\dagger_{\alpha\sigma''}c^\dagger_{\alpha'\sigma'}c_{\alpha\sigma'}c_{\alpha\sigma''} = -c^\dagger_{\alpha\sigma''}c_{\alpha\sigma'}c^\dagger_{\alpha'\sigma'}c_{\alpha'\sigma''} + \delta_{\alpha\alpha'}c^\dagger_{\alpha\sigma''}c_{\alpha\sigma''}$$

which has the mean value

$$-n_{\alpha\sigma'}n_{\alpha'\sigma'}\delta_{\sigma'\sigma''} + \delta_{\alpha\alpha'}n_{\alpha\sigma''}$$

When these are substituted in (10.74), the second terms cancel each other; the first ones give

$$\tfrac{1}{2}\sum n_{\alpha'\sigma''}n_{\alpha\sigma'}\langle\alpha'\alpha|v|\alpha\alpha'\rangle - \tfrac{1}{2}\sum n_{\alpha'\sigma''}n_{\alpha\sigma'}\delta_{\sigma'\sigma''}\langle\alpha\alpha'|v|\alpha\alpha'\rangle \quad (10.75)$$

It is easily verified that this is exactly half the sum of the interaction parts of the E_α's. This is because, in summing the E_α's, all interactions are taken into account *twice*; once in the equation for each of the two particles.

5. Hartree–Fock for Free Interacting Particles

Consider a system of identical particles which are free, i.e., not subject to an external potential but which interact with the potential $v(|\mathbf{r}-\mathbf{r}'|)$. We show that, in this case, plane waves (momentum eigenstates) constitute a solution of the Hartree–Fock equations.

The fact that momentum eigenfunctions provide a solution is suggested by the fact that the Hamiltonian of (10.72) with $V = 0$ is invariant under a spatial translation. It may be verified directly by putting

$$\varphi_\alpha, \varphi_\beta \rightarrow \frac{1}{\sqrt{\Omega}}e^{i\mathbf{k}\cdot\mathbf{r}}, \frac{1}{\sqrt{\Omega}}e^{i\mathbf{k}'\cdot\mathbf{r}} \quad (10.76)$$

respectively, where the system is assumed to be enclosed in a volume Ω. Substitution in (10.72) gives

$$\frac{\hbar^2 k^2}{2m}e^{i\mathbf{k}\cdot\mathbf{r}} + \sum_{\mathbf{k}',\sigma'} n_{\mathbf{k}'\sigma'}\frac{1}{\Omega}\int v(|\mathbf{r}-\mathbf{r}'|)\,d^3r' \cdot e^{i\mathbf{k}\cdot\mathbf{r}}$$

$$-\sum_{\mathbf{k}'} n_{\mathbf{k}'\sigma}\frac{1}{\Omega}\int e^{-i\mathbf{k}'\cdot\mathbf{r}'}v(|\mathbf{r}-\mathbf{r}'|)e^{i\mathbf{k}\cdot\mathbf{r}'}\,d^3r' \cdot e^{i\mathbf{k}'\cdot\mathbf{r}} = E_{\mathbf{k}\sigma}e^{i\mathbf{k}\cdot\mathbf{r}} \quad (10.77)$$

Taking out $e^{i\mathbf{k}\cdot\mathbf{r}}$ as a common factor, we find an expression for $E_{\mathbf{k}\sigma}$,

$$E_{\mathbf{k}\sigma} = \frac{\hbar^2 k^2}{2m} + \frac{N}{\Omega}\int v(|\mathbf{r}-\mathbf{r}'|)\,d^3r'$$

$$-\sum_{\mathbf{k}'} n_{\mathbf{k}'\sigma}\frac{1}{\Omega}\int e^{-i(\mathbf{k}-\mathbf{k}')\cdot\mathbf{R}}v(R)\,d^3R \quad (10.78)$$

where

$$\mathbf{R} = \mathbf{r}-\mathbf{r}' \quad (10.79)$$

and N is the total number of particles. (We let $\Omega \to \infty$, $N \to \infty$ in such a way that the particle density N/Ω is a constant.) As $\Omega \to \infty$ the second term becomes constant; we therefore pay no more attention to it, since it does not affect the dynamical properties of the particle. If we write

$$\tilde{v}(\mathbf{k}) = \frac{1}{\Omega} \int e^{-i\mathbf{k}\cdot\mathbf{r}} v(\mathbf{r}) \, d^3\mathbf{r} \tag{10.80}$$

the second term is $N\tilde{v}(0)$ and the third, the exchange term, is

$$\frac{1}{\Omega} \sum_{\mathbf{k}'} n_{\mathbf{k}'\sigma} \tilde{v}(\mathbf{k}-\mathbf{k}')$$

Let us consider the last term for the case of a Yukawa potential

$$v(\mathbf{r}) = \frac{g^2}{r} e^{-\lambda r} \tag{10.81}$$

It is then a straightforward calculation to show that

$$\tilde{v}(\mathbf{k}-\mathbf{k}') = \frac{4\pi g^2}{\Omega} \frac{1}{\lambda^2 + |\mathbf{k}-\mathbf{k}'|^2} \tag{10.82}$$

The sum over $\mathbf{k}'$ in (10.78) may now be replaced by an integral. The number of states per unit volume of $\mathbf{k}$ space is $\Omega/8\pi^3$;[1] therefore, the exchange energy of the particle in state $\mathbf{k}$ is

$$v_{\text{ex}}(\mathbf{k}) = \frac{g^2}{2\pi^2} \int \frac{1}{\lambda^2 + |\mathbf{k}-\mathbf{k}'|^2} \, d^3\mathbf{k}' \tag{10.83}$$

where the integral is over the occupied part of $\mathbf{k}'$ space.

If the cosine of the angle between $\mathbf{k}$ and $\mathbf{k}'$ is put equal to μ, (10.83) can be expressed as

$$v_{\text{ex}}(\mathbf{k}) = \frac{g^2}{2\pi^2} 2\pi \int_0^{k_F} k'^2 \, dk' \int_{-1}^1 \frac{1}{\lambda^2 + k^2 + k'^2 - 2kk'\mu} d\mu$$

where we have assumed that the states are occupied up to $k = k_F$. The value of k_F is determined by integrating the density of states up to $k = k_F$ and putting it equal to the number of particles of the appropriate spin.

[1]To see this consider a rectangular "box" of dimensions $a \times b \times c$. Free-particle solutions in it have the form $\sin(l_1\pi x/a) \sin(l_2\pi y/b) \sin(l_3\pi z/c)$ so the possible $\mathbf{k}$ vectors are $(l_1\pi/a, l_2\pi/b, l_3\pi/c)$. The volume of $\mathbf{k}$ space occupied by each is $\pi^3/abc = \pi^3/\Omega$, so that the density of states is Ω/π^3. But *identical* states in which any of the components may be negative are found in the other eight octants of $\mathbf{k}$ space. Therefore, the density of independent states is $\frac{1}{8}\Omega/\pi^3$.

Thus

$$\frac{\Omega}{8\pi^3}\frac{4}{3}\pi k_{F\sigma}^3 = N_\sigma$$

so that

$$k_{F\sigma} = \left(\frac{6\pi^2 N_\sigma}{\Omega}\right)^{1/3} \tag{10.84}$$

On integrating over μ, (10.83) becomes

$$v_{\text{ex}}(\mathbf{k}) = \frac{g^2}{2\pi k}\int^{k_F} k' \ln\frac{\lambda^2+(k+k')^2}{\lambda^2+(k-k')^2}$$

By evaluating the integral it may be verified that

$$v_{\text{ex}}(\mathbf{k}) = \frac{g^2}{2\pi}\left\{\frac{1}{2k}\,(k_F^2+\lambda^2-k^2)\,\ln\frac{\lambda^2+(k_F+k)^2}{\lambda^2+(k_F-k)^2}\right.$$

$$\left.+2k_F-2\lambda\,\tan^{-1}\frac{2k\lambda}{\lambda^2+k_F^2-k^2}\right\} \tag{10.85}$$

Problem 10-6:   Prove the formula (10.85).

In the case of Coulomb forces $g^2 = e^2$ and $\lambda = 0$, so that

$$v_{\text{ex}}(\mathbf{k}) = \frac{e^2}{2\pi}\left\{\frac{1}{k}\,(k_F^2-k^2)\,\ln\frac{k_F+k}{|k_F-k|}+2k_F\right\} \tag{10.86}$$

Problem 10-7:   Defining the effective mass m^* by

$$\frac{1}{m^*} = \frac{1}{\hbar^2}\frac{d^2E(\mathbf{k})}{dk^2}$$

show that at $k = k_F$ the effective mass of an electron in an electron gas is zero in the Hartree–Fock approximation.

Problem 10-8:   Prove that the exchange potential energy of an electron in an electron gas is twice as great at $k = 0$ as at $k = k_F$, and that the average value over the occupied levels is halfway between that at $k = 0$ and that at $k = k_F$.

---

6.   Kinetic and Exchange Energies of a "Fermi Gas" of Electrons

A many-body system of electrons is often referred to as a "Fermi gas" of electrons. In this section we calculate the kinetic and exchange energies of such a system. We ignore the direct Coulomb energy. If the electron system were considered to be embedded in a uniform distribution of positive charge, so as to ensure over-all charge neutrality, the Coulomb energies would cancel. In an actual solid, which is a practical realization of a many-electron system, the positive charges are not uniformly distributed, but rather appear as a lattice of nuclei (or ion cores). There are then Coulomb energies of quadrupole or higher order. We confine our attention here, however, to the idealized system.

We first calculate the kinetic and exchange energies of the spin-up electrons and the spin-down ones separately. Let there be N_+ of the first, N_- of the second, where the total number is

$$N = N_+ + N_-$$

We then consider two distinct problems:

(1) That of finding the condition that the ground state energy correspond to $N_+ = N_-$; this, we see, occurs at sufficiently high electron densities.

(2) That of finding the total kinetic and exchange energies when $N_+ = N_-$.

Consider first the *kinetic energy* of N_σ electrons of a given spin. Assuming the lowest electron states to be filled, it is

$$T = \frac{\Omega}{8\pi^3} \frac{\hbar^2}{2m} \int_0^{k_{F\sigma}} k^2 4\pi k^2 \, dk$$

$$= \Omega \frac{\hbar^2 k_{F\sigma}^5}{20\pi^2 m}$$

$$= \Omega \frac{\hbar^2}{20\pi^2 m} \left(\frac{6\pi^2 N_\sigma}{\Omega} \right)^{5/3} \tag{10.87}$$

by (10.84). Putting $\Omega = 1$, so that we calculate energy per unit volume, we find that

$$T = A N_\sigma^{5/3} \tag{10.88}$$

where

$$A = \frac{\pi^{4/3} 2^{-1/3} 3^{5/3}}{5} \frac{\hbar^2}{m} \tag{10.89}$$

Next, let us calculate the *exchange energy* for the particles of each spin. A general formula for the exchange energy can be evaluated from (10.82). If we put $g = e$, $\lambda = 0$,

$$-\bar{v}(\mathbf{k} - \mathbf{k}') = -\frac{4\pi e^2}{\Omega} \frac{1}{|\mathbf{k} - \mathbf{k}'|^2} \tag{10.90}$$

is the exchange energy of interaction of particles with momenta $\hbar\mathbf{k}$, $\hbar\mathbf{k}'$

and having the same spin. It follows that the exchange energy is

$$\epsilon = -\frac{2\pi e^2}{\Omega}\left(\frac{\Omega}{8\pi^3}\right)^2 \int d^3k \int d^3k' \frac{1}{|\mathbf{k}-\mathbf{k}'|^2} \tag{10.91}$$

where the $\mathbf{k}$ integrals are over spheres of radius $k_{F\sigma}$. The integral may be evaluated to give $4\pi^2 k_F{}^4$. The exchange energy is, then,

---

Problem 10-9:   Show that

$$\int_{(k_F)} d^3k' \, d^3k \frac{1}{|\mathbf{k}-\mathbf{k}'|^2} = 4\pi^2 k_F{}^4$$

[Hint: Use the fact that

$$\frac{1}{|\mathbf{k}-\mathbf{k}'|} = \sum \frac{k_<{}^l}{k_>{}^{l+1}} P_l(\cos\theta)$$

where $k_<$, $k_>$ are, respectively, the lesser and greater of $\mathbf{k}$, $\mathbf{k}'$, and θ is the angle between the vectors $\mathbf{k}$ and $\mathbf{k}'$.]

Problem 10-10:   As an alternative to the procedure employed above, use (10.86) to calculate the exchange energy of N particles of a given spin.

$$\epsilon = -\frac{2\pi e^2}{\Omega}\frac{\Omega^2}{64\pi^6}4\pi^2\left(\frac{6\pi^2 N_\sigma}{\Omega}\right)^{4/3} \tag{10.92}$$

Again putting $\Omega = 1$, this becomes

$$\epsilon = -BN_\sigma{}^{4/3} \tag{10.93}$$

where

$$B = 2^{-5/3}\pi^{-1/3}3^{4/3}e^2 \tag{10.94}$$

7.   Condition of Instability of the Zero-Spin State

It may be noted that for high enough densities the kinetic energy dominates the exchange, while for low density the reverse is the case. Now, so far as kinetic energy is concerned, it is least when there are equal number of particles of the two spins. Therefore, at high density we expect the lowest-energy state to correspond to equipartition of the electrons between the spin states.

With respect to exchange energy, on the other hand, it tends to favor lining up the electron spins, since the (negative) exchange energy operates only between particles of parallel spin. Therefore, at low enough density we expect that the state of equipartition of energy will be unstable, and that the energy can be lowered by having more electrons in one spin state than in the other. These considerations might constitute a basis for a model theory of ferromagnetism (spontaneous magnetization). We therefore investigate the conditions under which the equal or unequal spin occupancies correspond to the lowest energy.

The total energy is

$$U = T + \epsilon = AN_+^{5/3} + AN_-^{5/3} - BN_+^{4/3} - BN_-^{4/3} \qquad (10.95)$$

The energy is a maximum or minimum when $\partial U / \partial N_+ = 0$. Remembering that, since $N_- = N - N_+, \partial / \partial N_- = - \partial / \partial N_+$, we differentiate U to get

$$\frac{\partial U}{\partial N_+} = \tfrac{5}{3} A (N_+^{2/3} - N_-^{2/3}) - \tfrac{4}{3} B (N_+^{1/3} - N_-^{1/3}) \qquad (10.96)$$

This is zero for $N_+ = N_-$, but to determine whether it corresponds to a maximum or minimum of the energy, we must calculate the *second* derivative for $N_+ = N_-$:

$$\frac{\partial^2 U}{\partial N_+^2} = \frac{10}{9} A (N_+^{-1/3} + N_-^{-1/3}) - \frac{4}{9} B (N_+^{-2/3} N_-^{-2/3})$$

$$= \frac{20}{9} A \left(\frac{N}{2}\right)^{-1/3} - \frac{8}{9} B \left(\frac{N}{2}\right)^{-2/3} \qquad (10.97)$$

for the values in question. If this is > 0, the equal distribution gives minimum energy; if it is < 0, a maximum. The condition for unequal spin distribution is, therefore,

$$2B \left(\frac{N}{2}\right)^{-2/3} > 5A \left(\frac{N}{2}\right)^{-1/3}$$

or

$$\left(\frac{N}{2}\right)^{1/3} < \frac{2B}{5A} \qquad (10.98)$$

A convenient and commonly used parameter is the dimensionless constant which is the radius of the sphere which, in units of Bohr radii, contains on the average one electron. If this is designated r_s, it may be expressed in terms of N by the formula

$$\frac{\Omega}{N} = \frac{4}{3} \pi r_s^3 a_0^3 \qquad (10.99)$$

or

$$\left(\frac{N}{\Omega}\right)^{1/3} = \left(\frac{3}{4\pi}\right)^{1/3} \frac{1}{r_s a_0} \qquad (10.100)$$

The condition (10.98) may then be written (for $\Omega = 1$)

$$\left(\frac{3}{4\pi}\right)^{1/3} \frac{1}{r_s a_0} < 2^{1/3} \frac{2B}{5A}$$

or

$$r_s > \pi^{4/3} 3^{2/3} 2^{-2/3}$$

$$= 6.14 \tag{10.101}$$

This corresponds to a density a little lower than that found in any metal in which the electrons can reasonably be represented as free.

The concepts behind this calculation, however, form the basis of a quite reasonable approach to the problem of ferromagnetism. The necessary modification is to take account of the fact that the electrons are not free, and must be represented by more complicated wave functions, with a consequent modification in their energy spectrum.

It is particularly worth noting that the calculation given would not be modified by taking account of the mutual Coulomb repulsion of electrons, since the Coulomb potential operates between *all* pairs of electrons, and so is independent of the electron spins.

8. Total Kinetic and Exchange Energies of the Electron Gas

It is instructive to write the total kinetic and exchange energies in terms of the parameter r_s, in the case in which $N_+ = N_- = N/2$. The kinetic energy is

$$T_0 = 2A \left(\frac{N}{2}\right)^{5/3} \frac{1}{\Omega^{2/3}} \tag{10.102}$$

The kinetic energy per particle is

$$t_0 = \frac{1}{N} T_0 = 2^{-2/3} A \left(\frac{N}{\Omega}\right)^{2/3} \tag{10.103}$$

This may be written in terms of r_s by using (10.100):

$$t_0 = 2^{-2/3} A \left(\frac{3}{4\pi}\right)^{2/3} \frac{1}{r_s^2 a_0^2} \tag{10.104}$$

If we substitute for A from (10.89), we find that

$$t_0 = \frac{2^{-4/3} 3^{7/3} \pi^{2/3}}{5} \left(\frac{me^4}{2\hbar^2}\right) \frac{1}{r_s^2}$$

$$= \frac{2.21}{r_s^2} \text{ Rydbergs} \tag{10.105}$$

since $me^4/2\hbar^2$ is the Rydberg, the energy of the ground state of the hydrogen atom.

The exchange energy per particle is

$$\epsilon_0 = \frac{\epsilon}{N} = -2^{-1/3}B\left(\frac{N}{\Omega}\right)^{1/3}$$

$$= -2^{-5/3}\pi^{-2/3}3^{5/3}\frac{1}{r_s}\frac{e^2}{2a_0} \tag{10.106}$$

$$= -\frac{0.916}{r_s} \text{ Rydbergs} \tag{10.107}$$

If one goes beyond the Hartree–Fock approximation, it is possible to develop the energy per particle as an expansion in r_s. The expansion is one which converges more rapidly the smaller r_s, or the larger the density. The next term in the expansion has been shown to be proportional to $\ln r_s$.

9. Excited States of Many-Body Systems

So far, we have been considering only the *ground* state of many-body systems. If such systems are in a "temperature bath" they will be in excited states. For a system of many particles, there are many states within a very small energy range. Suppose, for instance, we consider N free, noninteracting particles in a volume Ω. The total number of states in the $3N$-dimensional volume $d^3k_1 \, d^3k_2 \ldots d^3k_N$ of the space of all their $\mathbf{k}$ vectors is $(2\Omega/8\pi^3)^N d^3k_1 \, d^3k_2 \ldots d^3k_N$. The number of states in an energy range $(E, E + \Delta E)$ is determined by the volume between the surfaces

$$\sum_{n=1}^{N} \frac{\hbar^2 k_n^2}{2m} = (E, E + \Delta E).$$

This was calculated in Chapter 7 and found to be

$$\left(\frac{2m\pi}{\hbar^2}\right)^{3N/2} \frac{1}{\Gamma(3N/2)} E^{(3N/2)-1}\Delta E$$

Therefore the number of states per unit energy range is

$$\nu(E) = \left(\frac{2\Omega}{8\pi^3}\right)^N \left(\frac{2m\pi}{\hbar^2}\right)^{3N/2} \frac{1}{\Gamma(3N/2)} E^{(3N/2)-1} \tag{10.108}$$

If the average energy per particle is E_0, so that $E = NE_0$, and if we use Stirling's formula for the Γ function, the last two factors can be combined into

$$\sqrt{\frac{3}{4\pi N}} \, e^{3N/2} \left(\frac{2}{3}\right)^{3N/2} E_0^{(3N/2)-1}$$

for large N. Thus

$$\nu(E) \approx \left(\frac{2\Omega}{8\pi^3}\right)^N \left(\frac{4\pi m e E_0}{3\hbar^2}\right)^{3N/2} \sqrt{\frac{3}{4\pi N}} \frac{1}{E_0} \tag{10.109}$$

$$\doteq \left(\frac{2\Omega}{8\pi^3}\right)^N \left(\frac{E_0(\mathrm{eV})}{0.55}\right)^{3N/2} 10^{24N} \sqrt{\frac{3}{4\pi N}} \frac{1}{E_0} \tag{10.110}$$

E_0 (eV) being the average electron energy in electron volts. We see that, if Ω is a macroscopic volume, the density of states is very large indeed.

In fact, due to collisions, the width of the many-particle states is much greater than the distance between them.

Therefore, it is not reasonable to try to determine the many-particle states. A macroscopic specimen with a given total energy therefore is not subject to an exact quantum treatment, but may be treated *statistically*. Such a statistical treatment may be formulated in terms of the density matrix, the theory of which we now develop.

10. Density Matrix

There are two ways of developing a theory of the density matrix. In one, the existence of quantum states for the system is assumed, and it is described in terms of the probability of occupancy of these states. In the other, due to Fano,[2] a more operational approach is adopted — the "*statistical operator*" (density matrix) is defined in terms of the statistical averages of dynamical operators. We outline both approaches, and show that they lead to identical results.

FIRST APPROACH

Suppose that the energy states of the system are designated by $|\alpha\rangle$. Let us imagine an ensemble of macroscopically "identical" systems. By virtue of the remarks above, they are not all in the same state (and in fact any single system is continually undergoing transitions from one state to another, so that a single system at sufficiently large intervals of time also effectively constitutes an ensemble). We therefore let the statistical probability that the system be in state $|\alpha\rangle$ be p_α. The statistical operator ρ is then defined as

$$\rho = \sum_\alpha |\alpha\rangle p_\alpha \langle\alpha| \tag{10.111}$$

[2]Fano, U., *Rev. Mod. Phys.* **29**, No. 1, 74–93 (1957).

Let us enumerate some of its properties:

(a) $\operatorname{Tr}\rho=1$. This follows from the fact that, if $|n\rangle$ is a complete set of states,

$$\operatorname{Tr}\rho = \sum_n \langle n|\rho|n\rangle$$

$$= \sum_{n,\alpha} \langle n|\alpha\rangle p_\alpha\langle\alpha|n\rangle$$

Since

$$\sum |n\rangle\langle n| = 1$$

$$\operatorname{Tr}\rho = \sum \langle\alpha|\alpha\rangle p_\alpha = \sum p_\alpha = 1 \tag{10.112}$$

(b) The ensemble average of a dynamical variable K is

$$\langle K\rangle = \operatorname{Tr}\rho K \tag{10.113}$$

Proof

The average of K in a *given* quantum state $|\alpha\rangle$ is $\langle\alpha|K|\alpha\rangle$. Therefore,

$$\langle K\rangle = \sum p_\alpha\langle\alpha|K|\alpha\rangle$$

But using an arbitrary basis $|n\rangle$,

$$\operatorname{Tr}(\rho K) = \sum_n \langle n|\rho K|n\rangle$$

$$= \sum_{n,n'} \langle n|\rho|n'\rangle\langle n'|K|n\rangle$$

$$= \sum_{\alpha,n,n'} \langle n|\alpha\rangle p_\alpha\langle\alpha|n'\rangle\langle n'|K|n\rangle$$

$$= \sum_{\alpha,n} \langle n|\alpha\rangle p_\alpha\langle\alpha|K|n\rangle$$

$$= \sum_\alpha p_\alpha\langle\alpha|K|\alpha\rangle$$

$$= \langle K\rangle \tag{10.114}$$

This result is clearly independent of the representation $|n\rangle$.

(c) The time dependence of ρ is given by

$$i\hbar\frac{\partial\rho}{\partial t} = -[\rho, H] \tag{10.115}$$

To prove this, we start from the time-dependent states

$$|\alpha(t)\rangle = e^{-iHt}|\alpha\rangle$$

where we have put $\hbar = 1$. Then we define

$$\rho(t) = \sum_\alpha |\alpha(t)\rangle p_\alpha \langle \alpha(t)|$$

$$= \sum_\alpha e^{-iHt}|\alpha\rangle p_\alpha \langle \alpha| e^{iHt}$$

$$= e^{-iHt}\rho(0)\, e^{iHt}$$

Therefore,

$$\frac{\partial \rho(t)}{\partial t} = -ie^{-iHt}[H, \rho(0)]\, e^{iHt}$$

so that

$$i\frac{\partial \rho}{\partial t} = [H, \rho(t)] = -[\rho(t), H] \tag{10.116}$$

(d) We can transform the matrix elements of the density matrix from one basis to another by a unitary transformation.

In a basis $|n\rangle$, the matrix elements of ρ are

$$\langle n'|\rho|n\rangle = \sum_\alpha \langle n'|\alpha\rangle\langle\alpha|n\rangle p_\alpha \tag{10.117}$$

To transform to another basis $|\nu\rangle$, it is necessary to write the vectors $|n\rangle$ in terms of the $|\nu\rangle$'s,

$$\langle n'|\rho|n\rangle = \sum_{\nu,\nu'} \langle n'|\nu'\rangle \langle\nu'|\alpha\rangle \langle\alpha|\nu\rangle\langle\nu|n\rangle p_\alpha$$

$$= \sum_{\nu,\nu'} \langle n'|\nu'\rangle \langle\nu'|\rho|\nu\rangle \langle\nu|n\rangle \tag{10.118}$$

Define a transformation matrix

$$S_{\nu n} = \langle\nu|n\rangle \tag{10.119}$$

Then we may write the matrix equation

$$\rho = S\dagger\rho'S = S^{-1}\rho'S$$

or, alternatively

$$\rho' = S\rho S^{-1} \tag{10.120}$$

That S is unitary follows from the fact that

$$(S\dagger S)_{nn'} = \sum_n S\dagger_{n\nu}S_{\nu n'}$$

$$= \sum_n \langle n|\nu\rangle\langle\nu|n'\rangle$$

$$= \langle n|n'\rangle$$

$$= \delta_{nn'} \tag{10.121}$$

11. Alternative Method (Fano)

The alternative procedure is to define ρ operationally as that operator with the property that the statistical expectation value of K is

$$\langle K \rangle = \text{Tr}\,\rho K \qquad (10.122)$$

This equation now provides a *definition* of ρ, and does not follow from other postulates. From this definition we can derive other properties of ρ, as follows.

(a) $\text{Tr}\,\rho = 1$. This follows immediately by taking K equal to the unit operator.

(b) ρ is Hermitian. This follows if it can be proven that

$$\langle n|\rho|n' \rangle = \langle n'|\rho|n \rangle^* \qquad (10.123)$$

To prove (10.123), we make use of the fact that $\langle K \rangle = \text{Tr}\,(\rho K)$ must be real, and must therefore be equal to its complex conjugate. Therefore,

$$\text{Tr}\,(\rho K) = \sum_{n,n'} \langle n|\rho|n' \rangle \, \langle n'|K|n \rangle$$

But the complex conjugate of $\langle K \rangle$ is then

$$\langle K \rangle = \sum_{n,n'} \langle n|\rho|n' \rangle^* \, \langle n|K|n' \rangle$$

since K is Hermitian. Now, in the equation

$$\sum \langle n|\rho|n' \rangle \langle n'|K|n \rangle = \sum \langle n'|\rho|n \rangle^* \langle n'|K|n \rangle \qquad (10.124)$$

compare coefficients of the arbitrary matrix elements $\langle m'|K|m \rangle$ on the two sides; it follows that

$$\langle m|\rho|m' \rangle = \langle m'|\rho|m \rangle^* \qquad (10.125)$$

as required.

(c) Time variation. If, in the Heisenberg representation, we introduce the time-variable operator

$$K(t) = e^{iHt}Ke^{-iHt} \qquad (10.126)$$

the mean value of K as a function of time is given by

$$\text{Tr}\,\rho K(t) = \text{Tr}\,\rho\, e^{iHt}Ke^{-iHt}$$
$$= \text{Tr}\,e^{-iHt}\rho\, e^{iHt}K$$

since the order of factors in a trace is irrelevant. Therefore,

$$\langle K(t) \rangle = \text{Tr}\,\rho(t)K \qquad (10.127)$$

where

$$\rho(t) = e^{-iHt} \rho e^{iHt} \qquad (10.128)$$

so that

$$\frac{\partial \rho(t)}{\partial t} = -i[H, \rho(t)] \qquad (10.129)$$

or

$$i\frac{\partial \rho(t)}{\partial t} = -[\rho(t), H]$$

as before.

If a system is in a stationary (i.e., equilibrium) state, so that ρ does not depend on time,

$$[H, \rho] = 0$$

or

$$\sum_n [\langle n'|H|n\rangle\langle n|\rho|n''\rangle - \langle n'|\rho|n\rangle\langle n|H|n''\rangle] = 0$$

If the basis states $|n\rangle$ are energy eigenstates, this becomes

$$(E_{n'} - E_{n''})\langle n'|\rho|n''\rangle = 0$$

Thus the matrix elements of ρ between different energy states are zero. If the energy spectrum is nondegenerate, ρ is diagonal.

(d) If, for any basis, ρ is diagonalized, then the diagonal matrix elements are the probabilities of occupancy of the states of that basis.

Assume ρ to be diagonalized in the basis $|n\rangle$, and calculate for an arbitrary operator K

$$\langle K \rangle = \sum_{n,n'} \langle n|\rho|n'\rangle \langle n'|K|n\rangle$$

$$= \sum_n \langle n|\rho|n\rangle\langle n|K|n\rangle \qquad (10.130)$$

$\langle n|K|n\rangle$ is the quantum mean value of the operator when the system is in the state $|n\rangle$. To get the ensemble average, we must multiply the quantum mean values in each state by the probabilities of occupancy of those states. Thus $\langle n|\rho|n\rangle$ must be the probability that systems of the ensemble be in state $|n\rangle$.

EXAMPLE: MEAN VALUE OF PARTICLE DENSITY

Let us calculate the mean value of the particle density in a system with statistical operator ρ. We use the second-quantization formalism. Then the density operator is

$$\int \psi^\dagger(\mathbf{x}')\delta(\mathbf{x}' - \mathbf{x})\psi(\mathbf{x}')\, d^3\mathbf{x}' = \psi^\dagger(\mathbf{x})\psi(\mathbf{x}) \qquad (10.131)$$

where **x** is the position vector of an arbitrary point. The mean density is, then,

$$\rho(\mathbf{x}) = \mathrm{Tr}\, \rho\psi^\dagger(\mathbf{x})\psi(\mathbf{x})$$

$$= \mathrm{Tr}\, \psi(\mathbf{x})\rho\psi^\dagger(\mathbf{x})$$

Let us expand $\psi(\mathbf{x})$, $\psi^\dagger(\mathbf{x})$ in plane waves (momentum eigenfunctions), and use as basis the set of states of occupation of these momentum states. Then

$$\rho(\mathbf{x}) = \sum_{\mathbf{k},\mathbf{k}'} e^{i(\mathbf{k}-\mathbf{k}')\cdot\mathbf{x}}\, \mathrm{Tr}\, c_k \rho c_{k'}^\dagger \tag{10.132}$$

Considering for the moment only the occupancy of the two states **k** and **k'**,

$$c_{k'}^\dagger|n_k n_{k'}\rangle = (1-n_{k'})|n_k, n_{k'}+1\rangle \tag{10.133}$$

and

$$\langle n_k n_{k'}|c_k = (1-n_k)\langle n_k+1, n_{k'}| \tag{10.134}$$

Therefore, on substituting in (10.132),

$$\rho(\mathbf{x}) = \sum_{\mathrm{all}\,n} \sum_{\mathbf{k},\mathbf{k}'} e^{i(\mathbf{k}-\mathbf{k}')\cdot\mathbf{x}}(1-n_k)(1-n_{k'})$$

$$\langle \ldots (n_k+1)\ldots n_{k'}\ldots |\rho| \ldots n_k \ldots (n_{k'}+1)\ldots\rangle$$

$$= \sum_{n_{k''},k''\neq k,k'} \sum_{\mathbf{k},\mathbf{k}'} e^{i(\mathbf{k}-\mathbf{k}')\cdot\mathbf{x}}\langle \ldots 1_k \ldots 0_{k'}|\rho| \ldots 0_k \ldots 1_{k'}\ldots\rangle \tag{10.135}$$

The outer sum is over all states of occupancy of all single-particle states *except* **k** and **k'**. We see, then, that the Fourier component of density with wave number **q** is associated with the annihilation of particles of momentum **k'** and the creation of those with momentum **k' + q**, whatever **k'** may be.

12. Density Matrix of the Canonical Ensemble

The canonical ensemble is an ensemble in thermal equilibrium with its surroundings, but containing a *fixed* number of particles. It can, therefore, exchange energy, but not particles, with its surroundings. It is part of a larger system. The remainder of the system may be designated the "heat reservoir."

Let the number of energy states of N particles with a total energy in the range $(E, E+\Delta E)$ be $\nu(E, N)\Delta E$. It has been calculated earlier for free particles [see formula (10.110)]. Suppose that our large system has N_0 particles and energy E_0, and our subsystem N particles; we are interested in the probability distribution of each of its energy states.

That for the states in the energy range $(E, E + dE)$ is $\propto [\nu(E, N)\nu(E_0 - E, N_0 - N)dE/\nu(E_0, N_0)]$. Therefore, the probability of occupancy per energy state in the range $(E, E + dE)$ is

$$p(E, N) = \frac{\nu(E_0 - E, N_0 - N)}{\nu(E_0, N_0)} \times \text{constant} \qquad (10.136)$$

For the canonical ensemble, the N's are fixed. Now

$$\ln p(E, N) = \ln \nu(E_0 - E, N_0 - N) - \ln \nu(E_0, N_0)$$

If we expand

$$\nu(E_0 - E, N_0 - N) = \nu(E_0, N_0) - E\left(\frac{\partial \nu}{\partial E}\right)_{E_0, N_0} - N\left(\frac{\partial \nu}{\partial N}\right)_{E_0, N_0}$$

$$= \nu(E_0, N_0)\left[1 - E\left(\frac{\partial \ln \nu}{\partial E}\right)_0 - N\left(\frac{\partial \ln \nu}{\partial N}\right)_0 + \cdots\right]$$

$$(10.137)$$

we may write

$$\ln \nu(E_0 - E, N_0 - N) \approx \ln \nu(E_0, N_0) - E\left(\frac{\partial \ln \nu}{\partial E}\right)_0 - N\left(\frac{\partial \ln \nu}{\partial N}\right)_0 \qquad (10.138)$$

The expansion to linear terms in E and N is valid in the limit of large systems of which the subsystem is a very small part, since $E(\partial \ln \nu/\partial N)_{E_0, N_0}$ is the fraction of the number of states of the whole system which lie in the range of energy of the subsystem, and the series is essentially an expansion in powers of this fraction.

From (10.136) and (10.138) it follows that

$$p(E, N) \approx \exp\left\{-E\left(\frac{\partial \ln \nu}{\partial E}\right)_{E_0, N_0} - N\left(\frac{\partial \ln \nu}{\partial N}\right)_{E_0, N_0}\right\} \qquad (10.139)$$

The coefficient $(\partial \ln \nu/\partial E)_{E_0, N_0}$ may be identified with $1/KT$, K being the Boltzmann constant and T the temperature. Thus

$$p(E, N) = \frac{1}{Z} e^{-E/KT} \qquad (10.140)$$

where Z is a constant to be chosen so that $\text{Tr}\, \rho = 1$.

The quantities $p(E_\alpha, N)$ are now the probabilities p_α entering into the definition (10.111) of the density matrix. The statistical operator may be written

$$\rho = \frac{1}{Z} e^{-\beta H} = \frac{e^{-\beta H}}{\text{Tr}\, e^{-\beta H}} \qquad (10.141)$$

where H is the Hamiltonian for the system and $\beta = 1/kT$. Matrix elements may then be calculated using an *arbitrary* basis. Mean values may also be calculated with an arbitrary complete set of states as basis, since the trace is independent of the basis used.

13. Grand Canonical Ensemble

The above calculation remains valid, except that N may now vary. $p(E, N)$ is given by (10.139). The quantity $1/\beta(\partial \ln \nu/\partial N)_{E_0, N_0}$ is known as the "chemical potential" and is designated as μ. The statistical operator has the form

$$\rho = \frac{e^{-\beta(H-\mu N)}}{\text{Tr }e^{-\beta(H-\mu N)}} \tag{10.142}$$

where N is now the number operator for the system.

The exponential operators in both cases (canonical and grand canonical ensemble) are to be understood in the sense of series expansions. It is sometimes convenient to take the subsystems to be single particles (e.g., single electrons in an electron gas, or a metal, or molecules in a gas). We again define the denominator, $\text{Tr }e^{-\beta(H-\mu N)}$, as Z.

14. Bose–Einstein and Fermi–Dirac Distributions

Let us calculate, for a many-particle system, the mean number of particles in a given single-particle state $|\alpha\rangle$, that is,

$$\langle n_\alpha \rangle = \text{Tr }\rho c_\alpha{}^\dagger c_\alpha \tag{10.143}$$

We of course use as a basis the states of occupation of the single-particle states of which $|\alpha\rangle$ is one. Now if we work with the grand canonical ensemble, so that we do not assume the total number of particles to be known,

$$\text{Tr }\rho c_\alpha{}^\dagger c_\alpha = \frac{1}{Z}\text{Tr }c_\alpha e^{-\beta(H-\mu N)}c_\alpha{}^\dagger \tag{10.144}$$

But

$$e^{-\beta(H-\mu N)}c_\alpha{}^\dagger = e^{-\beta(E_\alpha - \mu)}c_\alpha{}^\dagger e^{-\beta(H-\mu N)} \tag{10.145}$$

since the state created by operating with $c_\alpha{}^\dagger$ first has one additional particle and an extra energy E_α more than the one on which it operates. Therefore,

$$\langle n_\alpha \rangle = \frac{1}{Z}e^{-\beta(E_\alpha - \mu)}\text{Tr }c_\alpha c_\alpha{}^\dagger e^{-\beta(H-\mu N)}$$

Using the commutation rule

$$c_\alpha c_\alpha{}^\dagger + \eta c_\alpha{}^\dagger c_\alpha = 1$$

where $\eta = +1$ or -1 according as the particles are fermions or bosons, we obtain

$$\langle n_\alpha \rangle = \frac{1}{Z} e^{-\beta(E_\alpha - \mu)} \, \text{Tr} \, (1 - \eta c_\alpha^\dagger c_\alpha) e^{-\beta(H - \mu N)}$$

$$= e^{-\beta(E_\alpha - \mu)} - \eta e^{-\beta(E_\alpha - \mu)} \langle n_\alpha \rangle$$

It follows that

$$\langle n_\alpha \rangle = \frac{e^{-\beta(E_\alpha - \mu)}}{1 + \eta e^{-\beta(E_\alpha - \mu)}}$$

$$= \frac{1}{e^{\beta(E_\alpha - \mu)} + \eta} \qquad (10.146)$$

For $\eta = 1$ this is the Fermi–Dirac distribution; for $\eta = -1$, the Bose–Einstein.

The chemical potential μ is determined by the condition

$$\sum \langle n_\alpha \rangle = N \qquad (10.147)$$

where N is the total number of particles. (Note that the chemical potential is determined by the *mean* number of particles in the systems of the ensemble.) If now $N(E_\alpha)$ is the density of single-particle states, this condition becomes

$$\int N(E_\alpha) \langle n_\alpha \rangle \, dE_\alpha = N \qquad (10.148)$$

For fermions, $\langle n_\alpha \rangle$ must always be ≤ 1; it is therefore the *probability* of occupancy of a state and is designated $f(E_\alpha)$, which is called the "Fermi function."

It is important to distinguish here between the p_α of (10.111) and the Fermi function $f(E_\alpha)$, particularly when the subsystem is a single particle. The probability p_α is an a priori probability, while $f(E_\alpha)$ is the probability of occupancy of state E_α in a many-body system, taking account of the commutation properties of the particle operators which embody the assumption of indistinguishability of particles and, in the case of fermions, of the exclusion principle.

Problem 10-11: Show that, for the Fermi–Dirac distribution

$$\Delta n_\alpha = \frac{e^{(1/2)\beta(E_\alpha - \mu)}}{e^{\beta(E_\alpha - \mu)} + 1}$$

and for the Bose–Einstein distribution

$$\Delta n_\alpha = \frac{e^{(1/2)\beta(E_\alpha - \mu)}}{e^{\beta(E_\alpha - \mu)} - 1}$$

---

15. Linear Response Theory

We consider in this section a problem analogous to certain classical problems which have been dealt with in earlier chapters; that is, the problem of the *response* of a system to an external stimulus. The systems we consider are many-body ones, described by the statistical operator.

If the Hamiltonian of the system can be written

$$H = H_0 + H_1 F(t) \qquad (10.149)$$

where H_0 is taken to describe the unperturbed system and $H_1 F(t)$ the external stimulus, it is possible to write ρ in the form

$$\rho = \rho_0 + \rho_1 \qquad (10.150)$$

where

$$i\frac{\partial \rho_0}{\partial t} = [H_0, \rho_0] \qquad (10.151)$$

If the system is in a steady state

$$[H_0, \rho_0] = 0 \qquad (10.152)$$

The essence of the theory of *linear* response is that we treat the perturbation only to first order; the response ρ_1 is therefore also kept only to first order. The equation for ρ_1, is, then,

$$i\frac{\partial \rho_1}{\partial t} = [H_0, \rho_1] + F(t)[H_1, \rho_0] \qquad (10.153)$$

The substitution

$$\rho_1 = e^{-iH_0 t} f e^{iH_0 t} \qquad (10.154)$$

transforms the equation to

$$e^{-iH_0 t} i\frac{\partial f}{\partial t} e^{iH_0 t} = F(t)[H_1, \rho_0]$$

or

$$i\frac{\partial f}{\partial t} = F(t)[H_1(t), \rho_0] \qquad (10.155)$$

$H_1(t)$ being the time-dependent operator

$$H_1(t) = e^{iH_0 t} H_1 e^{-iH_0 t} \qquad (10.156)$$

Integration of Eq. (10.155) gives

$$f = -i \int_{-\infty}^{t} F(t')[H_1(t'), \rho_0]\, dt' \qquad (10.157)$$

where it has been assumed that the perturbation was zero at time $-\infty$.

Consider now any dynamical variable A whose value is zero in the unperturbed state; its value in the system subject to the external stimulus is

$$\langle A \rangle = \operatorname{Tr} \rho_1 A$$

$$= \operatorname{Tr} e^{-iH_0 t} f e^{iH_0 t} A$$

$$= \operatorname{Tr} f A(t) \tag{10.158}$$

where

$$A(t) = e^{iH_0 t} A e^{-iH_0 t} \tag{10.159}$$

To simplify (10.158), we note that

$$\operatorname{Tr} [H_1(t'), \rho_0] A(t) = \operatorname{Tr} H_1(t') \rho_0 A(t) - \operatorname{Tr} \rho_0 H_1(t') A$$

$$= \operatorname{Tr} \rho_0 [A(t), H_1(t')] \tag{10.160}$$

Therefore, finally

$$\langle A \rangle = -i \operatorname{Tr} \rho_0 \int_{-\infty}^{t} F(t')[A(t), H_1(t')] \, dt' \tag{10.161}$$

This is the linear response formula, originally due to Kubo.

The question of the convergence of (10.161) at the lower limit $-\infty$ is worthy of consideration. If $F(t)$ is "turned on" suddenly at some finite time (say $t = 0$) the use of (10.161) is straightforward. Since the relation between $\langle A \rangle$ and $F(t')$ is linear, if we express F in terms of its Fourier components

$$F(t') = \int \mathscr{F}(\omega) e^{i\omega t'} \, d\omega \tag{10.162}$$

the contributions from the different frequencies ω are additive. Consider then the case of a single frequency

$$F(t') = e^{i\omega t'} \tag{10.163}$$

In this case the integral (10.161) does *not* converge at the lower limit. Suppose, however, that the perturbation has been turned on slowly from some time in the distant past. This may be represented mathematically by multiplying (10.163) by $e^{\epsilon t'}$, where we may if we wish take ϵ vanishingly small. By this device, convergence is assured.

Making use of the fact that ρ_0 commutes with H_0 and that the order of two terms in the trace of a product may be reversed, we may write alternative forms for $\langle A \rangle$:

$$\langle A \rangle = -i \lim_{\epsilon \to 0} e^{i(\omega - i\epsilon)t} \operatorname{Tr} \int_{-\infty}^{t} e^{i(\omega - i\epsilon)(t' - t)} e^{iH_0 t'} \rho_0 e^{iH_0 t'} [A(t), H_1(t')] \, dt'$$

$$= -i \lim_{\epsilon \to 0} e^{i(\omega - i\epsilon)t} \operatorname{Tr} \rho_0 \int_{-\infty}^{t} e^{i(\omega - i\epsilon)(t' - t)} [A(t - t'), H_1] \, dt'$$

$$= -i e^{i\omega t} \operatorname{Tr} \rho_0 \lim_{\epsilon \to 0} \int_{0}^{\infty} e^{-i(\omega - i\epsilon)\tau} [A(\tau), H_1] \, d\tau \tag{10.164}$$

where, in the last line, the new variable $\tau = t - t'$ has been introduced. We see, then, that the response has the same periodic time variation as the applied perturbation.

16. An Example: Transverse Conductivity of an Electron Gas

As an illustration of the linear response formalism, we calculate the conductivity of an electron gas in a transverse field

$$\mathbf{E} = E_0 \boldsymbol{\xi} \, e^{i(\mathbf{q}\cdot\mathbf{r}-\omega t)} \tag{10.165}$$

where $\mathbf{q} \cdot \boldsymbol{\xi} = 0$. The field may be expressed in terms of a vector potential

$$\mathbf{A} = -\frac{ic}{\omega} E_0 \boldsymbol{\xi} \, e^{i(\mathbf{q}\cdot\mathbf{r}-\omega t)} \tag{10.166}$$

where $\nabla \cdot \mathbf{A} = i\mathbf{q} \cdot \mathbf{A} = 0$. Since the one-particle Hamiltonian of an electron of charge $-e$ in a field described by vector potential $\mathbf{A}$ is

$$H = \frac{1}{2m}\left(\mathbf{p} + \frac{e}{c}\mathbf{A}\right)^2$$

$$= \frac{p^2}{2m} + \frac{e}{mc}(\mathbf{p}\cdot\mathbf{A} + \mathbf{A}\cdot\mathbf{p}) \tag{10.167}$$

to the first order in the field, the many-body Hamiltonian splits into two parts, an unperturbed one

$$H_0 = \int \psi^\dagger(\mathbf{r}') \frac{p'^2}{2m} \psi(\mathbf{r}') \, d^3r'$$

$$= \frac{1}{\Omega} \sum_{\mathbf{k}} c_{\mathbf{k}}^\dagger c_{\mathbf{k}} \frac{\hbar^2 k^2}{2m}$$

$$= \frac{1}{\Omega} \sum_{\mathbf{k}} c_{\mathbf{k}}^\dagger c_{\mathbf{k}} \epsilon_{\mathbf{k}} \tag{10.168}$$

and the perturbation

$$H_1 F(t) = -\frac{ie\hbar}{2m\omega\Omega} E_0 \, e^{-i\omega t} \sum_{\mathbf{K},\mathbf{K}'} c_{\mathbf{K}'}^\dagger c_{\mathbf{K}} \int e^{-i\mathbf{K}'\cdot\mathbf{r}'} (2\mathbf{K}+\mathbf{q})\cdot\boldsymbol{\xi}\, e^{i(\mathbf{K}+\mathbf{q})\cdot\mathbf{r}'} d^3r'$$

$$= -\frac{ie\hbar}{2m\omega} E_0 \, e^{-i\omega t} \sum_{\mathbf{K}} c_{\mathbf{K}+\mathbf{q}}^\dagger c_{\mathbf{K}} (2\mathbf{K}+\mathbf{q})\cdot\boldsymbol{\xi} \tag{10.169}$$

Problem 10-12: Kubo has developed a general formula for electrical conductivity along the following lines: If $\mathbf{X}$ is the electrical polarization of the electrons, we may take

$$H_1 = - \mathbf{E} \cdot \mathbf{X} \, e^{i\omega t}$$

From (10.157) and (10.158),

$$\langle j_\mu(t) \rangle = -i \, \mathrm{Tr} \int_{-\infty}^{t} F(t') [H_1(t'), \rho_0] j_\mu(t) \, dt'$$

Now

(i) Show that

$$[H_1(t'), \rho_0] = -i\rho_0 \int_{0}^{\beta} \dot{H}_1(t' - i\lambda) \, d\lambda$$

where the dot indicates a time derivative.

(ii) Noting that $\dot{X}_\nu = j_\nu$, and making a change of variable of integration, show that the conductivity tensor is given by

$$\langle j_\mu(t) \rangle = \mathrm{Tr} \int_{0}^{\beta} d\lambda \int_{-\infty - i\lambda}^{-i\lambda} F(t + \tau + i\lambda) \rho_0 j_\nu(\tau) j_\mu(0) \, d\tau \, E_\nu$$

(iii) Putting $F(t) = e^{i\omega t}$, show that

$$\sigma_{\mu\nu} = \frac{1 - e^{-\beta\omega}}{2\omega} \int_{-\infty}^{\infty} \langle j_\nu(\tau) j_\mu(0) \rangle \, e^{i\omega\tau} \, d\tau$$

where

$$\langle j_\nu(\tau) j_\mu(0) \rangle = \mathrm{Tr} \, \rho_0 j_\nu(\tau) j_\mu(0)$$

where Ω is the volume within which the system is enclosed. Also, to be better able to check the dimensionality of our formulas, we have reintroduced the quantum constant $\hbar$ throughout. Now the many-particle operator for the current density is

$$\mathbf{J} = -\frac{e}{2m} \int \psi^\dagger(\mathbf{r}') \left[\left(\mathbf{p}' + \frac{e}{c} \mathbf{A}(\mathbf{r}') \right) \delta(\mathbf{r}' - \mathbf{r}) + \delta(\mathbf{r}' - \mathbf{r}) \right.$$

$$\left. \times \left(\mathbf{p}' + \frac{e}{c} \mathbf{A}(\mathbf{r}') \right) \right] \psi(\mathbf{r}') \, d^3\mathbf{r}'$$

$$= \mathbf{J}_p + \mathbf{J}_d \tag{10.170}$$

where $\mathbf{J}_p$ is given by

$$\mathbf{J}_p = -\frac{e}{2m} \int \psi^\dagger [\mathbf{p}' \delta(\mathbf{r}' - \mathbf{r}) + \delta(\mathbf{r}' - \mathbf{r}) \mathbf{p}'] \psi \, d^3\mathbf{r}' \tag{10.171}$$

and

$$\mathbf{J}_d = -\frac{e^2}{mc} \psi^\dagger(\mathbf{r}) \mathbf{A}(\mathbf{r}) \psi(\mathbf{r}). \tag{10.172}$$

600   Quantum Mechanics of Many-Body Problems

Again, putting

$$\psi = \frac{1}{\sqrt{\Omega}} \sum_k c_k e^{-i k' \cdot r'}, \text{etc.},$$

these become

$$\mathbf{J}_p = -\frac{e}{2m\Omega} \sum c_{k'}^\dagger c_k \int e^{i k \cdot r'} \; [\mathbf{p}'\delta(\mathbf{r}'-\mathbf{r}) + \delta(\mathbf{r}'-\mathbf{r})\mathbf{p}'] \, e^{i k \cdot r'} \, d^3 \mathbf{r}'$$

$$= -\frac{e\hbar}{2m\Omega} \sum c_{k'}^\dagger c_k (\mathbf{k}+\mathbf{k}') \, e^{i(\mathbf{k}-\mathbf{k}')\cdot\mathbf{r}} \tag{10.173}$$

and

$$\mathbf{J}_d = \frac{ie^2}{m\omega} E_0 \boldsymbol{\xi} \sum c_{k'}^\dagger c_k \, e^{-i(\mathbf{k}'-\mathbf{k})\cdot\mathbf{r}} \, e^{i(\mathbf{q}\cdot\mathbf{r}-\omega t)} \tag{10.174}$$

The mean value of the current density is now seen to be made up of two parts. The contribution from $\mathbf{J}_p$ must be calculated from linear response theory to first order in the field, using (10.161). Since $\mathbf{J}_d$ is *already* of first order in the field, its contribution to the mean current density is

$$\text{Tr}\,\rho_0 \mathbf{J}_d = \frac{ie^2}{m\omega} E_0 \boldsymbol{\xi} e^{i(\mathbf{q}\cdot\mathbf{r}-\omega t)} \sum \langle n_k \rangle$$

$$= \frac{iNe^2}{m\omega} E_0 \boldsymbol{\xi} e^{i(\mathbf{q}\cdot\mathbf{r}-\omega t)}$$

where N is the total number of electrons. This gives a contribution to the conductivity

$$\sigma_0 = \frac{iNe^2}{m\omega} \tag{10.175}$$

By (10.161), the mean value of the other part of the current density is

$$\langle \mathbf{J}_p \rangle = -\frac{i}{\hbar} \text{Tr}\,\rho_0 \int_{-\infty}^t e^{-i\omega t'} [\mathbf{J}_p(t), H_1(t')] \, dt' \tag{10.176}$$

But

$$\mathbf{J}_p(t) = -\frac{e\hbar}{2m\Omega} \sum e^{iH_0 t/\hbar} c_{k'}^\dagger c_k e^{-iH_0 t/\hbar} (\mathbf{k}+\mathbf{k}') e^{i(\mathbf{k}-\mathbf{k}')\cdot\mathbf{r}}$$

$$= -\frac{e\hbar}{2m\Omega} \sum e^{-i(\epsilon_{k'}-\epsilon_k)t/\hbar} c_{k'}^\dagger c_k (\mathbf{k}+\mathbf{k}') e^{i(\mathbf{k}-\mathbf{k}')\cdot\mathbf{r}} \tag{10.177}$$

and

$$H_1(t') = -\frac{ie\hbar}{2m\omega} E_0 \sum_K e^{iH_0 t'/\hbar} c_{K+q}^\dagger c_K e^{-iH_0 t'/\hbar} (2\mathbf{K}+\mathbf{q}) \cdot \boldsymbol{\xi}$$

$$= -\frac{ie\hbar}{2m\omega} E_0 \sum_K e^{-i(\epsilon_{K+q}-\epsilon_K)t/\hbar} c_{K+q}^\dagger c_K (2\mathbf{K}+\mathbf{q}) \cdot \boldsymbol{\xi} \tag{10.178}$$

Therefore,

$$\langle \mathbf{J}_p \rangle = \frac{e^2\hbar}{4m^2\omega\Omega} E_0 \sum_{\mathbf{k},\mathbf{k}',\mathbf{K}} (\mathbf{k}+\mathbf{k}') [(2\mathbf{K}+\mathbf{q})\cdot\boldsymbol{\xi}] e^{i(\mathbf{k}-\mathbf{k}')\cdot\mathbf{r}}$$

$$\times \operatorname{Tr} \rho_0 [c_{\mathbf{k}'}{}^\dagger c_{\mathbf{k}}, c^\dagger{}_{\mathbf{K}+\mathbf{q}} c_{\mathbf{K}}] e^{-i(\epsilon_{\mathbf{k}'}-\epsilon_{\mathbf{k}})t/\hbar}$$

$$\times \lim_{\mu\to 0} \int_{-\infty}^{t} e^{-i\omega t'} e^{-i(\epsilon_{\mathbf{K}+\mathbf{q}}-\epsilon_{\mathbf{K}})t'/\hbar} e^{\mu t'} \, dt' \qquad (10.179)$$

By simple algebra, using the anticommutation relations

$$[c_{\mathbf{k}'}{}^\dagger c_{\mathbf{k}}, c^\dagger{}_{\mathbf{K}+\mathbf{q}} c_{\mathbf{K}}] = \delta_{\mathbf{k},\mathbf{K}+\mathbf{q}} c_{\mathbf{k}'}{}^\dagger c_{\mathbf{K}} - \delta_{\mathbf{K}\mathbf{k}'} c^\dagger{}_{\mathbf{K}+\mathbf{q}} c_{\mathbf{k}} \qquad (10.180)$$

The trace of ρ_0 times this is

$$\delta_{\mathbf{k},\mathbf{K}+\mathbf{q}} \delta_{\mathbf{k}'\mathbf{K}} [f(\epsilon_{\mathbf{k}-\mathbf{q}}) - f(\epsilon_{\mathbf{k}})] \qquad (10.181)$$

$f(\epsilon)$ being the Fermi distribution function defined above. If this is substituted in (10.179), the equation for the mean current density becomes

$$\langle \mathbf{J} \rangle = E_0 e^{i(\mathbf{q}\cdot\mathbf{r}-\omega t)} \left\{ \frac{iNe^2}{m\omega} \boldsymbol{\xi} + \frac{ie^2\hbar}{4m^2\omega\Omega} \sum_{\mathbf{k}} [f(\epsilon_{\mathbf{k}-\mathbf{q}}) - f(\epsilon_{\mathbf{k}})] \right.$$

$$\left. \frac{(2\mathbf{k}-\mathbf{q})[(2\mathbf{k}-\mathbf{q})\cdot\boldsymbol{\xi}]}{\omega + (1/\hbar)(\epsilon_{\mathbf{k}}-\epsilon_{\mathbf{k}-\mathbf{q}})} \right\} \qquad (10.182)$$

By virtue of the fact that $\mathbf{q}\cdot\boldsymbol{\xi} = 0$ the second term may be simplified. For the sum to be different from zero, the summand must be even in $\mathbf{k}\cdot\boldsymbol{\xi}$. Therefore only the component of $\langle \mathbf{J} \rangle$ in the direction of $\boldsymbol{\xi}$ is nonzero, the current is in the direction of the field, and the conductivity is a scalar. Making use of this fact and replacing the sum by an integral, we obtain for the conductivity the formula

$$\sigma = \frac{iNe^2}{m\omega} + 2\frac{ie^2\hbar}{m^2\omega}\left(\frac{1}{2\pi}\right)^3 \int [f(\epsilon_{\mathbf{k}-\mathbf{q}}) - f(\epsilon_{\mathbf{k}})]$$

$$\times \frac{(\mathbf{k}\cdot\boldsymbol{\xi})^2}{\omega + (1/\hbar)(\epsilon_{\mathbf{k}}-\epsilon_{\mathbf{k}-\mathbf{q}})} \, d^3k \qquad (10.183)$$

The second term goes to zero as $\mathbf{q} \to 0$, that is, in the long-wavelength limit. It is in general very complicated, though it is not too difficult to get the leading term in an expansion in powers of $\mathbf{q}$, which is of order q^2.

If (10.183) is broken into two integrals, and in the first one makes the substitution

$$\mathbf{k} = -\mathbf{k}' + \mathbf{q}$$

and then drops the prime on the variable of integration, the two integrals combine into

$$\frac{2}{\hbar} \int f(\epsilon_{\mathbf{k}}) (\mathbf{k},\boldsymbol{\xi})^2 \frac{\epsilon_{\mathbf{k}}-\epsilon_{\mathbf{k}-\mathbf{q}}}{\omega^2 - (1/\hbar^2)(\epsilon_{\mathbf{k}}-\epsilon_{\mathbf{k}-\mathbf{q}})^2} \, d^3k \qquad (10.184)$$

Expanding the numerator, we see that the term $(\hbar^2/m)\mathbf{k}\cdot\mathbf{q}$, which is the only term of order q, vanishes by symmetry. The remaining term, $\hbar^2 q^2/2m$, is already of order q^2, so that to this order the denominator may be replaced simply by ω^2. The expression (10.184) then takes the approximate form

$$\frac{\hbar q^2}{m\omega^2}\int f(\epsilon_{\mathbf{k}})\tfrac{1}{3}k^2 4\pi k^2\,dk \qquad (10.185)$$

At zero temperature, $f=1$ for $k \leq k_F$, zero for $k \geq k_F$. In that case (10.185) becomes

$$\frac{\hbar q^2}{m\omega^2}\frac{4\pi}{15}k_F{}^5$$

Finally, substituting in (10.183), we obtain the conductivity up to quadratic terms in q^2:

$$\sigma = \frac{iNe^2}{m\omega}\left[1 + \frac{1}{5}\frac{\hbar^2 k_F^2}{m^2\omega^2}q^2\right] \qquad (10.186)$$

17. Propagation of Electromagnetic Waves in an Electron Gas

In Chapter 5, Eq. (5.550), we gave an equation for the dielectric function of a material consisting of free electrons and other polarizable components (ions, valence electrons). If we consider only frequencies low enough that the variation with ω of the contribution ϵ_L from these other components is negligible, and substitute the conductivity (10.186), we obtain for the frequency-dependent transverse dielectric response

$$\epsilon = \epsilon_L + \frac{i}{\omega\epsilon_0}\frac{iNe^2}{m\omega}\left(1 + \frac{1}{5}\frac{\hbar^2 k_F^2 q^2}{m^2\omega^2}\right)$$

$$= \epsilon_L\left\{1 - \frac{\omega_p^2}{\omega^2}\left(1 + \frac{1}{5}\frac{\hbar^2 k_F^2 q^2}{m^2\omega^2}\right)\right\} \qquad (10.187)$$

where $\omega_p{}^2$ is as defined in (5.576). The dispersion relation for waves in the medium is then given by (5.580),

$$\frac{c^2 q^2}{\omega^2} = \epsilon_L\left\{1 - \frac{\omega_p^2}{\omega^2}\left(1 + \frac{1}{5}\frac{\hbar^2 k_F^2 q^2}{m^2\omega^2}\right)\right\} \qquad (10.188)$$

or

$$c^2 q^2 = \frac{\epsilon_L(\omega^2 - \omega_p^2)}{1 + (1/5)\epsilon_L(\omega_p^2/\omega^2)(\hbar^2 k_F^2/m^2 c^2)} \qquad (10.189)$$

It is clear, since $\hbar k_F \ll mc$ for real electron densities, that the second term in the denominator is very small for $\omega \lesssim \omega_p$.

It follows also that waves can only be propagated for ω greater than the plasma frequency ω_p. Below that frequency, electromagnetic energy is dissipated in creating individual electron excitations. Above it, the wave is a coupled mode of particles and electromagnetic field; the motion of the particles *creates* the field which in turn governs their motion.

INDEX

INDEX

A CATALOGUE OF
SELECTED DOVER BOOKS
IN ALL FIELDS OF INTEREST

A CATALOGUE OF SELECTED DOVER
BOOKS IN ALL FIELDS OF INTEREST

CELESTIAL OBJECTS FOR COMMON TELESCOPES, T. W. Webb. The most used book in amateur astronomy: inestimable aid for locating and identifying nearly 4,000 celestial objects. Edited, updated by Margaret W. Mayall. 77 illustrations. Total of 645pp. 5⅜ x 8½.
20917-2, 20918-0 Pa., Two-vol. set $10.00

HISTORICAL STUDIES IN THE LANGUAGE OF CHEMISTRY, M. P. Crosland. The important part language has played in the development of chemistry from the symbolism of alchemy to the adoption of systematic nomenclature in 1892. ". . . wholeheartedly recommended,"—Science. 15 illustrations. 416pp. of text. 5⅝ x 8¼.   63702-6 Pa. $7.50

BURNHAM'S CELESTIAL HANDBOOK, Robert Burnham, Jr. Thorough, readable guide to the stars beyond our solar system. Exhaustive treatment, fully illustrated. Breakdown is alphabetical by constellation: Andromeda to Cetus in Vol. 1; Chamaeleon to Orion in Vol. 2; and Pavo to Vulpecula in Vol. 3. Hundreds of illustrations. Total of about 2000pp. 6⅛ x 9¼.
23567-X, 23568-8, 23673-0 Pa., Three-vol. set $32.85

THEORY OF WING SECTIONS: INCLUDING A SUMMARY OF AIR-FOIL DATA, Ira H. Abbott and A. E. von Doenhoff. Concise compilation of subatomic aerodynamic characteristics of modern NASA wing sections, plus description of theory. 350pp. of tables. 693pp. 5⅜ x 8½.
60586-8 Pa. $9.95

DE RE METALLICA, Georgius Agricola. Translated by Herbert C. Hoover and Lou H. Hoover. The famous Hoover translation of greatest treatise on technological chemistry, engineering, geology, mining of early modern times (1556). All 289 original woodcuts. 638pp. 6¾ x 11.
60006-8 Clothbd. $19.95

THE ORIGIN OF CONTINENTS AND OCEANS, Alfred Wegener. One of the most influential, most controversial books in science, the classic statement for continental drift. Full 1966 translation of Wegener's final (1929) version. 64 illustrations. 246pp. 5⅜ x 8½.(EBE)61708-4 Pa. $5.00

THE PRINCIPLES OF PSYCHOLOGY, William James. Famous long course complete, unabridged. Stream of thought, time perception, memory, experimental methods; great work decades ahead of its time. Still valid, useful; read in many classes. 94 figures. Total of 1391pp. 5⅜ x 8½.
20381-6, 20382-4 Pa., Two-vol. set $17.90

YUCATAN BEFORE AND AFTER THE CONQUEST, Diego de Landa. First English translation of basic book in Maya studies, the only significant account of Yucatan written in the early post-Conquest era. Translated by distinguished Maya scholar William Gates. Appendices, introduction, 4 maps and over 120 illustrations added by translator. 162pp. 5⅜ x 8½.
23622-6 Pa. $3.00

THE MALAY ARCHIPELAGO, Alfred R. Wallace. Spirited travel account by one of founders of modern biology. Touches on zoology, botany, ethnography, geography, and geology. 62 illustrations, maps. 515pp. 5⅜ x 8½.
20187-2 Pa. $6.95

THE DISCOVERY OF THE TOMB OF TUTANKHAMEN, Howard Carter, A. C. Mace. Accompany Carter in the thrill of discovery, as ruined passage suddenly reveals unique, untouched, fabulously rich tomb. Fascinating account, with 106 illustrations. New introduction by J. M. White. Total of 382pp. 5⅜ x 8½. (Available in U.S. only)   23500-9 Pa. $5.50

THE WORLD'S GREATEST SPEECHES, edited by Lewis Copeland and Lawrence W. Lamm. Vast collection of 278 speeches from Greeks up to present. Powerful and effective models; unique look at history. Revised to 1970. Indices. 842pp. 5⅜ x 8½.   20468-5 Pa. $9.95

THE 100 GREATEST ADVERTISEMENTS, Julian Watkins. The priceless ingredient; His master's voice; 99 44/100% pure; over 100 others. How they were written, their impact, etc. Remarkable record. 130 illustrations. 233pp. 7⅞ x 10 3/5.   20540-1 Pa. $6.95

CRUICKSHANK PRINTS FOR HAND COLORING, George Cruickshank. 18 illustrations, one side of a page, on fine-quality paper suitable for watercolors. Caricatures of people in society (c. 1820) full of trenchant wit. Very large format. 32pp. 11 x 16.   23684-6 Pa. $6.00

THIRTY-TWO COLOR POSTCARDS OF TWENTIETH-CENTURY AMERICAN ART, Whitney Museum of American Art. Reproduced in full color in postcard form are 31 art works and one shot of the museum. Calder, Hopper, Rauschenberg, others. Detachable. 16pp. 8¼ x 11.
23629-3 Pa. $3.50

MUSIC OF THE SPHERES: THE MATERIAL UNIVERSE FROM ATOM TO QUASAR SIMPLY EXPLAINED, Guy Murchie. Planets, stars, geology, atoms, radiation, relativity, quantum theory, light, antimatter, similar topics. 319 figures. 664pp. 5⅜ x 8½.
21809-0, 21810-4 Pa., Two-vol. set $11.00

EINSTEIN'S THEORY OF RELATIVITY, Max Born. Finest semi-technical account; covers Einstein, Lorentz, Minkowski, and others, with much detail, much explanation of ideas and math not readily available elsewhere on this level. For student, non-specialist. 376pp. 5⅜ x 8½.
60769-0 Pa. $5.00

CATALOGUE OF DOVER BOOKS

THE SENSE OF BEAUTY, George Santayana. Masterfully written discussion of nature of beauty, materials of beauty, form, expression; art, literature, social sciences all involved. 168pp. 5⅜ x 8½.   20238-0 Pa. $3.50

ON THE IMPROVEMENT OF THE UNDERSTANDING, Benedict Spinoza. Also contains *Ethics, Correspondence,* all in excellent R. Elwes translation. Basic works on entry to philosophy, pantheism, exchange of ideas with great contemporaries. 402pp. 5⅜ x 8½.   20250-X Pa. $5.95

THE TRAGIC SENSE OF LIFE, Miguel de Unamuno. Acknowledged masterpiece of existential literature, one of most important books of 20th century. Introduction by Madariaga. 367pp. 5⅜ x 8½.
                                                            20257-7 Pa. $6.00

THE GUIDE FOR THE PERPLEXED, Moses Maimonides. Great classic of medieval Judaism attempts to reconcile revealed religion (Pentateuch, commentaries) with Aristotelian philosophy. Important historically, still relevant in problems. Unabridged Friedlander translation. Total of 473pp. 5⅜ x 8½.   20351-4 Pa. $6.95

THE I CHING (THE BOOK OF CHANGES), translated by James Legge. Complete translation of basic text plus appendices by Confucius, and Chinese commentary of most penetrating divination manual ever prepared. Indispensable to study of early Oriental civilizations, to modern inquiring reader. 448pp. 5⅜ x 8½.   21062-6 Pa. $6.00

THE EGYPTIAN BOOK OF THE DEAD, E. A. Wallis Budge. Complete reproduction of Ani's papyrus, finest ever found. Full hieroglyphic text, interlinear transliteration, word for word translation, smooth translation. Basic work, for Egyptology, for modern study of psychic matters. Total of 533pp. 6½ x 9¼. (USCO)   21866-X Pa. $8.50

THE GODS OF THE EGYPTIANS, E. A. Wallis Budge. Never excelled for richness, fullness: all gods, goddesses, demons, mythical figures of Ancient Egypt; their legends, rites, incarnations, variations, powers, etc. Many hieroglyphic texts cited. Over 225 illustrations, plus 6 color plates. Total of 988pp. 6⅛ x 9¼. (EBE)
                            22055-9, 22056-7 Pa., Two-vol. set $20.00

THE STANDARD BOOK OF QUILT MAKING AND COLLECTING, Marguerite Ickis. Full information, full-sized patterns for making 46 traditional quilts, also 150 other patterns. Quilted cloths, lame, satin quilts, etc. 483 illustrations. 273pp. 6⅞ x 9⅝.   20582-7 Pa. $5.95

CORAL GARDENS AND THEIR MAGIC, Bronsilaw Malinowski. Classic study of the methods of tilling the soil and of agricultural rites in the Trobriand Islands of Melanesia. Author is one of the most important figures in the field of modern social anthropology. 143 illustrations. Indexes. Total of 911pp. of text. 5⅝ x 8¼. (Available in U.S. only)
                                                            23597-1 Pa. $12.95

THE PHILOSOPHY OF HISTORY, Georg W. Hegel. Great classic of Western thought develops concept that history is not chance but a rational process, the evolution of freedom. 457pp. 5⅜ x 8½. 20112-0 Pa. $6.00

LANGUAGE, TRUTH AND LOGIC, Alfred J. Ayer. Famous, clear introduction to Vienna, Cambridge schools of Logical Positivism. Role of philosophy, elimination of metaphysics, nature of analysis, etc. 160pp. 5⅜ x 8½. (USCO) 20010-8 Pa. $2.50

A PREFACE TO LOGIC, Morris R. Cohen. Great City College teacher in renowned, easily followed exposition of formal logic, probability, values, logic and world order and similar topics; no previous background needed. 209pp. 5⅜ x 8½. 23517-3 Pa. $4.95

REASON AND NATURE, Morris R. Cohen. Brilliant analysis of reason and its multitudinous ramifications by charismatic teacher. Interdisciplinary, synthesizing work widely praised when it first appeared in 1931. Second (1953) edition. Indexes. 496pp. 5⅜ x 8½. 23633-1 Pa. $7.50

AN ESSAY CONCERNING HUMAN UNDERSTANDING, John Locke. The only complete edition of enormously important classic, with authoritative editorial material by A. C. Fraser. Total of 1176pp. 5⅜ x 8½. 20530-4, 20531-2 Pa., Two-vol. set $16.00

HANDBOOK OF MATHEMATICAL FUNCTIONS WITH FORMULAS, GRAPHS, AND MATHEMATICAL TABLES, edited by Milton Abramowitz and Irene A. Stegun. Vast compendium: 29 sets of tables, some to as high as 20 places. 1,046pp. 8 x 10½. 61272-4 Pa. $17.95

MATHEMATICS FOR THE PHYSICAL SCIENCES, Herbert S. Wilf. Highly acclaimed work offers clear presentations of vector spaces and matrices, orthogonal functions, roots of polynomial equations, conformal mapping, calculus of variations, etc. Knowledge of theory of functions of real and complex variables is assumed. Exercises and solutions. Index. 284pp. 5⅝ x 8¼. 63635-6 Pa. $5.00

THE PRINCIPLE OF RELATIVITY, Albert Einstein et al. Eleven most important original papers on special and general theories. Seven by Einstein, two by Lorentz, one each by Minkowski and Weyl. All translated, unabridged. 216pp. 5⅜ x 8½. 60081-5 Pa. $3.50

THERMODYNAMICS, Enrico Fermi. A classic of modern science. Clear, organized treatment of systems, first and second laws, entropy, thermodynamic potentials, gaseous reactions, dilute solutions, entropy constant. No math beyond calculus required. Problems. 160pp. 5⅜ x 8½. 60361-X Pa. $4.00

ELEMENTARY MECHANICS OF FLUIDS, Hunter Rouse. Classic undergraduate text widely considered to be far better than many later books. Ranges from fluid velocity and acceleration to role of compressibility in fluid motion. Numerous examples, questions, problems. 224 illustrations. 376pp. 5⅝ x 8¼. 63699-2 Pa. $7.00

THE AMERICAN SENATOR, Anthony Trollope. Little known, long unavailable Trollope novel on a grand scale. Here are humorous comment on American vs. English culture, and stunning portrayal of a heroine/villainess. Superb evocation of Victorian village life. 561pp. 5⅜ x 8½.
23801-6 Pa. $7.95

WAS IT MURDER? James Hilton. The author of *Lost Horizon* and *Goodbye, Mr. Chips* wrote one detective novel (under a pen-name) which was quickly forgotten and virtually lost, even at the height of Hilton's fame. This edition brings it back—a finely crafted public school puzzle resplendent with Hilton's stylish atmosphere. A thoroughly English thriller by the creator of Shangri-la. 252pp. 5⅜ x 8. (Available in U.S. only)
23774-5 Pa. $3.00

CENTRAL PARK: A PHOTOGRAPHIC GUIDE, Victor Laredo and Henry Hope Reed. 121 superb photographs show dramatic views of Central Park: Bethesda Fountain, Cleopatra's Needle, Sheep Meadow, the Blockhouse, plus people engaged in many park activities: ice skating, bike riding, etc. Captions by former Curator of Central Park, Henry Hope Reed, provide historical view, changes, etc. Also photos of N.Y. landmarks on park's periphery. 96pp. 8½ x 11. 23750-8 Pa. $4.50

NANTUCKET IN THE NINETEENTH CENTURY, Clay Lancaster. 180 rare photographs, stereographs, maps, drawings and floor plans recreate unique American island society. Authentic scenes of shipwreck, lighthouses, streets, homes are arranged in geographic sequence to provide walking-tour guide to old Nantucket existing today. Introduction, captions. 160pp. 8⅞ x 11¾. 23747-8 Pa. $7.95

STONE AND MAN: A PHOTOGRAPHIC EXPLORATION, Andreas Feininger. 106 photographs by *Life* photographer Feininger portray man's deep passion for stone through the ages. Stonehenge-like megaliths, fortified towns, sculpted marble and crumbling tenements show textures, beauties, fascination. 128pp. 9¼ x 10¾. 23756-7 Pa. $5.95

CIRCLES, A MATHEMATICAL VIEW, D. Pedoe. Fundamental aspects of college geometry, non-Euclidean geometry, and other branches of mathematics: representing circle by point. Poincare model, isoperimetric property, etc. Stimulating recreational reading. 66 figures. 96pp. 5⅝ x 8¼.
63698-4 Pa. $3.50

THE DISCOVERY OF NEPTUNE, Morton Grosser. Dramatic scientific history of the investigations leading up to the actual discovery of the eighth planet of our solar system. Lucid, well-researched book by well-known historian of science. 172pp. 5⅜ x 8½. 23726-5 Pa. $3.50

THE DEVIL'S DICTIONARY. Ambrose Bierce. Barbed, bitter, brilliant witticisms in the form of a dictionary. Best, most ferocious satire America has produced. 145pp. 5⅜ x 8½. 20487-1 Pa. $2.50

CATALOGUE OF DOVER BOOKS

HISTORY OF BACTERIOLOGY, William Bulloch. The only comprehensive history of bacteriology from the beginnings through the 19th century. Special emphasis is given to biography-Leeuwenhoek, etc. Brief accounts of 350 bacteriologists form a separate section. No clearer, fuller study, suitable to scientists and general readers, has yet been written. 52 illustrations. 448pp. 5⅝ x 8¼. 23761-3 Pa. $6.50

THE COMPLETE NONSENSE OF EDWARD LEAR, Edward Lear. All nonsense limericks, zany alphabets, Owl and Pussycat, songs, nonsense botany, etc., illustrated by Lear. Total of 321pp. 5⅜ x 8½. (Available in U.S. only) 20167-8 Pa. $4.50

INGENIOUS MATHEMATICAL PROBLEMS AND METHODS, Louis A. Graham. Sophisticated material from Graham *Dial*, applied and pure; stresses solution methods. Logic, number theory, networks, inversions, etc. 237pp. 5⅜ x 8½. 20545-2 Pa. $4.50

BEST MATHEMATICAL PUZZLES OF SAM LOYD, edited by Martin Gardner. Bizarre, original, whimsical puzzles by America's greatest puzzler. From fabulously rare *Cyclopedia*, including famous 14-15 puzzles, the Horse of a Different Color, 115 more. Elementary math. 150 illustrations. 167pp. 5⅜ x 8½. 20498-7 Pa. $3.50

THE BASIS OF COMBINATION IN CHESS, J. du Mont. Easy-to-follow, instructive book on elements of combination play, with chapters on each piece and every powerful combination team—two knights, bishop and knight, rook and bishop, etc. 250 diagrams. 218pp. 5⅜ x 8½. (Available in U.S. only) 23644-7 Pa. $4.50

MODERN CHESS STRATEGY, Ludek Pachman. The use of the queen, the active king, exchanges, pawn play, the center, weak squares, etc. Section on rook alone worth price of the book. Stress on the moderns. Often considered the most important book on strategy. 314pp. 5⅜ x 8½. 20290-9 Pa. $5.00

LASKER'S MANUAL OF CHESS, Dr. Emanuel Lasker. Great world champion offers very thorough coverage of all aspects of chess. Combinations, position play, openings, end game, aesthetics of chess, philosophy of struggle, much more. Filled with analyzed games. 390pp. 5⅜ x 8½. 20640-8 Pa. $5.95

500 MASTER GAMES OF CHESS, S. Tartakower, J. du Mont. Vast collection of great chess games from 1798-1938, with much material nowhere else readily available. Fully annotated, arranged by opening for easier study. 664pp. 5⅜ x 8½. 23208-5 Pa. $8.50

A GUIDE TO CHESS ENDINGS, Dr. Max Euwe, David Hooper. One of the finest modern works on chess endings. Thorough analysis of the most frequently encountered endings by former world champion. 331 examples, each with diagram. 248pp. 5⅜ x 8½. 23332-4 Pa. $3.95

CATALOGUE OF DOVER BOOKS

THE COMPLETE BOOK OF DOLL MAKING AND COLLECTING, Catherine Christopher. Instructions, patterns for dozens of dolls, from rag doll on up to elaborate, historically accurate figures. Mould faces, sew clothing, make doll houses, etc. Also collecting information. Many illustrations. 288pp. 6 x 9. 22066-4 Pa. $4.95

THE DAGUERREOTYPE IN AMERICA, Beaumont Newhall. Wonderful portraits, 1850's townscapes, landscapes; full text plus 104 photographs. The basic book. Enlarged 1976 edition. 272pp. 8¼ x 11¼. 23322-7 Pa. $7.95

CRAFTSMAN HOMES, Gustav Stickley. 296 architectural drawings, floor plans, and photographs illustrate 40 different kinds of "Mission-style" homes from The Craftsman (1901-16), voice of American style of simplicity and organic harmony. Thorough coverage of Craftsman idea in text and picture, now collector's item. 224pp. 8⅛ x 11. 23791-5 Pa. $6.50

PEWTER-WORKING: INSTRUCTIONS AND PROJECTS, Burl N. Osborn. & Gordon O. Wilber. Introduction to pewter-working for amateur craftsman. History and characteristics of pewter; tools, materials, step-by-step instructions. Photos, line drawings, diagrams. Total of 160pp. 7⅞ x 10¾. 23786-9 Pa. $3.50

THE GREAT CHICAGO FIRE, edited by David Lowe. 10 dramatic, eyewitness accounts of the 1871 disaster, including one of the aftermath and rebuilding, plus 70 contemporary photographs and illustrations of the ruins—courthouse, Palmer House, Great Central Depot, etc. Introduction by David Lowe. 87pp. 8¼ x 11. 23771-0 Pa. $4.00

SILHOUETTES: A PICTORIAL ARCHIVE OF VARIED ILLUSTRATIONS, edited by Carol Belanger Grafton. Over 600 silhouettes from the 18th to 20th centuries include profiles and full figures of men and women, children, birds and animals, groups and scenes, nature, ships, an alphabet. Dozens of uses for commercial artists and craftspeople. 144pp. 8⅜ x 11¼. 23781-8 Pa. $4.50

ANIMALS: 1,419 COPYRIGHT-FREE ILLUSTRATIONS OF MAMMALS, BIRDS, FISH, INSECTS, ETC., edited by Jim Harter. Clear wood engravings present, in extremely lifelike poses, over 1,000 species of animals. One of the most extensive copyright-free pictorial sourcebooks of its kind. Captions. Index. 284pp. 9 x 12. 23766-4 Pa. $8.95

INDIAN DESIGNS FROM ANCIENT ECUADOR, Frederick W. Shaffer. 282 original designs by pre-Columbian Indians of Ecuador (500-1500 A.D.). Designs include people, mammals, birds, reptiles, fish, plants, heads, geometric designs. Use as is or alter for advertising, textiles, leathercraft, etc. Introduction. 95pp. 8¾ x 11¼. 23764-8 Pa. $4.50

SZIGETI ON THE VIOLIN, Joseph Szigeti. Genial, loosely structured tour by premier violinist, featuring a pleasant mixture of reminiscences, insights into great music and musicians, innumerable tips for practicing violinists. 385 musical passages. 256pp. 5⅝ x 8¼. 23763-X Pa. $4.00

CATALOGUE OF DOVER BOOKS

TONE POEMS, SERIES II: TILL EULENSPIEGELS LUSTIGE STREICHE, ALSO SPRACH ZARATHUSTRA, AND EIN HELDEN-LEBEN, Richard Strauss. Three important orchestral works, including very popular *Till Eulenspiegel's Marry Pranks,* reproduced in full score from original editions. Study score. 315pp. 9⅜ x 12¼. (Available in U.S. only)
23755-9 Pa. $8.95

TONE POEMS, SERIES I: DON JUAN, TOD UND VERKLARUNG AND DON QUIXOTE, Richard Strauss. Three of the most often performed and recorded works in entire orchestral repertoire, reproduced in full score from original editions. Study score. 286pp. 9⅜ x 12¼. (Available in U.S. only)
23754-0 Pa. $8.95

11 LATE STRING QUARTETS, Franz Joseph Haydn. The form which Haydn defined and "brought to perfection." (*Grove's*). 11 string quartets in complete score, his last and his best. The first in a projected series of the complete Haydn string quartets. Reliable modern Eulenberg edition, otherwise difficult to obtain. 320pp. 8⅜ x 11¼. (Available in U.S. only)
23753-2 Pa. $8.95

FOURTH, FIFTH AND SIXTH SYMPHONIES IN FULL SCORE, Peter Ilyitch Tchaikovsky. Complete orchestral scores of Symphony No. 4 in F Minor, Op. 36; Symphony No. 5 in E Minor, Op. 64; Symphony No. 6 in B Minor, "Pathetique," Op. 74. Bretikopf & Hartel eds. Study score. 480pp. 9⅜ x 12¼. 23861-X Pa. $10.95

THE MARRIAGE OF FIGARO: COMPLETE SCORE, Wolfgang A. Mozart. Finest comic opera ever written. Full score, not to be confused with piano renderings. Peters edition. Study score. 448pp. 9⅜ x 12¼. (Available in U.S. only) 23751-6 Pa. $12.95

"IMAGE" ON THE ART AND EVOLUTION OF THE FILM, edited by Marshall Deutelbaum. Pioneering book brings together for first time 38 groundbreaking articles on early silent films from *Image* and 263 illustrations newly shot from rare prints in the collection of the International Museum of Photography. A landmark work. Index. 256pp. 8¼ x 11.
23777-X Pa. $8.95

AROUND-THE-WORLD COOKY BOOK, Lois Lintner Sumption and Marguerite Lintner Ashbrook. 373 cooky and frosting recipes from 28 countries (America, Austria, China, Russia, Italy, etc.) include Viennese kisses, rice wafers, London strips, lady fingers, hony, sugar spice, maple cookies, etc. Clear instructions. All tested. 38 drawings. 182pp. 5⅜ x 8.
23802-4 Pa. $2.75

THE ART NOUVEAU STYLE, edited by Roberta Waddell. 579 rare photographs, not available elsewhere, of works in jewelry, metalwork, glass, ceramics, textiles, architecture and furniture by 175 artists—Mucha, Seguy, Lalique, Tiffany, Gaudin, Hohlwein, Saarinen, and many others. 288pp. 8⅜ x 11¼. 23515-7 Pa. $8.95

CATALOGUE OF DOVER BOOKS

THE CURVES OF LIFE, Theodore A. Cook. Examination of shells, leaves, horns, human body, art, etc., in *"the* classic reference on how the golden ratio applies to spirals and helices in nature . . . ."—Martin Gardner. 426 illustrations. Total of 512pp. 5⅜ x 8½. 23701-X Pa. $6.95

AN ILLUSTRATED FLORA OF THE NORTHERN UNITED STATES AND CANADA, Nathaniel L. Britton, Addison Brown. Encyclopedic work covers 4666 species, ferns on up. Everything. Full botanical information, illustration for each. This earlier edition is preferred by many to more recent revisions. 1913 edition. Over 4000 illustrations, total of 2087pp. 6⅛ x 9¼. 22642-5, 22643-3, 22644-1 Pa., Three-vol. set $28.50

MANUAL OF THE GRASSES OF THE UNITED STATES, A. S. Hitchcock, U.S. Dept. of Agriculture. The basic study of American grasses, both indigenous and escapes, cultivated and wild. Over 1400 species. Full descriptions, information. Over 1100 maps, illustrations. Total of 1051pp. 5⅜ x 8½. 22717-0, 22718-9 Pa., Two-vol. set $17.00

THE CACTACEAE,, Nathaniel L. Britton, John N. Rose. Exhaustive, definitive. Every cactus in the world. Full botanical descriptions. Thorough statement of nomenclatures, habitat, detailed finding keys. The one book needed by every cactus enthusiast. Over 1275 illustrations. Total of 1080pp. 8 x 10¼. 21191-6, 21192-4 Clothbd., Two-vol. set $50.00

AMERICAN MEDICINAL PLANTS, Charles F. Millspaugh. Full descriptions, 180 plants covered: history; physical description; methods of preparation with all chemical constituents extracted; all claimed curative or adverse effects. 180 full-page plates. Classification table. 804pp. 6½ x 9¼. 23034-1 Pa. $13.95

A MODERN HERBAL, Margaret Grieve. Much the fullest, most exact, most useful compilation of herbal material. Gigantic alphabetical encyclopedia, from aconite to zedoary, gives botanical information, medical properties, folklore, economic uses, and much else. Indispensable to serious reader. 161 illustrations. 888pp. 6½ x 9¼. (Available in U.S. only) 22798-7, 22799-5 Pa., Two-vol. set $15.00

THE HERBAL or GENERAL HISTORY OF PLANTS, John Gerard. The 1633 edition revised and enlarged by Thomas Johnson. Containing almost 2850 plant descriptions and 2705 superb illustrations, Gerard's *Herbal* is a monumental work, the book all modern English herbals are derived from, the one herbal every serious enthusiast should have in its entirety. Original editions are worth perhaps $750. 1678pp. 8½ x 12¼. 23147-X Clothbd. $75.00

MANUAL OF THE TREES OF NORTH AMERICA, Charles S. Sargent. The basic survey of every native tree and tree-like shrub, 717 species in all. Extremely full descriptions, information on habitat, growth, locales, economics, etc. Necessary to every serious tree lover. Over 100 finding keys. 783 illustrations. Total of 986pp. 5⅜ x 8½. 20277-1, 20278-X Pa., Two-vol. set $12.00

GREAT NEWS PHOTOS AND THE STORIES BEHIND THEM, John Faber. Dramatic volume of 140 great news photos, 1855 through 1976, and revealing stories behind them, with both historical and technical information. Hindenburg disaster, shooting of Oswald, nomination of Jimmy Carter, etc. 160pp. 8¼ x 11. 23667-6 Pa. $6.00

CRUICKSHANK'S PHOTOGRAPHS OF BIRDS OF AMERICA, Allan D. Cruickshank. Great ornithologist, photographer presents 177 closeups, groupings, panoramas, flightings, etc., of about 150 different birds. Expanded *Wings in the Wilderness*. Introduction by Helen G. Cruickshank. 191pp. 8¼ x 11. 23497-5 Pa. $7.95

AMERICAN WILDLIFE AND PLANTS, A. C. Martin, et al. Describes food habits of more than 1000 species of mammals, birds, fish. Special treatment of important food plants. Over 300 illustrations. 500pp. 5⅜ x 8½. 20793-5 Pa. $6.50

THE PEOPLE CALLED SHAKERS, Edward D. Andrews. Lifetime of research, definitive study of Shakers: origins, beliefs, practices, dances, social organization, furniture and crafts, impact on 19th-century USA, present heritage. Indispensable to student of American history, collector. 33 illustrations. 351pp. 5⅜ x 8½. 21081-2 Pa. $4.50

OLD NEW YORK IN EARLY PHOTOGRAPHS, Mary Black. New York City as it was in 1853-1901, through 196 wonderful photographs from N.-Y. Historical Society. Great Blizzard, Lincoln's funeral procession, great buildings. 228pp. 9 x 12. 22907-6 Pa. $8.95

MR. LINCOLN'S CAMERA MAN: MATHEW BRADY, Roy Meredith. Over 300 Brady photos reproduced directly from original negatives, photos. Jackson, Webster, Grant, Lee, Carnegie, Barnum; Lincoln; Battle Smoke, Death of Rebel Sniper, Atlanta Just After Capture. Lively commentary. 368pp. 8⅜ x 11¼. 23021-X Pa. $11.95

TRAVELS OF WILLIAM BARTRAM, William Bartram. From 1773-8, Bartram explored Northern Florida, Georgia, Carolinas, and reported on wild life, plants, Indians, early settlers. Basic account for period, entertaining reading. Edited by Mark Van Doren. 13 illustrations. 141pp. 5⅜ x 8½. 20013-2 Pa. $6.00

THE GENTLEMAN AND CABINET MAKER'S DIRECTOR, Thomas Chippendale. Full reprint, 1762 style book, most influential of all time; chairs, tables, sofas, mirrors, cabinets, etc. 200 plates, plus 24 photographs of surviving pieces. 249pp. 9⅞ x 12¾. 21601-2 Pa. $8.95

AMERICAN CARRIAGES, SLEIGHS, SULKIES AND CARTS, edited by Don H. Berkebile. 168 Victorian illustrations from catalogues, trade journals, fully captioned. Useful for artists. Author is Assoc. Curator, Div. of Transportation of Smithsonian Institution. 168pp. 8½ x 9½. 23328-6 Pa. $5.00

SECOND PIATIGORSKY CUP, edited by Isaac Kashdan. One of the greatest tournament books ever produced in the English language. All 90 games of the 1966 tournament, annotated by players, most annotated by both players. Features Petrosian, Spassky, Fischer, Larsen, six others. 228pp. 5⅜ x 8½. 23572-6 Pa. $3.50

ENCYCLOPEDIA OF CARD TRICKS, revised and edited by Jean Hugard. How to perform over 600 card tricks, devised by the world's greatest magicians: impromptus, spelling tricks, key cards, using special packs, much, much more. Additional chapter on card technique. 66 illustrations. 402pp. 5⅜ x 8½. (Available in U.S. only) 21252-1 Pa. $5.95

MAGIC: STAGE ILLUSIONS, SPECIAL EFFECTS AND TRICK PHO-TOGRAPHY, Albert A. Hopkins, Henry R. Evans. One of the great classics; fullest, most authorative explanation of vanishing lady, levitations, scores of other great stage effects. Also small magic, automata, stunts. 446 illustrations. 556pp. 5⅜ x 8½. 23344-8 Pa. $6.95

THE SECRETS OF HOUDINI, J. C. Cannell. Classic study of Houdini's incredible magic, exposing closely-kept professional secrets and revealing, in general terms, the whole art of stage magic. 67 illustrations. 279pp. 5⅜ x 8½. 22913-0 Pa. $4.00

HOFFMANN'S MODERN MAGIC, Professor Hoffmann. One of the best, and best-known, magicians' manuals of the past century. Hundreds of tricks from card tricks and simple sleight of hand to elaborate illusions involving construction of complicated machinery. 332 illustrations. 563pp. 5⅜ x 8½. 23623-4 Pa. $6.95

THOMAS NAST'S CHRISTMAS DRAWINGS, Thomas Nast. Almost all Christmas drawings by creator of image of Santa Claus as we know it, and one of America's foremost illustrators and political cartoonists. 66 illustrations. 3 illustrations in color on covers. 96pp. 8⅜ x 11¼. 23660-9 Pa. $3.50

FRENCH COUNTRY COOKING FOR AMERICANS, Louis Diat. 500 easy-to-make, authentic provincial recipes compiled by former head chef at New York's Fitz-Carlton Hotel: onion soup, lamb stew, potato pie, more. 309pp. 5⅜ x 8½. 23665-X Pa. $3.95

SAUCES, FRENCH AND FAMOUS, Louis Diat. Complete book gives over 200 specific recipes: bechamel, Bordelaise, hollandaise, Cumberland, apricot, etc. Author was one of this century's finest chefs, originator of vichyssoise and many other dishes. Index. 156pp. 5⅜ x 8. 23663-3 Pa. $2.75

TOLL HOUSE TRIED AND TRUE RECIPES, Ruth Graves Wakefield. Authentic recipes from the famous Mass. restaurant: popovers, veal and ham loaf, Toll House baked beans, chocolate cake crumb pudding, much more. Many helpful hints. Nearly 700 recipes. Index. 376pp. 5⅜ x 8½. 23560-2 Pa. $4.95

ILLUSTRATED GUIDE TO SHAKER FURNITURE, Robert Meader. Director, Shaker Museum, Old Chatham, presents up-to-date coverage of all furniture and appurtenances, with much on local styles not available elsewhere. 235 photos. 146pp. 9 x 12.   22819-3 Pa. $6.95

COOKING WITH BEER, Carole Fahy. Beer has as superb an effect on food as wine, and at fraction of cost. Over 250 recipes for appetizers, soups, main dishes, desserts, breads, etc. Index. 144pp. 5⅜ x 8½. (Available in U.S. only)   23661-7 Pa. $3.00

STEWS AND RAGOUTS, Kay Shaw Nelson. This international cookbook offers wide range of 108 recipes perfect for everyday, special occasions, meals-in-themselves, main dishes. Economical, nutritious, easy-to-prepare: goulash, Irish stew, boeuf bourguignon, etc. Index. 134pp. 5⅜ x 8½.
23662-5 Pa. $3.95

DELICIOUS MAIN COURSE DISHES, Marian Tracy. Main courses are the most important part of any meal. These 200 nutritious, economical recipes from around the world make every meal a delight. "I . . . have found it so useful in my own household,"—N.Y. Times. Index. 219pp. 5⅜ x 8½.   23664-1 Pa. $3.95

FIVE ACRES AND INDEPENDENCE, Maurice G. Kains. Great back-to-the-land classic explains basics of self-sufficient farming: economics, plants, crops, animals, orchards, soils, land selection, host of other necessary things. Do not confuse with skimpy faddist literature; Kains was one of America's greatest agriculturalists. 95 illustrations. 397pp. 5⅜ x 8½.
20974-1 Pa. $4.95

A PRACTICAL GUIDE FOR THE BEGINNING FARMER, Herbert Jacobs. Basic, extremely useful first book for anyone thinking about moving to the country and starting a farm. Simpler than Kains, with greater emphasis on country living in general. 246pp. 5⅜ x 8½.
23675-7 Pa. $3.95

PAPERMAKING, Dard Hunter. Definitive book on the subject by the foremost authority in the field. Chapters dealing with every aspect of history of craft in every part of the world. Over 320 illustrations. 2nd, revised and enlarged (1947) edition. 672pp. 5⅜ x 8½.   23619-6 Pa. $8.95

THE ART DECO STYLE, edited by Theodore Menten. Furniture, jewelry, metalwork, ceramics, fabrics, lighting fixtures, interior decors, exteriors, graphics from pure French sources. Best sampling around. Over 400 photographs. 183pp. 8⅜ x 11¼.   22824-X Pa. $6.95

ACKERMANN'S COSTUME PLATES, Rudolph Ackermann. Selection of 96 plates from the Repository of Arts, best published source of costume for English fashion during the early 19th century. 12 plates also in color. Captions, glossary and introduction by editor Stella Blum. Total of 120pp. 8⅜ x 11¼.   23690-0 Pa. $5.00

CATALOGUE OF DOVER BOOKS

THE ANATOMY OF THE HORSE, George Stubbs. Often considered the great masterpiece of animal anatomy. Full reproduction of 1766 edition, plus prospectus; original text and modernized text. 36 plates. Introduction by Eleanor Garvey. 121pp. 11 x 14¾. 23402-9 Pa. $8.95

BRIDGMAN'S LIFE DRAWING, George B. Bridgman. More than 500 illustrative drawings and text teach you to abstract the body into its major masses, use light and shade, proportion; as well as specific areas of anatomy, of which Bridgman is master. 192pp. 6½ x 9¼. (Available in U.S. only) 22710-3 Pa. $4.50

ART NOUVEAU DESIGNS IN COLOR, Alphonse Mucha, Maurice Verneuil, Georges Auriol. Full-color reproduction of Combinaisons ornementales (c. 1900) by Art Nouveau masters. Floral, animal, geometric, interlacings, swashes—borders, frames, spots—all incredibly beautiful. 60 plates, hundreds of designs. 9⅜ x 8-1/16. 22885-1 Pa. $4.50

FULL-COLOR FLORAL DESIGNS IN THE ART NOUVEAU STYLE, E. A. Seguy. 166 motifs, on 40 plates, from Les fleurs et leurs applications decoratives (1902): borders, circular designs, repeats, allovers, "spots." All in authentic Art Nouveau colors. 48pp. 9⅜ x 12¼. 23439-8 Pa. $6.00

A DIDEROT PICTORIAL ENCYCLOPEDIA OF TRADES AND IN-DUSTRY, edited by Charles C. Gillispie. 485 most interesting plates from the great French Encyclopedia of the 18th century show hundreds of working figures, artifacts, process, land and cityscapes; glassmaking, paper-making, metal extraction, construction, weaving, making furniture, clothing, wigs, dozens of other activities. Plates fully explained. 920pp. 9 x 12. 22284-5, 22285-3 Clothbd., Two-vol. set $50.00

HANDBOOK OF EARLY ADVERTISING ART, Clarence P. Hornung. Largest collection of copyright-free early and antique advertising art ever compiled. Over 6,000 illustrations, from Franklin's time to the 1890's for special effects, novelty. Valuable source, almost inexhaustible.
Pictorial Volume. Agriculture, the zodiac, animals, autos, birds, Christmas, fire engines, flowers, trees, musical instruments, ships, games and sports, much more. Arranged by subject matter and use. 237 plates. 288pp. 9 x 12. 20122-8 Clothbd. $15.00

Typographical Volume. Roman and Gothic faces ranging from 10 point to 300 point, "Barnum," German and Old English faces, script, logotypes, scrolls and flourishes, 1115 ornamental initials, 67 complete alphabets, more. 310 plates. 320pp. 9 x 12. 20123-6 Clothbd. $15.00

CALLIGRAPHY (CALLIGRAPHIA LATINA), J. G. Schwandner. High point of 18th-century ornamental calligraphy. Very ornate initials, scrolls, borders, cherubs, birds, lettered examples. 172pp. 9 x 13. 20475-8 Pa. $7.95

CATALOGUE OF DOVER BOOKS

GEOMETRY, RELATIVITY AND THE FOURTH DIMENSION, Rudolf Rucker. Exposition of fourth dimension, means of visualization, concepts of relativity as Flatland characters continue adventures. Popular, easily followed yet accurate, profound. 141 illustrations. 133pp. 5⅜ x 8½.
23400-2 Pa. $2.75

THE ORIGIN OF LIFE, A. I. Oparin. Modern classic in biochemistry, the first rigorous examination of possible evolution of life from nitrocarbon compounds. Non-technical, easily followed. Total of 295pp. 5⅜ x 8½.
60213-3 Pa. $5.95

PLANETS, STARS AND GALAXIES, A. E. Fanning. Comprehensive introductory survey: the sun, solar system, stars, galaxies, universe, cosmology; quasars, radio stars, etc. 24pp. of photographs. 189pp. 5⅜ x 8½. (Available in U.S. only)
21680-2 Pa. $3.75

THE THIRTEEN BOOKS OF EUCLID'S ELEMENTS, translated with introduction and commentary by Sir Thomas L. Heath. Definitive edition. Textual and linguistic. notes, mathematical analysis, 2500 years of critical commentary. Do not confuse with abridged school editions. Total of 1414pp. 5⅜ x 8½.
60088-2, 60089-0, 60090-4 Pa., Three-vol. set $19.50